SUPERCONDUCTIVITY AND ITS APPLICATIONS

AIP
CONFERENCE
PROCEEDINGS 219

SUPERCONDUCTIVITY
AND ITS
APPLICATIONS

BUFFALO, NY 1990

EDITORS:

YI-HAN KAO, PHILIP COPPENS, &
HOI-SING KWOK
UNIVERSITY OF NEW YORK, BUFFALO

AIP
American Institute of Physics   New York

L.C. Catalog Card No. 91-55020
ISBN 0-88318-835-X
DOE CONF-900990

Printed in the United States of America.

This book was put into production on 2 January 1991 and was published on 21 March 1991.

CONTENTS

THIN FILMS AND SYNTHESIS

PREFACE

This volume is a collection of papers presented at the Fourth Annual Conference on Superconductivity and Applications organized by the New York State Institute on Superconductivity (NYSIS). The Conference was held at the Buffalo Hilton Hotel on 18–20 September 1990. There were about 200 participants, including many scientists from Japan, China, U.S.S.R., Germany, and other countries.

The field of high-T_c superconductivity is growing so rapidly that it is difficult to keep up with the literature. An open meeting like the Buffalo Conference serves the purpose of helping to keep researchers abreast of important progress being made around the world. The Conference benefited from the enthusiastic participation of many leading experts in the field. It is hoped that our Conference has established a tradition for the future exchange of ideas.

We are grateful to many outstanding scientists and friends who have kindly agreed to give the plenary and invited talks. Our particular thanks go to Dr. R. S. Hamilton and Mrs. P. F. Barnett for their excellent contributions in planning, managing, and running the secretariat of this Conference.

Y.-H. Kao
P. Coppens
H.-S. Kwok

Plenary Reviews

THE s-CHANNEL THEORY OF SUPERCONDUCTIVITY

T. D. Lee

Columbia University, New York, N.Y. 10027

ABSTRACT

The essential features and new developments of the s-channel theory of super-conductivity are presented. Applications to the μsR and Hall number experiments are analyzed. The relations between small coherence length, Bose-Einstein condensation and high T_c are emphasized.

1. INTRODUCTION

In this talk I wish to discuss a new phenomenological theory of high temperature superconductivity, which has been developed in collaboration with R. Friedberg and H. C. Ren. Our starting point is the experimental observation of a small coherence length $\xi \approx 10\mathring{A}$ in high T_c superconductors[1,2]. This is in contrast with a much larger ξ in the usual cold superconductor, typically $\approx 10^4\mathring{A}$ for type I and $\approx 10^2\mathring{A}$ for type II. In addition, it is known that in all these superconductors the magnetic flux carried by each vortex filament is $2\pi\hbar c/2e$, showing the existence of a pairing state.

A small coherence length ξ in the high T_c superconductors indicates that the pairing between electrons, or holes, is reasonably localized in the coordinate space. Hence, the pair-state can be approximated by a phenomenological local boson field $\phi(\vec{r})$, whose mass m_b is $\approx 2m_e$ and whose elementary charge unit is $2e$, where m_e and e are the mass and charge of an electron. It follows then that the transition

$$2e \xrightarrow{\quad} \phi \xrightarrow{\quad} 2e \tag{1.1}$$

must occur, in which e denotes either an electron or a hole; furthermore, the localization of ϕ implies that phenomena at distances larger than the physical extension of ϕ (which is $< \xi$) are insensitive to the interior of ϕ. Since ξ is of the same order as the scale of a lattice unit cell, it becomes possible to develop a phenomenological theory of superconductivity based *only* on the local character of ϕ.

Of course, physics at large does depend on several overall properties: the spin of ϕ, the stability of an individual ϕ-quantum, the isotropicity and homogeneity (or their absence) of the space containing ϕ and so on. The situation is analogous to that in particle physics: the smallness of the radii of pions, ρ-mesons, kaons, $\cdots$ makes it possible for us to handle much of the dynamics without any reference to their internal structure, such as quark-antiquark pairs or bag models. Hence, the origin of their formation becomes a problem separate from the description of their mechanics. An important ingredient in this type of phenomenological approach is the selection of the basic interaction Hamiltonian that describes the underlying dominant process. In the usual low-temperature superconductors, the large ξ-value makes the corresponding pairing state ϕ too extended and ill-defined in the coordinate space; therefore, (1.1) does not play an important role. Instead, the BCS theory of superconductivity[5] is based on the emission and absorption of phonons,

$$2e \; \rightarrow \; 2e + \text{phonon} \; \rightarrow \; 2e. \tag{1.2}$$

In the language of particle physics, (1.1) is an s-channel process, while (1.2) is t-channel. The BCS theory may be called the t-channel theory, and the model that is based on (1.1) the *s-channel* theory.[6,7]

The use of a boson field for the superfluidity of Liquid $HeII$ has had a long history. However, there are some major differences in the following application to (high temperature) superconductors:

1. The ϕ-quantum is charged, carrying $2e$, while the helium atom is neutral.

2. We assume each individual ϕ-quantum to be *unstable*, with 2ν as its excitation energy.

In any microscopic attempt to construct ϕ out of $2e$, because of the short-range Coulomb repulsion it is very difficult to have ϕ stable. The explicit assumption of instability bypasses this difficulty; it also makes the present boson-fermion model different from the theory of Schafroth[8,9] and others.

In the rest frame of a single ϕ-quantum (in isolation), the decay

$$\phi \rightarrow 2e \tag{1.3}$$

occurs, in which each e carries an energy

$$\frac{k^2}{2m_e} = \nu \, .$$

Consequently, in a large system, there are macroscopic numbers of both bosons (the ϕ-quanta) and fermions (electrons or holes), distributed according to the principles of statistical mechanics.

At temperature $T < T_c$, there is always a macroscopic distribution of zero momentum bosons co-existing with a Fermi distribution of electrons (or holes). Take the simple example of *zero* temperature: Let ϵ_F be the Fermi energy. When $\epsilon_F = \nu$, the decay $\phi \rightarrow 2e$ cannot take place because of the exclusion principle; therefore, the bosons are present. Even when $\epsilon_F < \nu$, there is still a macroscopic number of (virtual) zero momentum bosons in the form of a static coherent field amplitude whose source is the fermion pairs. This then leads to the following essential features of the s-channel model.

Below the critical temperature T_c the long range order in the boson field can always be described by its zero-momentum bosonic amplitude B, as in the Bose-Einstein condensation (and therefore similar to liquid $HeII$). Because of the transition (1.1), the zero-momentum of the boson in the condensate forces the two e to have equal and opposite momenta, forming a Cooper pair. Therefore, the same long range order B also applies to the Cooper pairs of the fermions. Furthermore, as we shall see, the gap energy Δ of the fermion system is related to B by

$$\Delta^2 = |gB|^2 \tag{1.4}$$

where g is the coupling for $\phi \rightarrow 2e$.

Since in reality ϕ is a composite of $2e$, when the average distance between ϕ-quanta becomes less than the diameter of the composite the approximation of treating each ϕ as a single boson breaks down. However, for densities not that high, by representing the 2e-resonance as an independent ϕ-field, we may convert an otherwise strong interaction problem (which forms the resonance and exists at small distances) to one that can be handled by perturbative series in weak coupling (i.e., the residual interaction at relatively larger distances). This enables us to give a systematic analysis of such a theory; it also makes transparent the questions of gauge invariance and symmetry breaking.

2. μsR EXPERIMENTS

In the s-channel theory, the long-range order parameter B is due to Bose-Einstein condensation. Consequently, the phase transition can be of statistical origin, in contrast to the usual BCS theory. As we shall see, the critical temperature T_c may then be much higher. Let us first examine the evidence supporting such a picture.

Recently, Uemura *et al.*[10] discovered that in all (high temperature) cupric superconductors there is a universality law:

$$T_c \propto \rho^*/m^* \tag{2.1}$$

where ρ^* is the number density of superconducting charge carriers and m^* their effective mass; the proportionality constant is the same for all materials, about

$$40°K \qquad \text{to} \qquad 4 \times 10^{20} cm^{-3}/m_e \,, \tag{2.2}$$

assuming each carrier bears a charge e . In the s-channel theory, the ρ^* in the μ SR experiment[10] should be interpreted as due to bosons of charge $2e$; the proportionality constant would then be reduced by a factor 4 , and the experimentally determined proportionality constant (2.2) becomes

$$40°K \qquad \text{to} \qquad 10^{20} cm^{-3}/m_e \,. \tag{2.3}$$

In these cupric superconductors, the charge carriers concentrate on the two - dimensional CuO_2 plane; their tunneling between these planes gives rise to the three-dimensional character. The average separation c between CuO_2 planes is approximately constant for different materials:

$$c \cong 6\mathring{A} \,.$$

Let d be the average distance between neighboring bosons in the same CuO_2 plane; the boson density n_b is

$$n_b = (d^2 c)^{-1}. \tag{2.4}$$

At temperature T_c, each boson of mass m_b has a thermal deBroglie wavelength ($\hbar = 1$).

$$\lambda_T \equiv \sqrt{\frac{2\pi}{m_b \kappa T}}. \tag{2.5}$$

Setting

$$m^* = m_b \quad \text{and} \quad \rho^* = n_b \tag{2.6}$$

in (2.1), one sees that the product $m_b T_c$ is $\propto d^{-2}$; from the definition of λ_T, the same product is also $\propto \lambda_T^{-2}$. Eliminating $m_b T_c$ and using the $\mu s R$ experimental data we arrive at

$$(\lambda_T/d)_{\text{exp}} \cong 2.8 \tag{2.7}$$

for all cupric superconductors. Furthermore, this value is independent of m_b.

In the s-channel theory, the pair state is represented phenomenologically by a local field ϕ which can propagate relatively freely within each CuO_2 plane; let m_b denote its mass in the CuO_2 plane. Because the ϕ-quantum, in reality, is shaped like a flat disc with its face parallel to the plane, this makes it difficult to tunnel across different CuO_2. The effective boson mass M_b in the direction $\perp$ to the CuO_2 plane should therefore be much larger than m_b:

$$M_b \gg m_b. \tag{2.8}$$

The criterion of Bose-Einstein condensation for such a configuration is[11]

$$n_b \cong (\lambda_T^2 c)^{-1} \ln(2M_b c^2 \kappa T_c) \tag{2.9}$$

and, because of (2.4),

$$\lambda_T/d. \cong \left[\ln(2M_b c^2 \kappa T_c) \right]^{\frac{1}{2}}. \tag{2.10}$$

When $M_b \to \infty$, or $c \to \infty$, λ_T/d becomes infinite and $T_c = 0$; this gives the well-known result that there is no Bose-Einstein condensation in two dimensions. Because of the log-factor in (2.9)-(2.10), we expect that while λ_T is $O(d)$, the ratio λ_T/d can be ~ 2.8, larger than 1.

The theoretical formula (2.10) gives a slight variation of λ_T/d, consistent with experimental data, as shown in Figure 1.

3.   BOSE-EINSTEIN CONDENSATION AND HIGH T_c

An essential feature of the Bose-Einstein condensation is that the thermal de Broglie length λ_T should be comparable to the interparticle distance d, so that the effect of symmetric statistics becomes manifest. It is instructive to put side by side the ratios λ_T/d for the ideal bosons, liquid Helium II together with cupric superconductors:

$$\lambda_T/d = \begin{cases} 1.377 & \text{ideal bosons} \\ 1.65 & \text{He II} \\ 2.8 & \text{cupric superconductors.} \end{cases} \tag{3.1}$$

In the BCS theory, T_c depends sensitively on the interaction between electrons and phonons (or other excitations); $T_c \to 0$ when there is no interaction. In the Bose-Einstein condensation, T_c is determined by $\lambda_T \sim d$, which is of statistical origin and therefore can be much higher (T_c exists even without interaction). In the boson picture, on account of (2.5) and (3.1), we have

$$(m_b T_c)^{\frac{1}{2}} d \approx \text{constant} . \tag{3.2}$$

This product varies by only a factor less than, or ~ 2, from ideal boson to He, and from He II to cupric superconductors. For He, $d \cong 3.58\text{Å}$, $T_c \cong 2.2°K$ and $m_b \cong 8000 m_e$, whereas for cupric superconductors the relevant m_b is only a few times m_e, the electron mass. (See (8.7) below.) Thus, between He and cupric superconductors there is a change in m_b of *three* orders of magnitude. The relative constancy of the product $(m_b T_c)^{\frac{1}{2}} d$ naturally leads to a much higher T_c for cupric superconductors. In addition, if one could have smaller d, then T_c would increase accordingly. Of course, d must not be too small; otherwise, the pair-states overlap, and the boson approximation breaks down (as in the case of cold superconductors, because of their large coherence lengths).

4.   A PROTOTYPE s-CHANNEL MODEL

As a prototype of the s-channel theory of superconductivity, we assume ϕ to be of spin 0 and that the space containing ϕ is a three-dimensional homogeneous and isotropic continuum. For realistic applications[11,12], as emphasized before, a more appropriate approximation of the latter would be the product of a two-dimensional x, y-continuum (simulating the CuO_2 plane) and a discrete lattice of spacing c along the z-direction. The two-dimensional layer character of CuO_2 planes helps in the localization of the pair state in the z-direction, making the ϕ-quantum disc-shaped. The space that ϕ moves in becomes a three-dimensional continuum when $c \to 0$, but two-dimensional when $c \to \infty$.

Here we consider an idealized isotropic and homogeneous space; the system consisting of the local scalar field ϕ of mass M and the electron (or hole) field ψ_σ of mass m, with $\sigma = \uparrow$ or $\downarrow$ denoting the spin. The Hamiltonian is ($\hbar = 1$)

$$H = H_0 + H_1 \tag{4.1}$$

in which the free Hamiltonian is

$$H_0 = \int \left[\phi^\dagger (2\nu_0 - \frac{1}{2M} \nabla^2) \phi + \psi_\sigma^\dagger (-\frac{1}{2m} \nabla^2) \psi_\sigma \right] d^3r \tag{4.2}$$

with the repeated spin index σ summed over and $\dagger$ denoting the hermitian conjugate. The interaction H_1 can be simply

$$H_1 = g \int (\phi^\dagger \psi_\uparrow \psi_\downarrow + \text{h.c.}) \, d^3r \,. \tag{4.3}$$

Both ϕ and ψ_σ are the usual quantized field operators whose equal-time commutator and anticommutator are

$$[\phi(\vec{r}), \, \phi^\dagger(\vec{r}')] = \delta^3(\vec{r} - \vec{r}')$$

and

$$\{\psi_\sigma(\vec{r}), \, \psi_{\sigma'}^\dagger(\vec{r}') \} = \delta_{\sigma\sigma'} \, \delta^3(\vec{r} - \vec{r}') \,.$$

The total particle number operator is defined to be

$$N = \int (2\phi^\dagger \phi + \psi_\sigma^\dagger \psi_\sigma) \, d^3r \tag{4.4}$$

which commutes with H and is therefore conserved.

Expand the field operators in Fourier components inside a volume Ω with periodic boundary conditions:

$$\psi_\sigma(\vec{r}) \;=\; \sum_k \Omega^{-\frac{1}{2}} \, a_{\vec{k},\sigma} e^{i\vec{k}\cdot\vec{r}}$$

and   (4.5)

$$\phi(\vec{r}) \;=\; \sum_k \Omega^{-\frac{1}{2}} \, b_{\vec{k}} \, e^{i\vec{k}\cdot\vec{r}}$$

where the anticommutator $\{a_{\vec{k},\sigma}, a^\dagger_{\vec{k}',\sigma'}\} = \delta_{\vec{k}\vec{k}'}\,\delta_{\sigma\sigma'}$ and the commutator $[b_{\vec{k}}, b^\dagger_{\vec{k}'}] = \delta_{\vec{k}\vec{k}'}$. Equation (4.3) can then be written as

$$H_1 \;=\; \frac{g}{\sqrt{\Omega}} \sum_{p,k} \left[b^\dagger_p \, a_{\frac{\vec{p}}{2}+\vec{k},\uparrow} \, a_{\frac{\vec{p}}{2}-\vec{k},\downarrow} + \text{h.c.} \right] .$$ (4.6)

In (4.2), $2\nu_0$ is the "bare" excitation energy of ϕ. Because of the interaction, the "physical" (i.e., renormalized) excitation energy 2ν in reaction (1.3) is given by

$$2\nu \;=\; 2\nu_0 + \frac{g^2}{2\Omega} \sum_k P \, \frac{1}{\nu - \omega_k}$$ (4.7)

where P denotes the principal value and

$$\omega_k \;=\; \frac{k^2}{2m} .$$ (4.8)

The decay width Γ of a ϕ-quantum (in vacuum) is given by

$$\Gamma \;=\; (g^2/\pi)\, m^{\frac{3}{2}} \sqrt{\frac{\nu}{2}} .$$ (4.9)

5. GAP ENERGY

For $T < T_c$, the zero-momentum occupation number $b^\dagger_0 b_0$ of the boson field ϕ becomes macroscopic; hence, we may replace the operator b_0 by a macroscopic constant. Set in (4.5)

$$\Omega^{-\frac{1}{2}} b_0 \;=\; B \;=\; \text{c number}$$

and write

$$\phi = B + \phi_1 \tag{5.1}$$

where

$$\phi_1 = \sum_{k \neq 0} \Omega^{-\frac{1}{2}} b_{\vec{k}} \, e^{i\vec{k}\cdot\vec{r}} . \tag{5.2}$$

In the following, we shall treat the effects of ϕ_1 perturbatively. Let μ be the chemical potential. Introduce

$$\mathcal{H} \equiv H - \mu N = \mathcal{H}_0 + \mathcal{H}_1 \tag{5.3}$$

where

$$\mathcal{H}_0 = \sum_k \{ [\frac{k^2}{2M} + 2(\nu_0 - \mu)] \, b_{\vec{k}}^\dagger b_{\vec{k}} + (\omega_k - \mu) \, a_{\vec{k},\sigma}^\dagger \, a_{\vec{k},\sigma}$$

$$+ g[B^* a_{\vec{k},\uparrow} \, a_{-\vec{k},\downarrow} + B \, a_{-\vec{k},\downarrow}^\dagger \, a_{\vec{k},\uparrow}^\dagger] \} \tag{5.4}$$

and

$$\mathcal{H}_1 = g \int (\phi_1^\dagger \psi_\uparrow \psi_\downarrow + \text{h.c.}) \, d^3 r \tag{5.5}$$

is the perturbation.

The zeroth-order Hamiltonian $\mathcal{H}_0$ is quadratic in field operators, and can therefore be readily diagonalized. Its fermion-dependent part can be written as a sum of matrix products, each of the form:

$$(a_{\vec{k},\uparrow}, \, a_{-\vec{k},\downarrow}^\dagger) \cdot A \cdot \begin{pmatrix} a_{\vec{k},\uparrow}^\dagger \\ a_{-\vec{k},\downarrow} \end{pmatrix}$$

where A is a 2×2 matrix

$$A = \begin{pmatrix} -(\omega_k - \mu) & g B^* \\ g B & \omega_k - \mu \end{pmatrix} .$$

Because of Fermi statistics, the two diagonal elements of A have opposite signs. The eigenvalue E_k of A is determined by

$$E_k^2 - (\omega_k - \mu)^2 = g^2 |B|^2 ;$$

i.e.,

$$E_k = [(\omega_k - \mu)^2 + g^2 |B|^2]^{\frac{1}{2}} . \tag{5.6}$$

Thus, we establish the formula for the gap energy (1.4):

$$\Delta^2 = g^2 |B|^2.$$

6. THERMODYNAMICS

The zeroth-order grand canonical partition function is

$$\mathcal{Q} = \text{trace } e^{-\beta \mathcal{H}_0},$$

whose logarithm is $p\Omega/\kappa T$, where $\beta = (\kappa T)^{-1}$ and p is the pressure. By using the diagonalized form of $\mathcal{H}_0$, we find

$$p = -2(\nu_0 - \mu)|B|^2 + \Omega^{-1} \sum_k (E_k + \mu - \omega_k)$$

$$+2(\beta\Omega)^{-1} \sum_k ln(1 + e^{-\beta E_k})$$

$$-(\beta\Omega)^{-1} \sum_k ln\{1 - \exp \beta [2\mu - 2\nu - (k^2/2M)]\}. \qquad (6.1)$$

In accordance with the general thermodynamical principle, at constant T and μ, the function p should be a maximum with respect to any internal parameter, such as $|B|$. Setting $(\partial p/\partial |B|)_{\mu,T} = 0$, we have

$$\nu_0 - \mu - \Omega^{-1} \frac{g^2}{4} \sum_k \frac{1}{E_k} \tanh \frac{1}{2}\beta E_k = 0.$$

By using (4.7), we may express the above formula in terms of the physical excitation energy 2ν of the ϕ-quantum:

$$\nu - \mu = \Omega^{-1} \frac{g^2}{4} \sum_k \left[\frac{1}{E_k} \tanh \frac{1}{2}\beta E_k + P \frac{1}{\nu - \omega_k} \right], \qquad (6.2)$$

where P denotes the principal value, as before. The right-hand side is convergent in the ultra-violet region since the theory is renormalizable. The particle density ρ is given by $(\partial p/\partial \mu)_{T,B}$, which yields

$$\rho = 2\,|\,B\,|^2 + 2\Omega^{-1}\sum_k [\,e^{\beta(2\nu+(k^2/2M)-2\mu)} - 1\,]^{-1}$$

$$+\Omega^{-1}\sum_k [\,E_k(1+e^{-\beta E_k})\,]^{-1}[\,E_k+\mu-\omega_k+(E_k-\mu+\omega_k)\,e^{-\beta E_k}\,]. \quad (6.3)$$

From (6.2) and (6.3), μ and $|\,B\,|^2$ can be determined as functions of ρ and T. (Equation (6.2) is similar to the gap equation in the BCS theory, and Eq. (6.3) is the generalization of the density equation in the Bose-Einstein condensation.)

Regarding (6.1)-(6.3) as the zeroth approximation, one can develop a systematic expansion using $\mathcal{H}_1$ of (5.5) as the perturbation.

It is useful to introduce

$$\rho_\nu \equiv (3\pi^2)^{-1}(2m\nu)^{\frac{3}{2}}, \quad (6.4)$$

the fermionic density when the Fermi-energy equals ν, with the excitation energy of the ϕ-quantum $= 2\nu$. For $\rho << \rho_\nu$, one finds that the gap energy Δ_0 at zero temperature is related to the critical temperature T_c by, as in the BCS theory,

$$\frac{\Delta_0}{\kappa T_c} = \pi\,e^{-\gamma} = 1.7639 \quad (6.5)$$

where $\gamma =$ Euler's constant $= 0.5772$. For $\rho >> \rho_\nu$

$$\Delta_0^2 = (2.612)\,g^2\left(\frac{M\kappa T_c}{2\pi}\right)^{\frac{3}{2}} \quad (6.6)$$

and (at any temperature $T < T_c$)

$$\Delta(T)^2 = g^2\,|\,B(T)\,|^2 = \Delta_0^2\left[1 - \left(\frac{T}{T_c}\right)^{\frac{3}{2}}\right], \quad (6.7)$$

as in the Bose-Einstein condensation.

A detailed study of (6.2) and (6.3) shows that typical BCS and Bose-Einstein formulas can be analytically connected within one single expression. In this way, these two approaches become closely unified. The s-channel theory has an intrinsically simpler structure than the t-channel theory; this makes it possible to take a deductive approach, thereby rendering the analysis attractive on the pedagogical level.

7. COHERENCE LENGTH

Consider the case of a scalar ϕ interacting with an electron (or hole) field ψ through (1.1). Let $\vec{A}$ be the *transverse* electromagnetic field. Assume the space to be isotropic and homogeneous. Define the phase-angle variable $\theta(x)$ by

$$\phi(x) = R(x)\,e^{i\theta(x)} \tag{7.1}$$

with R and θ both hermitian. Write

$$\psi(x) = \psi'(x)\,e^{\frac{1}{2}i\theta(x)}$$

and

$$\vec{V}(x) \equiv \vec{A}(x) - (2e)^{-1}\,\vec{\nabla}\theta(x). \tag{7.2}$$

At very low temperature we have $R \cong B$, the long-range order parameter (chosen to be real). As shown in Ref. 7, the energy spectra for the transverse and longitudinal modes of $\vec{V}$ are (in units of $\hbar = c = 1$)

$$\omega_t(k) = (\lambda_L^{-2} + k^2)^{\frac{1}{2}} \tag{7.3}$$

and

$$\omega_\ell(k) = \left[\lambda_L^{-2} + k^2 v^2 + (k^2/2M)^2\right]^{\frac{1}{2}} \tag{7.4}$$

where k is the momentum (or wave nunber),

$$\lambda_L^{-2} = (2e\,B)^2/M \tag{7.5}$$

is the inverse square of the London length, $e^2 = 4\pi/137$, v is the "sound" velocity of the boson-fermion system and M the mass of ϕ.

Equations (7.3)-(7.4) also follow from general arguments: (i) At zero momentum $k = 0$, as in the Higgs mechanism,[13] the energies of the three spin-components of the massive vector field $\vec{V}$ become the same; i.e., they are all equal to the rest mass m_V, given by

$$m_V = \lambda_L^{-1}. \tag{7.6}$$

(ii) When $e = 0$, we have $m_V = 0$ and the transverse mode is the usual photon with $\omega_t = k$ (since the velocity of light c is 1). On the other hand, the longitudinal mode describes the Goldstone-Nambu boson[13] which, for $e = 0$, corresponds to the vibration of ϕ, propagating with the sound velocity $v << 1$ (i.e., $\omega_\ell \to kv$ as $k \to 0$). (iii) For very large k, the excitation of ϕ approaches the free boson spectrum $k^2/2M$,

$$\omega_\ell \to \frac{k^2}{2M} \qquad \text{for} \qquad k >> 2Mv \quad \text{and} \quad (2M/\lambda_L)^{\frac{1}{2}}. \qquad (7.7)$$

For $e \neq 0$, the Goldstone-Nambu boson joins with the transverse photon to form a massive vector field $\vec{V}$, which leads to the above formulas for ω_ℓ and ω_t, consistent with (i)-(iii).

For the coherence length ξ, we may set $\omega_\ell(k) = 0$ and k becomes complex, which gives a boson-amplitude, say $\exp(ikx)$, that varies exponentially with distance (e.g., along the radius of a vortex filament). The decay rate in x determines ξ. From (7.4), the root

$$k \equiv i\sqrt{2}\,\mu_\pm \qquad \text{for} \qquad \omega_\ell(k) = 0 \qquad (7.8)$$

satisfies

$$\mu_\pm^2 = (Mv)^2 \pm \left[(Mv)^4 - (M/\lambda_L)^2 \right]^{\frac{1}{2}}. \qquad (7.9)$$

The amplitude $\exp(ikx)$ becomes, then, $\exp(-\sqrt{2}\,\mu_\pm x)$. To conform to the usual definition, the coherence length ξ is given by $[Re(\mu_-)]^{-1}$, which is always $\geq [Re(\mu_+)]^{-1}$.

(1) For $v^2 > (M\lambda_L)^{-1}$, μ_+ and μ_- are real and

$$\xi = 1/\mu_-. \qquad (7.10)$$

(2) For $v^2 < (M\lambda_L)^{-1}$, $\mu_\pm$ are complex with

$$\mu_+ = \mu_-^* = (M/\lambda_L)^{\frac{1}{2}} e^{i\alpha} \qquad (7.11)$$

where

$$\cos 2\alpha = M \lambda_L v^2 \qquad (7.12)$$

and

$$\sin 2\alpha = [1 - (M \lambda_L v^2)^2]^{\frac{1}{2}}; \qquad (7.13)$$

correspondingly,

$$\xi = \sqrt{\frac{\lambda_L}{M}} \sec \alpha . \tag{7.14}$$

A complex $\mu_{\pm}$ implies the condensate amplitude inside a vortex filament also contains an oscillatory component, which may lead to new observational possiblities.

In the case $v^2 < (M\lambda_L)^{-1}$, according to (7.12) $\cos 2\alpha$ varies from 0 to 1; therefore, $\cos \alpha$ is between $\frac{1}{\sqrt{2}}$ and 1. Hence

$$\sqrt{\frac{2\lambda_L}{M}} \geq \xi \geq \sqrt{\frac{\lambda_L}{M}} . \tag{7.15}$$

(Recall that $\lambda_L^{-2} = (2eB)^2/Mc^2$. The product λ_L times the Compton wavelength $\hbar/Mc$ is independent of c, the velocity of light.) Assume a boson condensate density B^2 (at $T \ll T_c$) between $10^{20} - 10^{21}\text{cm}^{-3}$. On account of (7.5), $M \cong 2m_e$ and $e^2/4\pi = 1/137$, the London length is

$$\lambda_L \sim 1200\mathring{A} \qquad \text{for} \qquad B^2 \sim 10^{21}\text{cm}^{-3} \tag{7.16}$$

and $\lambda_L \sim 3800\mathring{A}$ for $B^2 \sim 10^{20}\text{cm}^{-3}$. Since the Compton wavelength M^{-1} is $\sim 2 \times 10^{-3}\mathring{A}$, we see that in case (2), (7.14)-(7.15) give

$$\xi \sim \text{few } \mathring{A} . \tag{7.17}$$

Case (1) holds only if v is larger than $(M\lambda_L)^{-\frac{1}{2}} \sim 10^{-3}$ times the velocity of light; hence, depending on v, $\xi \sim (Mv)^{-1} < \text{few } \mathring{A}$, or $\xi \sim \sqrt{2}v\lambda_L \ll \lambda_L$. In either case, the theory predicts a very small ξ, consistent with experimental observations. Because $\lambda_L \gg \xi$, the s-channel theory gives, in general, a type II superconductor.

The s-channel theory is based on the observation that the coherence length ξ is small for all recently discovered high T_c superconductors. It is indeed satisfying that within the s-channel theory, the smallness of ξ can in turn be calculated.

8. HALL ANOMALY

The Hall number $n_H(T_c, T)$ has been extensively measured[14-20] for a variety of cupric superconductors and for various T_c and $T > T_c$. In the s-channel theory, the fermion carries charge e and the boson $2e$. Let n_f, μ_f be the density and mobility of the fermion, and n_b, μ_b be those of the boson. In a simple two-carrier model, n_H is given by

$$n_H(T_c, T) = \frac{(n_f\mu_f + 2n_b\mu_b)^2}{n_f\mu_f^2 + 2n_b\mu_b^2} . \tag{8.1}$$

From either the μsR experiment (equations (2.1) and (2.6)) or the theoretical formula (2.9), we know that as $T_c \to 0$, $n_b \to 0$ and therefore

$$n_H \to n_f . \tag{8.2}$$

In addition, since the fermions form a degenerate Fermi sea with top energy $= \nu$, let its density at $T = 0$ be n_ν:

$$n_\nu \cong \frac{m_f\nu}{\pi c} \tag{8.3}$$

where m_f is the fermion mass and c is, as before, the average spacing between neighboring CuO_2 planes. In the temperature range of interest, we may neglect the small temperature variation of the fermion density n_f and set

$$n_f \cong n_\nu = \text{constant} , \tag{8.4}$$

which is the same parameter for all cupric superconductors.

For a given sample ($La - 214$, or $Y - 123$, or $\cdots$), the total charge density (in units of e)

$$n = n_f + 2n_b \cong n_\nu + 2n_b$$

is independent of T but, because of doping, varies with T_c. Therefore,

$$n_b \cong \frac{1}{2}(n - n_\nu) = n_b(T_c) \tag{8.5}$$

is also independent of T. Taking the $T = 77°$K Hall number measurements (for different T_c) on $La - 214$ by Ong et $al.$[14] and extrapolating to $T_c \to 0$ (which has a nonzero doping), we find[11], on account of (8.2)

$$n_\nu \cong 5.2 \times 10^{-4} \text{Å}^{-3} \,. \tag{8.6}$$

By using (8.5) with n from the stoichiometric measurements (for $x < 0.15$), substituting the ratio T_c/n_b into (2.9) and combining that with the μsR measurements[10], we find

$$m_b = 2m_f \cong 6.2m_e \,. \tag{8.7}$$

To analyze the large number of $n_H(T_c, T)$ measurements, we assume the mobilities μ_b and μ_f are insensitive to doping; hence, they depend only on T, so does their ratio

$$r(T) \equiv \frac{\mu_b(T)}{\mu_f(T)} \,. \tag{8.8}$$

Thus, (8.1) becomes

$$n_H(T_c, T) \cong \frac{[\,n_\nu + 2n_b(T_c)\,r(T)\,]^2}{n_\nu + 2n_b(T_c)\,r^2(T)} \,. \tag{8.9}$$

We then determine $r(T)$ by making a least-square fit of (8.9) to all $n_H(T_c, T)$ at a given T (but for different T_c and with $n_b(T_c)$ given by the theoretical formula (2.9)). The same $r(T)$ is then substituted back into (8.9) for calculating n_b from the observed n_H. The resulting T_c vs. n_b/n_ν (with n_ν given by (8.6)) is plotted[11] in Figure 2, with its details listed in Table 1. This is to be compared with Figure 1 which gives the same T_c vs. n_b/n_ν, but based on the μsR measurements[10]. In both figures the solid line is the theoretical curve (2.9). As one can see, these results are mutually consistent, and lend support to the theory.

9. HIGHER T_c?

The small coherence length in the (warm) cupric superconductors makes it possible for us to assume Bose-Einstein condensation as the underlying phase transition mechanism. In that case, T_c can be raised by increasing the bosonic density n_b. In practice, this is difficult; e.g., the average spacing between CuO_2 planes can hardly be changed, partly because these cupric layers are electrically charged. However, for Bose-Einstein condensation, there is no particular importance to the lattice structure that the bosons (i.e., ϕ-quanta) move in. In this section, we comment briefly on the possibility of non-crystalline high T_c superconductors.

Since the ϕ-quantum is well-localized in the coordinate space, it seems reasonable that its existence can be preserved on a microscopic plaquette made of a few (two or more) connected CuO_2 unit squares. These plaquettes may then be immersed in some suitable three-dimensional conducting liquid or amorphous medium. Two conditions must be fulfilled: (i) the plaquettes are structurally stable and (ii) the ϕ-quanta can propagate relatively freely in the medium, without moving the plaquettes. Bose-Einstein condensation occurs when the thermal wavelength λ_T is comparable to the interparticle distance d.

Assuming that the structural problem of stabilizing small plaquettes of CuO_2 squares can be solved, there remains the question of how small these plaquettes can be. Under the hypothesis that each ϕ-quantum is made of a pair of O^- holes, a prototype ϕ-quantum may simply reside on a single rectangular plaquette consisting of two CuO_2 squares, sharing a common side. Such a plaquette can contain six Cu^{2+}, five O^{2-} and two O^-, with the two O^- forming a ϕ-quantum. As shown in Figure 3, there are eight different configurations (arranged in ascending order from (i) to (viii) according to their Coulomb energies, which are calculated in the approximation of treating all ionic charges as point-like). Each can be a candidate to be an excited boson state. Since each plaquette is electrically neutral, one may hope to achieve a closer packing of these plaquettes. Their immersion in a suitable medium may enable the ϕ-quanta to move from plaquette to plaquette. If so, this could open new directions to even higher T_c superconducting materials, which might be liquid or plastic.

REFERENCES

This research was supported in part by the U.S. Department of Energy.

[1]. J. C. Bednorz and K. A. Müller, Z.Phys. **B64**, 189 (1986).

[2]. M. K. Wu, J. R. Ashburn, C. J. Torng, P. H. Hor, R. L. Meng, L. Gao, Z. J.

Huang, Y. Q. Wang and C. W. Chu, Phys.Rev.Lett. **58**, 908 (1987). Z. X. Zhao, L.Chen, Q. Yang, Y. Huang, G. Chen, R. Tang, G. Liu, C. Cui, L. Wang, S. Guo, S. Lin and J. Bi, Kexue Tongbao **6**, 412 (1987).

[3]. P. Chaudhari *et al.*, Phys.Rev. **B36**, 8903 (1987).

[4]. T. K. Worthington, W. J. Gallagher and T. R. Dinger, Phys.Rev.Lett. **59**, 1160 (1987).

[5]. J. Bardeen, L.N. Cooper and J.R. Schrieffer, Phys.Rev. **106**, 162 (1957); **108**, 1175 (1957).

[6]. T.D. Lee, "s-Channel Theory of Superconductivty," in *Symmetry in Nature* (Pisa, Scuola Normale Superiore, 1989), Vol. 2. R. Friedberg and T.D. Lee, Phys.Lett. **138A**, 423 (1989).

[7]. R. Friedberg and T.D. Lee, Phys.Rev. **B40**, 6745 (1989).

[8]. M.R. Schafroth, Phys.Rev. **100**, 463 (1955).

[9]. R. Friedberg, T.D. Lee and H.C. Ren, "A Correction to Schafroth's Superconductivity Solution of an Ideal Charged Boson System," Preprint CU-TP-460.

[10]. Y.J. Uemura *et al.*, Phys.Rev.Lett. **62**, 2317 (1989).

[11]. R. Friedberg, T.D. Lee and H.C. Ren, "Applications of the s-Channel Theory to the μsR and Hall Number Experiments on High T_c Superconductors," Columbia University Preprint CU-TP-483.

[12]. R. Friedberg, T.D. Lee and H.C. Ren, "The s-Channel Theory of Superconductivity on a One-Dimensional Lattice of Parallel (CuO_2) Planes," Columbia University Preprint CU-TP-482.

[13]. P.W. Higgs, Phys.Lett. **12**, 132 (1964). Y. Nambu, Phys.Rev.Lett. **4**, 350 (1960). J. Goldstone, Nuovo Cimento **19**, 154 (1961).

[14]. N.P. Ong *et al.*, Phys.Rev. **B35**, 8807 (1987).

[15]. M. Suzuki, Phys.Rev. **B39**, 2312 (1989).

[16]. Z.Z. Wang *et al.*, Phys.Rev. **B36**, 7222 (1987).

[17]. J. Clayhold *et al.*, Phys.Rev. **B39**, 7320 (1989).

T. D. Lee 21

[18]. H. Takagi *et al.*, Nature **332**, 236 (1988).

[19]. A. Maeda *et al.*, Jpn J.Appl.Phys. **28**, L576 (1989).

[20]. J. Clayhold *et al.*, Phys.Rev. B**38**, 7016 (1988).

[21]. V. Aharony *et al.*, Phys.Rev.Lett. **60**, 1330 (1988).

FIGURE CAPTIONS

FIGURE 1. Critical temperature T_c versus the boson density n_b in units of $n_\nu = 5.2 \times 10^{-4} \text{Å}^{-3}$, with points determined from the μsR measurements (Ref. 10) and the solid line according to the theoretical formula (2.9). The dots and solid triangles are from the $La - 214$ and $Y - 123$ data. From left to right, the two open triangles are from the $Bi - 2212$ and $Tl - 2212$ data; the first diamond refers to $Bi - 2223$ and the other two to $Tl - 2223$.

FIGURE 2. Critical temperature T_c versus the boson density n_b in units of $n_\nu = 5.2 \times 10^{-4} \text{Å}^{-3}$. The points are determined from the Hall number measurements (details given in Table 1). The solid line is according to the theoretical formula (2.9).

FIGURE 3. Examples of single neutral rectangular plaquettes, each containing six Cu^{2+} (squares), five O^{2-} (open circles) and two O^- (filled circles), with the two O^- forming a ϕ-quantum. For point-like ions, (i)-(viii) are arranged in ascending order of their Coulomb energies, with (i) the lowest. Each O^- forces its two neighboring Cu^{2+} spins to be parallel to each other. A linear $Cu^{2+} - O^{2-} - Cu^{2+}$ chain with two parallel Cu^{2+} spins is called a frustrated bond (Ref. 21). Figures (v) and (viii) have no frustrated bond; all others have one each.

TABLE CAPTION

TABLE 1. Mobility ratio μ_b/μ_f and $\hat{n}_b = n_b/n_\nu$ (boson density n_b in units of $n_\nu = 5.2 \times 10^{-4} \text{Å}^{-3}$) determined from Hall number measurements. Except for the 77°K measurement of $La_{1.9} Sr_{0.1} CuO_4$ and the 300°K measurement of Bi-compound (which are on single crystals), all data are taken from ceramics results. The symbols refer to those used in Figure 2.

Material	$T_{\text{measure}}(^0K)$	$\frac{\mu_b}{\mu_f}$	$\hat{n}_b$	$T_C(^0K)$	Symbol	References
$La_{1.95}Sr_{0.05}CuO_4$	77	0.5	0	0	•	14
$La_{1.90}Sr_{0.10}CuO_4$	77	0.5	0.723	28	•	15
$La_{1.86}Sr_{0.14}CuO_4$	77	0.5	1.00	37.5	•	14
$La_{1.85}Sr_{0.15}CuO_4$	77	0.5	1.73	38	•	14
$La_{1.14}Sr_{0.16}CuO_4$	77	0.5	2.79	35	•	14
$YBa_2Cu_3O_{7-y}$	77	0.262	1.26	51	▲	16
$YBa_2Cu_3O_{7-y}$	77	0.262	1.52	50	▲	16
$YBa_2Cu_3O_{7-y}$	95	0.256	1.70	50	▲	16
$YBa_2Cu_3O_{7-y}$	95	0.256	1.45	57	▲	16
$YBa_2Cu_3O_{7-y}$	95	0.256	1.33	51	▲	16
$YBa_2Cu_3O_{7-y}$	95	0.256	1.70	63	▲	16
$YBa_2Cu_3O_{7-y}$	95	0.256	3.94	92	▲	16
$YBa_2Cu_3O_{7-y}$	160	0.589	1.63	63	▲	16
$YBa_2Cu_3O_{7-y}$	160	0.589	1.56	50	▲	16
$Bi_4Sr_3Ca_3Cu_4O_{16+y}$	100	1.0	2.95	80	Δ	17
$Bi_2(Sr,Ca)_3Cu_2O_{8+y}$	300	1.0	2.38	80	Δ	18
$(Bi,Pb)_2Sr_2Ca_2Cu_3O_y$	130	0.232	3.36	110	♦	19
$Tl_2Ca_2Ba_2Cu_3O_x$	130	0.074	3.70	120	♦	20

TABLE 1.

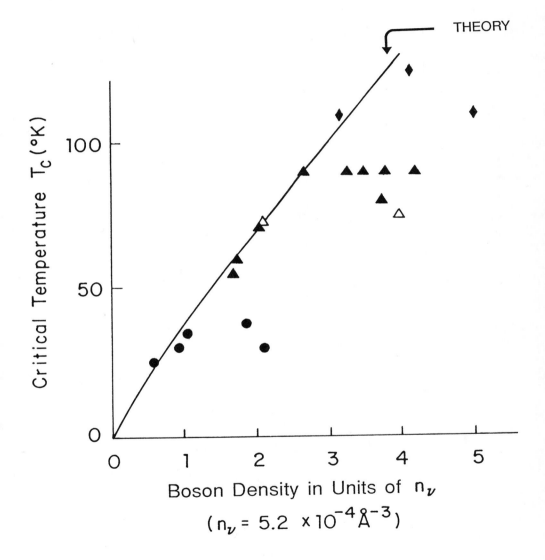

FIGURE 1.

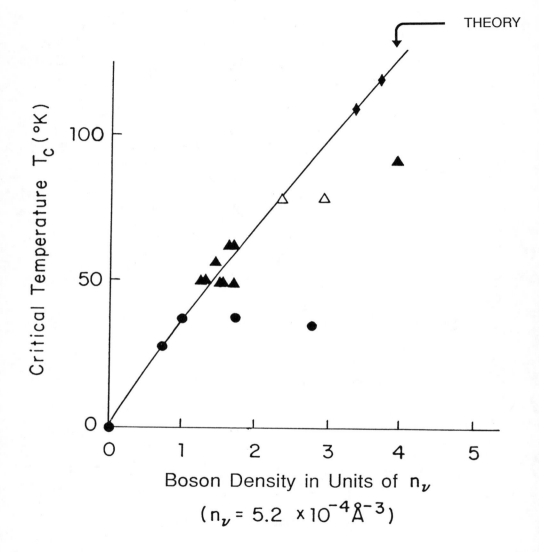

FIGURE 2.

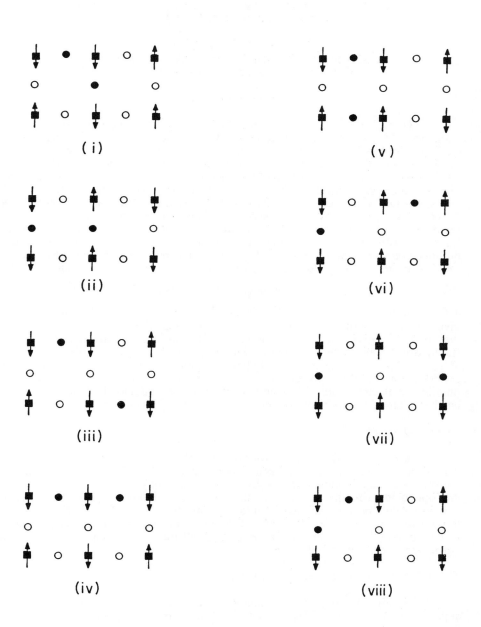

FIGURE 3.

CRYSTAL CHEMISTRY AND MECHANISM IN THE HIGH-T$_c$ COPPER OXIDES

John B. Goodenough
Center for Material Science and Engineering, ETC 5.160
University of Texas at Austin, Austin, Tx 78712-1084

ABSTRACT

Several structural, normal-state, and superconductive properties of the high-T$_c$ copper oxides are discussed as a prelude to the proposal of a correlation-bag model of Cooper-pair formation. All copper oxides have intergrowth structures, which introduce not only anisotropy -- both electronic and mechanical -- but also internal electric fields normal to the intergrowths in addition to the local electric fields at copper due to the oxygen coordination. The significance of these internal fields for the oxidation state of the superconductive CuO$_2$ layers is illustrated, and the need for all Cu of a CuO$_2$ layer to have the same oxygen coordination is emphasized. The role of interlayer bond-length mismatch is illustrated by the A$_2$CuO$_4$ compounds. Evidence for a first-order phase change on going from the antiferromagnetic-semiconductor to the superconductor phase is presented, and XPS data indicative of a change from more ionic to more covalent Cu-3d, O-2p$_\sigma$ bonding across this phase change is cited.

Evidence for intermediate correlation energies (U $\gtrsim$ W) in the superconductive phase and an equilibrium reaction Cu^{3+} + O^{2-} $\rightleftarrows$ Cu^{2+} + O$^-$ near cross-over are argued to give rise to instabilities that cannot be expressed as a negative-U charge-density wave; they therefore give rise to dynamic charge fluctuations that are manifest as spin fluctuations in the hole-poor regions and as paired holes in the hole-rich regions. Where the pair density is large enough for pair-pair interactions, a cooperative condensation into the superconductive state occurs. Pair formation is stabilized not only by the electron-phonon interactions, but also by modulation of the bandwidth -- and hence also the correlation energy -- within a charge fluctuation.

INTRODUCTION

The superconductive properties of the high-T$_c$ copper-oxide superconductors cannot be described by the Bardeen-Cooper-Schrieffer (BCS) theory applicable to conventional superconductors; they indicate a new physical phenomenon. In an attempt to identify the essential features of this phenomenon and the particular conditions that must be satisfied for it to be manifest in nature, I shall call attention to those distinctive properties of the high-T$_c$ copper-oxide superconductors that I believe to be particularly relevant. These properties fall into three categories: structural, normal-state, and superconductive.

STRUCTURAL CONSIDERATIONS

The high-T$_c$ copper-oxide superconductors all have intergrowth structures; superconductively "active" layers containing one or more CuO$_2$ sheets alternate with "inactive" layers of variable oxygen concentration. The CuO$_2$ sheets consist of a square network of copper atoms with oxygen atoms forming 180° (or nearly 180°) Cu-O-Cu bridges. Within these sheets the important interactions are Cu-O-Cu interactions and O-O interactions; the separation of the Cu atoms from one another is too great for any significant Cu-Cu interactions.

Structural intergrowth introduces three important properties [1]:

- Anisotropy -- both electronic and mechanical

- The possibility of an internal electric field normal to the intergrowth layers
  that shifts
  the electronic energy levels of adjacent layers in opposite directions.

- Bond-length mismatch across an interface between layers that places one
  layer under compression and the other under tension.

The anisotropic properties of the copper-oxide superconductors make difficult the processing of these materials into wires suitable for power applications; it is not yet established whether they play a fundamental role in the mechanism of high-T_c superconductivity.

In an ionic material, the individual intergrowth layers may carry formal charges of opposite sign; and such charges introduce internal electric fields that may shift the electronic energy levels of neighboring layers by more than 1 eV with respect to one another, as has been demonstrated by the stabilization of Mn(II) in the presence of Fe(IV) in $BaFe_{12-x}Mn_xO_{19}$ [2]. However, the oxygen coordination of the copper atoms appears to be even more critical; the site Madelung energy -- and hence the relative positions of the Cu-3d and O-2p states in an ionic model -- is sensitive to the number of oxygen near neighbors and the mean Cu-O distance. This situation is well illustrated by the $YBa_2Cu_3O_{6+x}$ structure of Fig. 1.

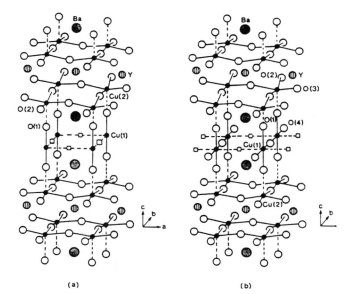

(a)   (b)

Fig. 1   Structures of (a) tetragonal $YBa_2Cu_3O_6$ and (b) ideal orthorhombic $YBa_2Cu_3O_7$.

In the structure of Fig. 1, the active layers contain two CuO_2 sheets, the inactive layers contain a CuO_x plane of variable oxygen content x located between two BaO rocksalt (001) planes. We represent the stacking sequence along the c-axis as

$$-CuO_2 \mid BaO\text{-}CuO_x\text{-}BaO \mid CuO_2\text{-}Y\text{-}CuO_2 \mid BaO- \qquad (1)$$

where the vertical lines denote interlayer interfaces on traversing the c-axis of the structure, *i.e.* moving to the right in (1). The Cu(2) atoms of a CuO_2 sheet are seen to be coordinated by five near-neighbor oxygen atoms; the Cu(1) atoms of the CuO_x planes have an oxygen coordination that varies with x and with the ordering of the oxygen atoms within the plane. Cava [3] was the first to show that careful extraction of oxygen at 400 °C from $YBa_2Cu_3O_{6.95}$ leads to the step-wise variation of the critical temperature T_c versus x shown in Fig. 2. However, introduction of oxygen to x > 1 in $YBa_{2-y}La_yCu_3O_{6+x}$ causes T_c to decrease linearly with x[4], see Fig. 4. Interpretation of these data follows from two observations and one assumption:

Observation 1: Where the CuO_2 sheets carry a formal charge $(CuO_2)^{2-}$ corresponding to a single valence state Cu(II), there each copper atom carries a localized magnetic moment of about 0.5 μ_B, and long-range antiferromagnetic order

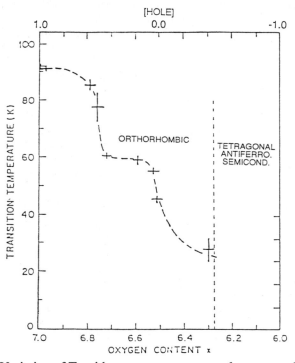

Fig. 2   Variation of T_c with oxygen content x on slow removal of oxygen from
$YBa_2Cu_3O_{6.95}$ at 400 °C, after ref. [3].

occurs below a Néel temperature T_N [5]; but the oxidation of the sheets to $(CuO_2)^{(2-p)-}$ within a range $p_c < p < p_m$ leads to superconductivity with a critical temperature that increases with p as $T_c \sim p/m^*$, where the effective mass m^* of the holes of concentration p per Cu(2) atom is probably not a sensitive function of p [6].

Observation 2: $YBa_2Cu_3O_{6+x}$ remains tetragonal and antiferromagnetic for $0 \leq x < 0.27$; it becomes orthorhombic due to oxygen ordering within the CuO_x plane into b-axis sites in the range $0.27 < x < 1$, see Fig. 1(b) for x = 1 with ideal ordering [7]. Moreover, a second level of order is found [8] for x = 0.5; the oxygen atoms order into alternate *b*-axis chains as illustrated in Fig. 3(a). Finally, a third level of order has been observed[9] for x = 0.75; the oxygen atoms of the partially filled chains tend to order into alternate *b*-axis sites as illustrated in Fig. 3(b).

Assumption: Twofold-coordinated Cu(1) atoms have the formal valence Cu(I) even in the presence of mobile holes p in an oxidized $(CuO_2)^{(2-p)-}$ plane, but Cu(1) atoms with threefold or fourfold coordination are oxidized to Cu(II) before mobile holes are introduced into the Cu(2) planes. Finally, fivefold-coordinated Cu(1) may compete with the Cu(2) for oxidation and therefore trap out mobile holes from the CuO_2 planes. These relationships are the consequence of a larger Madelung electric field at a copper site for a larger anion coordination given the constraint that the in-plane Cu-O distances are nearly the same.

Returning to Fig. 2, the introduction of disordered atoms into the CuO_x plane in the range $0 \leq x \leq 0.27$ can only oxidize Cu(I) to Cu(II). Consequently the Cu(2) atoms remain Cu(II) in antiferromagnetic $(CuO_2)^{2-}$ sheets. Within the range $0.27 < x < 0.5$, oxygen ordering onto *b*-axis sites converts the crystal symmetry from tetragonal

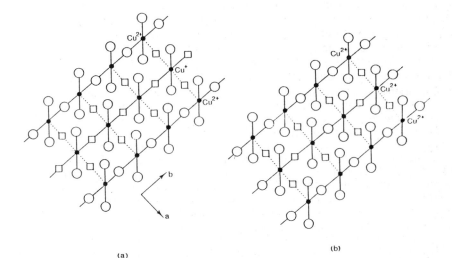

(a)   (b)

Fig. 3   (a) Intercahin ordering in $YBa_2Cu_3O_{6.5}$ and (b) interchain and intrachain ordering in fully ordered $YBa_2Cu_3O_{6.75}$.

to orthorhombic, and interchain ordering increases the concentration of twofold-coordinated Cu(I); at x = 0.5 ideal ordering would convert the Cu(1) from all Cu(II) for complete disorder to half Cu(I) and half Cu(II) for perfect order, which creates p = 0.25 holes per Cu(2) in the CuO_2 planes. At x = 0.5, a $p_c < p \approx 0.25 < p_m$ renders the CuO_2 sheets superconductive. Whether the order-disorder transition versus x is smooth or first-order has not yet been established for this system; the associated antiferromagnetic-superconductive transition probably makes it first-order and gives rise to a two-phase region within the compositional range 0.27 < x < 0.45. In the range 0.5 < x < 0.75, the Cu(1) atoms in partially filled chains are oxidized from Cu(I) to Cu(II), so -- with perfect order -- no additional oxidation of the CuO_2 planes occurs as x increases; consequently with ideal ordering $T_c \sim p/m^*$ would remain independent of x in this compositional range. Note that with T_c independent of x, inhomogeneities in x will not give rise to a broad transition region ΔT associated with T_c. A sharp transition does not require a distinguishable $T_c \approx 60$ K phase in the range 0.5 < x < 0.75. Finally, in the range $0.75 \leq x < 1$, all the Cu(1) are Cu(II) so the concentration p of mobile holes per Cu(2) increases with x. Consequently $T_c \sim p/m^*$ increases with x up to about 90 K at a $p \approx 0.45$ corresponding to x = 0.90. It appears that in the range 0.90 < x < 0.95 the value of T_c is again nearly independent of x because T_c saturates at $p \approx p_m \approx 0.45$.

Substitution of La for Ba in $YBa_{2-y}La_yCu_3O_{6+x}$ allows stabilization of compositions with x > 1, which means the introduction of a-axis oxygen atoms. Each a-axis oxygen in the $Cu(1)O_x$ plane creates two fivefold-coordinated copper, and a linear decrease in T_c with x > 1 for y > 0.1 corresponds to a trapping out of two holes per a-axis oxygen from the CuO_2 sheets [4]. The original interpretation [4, 10] of a trapping out at peroxide ions $(O_2)^{2-}$ appears to be incorrect; it is more probable that the holes are trapped within a Cu_2O_9 cluster. Interestingly, modulation of the c-axis electric field by co-doping Ca^{2+} for Y^{3+} in the system $Y_{1-z}Ca_zBa_{2-y}La_yCu_3O_{6+x}$ raises the critical concentration x_c at which holes are trapped out from the CuO_2 sheets by fivefold-coordinated Cu(1) atoms to $x_c = 1.12$ for z = 0.4 [11], see Fig. 4. Seebeck measurements showed an increase in the trapping energy with increasing $(x-x_c)$ [12].

An important general conclusion from this discussion is the following: *The potential energy for electrons at a copper site varies sensitively with the oxygen coordination and the mean Cu-O band length at that site.* It follows that narrow-band superconductive behavior within CuO_2 planes requires an identical oxygen coordination at each copper atom in the plane. The intergrowth structure of $YBa_2Cu_3O_{6+x}$ permits a variable oxygen concentration that leaves unchanged the oxygen coordination of the Cu(2) atoms in the CuO_2 sheets. On the other hand, substitution of La or Sr for Y leads to the incorporation of some oxygen between the CuO_2 sheets, which produces Cu(2) with sixfold as well as fivefold coordination; these materials are not superconductors. [13].

The problem of bond-length mismatch is well illustrated by the system $La_{2-x}Sr_xCuO_4$, which has the structure illustrated in Fig. 5. In this structure a single CuO_2 sheet alternates with two (001) LaO rocksalt planes; this stacking may be represented schematically as

$$-LaO \mid CuO_2 \mid LaO\text{-}LaO \mid CuO_2 \mid LaO- \qquad (2)$$

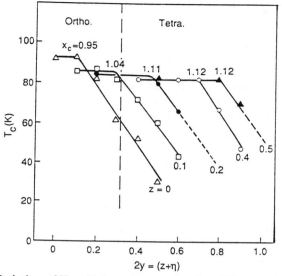

Fig. 4   Variation of T_c with La concentration for different values of Ca concentration z in the system $Y_{1-z}Ca_2Ba_{2(1-y)}La_{2y}Cu_3O_{b+x}$ in which $x=6.94 + 0.5(2y-z)$; x_c refers to the critical oxygen content above which hole trapping occurs.

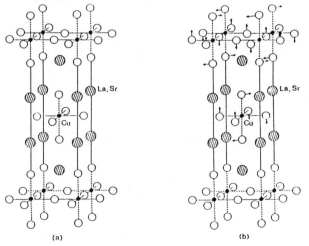

Fig. 5   Structures of (a) T-tetragonal and (b) O-orthorhombic La_2CuO_4.

A measure of the bond-length matching across the interlayer interface is the tolerance factor

$$t = (La-O)/\sqrt{2} \ (Cu-O) \tag{3}$$

where the bond-lengths La-O and Cu-O are temperature-dependent. Since the thermal expansion of the La-O bond is larger than that of the Cu-O bond, the tolerance factor t decreases with decreasing temperature. A $t < 1$ places the CuO_2 planes under compression, the $(LaO)_2$ layers under tension. With a $t \leq 1$ at the sintering temperature, the bondlength mismatch increases with decreasing temperature. As the temperature decreases, the increasing compressive stress on the CuO_2 planes is relieved by three successive mechanisms [1].

First, the single hole in the 3d shell of the Cu(II) ion is ordered into an antibonding orbital of d_{x2-y2} parentage, where the x and y axes are taken to be along the Cu-O bond axes of a CuO_2 plane. This antibonding orbital forms a narrow, strongly correlated band of one quasi-particle states designated σ^*_{x2-y2}; it includes covalent hybridization with the near-neighbor O-2s, $2p_x,2p_y$ orbitals. Ordering of the 3d hole strongly distorts the octahedral site of the Cu atoms to tetragonal ($c/a > 1$) symmetry with the long axis parallel to the c-axis.

Second, the structure tends to incorporate excess interstitial oxygen atoms; these occupy sites between the two LaO planes of a rocksalt layer where they are coordinated by four La and four O atoms. The c-axis stacking sequence for $La_2CuO_{4+\delta}$ may be represented as

$$-LaO \mid CuO_2 \mid LaO-O_\delta-LaO \mid CuO_2 \mid LaO- \tag{4}$$

The intersitial oxygen atoms capture $p = 2\delta$ electrons from the CuO_2 planes to reduce the charges on the layers -- and hence the c-axis electric field. For $\delta \leq 0.02$, the phase remains an antiferromagnetic semiconductor; for $\delta > 0.05$ the oxide is a superconductor [14]. A two-phase region $0.02 < \delta < 0.05$ separates the antiferromagnet and superconductor phases, which indicates that there is a first-order phase change on going from the antiferromagnetic to the superconductive phase. A first-order phase change with loss of oxygen occurs on heating the superconductive phase, prepared under high oxygen pressure, to above 225 °C [15]. Note that the interstitial oxygen do not change the oxygen coordination about a Cu atom of the oxidized $(CuO_2)^{(2-p)-}$ plane. At atmospheric oxygen pressure, only filamentary superconductivity is found [16]. As the temperature is lowered to where a $\delta > 0.02$ might be incorporated into the structure, the mobility of the oxygen atoms at atmospheric pressure becomes too low for any significant penetration of the superconductive phase into the bulk of the sample.

Third, the CuO_6 octahedra rotate cooperatively about a [110] axis so as to buckle the CuO_2 planes as indicated by the arrows in Fig. 5(b). This buckling gives rise to a tetragonal-to-orthorhombic transition on lowering the temperature through the transition temperature T_t. As indicated schematically in Fig. 6, the introduction of interstitial oxygen, which relieves the compressive stress, lowers T_t from above room temperature to below it at $\delta \approx 0.02$. The superconductive phase ($\delta > 0.05$), on the other hand, exhibits an orthorhombic-to-tetragonal transition with increasing temperature at a termperature T'_t, and T'_t increases with δ. The distortion in the superconductive phase appears to be associated with elastic coupling between the local

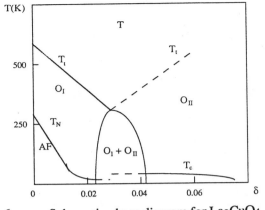

Fig. 6   Schematic phase diagram for $La_2CuO_4 + \delta$.

lattice deformations about the interstitial oxygen atoms [17]; these interactions may also be assocated with short-range ordering of the interstitial atoms [18]. At $\delta \approx 0.03$, a segregation into two phases, O_I and O_{II}, occurs below a temperature $T_s < T_t'$ [19].

If the La(III) ion is replaced by a smaller rare-earth ion such as Nd(III), then the tolerance factor is too small for stabilization of the structure of Fig. 5; the oxygen of the $(LaO)_2$ rocksalt layers are displaced from the c-axis to the tetrahedral-site positions between Nd planes as shown in Fig. 7(a). The resulting tetragonal structure has fluorite $Nd-O_2-Nd$ stacking along the c-axis; it is designated the T' structure to distinguish it from the T/O structure of La_2CuO_4. If solid solutions are made between La_2CuO_4 and Ln_2CuO_4, where La is Sm or a smaller rare-earth atom, then interlayer ordering of the La(III) and the smaller, e.g. Dy(III), rare-earth atom may give the alternating rocksalt and flourite layers of the T^* structure of Fig. 7(b). Where Ln=Pr or Nd, the size difference between La and the Ln atom is not sufficient to produce the ordering required to stabilize the T^* phase, but a T" phase appears within the two-phase domain separating the T' and T/O compositonal ranges, see Fig. 8 [20]; the T" phase has the T' structure but probably with ordering within the La_3Ln planes and/or some oxygen displaced back to the c-axis.

NORMAL-STATE PROPERTIES

It is instructive to consider the simplest of the high-T_c copper-oxide systems, $La_{2-x}Sr_xCuO_4$. As pointed out above, a $t < 1$ produces a compressive stress on the CuO_2 planes that is partially relieved by an ordering of the Cu-3d hole into orbitals of $3d_{x2-y2}$ parentage. This ordering distorts the octahedral interstice of the Cu atom to tetragonal $(c/a>1)$ symmetry so as to remove the degeneracy of the $3d_{x2-y2}$ and $3d_{z2}$ orbitals.

For the antiferromagnetic end number La_2CuO_4, removal of the orbital degeneracy simplifies the problem to that of a half-filled, nondegenerate σ^*_{x2-y2} band. The presence of antiferromagnetic order and a localized magnetic moment on the

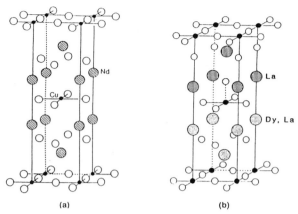

(a)   (b)

Fig. 7   Structures of (a) T'-tetragonal Nd_2CuO_4 and (b) T*-tetragonal
            $LaDyCuO_4$.

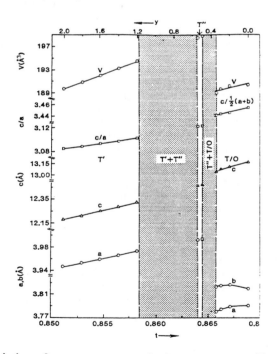

Fig. 8   Variation of room-temperature lattice parameters with tolerance factor t or
            compostion parameter y for the system $La_{2-y}Nd_yCuO_4$.

copper, $\mu_{Cu} \approx 0.5\mu_{\beta}$, signals two things. First, the crystalline electrostatic Madelung energy is large enough to bias the equilibrium

$$Cu^{2+} + O^{2-} \rightleftarrows Cu^+ + O^- \tag{5}$$

sufficiently far to the left that the covalent mixing associated with elecron transfer from oxygen to copper can be treated in second-order perturbation theory. In this situation, it is meaningful to construct crystal-field wavefunctions of the type

$$\psi_{x^2-y^2} = N_{\sigma}(f_{x^2-y^2} - \lambda_s\phi_s - \lambda_{\sigma}\phi_{\sigma}) \tag{6}$$

where $f_{x^2-y^2}$ is the Cu-3$d_{x^2-y^2}$ orbital, ϕ_s and ϕ_{σ} are appropriately symmetrized O-2s and O-2p_x,2p_y orbitals, and the

$$\lambda_i = b^{ca}/\Delta E_i \tag{7}$$

are the covalent-mixing parameters for a cation-anion resonance integral b^{ca} and an energy ΔE for electron transfer from oxygen to an empty Cu-3$d_{x^2-y^2}$ state   Second, the Cu-O-Cu interactions between crystal-field $\psi_{x^2-y^2}$ orbitals on neighboring Cu atoms, which is given by the resonance integral

$$b_{\sigma} = (\psi_i, H'\psi_i) \approx \varepsilon_{\sigma}(\lambda_{\sigma^2} + \lambda_{\sigma^2}) \tag{8}$$

is small enough that the $\sigma^*_{x^2-y^2}$ band electrons remain strongly correlated. Strong correlations in a half-filled band produce a spontaneous atomic moment μ_{Cu} and split the band in two (Mott-Hubbard correlation splitting). A measured bandgap of 2 eV [21] means that $(U - W) \gtrsim 2$ eV, where U is the on-site correlation energy and W is the width of the $\sigma^*_{x^2-y^2}$ band. In tight-binding approximation,

$$W \approx 2zb_{\sigma} \sim \lambda_{\sigma}^2 \tag{9}$$

where z=4 is the number of Cu nearest neighbors to a Cu atom.

Fig. 9 illustrates, for the case of an on-site U too large for stabilization of a negative-U charge-density wave, the evolution of the one-particle energies as a function of the interatomic interaction parameter b for a half-filled band and the associated phase diagram; in the phase diagram, T_{cs} is a BCS superconductive critical temperature [22]. The antiferromagnetic semiconductor La$_2$CuO$_4$ has a b_{σ} for the $\sigma^*_{x^2-y^2}$ band that lies to the left of b_g in this diagram.

We must now inquire into what happens on oxidation of the CuO$_2$ sheets. This inquiry is complicated by two factors. First, the correlation splitting may be (U - W) > 2 eV, in which case the lower $\sigma^*_{x^2-y^2}$ band lies below the top of the $(\pi^* + \sigma^*_{z^2})$ bands; oxidation of the CuO$_2$ planes would, in this case, introduce holes into the $(\pi^* + \sigma^*_{z^2})$ bands. Although this situation may apply on initial oxidation, an importnat reduction in (U-W) occurs -- see below -- on going from the antiferromagnetic to the superconductive state. Therefore it seems unnecessary to introduce this complication into the primary analysis. Second, Koopman's theorem is not appplicable to strongly correlated electronic energy bands; removal of electrons from the CuO$_2$ sheets must cause an important decrease in ΔE of equation (6) and hence an important increase in λ_{σ} and, from equations (7) and (8), b and W. Therefore, for b < b_g oxidation not only

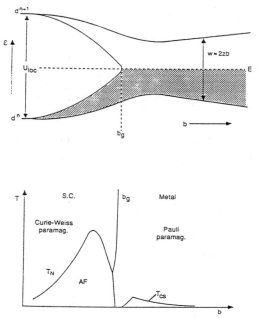

Fig. 9   Schematic variation with interatomic electron-transfer-energy integral b of
             (a) quasiparticle energies and (b) magnetic and superconductor ordering
             temperatures T_N and T_c for a half-filled band.

lowers the Fermi energy into the lower Hubbard band of Fig. 9 (with an
accompanying rearrangement of numbers of states in the two bands), it also increases
b_σ towards $b_\sigma \approx b_g$. In fact, although equation (4) may be considered to be biased to
the left, the equilibrium

$$Cu^{3+} + O^{2-} \rightleftarrows Cu^{2+} + O^- \qquad (10)$$

does not appear to be biased strongly to either the right or the left. As the electron-
transfer equilibrium between copper and oxygen arrays approaches cross-over, the
energy ΔE of equation (6) goes to zero and the perturbation expansion defining λ_σ
breaks down. Although a breakdown in our mathematical representation does not
necessarily represent a breakdown in the fundamental physics, nevertheless a change
from more ionic-type bonding to more covalent-type bonding could well occur via a
first-order phase change. In fact, a recent XPS study [23] indicates just such a first-
order change on going from the antiferromagnetic to the superconductive state.
        Fig. 10 shows four Cu ($2p_{3/2}$) spectra taken from this study. Curves (a) and
(b) are for $LaCuO_3$ and La_2CuO_4, respectively. The binding energy (BE) of the Cu
($2p_{3/2}$) electron in $LaCuO_3$ is shifted by 2.5 eV relative to its position in La_2CuO_4; it is
shifted by 1.2 eV relative to its postion in $NaCuO_2$ (not shown). The larger shift
associated with Cu(III) in the metallic perovskite compound $LaCuO_3$ compared to that

of the Cu(III) in the insulator $NaCuO_2$ is due to the delocalization of all the σ-bonding electrons at octahedral-site Cu(III) in $LaCuO_3$. In $NaCuO_2$, the Cu(III) have square-coplanar coordination with a localized $3d_{z^2}^2$ configuration (z-axis perpendicular to the CuO_4 plane) that provides stronger screening of the nuclear charge. Curves (b) and (d) of Fig. 10 correspond, respectively, to "clean" and somewhat "dirty" powder samples of $La_{1.85}Sr_{0.15}CuO_4$. Each sample was annealed at 1000 °C in air to remove carbonaceous species from the surface, slow cooled in O_2 to 250 °C, and quenched under He into a desiccator. Samples were reground in inert atmosphere to attain fresh surfaces and, within 20 minutes, transferred to the vacuum chamber of the spectrometer with less than five minutes exposure to air. Nevertheless, a rapid build-up of carbonaceous surface species occurs on exposure to air, and the surface carbon/oxygen ratio C/O remained finite. "Clean" samples had a surface C/O ratio below a fixed limit; "dirty" samples had a larger C/O ratio. The "clean" samples gave a Cu ($2p_{3/2}$) spectrum like that of $LaCuO_3$; the "dirty" samples like that of La_2CuO_4. The curve (d) represents a sample with a dirty surface layer and some contribution from the bulk having a spectrum like $LaCuO_3$. The La(3d) peaks showed a shake-up satellite for the "dirty" samples like that found in La_2CuO_4, but the "clean" samples had a shake-down satellite like that found in $LaCuO_3$; the remarkable shift in the Fermi energy that this result implies was also seen in the band region of the XPS spectrum. Moreover, clean samples of the n-type superconductors having the T' phase gave a Cu ($2p_{3/2}$) spectrum like that of $NaCuO_2$, as would be expected for delocalized $σ^*_{x^2-y^2}$

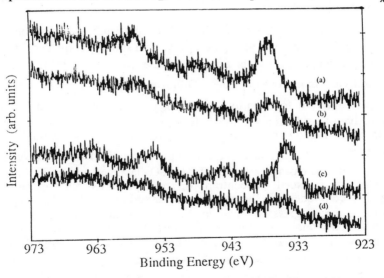

Fig. 10   Comparisons of the Cu($2p_{3/2}$) XPS spectra for (a) $LaCuO_3$, (b) "clean" $La_{1.85}Sr_{0.15}CuO_4$,(c) La_2CuO_4, and (d) somewhat "dirty" $La_{1.85}Sr_{0.15}CuO_4$.

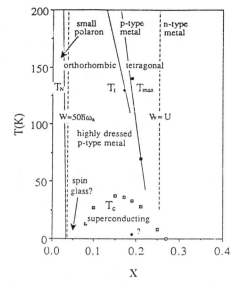

Fig. 11   Phase diagram for the system $La_{2-x}Sr_xCuO_4$ adopted from refs. 14, 33, 38.

electrons and a localized $d_{z^2}^2$ configuration. These data indicate a first-order phase change on going from the antiferromagnetic phase with strongly correlated 3d electrons and more ionic bonding to the superconductive phase with delocalized $\sigma_{z^2}^*$ as well as $\sigma_{x^2-y^2}^*$ electrons and more covalent Cu-3d, O-2p$_\sigma$ bonding.

Significantly, delocalization of the $\sigma_{x^2-y^2}^*$ and $\sigma_{z^2}^*$ electrons in the superconductive phase does not completely suppress the electron correlations. The phase diagram of Fig. 11 for the system $La_{2-x}Sr_xCuO_4$ shows several interesting features: (1) The Néel temperature T_N for long-range antiferromagnetic order drops precipitously with increasing oxidation of the CuO_2 layers; the holes in the oxidized CuO_2 planes are, for small concentrations, small ploarons, but they introduce a ferromagnetic component to the magnetic interactions that leads to spin-order "frustration" and spin-glass behavior in the semiconductive phases [14]. (2) The orthorhombic-tetragonal transition temperature T_t decreases with increasing Sr concentration x; the larger Sr^{2+} ion relieves the tensile stress in the rocksalt layer, and the introduction of more holes into the antibonding $\sigma_{x^2-y^2}^*$ band relieves the compressive stress on the CuO_2 sheets. (3) The temperature T_{max} at which the paramagnetic susceptibility shows a maximum also decreases with increasing x; at temperatures $T < T_{max}$ a two-dimensional system generally exhibits short-range magnetic ordering, and the presence of short-range spin fluctuations in the superconductive phases has been demonstrated by several techniques [24-26]. (4) Reliable data for x > 0.25 is difficult due to a loss of oxygen; deeper oxidation of the

CuO_2 planes becomes more difficult as the tolerance factor becomes larger. (5) The superconductors are p-type metals, which indicates retention of the correlation splitting (U - W) > 0, but at higher oxidation states the system appears to become an n-type metal with no correlation splitting of the σ^*_{x2-y2} band.

These features suggest an evolution with x of the single-particle energy-density of states $N(\varepsilon)$ as shown schematically in Fig. 12. Whether the small polarons associated with hole concentrations in the semiconductive phase occupy the $\pi^* + \sigma^*_{z2}$ or the lower σ^*_{x2-y2} band has not been clarified; but spectroscopic data supports holes in the lower σ^*_{x2-y2} band in the superconductive phase [27, 28]. Stabilization of an oxide phase with W ≈ U is unusual; commonly a semiconductor-metal transition occurs in association with the formation of a charge-density wave (CDW) at low temperatures. In the CuO_2 sheets, Cu-O-Cu interactions restrict formation of a CDW to a negative-U type. But in an oxide, the on-site correlation energy U at a Cu^{2+} ion is too large to permit stabilization of the reaction

$$2Cu^{2+} \rightarrow Cu^+ + Cu^{3+} \tag{11}$$

Nevertheless an inherent instability associated with the condition W ≈ U can be anticipated. Given the clear indication from Fig. 11 that the ratio U/W decreases sensitively with increasing hole concentration $p \approx x$ in $La_{2-x}Sr_xCuO_4$ -- as is also anticipated for a ΔE in equation (6) that approaches zero as x increases toward x = 1, a spinodal separation into phases rich and poor in holes could be expected. However, such a phase separation would require atomic diffusion -- in this system of Sr^{2+} ions, which would not occur at lower temperatures. The only alternative left to the system to express its inherent instability is the formation of dynamic charge fluctuations via

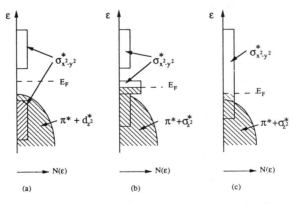

Fig. 12   Schematic variation with increasing hole concentration p of the energy density of quasiparticle σ^*_{x2-y2} states $N(\varepsilon)$ versus energy ε of a CuO_2 sheet: (a) antiferromagnetic La_2CuO_4, (b) superconductive p-type, $La_{1.85}Sr_{0.15}CuO_4$, and (c) n-type, metallic $La_{1.7}Sr_{0.3}CuO_4$.

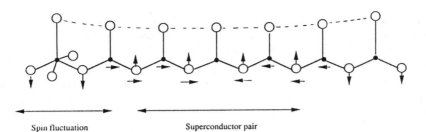

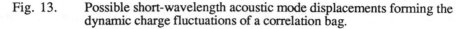

Spin fluctuation   Superconductor pair

Fig. 13.   Possible short-wavelength acoustic mode displacements forming the
dynamic charge fluctuations of a correlation bag.

atomic displacements. Charge fluctuations poor in holes would have a larger ratio
U/W; the spin fluctuations below T_c would be associated with these hole-poor
fluctuations. Charge fluctuations rich in holes would have a smaller ratio U/W; the
formation of superconductive pairs could occur within these hole-rich fluctuations.
The atomic displacements that would capture mobile hole pairs within a real-space,
non-retarded potential would be short-wavelength acoustic modes that created shorter
mean Cu-O distances within a hole-rich fluctuation and longer mean Cu-O distances
within a hole-poor fluctuation as indicated schematically in Fig. 13.

Evidence for the onset of charge fluctuations just above T_c has recently come
from Seebeck data on single crystals [29]; ion-channeling studies of the $YBa_2Cu_3O_{6+x}$
superconductors [30] indicate the presence of dynamic Cu-atom fluctuations within the
CuO_2 planes below T_c. Positron annihilation studies also support the presence of
charge fluctuations [31].

SUPERCONDUCTIVE PROPERTIES

Four superconductive properties of the high-T_c copper oxides are particularly
relevant to our discussion;

- the coherence length $\xi \sim 10$ Å is much smaller that the $\xi \sim 10^3$ Å of
  conventional superconductors [32];
- the opening of an energy gap 2Δ at the Fermi energy is about a factor 2 larger
  than predicted by BCS theory, but with the apparent retention of electronic
  states within the gap [33];
- a $T_c \sim p/m^*$ in the compositional range $p_c < p < p_m$ followed by saturation
  and a subsequent decrease in T_c with increasing p [6, 34]; and
- a pressure dependence of T_c that varies as $dT_c/dP > 0$ and increases to a
  maximum value at $p \approx p_m$, changing to $dT_c/dP < 0$ as T_c drops to zero with
  increasing p [35].

A $\xi \approx 10$ Å implies, first of all, that the superconductive hole pairs are bound
by a non-retarded potential -- not the retarded potential of BCS theory; if this potential
is due to electron-phonon interactions, it should be observable as a dynamic atomic-
displacement fluctuation in real space.

The opening up of an energy gap at the Fermi energy below T_c implies the cooperative formation of cooper pairs rather than a Bose condensation of bipolarons. The critical temperature T_c for the condensation of superconductive Cooper pairs has the general form

$$kT_c \approx 1.14\ \hbar\omega \exp(-1/\Lambda) \qquad (12)$$

where $\hbar\omega = \hbar\omega_D$ is the Debye energy and $\Lambda=\lambda$ is the electron-phonon coupling parameter in the BCS weak-coupling limit. From the Uncertainty Principle, a small coherence length ξ implies a large range of momenta $\Delta(\hbar k)$ of states about ε_F that participate in Cooper-pair formation. With $\xi \sim 10$ Å, an $\hbar\omega \gg \hbar\omega_D$ corresponds to an $\hbar\omega_{max} \approx 0.1W$, where W is the width of the lower $\sigma^*_{x^2-y^2}$ band. If the Fermi energy ε_F, as measured from the top of the lower $\sigma^*_{x^2-y^2}$ band, is $\varepsilon_F < \hbar\omega_{max}$, then all the holes should participate in pair formation to make $\hbar\omega \approx \varepsilon_F$. However, $\hbar\omega = \hbar\omega_{max}$ holds it $\varepsilon_F > \hbar\omega_{max}$. For a two-dimensional system, $N(\varepsilon)$ is constant over the bandwidth, so the hole concentration is $p \sim \varepsilon_F$, which makes $T_c \sim p$ for $\varepsilon_F < \hbar\omega_{max}$. Thus a small ξ provides a rationale for a $T_c \sim p$ over a compositional range $p_c < p < p_1$, where $p = p_1$ at $\varepsilon_F \approx \hbar\omega_{max}$.

In order to obtain an expression for $\hbar\omega_{max}$, it is necessary to have a model of the hole-pairing mechanism. For this model, we turn to the idea of a charge-fluctuation instability. Since electron-phonon interactions alone, unless strongly screened by a large dielectric constant, cannot overcome the electrostatic repulsion between paired electrons in a non-retarded potential well [36], an additional term must be added to the binding energy; this term is a modulation of the electron-electron interactions by the phonons in the creation of hole-rich and hole-poor charge fluctuations. Within a hole-rich fluctuation, the ratio U/W is smaller because U is decreased to $\tilde{U}$ and W is increased to $\tilde{W}$. In the region $W \approx U$, the electron-electron electrostatic energy U is a particularly sensitive function of W.

THE CORRELATION-BAG MODEL

The physical reasoning outlined to this point led my student, J.-S. Zhou, and I to propose a "correlation-bag" model of superconductive-pair formation [37]. The physical picture is analogous to the "spin-bag" model of Schrieffer et al [38] except that we invoke charge fluctuations rather than spin-density fluctuations. Therefore we utilize their mathematical formalism to obtain an

$$\hbar\omega_{max} = [\varepsilon_F(U-\tilde{U}')]^{1/2} \qquad (13)$$

Substitution of $\hbar\omega = \varepsilon_F$ for $\varepsilon_F < \hbar\omega_{max}$ and $\hbar\omega = \hbar\omega_{max}$ for $\varepsilon_F \gtrsim \hbar\omega_{max}$ leads to the qualitative variation of T_c versus x shown in Fig. 14. An $\varepsilon_F < \hbar\omega_{max}$ for $p < p$ leads to a linear increase in T_c with hole concentration p in the range $p_c < p < p_1$, where p_c is the critical hole concentration for Cooper-pair formation; for $p < p_c$ the system is either antiferromagnetic or stabilizes isolated bipolarons that do not superconduct. In the

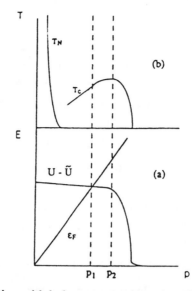

Fig. 14.   Variation with hole concentration p in a CuO_2 sheet of (a) (U - Ũ and ε_F and (b) T_N and T_c.

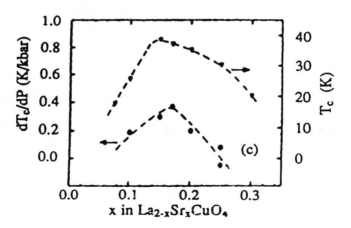

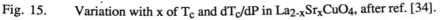

Fig. 15.   Variation with x of T_c and dT_c/dP in $La_{2-x}Sr_xCuO_4$, after ref. [34].

range $p_1 < p < p_2$, the correlation energies U and $\tilde{U}$ vary nearly equally with p, so $\hbar\omega_{max}$ is nearly constant and T_c saturates. For $p > p_2$, the correlation splitting (U-W) approaches zero, which makes not only U-$\tilde{U}$ decrease, but also the meaning of the hole concentration ambiguous.

Finally, it is interesting to note that the application of hydrostatic pressure P increases the bandwidth and hence the magnitude of ϵ_F for a given hole concentration p. Moreover, the magnitude of the increase in ϵ_F increases with P, but the critical concentration p_h does not increase with P so long as the coherence length ξ remains invariant with p. Thus $dT_c/dP > 0$ should increase to a maximum in the range $p_1 < p < p_2$. The concentration p_2 beyond which T_c decreases with increasing hole concentration is reduced by P since (U-W) for a given p decreases with increasing P. Therefore we can also anticipate that dT_c/dP becomes negative as p approaches the value where T_c drops to zero. These predictions are in accord with the experiments of Schirber et al [35], Fig. 15

ACKNOWLEDGEMENTS

This work was supported by the National Science Foundation, the Texas Advanced Research Program, and the Robert A. Welch Foundation, Houston, Texas.

REFERENCES

1.   J.B. Goodenough, Supercond. Sci. Technol. 3, 26 (1990).

2.   J. Fontcuberta, X. Obradors, and J.B. Goodenough, J.Phys. C. : Solid State Phys. 20, 441 (1987).

3.   R.J. Cava, B. Batlogg, C.H. Chen, E.A. Rietman, S.M. Zahurak, and D. Weider, Nature (London) 329, 423 (1987); Phys Rev B 36, 5619 (1987).

4.   A. Manthiram, X.X. Tang, and J.B. Goodenough, Phys. Rev. B 37, 3734 (1988).

5.   T. Freltoft, J.E. Fischer, G. Shirane, D.E. Moncton, S.K. Sinha, D. Vaknin, J.P. Remeika, A.S. Cooper, and D. Harshman, Phys. Rev B 36, 826 (1987).

6.   Y.J. Uemura et al, Phys. Rev.Lett. 62, 2317 (1989).

7.   A. Manthiram, J.S. Swinnea, Z.T. Sui, H Steinfink, and J.B. Goodenough, J. Am. Chem Soc. 109, 6667 (1987); W.I.F. David et al Nature 327, 310 (1987).

8.   M.A. Alario-Franco, C. Chaillout, J.J. Capponi, and J.Chenavas, Mater. Res. Bull. 22, 1685 (1987).

9.   C.J. Hou, A. Manthiram, L. Rabenberg, and J.B. Goodenough, J. Mater Res. 5, 9 (1990).

10.   Y. Dai, A. Manthiram, A. Campion, and J.B. Goodenough, Phys. Rev. B 38, 5091 (1988).

11.   A. Manthiram and J.B. Goodenough, Physica C 162-164, 69(1989); C 159, 760 (1989).

12.   F. Devaux, A. Manthiram, and J.B. Goodenough, Phys. Rev. B 41, 8723 (1990).

13.   P. Lightfoot, S. Pei, J.D. Jorgensen, X.-X. Tang, A. Manthiram, and J.B. Goodenough, Physica C 169, 464 (1990).

14.   S. Uchida, in "Proc. IX Winter Meeting on Low Temperature Physics: Progress in High-Temperature Superconductivity", Vol. 5, J. Heiras, R.A. Barrio, T Akathi, and J.Taguena, eds, (World Scientific, Singapore, 1988) p. 13.

15.   J.-S. Zhou, Sanjai Sinha, and J.B. Goodenough, Phys. Rev. B 39, 12331 (1989).

16.   P.M. Grant et al, Phys Rev. Lett. 58, 2482 (1987).

17.   P. Lightfoot, Shinyou Pei, J.D. Jorgensen, X.-X. Tang, A. Manthiram, and J.B. Goodenough, Physica C 169, 15 (1990).

18.   Z. Hiroi, T. Obata, M. Takano, and Y. Bando, Phys. Rev. B 41, 11665 (1990).

19.   J.D. Jorgensen, et al, Phys. Rev. B. 38, 11337 (1988).

20.   A. Manthiram and J.B. Goodenough, J. Solid State Chem 87, 402 (1990).

21.   M. Suzuki, Phys. Rev. B 39, 2312 (1989).

22.   J.B. Goodenough, Progress in Solid State Chemistry 5, 145 (1971).

23.   J.-S. Zhou, J.B. Goodenough, K. Allan, and A. Campion, (unpublished).

24.   R. Birgeneau et al, Phys Rev. B 38, 6614 (1988).

25.   A. Weidinger et al, Phys. Rev. Lett 62, 102 (1988).

26.   J. Tsuda, T. Shimizu, H. Yasuoka, K. Kishio, and K. Kitazawa, J. Phys. Soc. Japan 57, 2908 (1987).

27.   K. Kumazi and Y. Nakamura, Physica C 157, 307 (1989).

28.   N. Nücker, J.Fink, J.C. Fuggle, P.J. Durham, and W.M. Temmerman, Phys. Rev. B 37, 5158 (1988); F.J. Himpsel, F.V. Chandreshakar, A.B. McLean, and M.W. Shafer, Phys Rev. B 38, 11946 (1988).

29.   M.A. Howson, M.B. Salamon, T.A. Friedmann, J.P. Rice, and D. Ginsberg, Phys. Rev. B 41, 300 (1990).

30.   L.E. Rehn, R.P. Sharma, P.M. Baldo, Y.C. Chang. and P.Z. Jiang, Phys. Rev B 42, 4175 (1990).

31.   L. Chen, G. Wang. and M. Teng. Phys. Stat. Sol. (b) 157, 411 (1990).

32.   W.C. Lee, R.A. Klemm, and D.C. Johnston, Phys. Rev.Lett. 62, 1012 (1989).

33.   O.G. Olson *et al* Science 245, 731 (1989); R. Manzke, T. Buslaps, R. Cleassen. and J.Fink, Europhys. Lett. 9, 477 (1989).

34.   J.B. Torrance *et al*, Phys. Rev B 40, 8872 (1989).

35.   J.E. Schirber, E.L. Vanturini, J.F. Kwak, D.S. Ginley, and B. Morosin, J. Mater. Res. 2, 421 (1987).

36.   L.J. DeJongh, Physica C 152, 171 (1988).

37.   J.B. Goodenough and J.-S. Zhou, Phys. Rev. B 42, 4276 (1990).

38.   J.R. Shcrieffer, X.-G. Wen, and S.-C. Zhang, Phys Rev. Lett. 60, 944 (1988); Phys. Rev B 39, 11663 (1989).

39.   R.M. Fleming, B. Batlogg, R.J. Cava, and E.A. Rietman, Phys. Rev. B 35, 7191 (1987).

Electron Tunneling into High-T$_c$ Oxide Superconductors

R. C. Dynes
AT&T Bell Laboratories
Murray Hill, NJ 07974

Abstract

In this paper the results of electron tunneling measurements in two classes of high T$_c$ superconductors are reported. In the cubic BiO$_3$ materials (notably BaPb$_{1-x}$Bi$_x$O$_3$ and Ba$_{1-x}$K$_x$BiO$_3$) the results show these materials to be BCS superconductors. Indications point to electron-phonon coupling as the pair binding mechanism but quantitative verification still eludes experiment. In the cuprates (most of the data is on YBa$_2$Cu$_3$O$_7$) "gap-like" structure is observed but it is not BCS-like. Potential reasons for this deviation are discussed and some indications in the tunneling data for the pairing mechanism are revealed.

Introduction

Electron tunneling measurements have proven enormously valuable in studies of conventional superconductors. Very early measurements confirmed, in an especially convincing way, the existence of the superconducting energy gap and more detailed studies demonstrated the spectral form of the gap and the temperature dependence of it. These measurements were instrumental in confirming in some detail the predictions of the Bardeen, Cooper, Schrieffer (BCS) theory of superconductivity in simple metals. For example, it was shown very clearly that the ratio of the energy gap (2Δ) and critical temperature T$_c$ was close to the BCS value $\left(\dfrac{2\Delta}{kT_c} = 3.5 \right)$. As the sophistication of the technique improved, deviations from this BCS weak coupling limit became apparent $\left(\dfrac{2\Delta}{kT_c} \right.$ was measured to be >4 in materials like Pb for example$\left.\right)$ and subtle structure in the current-voltage characteristics of tunnel junctions unearthed a signature of the electron-phonon interaction - the microscopic mechanism responsible for superconductivity in these traditional materials. Through a quantitative analysis of this structure, people were able to extract a function $\alpha^2(\omega)F(\omega)$ (which is the phonon density of states ($F(\omega)$) modulated by the electron-phonon coupling function ($\alpha^2(\omega)$)). This function gave a quantitative description of the electron phonon interaction and confirmed beyond a doubt that the electron-phonon interaction was responsible for superconductivity.

In view of this success, it is not surprising that investigators have been vigorously exploring the possibility of a similar quantitative investigation of the recently discovered high T_c superconducting oxides. The existence and nature of the superconducting energy gap as well as a signature of the detailed binding mechanism would go a long way to reveal the similarities and differences of these remarkable materials with the more conventional superconductors.

In this paper, we will discuss some of what has been learned from tunneling measurements in the high T_c oxides. We take the point of view that the $BaBiO_3$ class of materials are also high T_c superconductors. We will concentrate on measurements in $BaPb_{1-x}Bi_xO_3$, $Ba_{1-x}K_xBiO_3$ and $YBa_2Cu_3O_7$, as it is these compounds which have been most extensively studied and reproducible results exist.

As a practical matter, investigators have been attempting to fabricate an all-high T_c Josephson junction which would have applications in detectors and digital electronics. We will also report results of the first of these junctions which have been achieved in the $Ba_{1-x}K_xBiO_3$ system.

Bismuth Oxides

Extensive tunneling studies have been performed on the compound $BaPb_{1-x}Bi_xO_3$ in the region $0<x<.25$ where the material is a superconductor.[2] By far, the cleanest data is obtained at x = .25.[1] For measurements x ≠ .25 it is difficult to convince oneself that the material is single phase as the superconducting resistive transition is broad[1] and the tunneling measurements show either very broad gap structure or multiple gaps.[3]

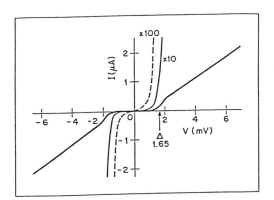

Fig. 1 Current vs voltage for a typical $BaPb_{.25}Bi_{.75}O_3$/In junction. From this plot, we obtain $\Delta \simeq 1.65$ meV.

At x = .25 high quality tunnel junctions are routinely formed by either condensing a thin film on a freshly cleaved surface or bringing a metal into contact with the surface (point contact, solder blob, pressed metal contact etc.). The tunnel barrier that forms occurs naturally and may be either a natural Shottky barrier or a region of depleted oxygen, for example. At any rate the junctions are of extremely high quality as is evidenced by the energy gap structure of both the $BaPb_{.25}Bi_{.25}O_3$ and the counter-electrode (Pb, Sn, In etc.).

A typical I-V characteristic of a $BaPb_{.75}Bl_{.25}O_3/In$ junction where the In is driven normal by a modest (1 kilogauss) magnetic field is shown in Fig. 1. The energy gap edge is clearly evident as the point where current begins to flow. Studies of over 60 junctions of this type have shown similar behavior with very little variation in the extracted value of the energy gap 2Δ.

For finite temperatures, the I-V characteristic of a superconductor insulator-normal metal tunnel junction is given by

$$I(V) = C\,N_o \int_{-\infty}^{\infty} N_S(E) \Big[f(E,T) - f(E + V,T) \Big] dE \qquad (1)$$

where

> V is the applied voltage
> N_o is the normal metal density of states (assumed constant)
> C is a constant which contains all the tunneling probabilities
> f is the Fermi function at temperature T
> $N_S(E)$ is the superconducting density of states

In the weak coupled BCS approximation,

$$N_S(E) = \left\{ \frac{E}{\sqrt{E^2 - \Delta^2}} \right\} \qquad (2)$$

where Δ is the superconducting energy gap. At $T = 0$, this expression allows no current flow until $eV = \Delta$, a condition approximately satisfied ($T = 1.1K$) in Fig. 1. The first derivative $\left(\dfrac{dI}{dV} \right)$ of this I-V characteristic, at $T = 0$, yields directly $N_S(E)$ and so it is commonplace to study the derivative $\dfrac{dI}{dV}$, even at finite temperatures where a thermally smeared measure of $N_s(E)$ results. The results of a $\dfrac{dI}{dV}$ measurement of a $BaPb_{.75}Bi_{.25}O_3/normal$ In tunnel junction are shown in Fig. 2. Also shown is a $\dfrac{dI}{dV}$

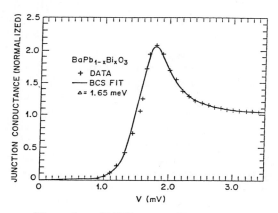

Fig. 2   dI/dV   vs   V   for   a $BaPb_{.25}Bi_{.75}O_3/In$ junction. The data are the crosses and the solid line is Eq. 1 assuming a BCS form for the density of states and delta=1.65 meV.

determined from Eq. 1 assuming the BCS form for $N_S(E)$ with $\Delta = 1.65$ meV. The fit is excellent and gives us confidence in the BCS form of the density of states. Also $2\Delta/kT_c = 3.5$ which is equal to the BCS value.

From these studies this material appears to be a BCS superconductor. That is to say the energy gap and the excitation spectrum are very much the form expected for a BCS superconductor. This implies the usual s-wave electron pairing but does not necessarily imply electron-phonon coupling.

We have performed similar studies on the compound $Ba_{1-x}K_xBiO_3$[4] and the results and conclusions are quantitavely similar to the case of $BaPb_{1-x}Bi_xO_3$. In this superconductor with T_c up to 30K depending on the K concentration, we also find extremely good fits to a simple BCS expression. The results of a measurement and fit are shown

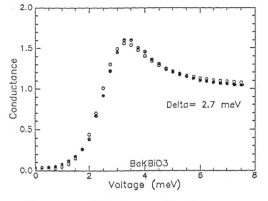

Fig. 3   dI/dV   vs   V   for   a $Ba_{1-x}K_xBiO_3$/Au junction. The data are the crosses and the fit to Eq. 1 is the open circles assuming an energy gap $\Delta = 2.7$ meV.

in Fig. 3 where again we see the fit is quite good and we extract the ratio $\dfrac{2\Delta}{kT_c} = 3.5$ from these measurements. Again, the straightforward conclusion from this work is that the class of superconductors that are BiO_3 compounds, have a conventional energy gap and look very much like BCS superconductors.

Given the quality of the junctions of this type, further detailed studies at higher voltages to investigate structure associated with the mechanism responsible for superconductivity

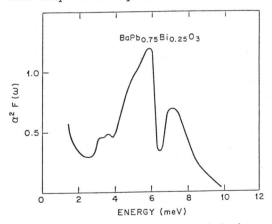

Fig. 4 $\alpha^2(\omega)F(\omega)$ function derived from tunneling into $BaPb_{1-x}Bi_xO_3$.

are justified.[5] Structure in $\dfrac{dI}{dV}$ vs V, typical of electron-phonon interactions are routinely observed in these junctions and an elaborate analysis of this structure via the Eliashberg equations[6] results in an electron-phonon coupling function $\alpha^2(\omega)F(\omega)$ shown in Fig. 4.

From these studies several difficulties have been apparent. We see that the function $\alpha^2(\omega)F(\omega)$ is uncertain at low energies. This suggests that there are some strongly coupled low frequency modes that are difficult to extract from the raw data and so some of the spectral weight at low energy may be missing. It is also clear from the data that we do not observe the anticipated high frequency (oxygen) optical modes[7] in this experiment. There are two possibilities for this. First, the strong sloping background in the tunnel junction conductance[8] can make the observation of weak structure at high voltages more difficult. Secondly, a model calculation of the coupling strength necessary to account for the superconductivity by way of the electron-phonon interaciton suggests we have not yet fabricated junctions of sufficient quality to resolve the weak features at high energies. Although some investigations claim to have resolved this issue,[9] it is not yet generally accepted.

Summarizing the bismuth oxides, it appears that a superconducting energy gap exists in these compounds. The ratio $\dfrac{2\Delta}{kT_c} \sim 3.5$, which is the BCS weak coupling limit and there is evidence for electron-phonon coupling as the responsible mechanism. While it has not been shown that there is sufficient electron-phonon coupling to account for all of the superconductivity, the quantitative role of the optic phonons awaits study.

Copper Oxides

Now we turn to a discussion of tunneling data on the cuprates concentrating on YBCO for which there have been extensive studies. Many different tunneling techniques have been attempted with variable results.[10] More recently several groups studying thin films[11,12] and single crystals[12,13,14,15] are beginning to show a reproducibility and consistency in the data which must be taken seriously. In Fig. 5 we show the tunneling conductance data for a single crystal YBCO/Pb tunnel junction for various temperatures above and below $T_c (\simeq 89K)$. There are several aspects to this data which are notable. First, there is a background conductance at higher bias which is approximately linear with voltage. Second, there is structure which is generally gap-like which appears below T_c and becomes more pronounced and sharper at lower temperature. Unlike the bismuthates, however, this structure does not fit a BCS gap shape. Superimposed on Fig. 5 is the shape one would expect for T = 0 with a BCS gap given by $\dfrac{2\Delta}{kT_c} = 3.5$. The energy scale is approximately correct, but the shape is decidedly different. At the lowest temperatures studied (down to 50 mK) there are states at E_F with no real "gap" appearing. At higher voltages (30-40 mV) there is additional structure reminiscent of the phonon renormalization effects observed in conventional superconductors and unfolded to yield the strength of the electron-phonon interaction. If the presence of spin fluctuations is an issue then the observed states in the "gap" could be due to spin flip scattering resulting in gaplessness. In conventional superconductors if the spin-flip scattering rate $\dfrac{1}{\tau_{SF}}$ is such that $\Delta\tau_{SF} \sim \hbar$, gaplessness occurs.[16] Tunneling studies of conventional superconductors with dilute magnetic impurities show behavior which is qualitatively similar to that shown in Fig. 5. Depending upon the strength of the electron-local moment interaction, only 1 impurity in 10^2 is necessary for such effects.[17] A small density of defects in the copper oxide planes might result in a moment and subsequent magnetic scattering. A problem with this explanation is the apparent reproducibility of the tunneling results in thin film and single crystal samples. It is difficult to imagine all samples having approximately the same density of spin-flip scattering centers unless such defects are intrinsic.

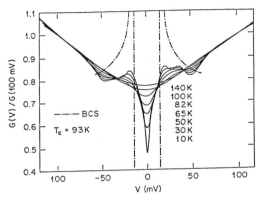

It is well known that superconducting tunneling probes a depth into the sample of the coherence length ξ. In the cuprates it is now agreed that ξ is quite anisotropic with $\xi_{ab} \sim 10-20\text{Å}$ and $\xi_c \approx 5\text{Å}$. With $\xi_c <$ lattice constant a serious spatial

Fig. 5 dI/dV vs V for a $YBa_2Cu_3O_7$ single crystal/Pb junction normalized to dI(100 mV)/dV for a range of temperatures. The polarity refers to the $YBa_2Cu_3O_7$ electrode. The heavy dashed-dotted line gives the density of states for a weak-coupled BCS superconductor with $\dfrac{2\Delta}{k_B T_c} = 3.5$.

variation of Δ is possible in the c direction and surface effects could seriously alter both the energy gap and the shape of the excitation spectrum, modifying it substantially from its BCS form. Again, such an explanation would suggest a wide variation from sample to sample as the detailed nature of the surface would be expected to vary. The reproducibility of these results from experiment to experiment and from laboratory to laboratory argues against this as an explanation unless there is an intrinsic nature to it.

Such an intrinsic possibility has been suggested by Takahashi and Tachiki.[18] Using a model of strongly and weakly superconducting layers in an "internal proximity effect" they have reproduced the results shown in Fig. 5. Physically they envision multiple gaps due to the relative decoupling of the various planes and chains and the very short coherence length. If this explanation is correct, the results of Fig. 5 are intrinsic to a material where the coherence length is $\lesssim$ lattice spacing and there are decoupled chains and planes.

An intriguing feature of the data in Fig. 5 is the structure observed at voltages in the range $\sim$ 40 meV. In a conventional superconductor this would be identified as a reflection of the electron-phonon interaction and using the technique of McMillan and Rowell[5] the electron-phonon coupling function $\alpha^2(\omega)F(\omega)$ extracted. The situation here is not so clear as there are real differences between these cuprates and conventional superconductors. In addition, careful inspection of Fig. 5 shows that there is some assymetry about zero bias in this structure; a result not expected normally. Nevertheless, it is interesting to take this data and in a

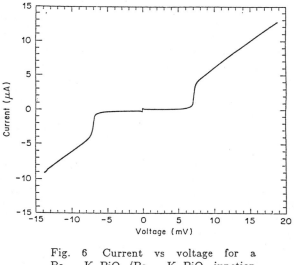

Fig. 6   Current vs voltage for a $Ba_{1-x}K_xBiO_3/Ba_{1-x}K_xBiO_3$ junction taken at T = 1.1K.

semiquantitative way, extract an $\alpha^2(\omega)F(\omega)$ assuming an Eliashberg analysis on whatever these structures reflect. Doing this we extract an $\alpha^2(\omega)F(\omega)$ function which has a single, broad peak at $\approx$ 23 meV and a peak height in dimensionless units $\approx$ 2. A traditional measure of the coupling strength is given by

$$\lambda = 2 \int\limits_0^\infty \frac{\alpha^2(\omega)F(\omega)}{\omega} \, d\omega.$$

For this data we calculate a $\lambda \approx 2$. Within the strong coupling calculations of Allen and Dynes[19] this is enough strength to account for a superconductor with $T_c \approx 60$K. We caution that this result at the moment is only semi-quantitative in that several assumptions were made in analyzing the data. Nevertheless, it is an intriguing result and motivates us to pursue further whether this structure is signaling the coupling mechanism.

Josephson Junctions

On an applications side, one of the important accomplishments would be an all-high T_c Josephson junction. Recently we have accomplished this in the $Ba_{1-x}K_xBiO_3$ system.[20] We have fabricated tunnel junctions of the type $Ba_{1-x}K_xBiO_3$ which show high quality quasiparticle tunneling and a well characterized Josephson effect. A typical I-V characteristic is shown in Fig. 6. An extensive set of measurements[20] indicate that the Josephson devices behave in a conventional manner and offer a Josephson technology at 30K.

Summary

Electron tunneling measurements in the high T_c oxide superconductors are yielding quantitative information and insight into these fascinating materials. In the case of the BiO_3 compounds, measurements of the superconducting energy gap Δ reveal conventional BCS behavior. We have even achieved fabrication of an all-high T_c Josephson junction in these materials. There are strong indications that electron-phonon coupling contributes substantially to the electron pairing and the question of whether it accounts for all of the superconductivity remains open.

In the case of the copper oxides, we do not observe a "true" gap while structure is observed reminescent of an energy gap. States persist inside the gap and whether these states are due to spin flip scattering, surface effects or an internal proximity effect remains a crucial question. Higher energy structure in the case of $YBa_2Cu_3O_7$ may be giving us a glimpse of the culpable mechanism for superconductivity in these remarkable materials.

Acknowledgments

I have benefited enormously over the years from collaborations and discussions with P. W. Anderson, J. M. Rowell, J. M. Valles, Jr., M. Gurvitch, J. P. Garno, B. Batlogg, J. P. Remeika, R. J. Cava, A. M. Cucolo, L. Schneemeyer, F. Sharifi, A. Pargellis, E. S. Hellman, E. H. Hartford, Jr., B. Miller and J. M. Rosamilia.

References

[1] B. Batlogg et al., in "Proceedings of the International Conference on d and f band Superconductors", edited by W. Buckel and W. Werber (Kernforschungszentrum Karlsruhe, Karlsruhe, Federal Republic of Germany, 1982), pp. 401-403

[2] A. W. Sleight, J. L. Gillson, and P. E. Bierstedt, Solid State Commun. **17**, (1975) 27.

[3] J. Akimitsu, T. Ekino and K. Kobayashi, Jpn, J. Appl. Phys. Suppl. **26**, (1987) 995.

[4] See, for example, B. Batlogg, R. J. Cava, L. F. Schneemeyer and G. P. Espinosa, IBM Journal of Research and Development, **33**, 2081

(1989).

[5]   W. L. McMillan and J. M. Rowell, in "Superconductivity", edited by R. D. Parks, (Dekker, New York, 1969).

[6]   W. L. McMillan, Phys. Rev. 167, (1968) 331.

[7]   L. F. Matheiss and D. R. Hamann, Phys. Rev. B28, (1983) 4227.

[8]   J. M. Valles, Jr. and R. C. Dynes, MRS Bulletin, XV, 44 (1990).

[9]   J. F. Zasadzinski et al., Physica C 158, (1989) 519 and J. F. Zasadzinski et al., in "Proceedings of the Conference on Materials and Mechanisms of Superconductivity ", Stanford, CA (1989).

[10] J. Moreland et al., Phys. Rev. B35, (1987) 8856; H. F. C. Hoevers et al., Physica (Amsterdam) 152C, (1988) 105; J. R. Kirtley et al., Phys. Rev. B35, (1987) 8846.

[11] J. Geerk, X. X. Xi, and G. Linker, Z. Phys. B73, Conference on Materials and Mechanisms of Superconductivity", Stanford, CA (1989).

[12] A. M. Cucolo et al., Physics C 161, (1989) 351.

[13] A. Fournel et al., Europhys. Lett. 6, (1988) 653.

[14] M. Gurvitch et al., Phys. Rev. Lett. 63, (1989) 1008; J. M. Valles, Jr. et al. "Proceedings of the Fall MRS Meeting", 1989.

[15] M. Lee, A. Kapitulnik and M. R. Beasley in Mechanisms of High-Temperature Superconductivity, edited by H. Kamimura and A. Oshiyama, Springer Series in Materials Science Vol. 11 (Springer-Verlag, Heidelberg, 1989), p. 220; J. S. Tsai et al., ibid., p. 229.

[16] K. Maki, in "Superconductivity", edited by R. D. Parks, (Dekker, New York, 1969).

[17] A. S. Edelstein, Phys. Rev. 180, (1968) 505.

[18] S. Takahashi and M. Tachiki, Physica B 165&166, 1067 (1990).

[19] P. B. Allen and R. C. Dynes, Phys. Rev. B12, 905 (1975).

[20] A. N. Pargellis, F. Sharifi, R. C. Dynes, B. Miller, E. S. Hellman, J. M. Rosamilia and E. H. Hartford, Jr., submitted to Appl. Phys. Lett.

LOOKING FOR NEW HIGH-T$_c$ SUPERCONDUCTORS

R. J. Cava
B. Batlogg
R. B. van Dover
P. Gammel
J. J. Krajewski
W. F. Peck, Jr.
L. W. Rupp, Jr.

AT&T Bell Laboratories
Murray Hill, New Jersey 07974

ABSTRACT

A major driving force in the field of high T$_c$ superconductivity during the first few years has been the search for new materials. "Why copper and bismuth oxides?" is a question that is often raised. This paper summarizes recent results on nonsuperconducting $BaSn_{1-x}Sb_xO_3$ and $La_{1-x}Sr_xNiO_4$, which were studied in an attempt to address that question. Recent results on the new 60 K superconductor $La_{1.6}Sr_{0.4}CaCu_2O_6$ are also described.

INTRODUCTION

Since the initial discovery of superconductivity in copper oxides,[1] many new oxide superconductors have been discovered. Nearly all have been complex compounds of copper and oxygen. Many thousands of papers have been published on these materials in the past four years. Their crystallography and crystalchemistry have been reviewed several times (see for instance references 2 and 3). High T$_c$ superconductors have also been found in the perovskites based on atomic substitutions in $BaPbO_3$ and $BaBiO_3$ but they have been studied considerably less than have the copper oxides, no doubt because their T$_c$'s are not yet comparable to those of the best copper oxides. At this time there remains considerable controversy over whether the copper oxides and bismuth/lead oxides are superconducting for the same microscopic reason. We have recently been studying conductive oxides which are not based on either copper or lead/bismuth in an attempt to understand what makes the superconductors so special. The results on the Sn/Sb analog of $Ba(Pb,Bi)O_3$ and the Ni analog of $(La,Sr)CuO_4$ are described as examples. Finally, the new 60 K superconductor $(La,Sr)_2CaCu_2O_6$ is described.

$BaSn_{1-x}Sb_xO_3$

The most unusual of the oxide superconductors known before 1986 is $BaPb_{.75}Bi_{.25}O_3$,[4] with a T$_c$ of 12 K. This material has the simple perovskite structure, with a small tilting of the $(Pb/Bi)O_6$ octahedra, and Pb and Bi

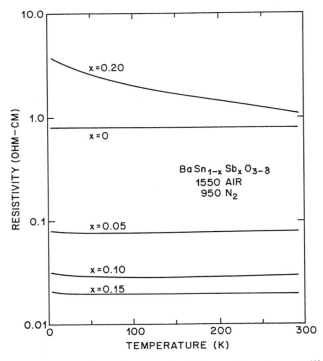

Fig. 1:   Temperature dependent resistivities of polycrystalline samples of BaSn$_{1-x}$Sb$_x$O$_3$ from 300-4.2K.

disordered among the octahedra. BaPbO$_3$ is a metal because of the covalent Pb-O bonding gives rise to a broad Pb 6s–O$_{2p\parallel}$ band which overlaps in energy with the narrower O$_{2p\perp}$ band.[5] When Bi or Sb is added, electrons are put into the broad 6s-2p$\parallel$ band, and superconductivity is induced with T_c's of 12 and 3.5 K, respectively. Alternately, a T_c of 30 K is found for K doping of the charge density wave (CDW) insulator BaBiO$_3$, Ba$_{1-x}$K$_x$BiO$_3$, for x = 0.4. The CDW state of BaBiO$_3$ has been much discussed. It is driven by the energetic instability of Bi in the 4+ state, which would result in an unpaired 6s electron. Bi has one more valence electron than Pb, so the K for Ba partial substitution drives the electron count toward that of BaPbO$_3$.

In order to investigate the importance of the metal-oxygen band for superconductivity in this class of materials we have recently investigated the properties of the BaSn$_{1-x}$Sb$_x$O$_3$ perovskite.[6] For this material the metal valence electrons are solely of the 5s 5p type. We expect the same kind of inherent instability of Sb^{4+} as occurs for Bi^{4+}, this time because a single 5s electron would be energetically unfavorable. Thus the BaSn$_{1-x}$Sb$_x$O$_3$ perovskite is an exact 5s 5p analog of the totally 6s 6p superconducting system BaPb$_{1-x}$Bi$_x$O$_3$.

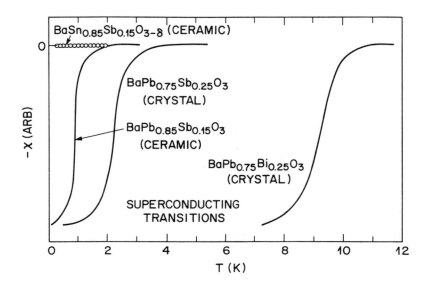

Fig. 2: Comparison of the superconducting transitions for $BaPb_{.75}B_{.25}O_3$, $BaPb_{.85}Sb_{.15}O_3$, $BaPb_{.75}Sb_{.25}O_3$, and the susceptibility data for $BaSn_{.85}Sb_{.15}O_3$ between 2 and 0.05K.

Details of the preparation and characterization of the materials are presented elsewhere.[6] Under optimal conditions, $BaSn_{1-x}Sb_xO_3$ exist for $0 \leq x \leq 0.2$, and can have a very small range of oxygen deficiency. Although less than 0.01 per formula unit, this oxygen deficiency strongly influences the electrical properties. Figure 1 shows the temperature dependent resistivities for single phase polycrystalline pellets of $BaSn_{1-x}Sb_xO_{3-\delta}$ prepared in air at 1550C and later annealed in N_2 at 950°C. Although the resistivities are much larger than are expected for the donation of 1 electron per Sb, they do not show the localization behavior (until $x = 0.20$) one would expect at these resistivity values. $BaSnO_{3-\delta}$, for instance, is a light grey material, with a virtually temperature independent resistivity of nearly 1 ohm-cm below 300K.

Figure 2 shows the low temperature magnetic susceptibility data for $BaSn_{.85}Sb_{.15}O_3$, the most conductive of the series, compared to the results of our measurements for optimal ($x = 0.25$) and non-optimal ($x = 0.15$) compositions of polycrystalline and single crystal $BaPb_{1-x}(Sb,Bi)_xO_3$. The tin-antimony compound is not superconducting down to 0.05K.

$BaPbO_3$ has a metallic resistivity due to an "accident" of the band structure. Like white insulating $BaSnO_3$, by electron count alone it would not be expected to be conductive. Conductivity arises because of the overlap of the narrow oxygen $p_\perp$ band (from oxygen 2p states perpendicular to the metal-oxygen bonds) with the broad antibonding $sp_\parallel$ σ^* metal-oxygen band (from metal-oxygen states parallel to

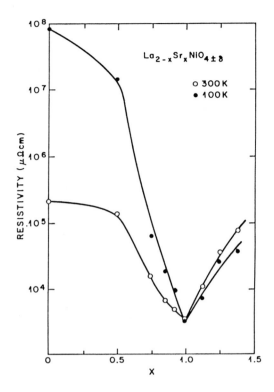

Fig. 3:   Room temperature (open circles) and 100K (closed circles) resistivities for $(La, Sr)_2NiO_4$ polycrystalline pellets. The apparent increase in ρ for $x > 1$ may be due to the lower density of these samples.

the metal-oxygen bond). When electrons are added to $BaPbO_3$ by additions of Bi or Sb they go into the antibonding $sp_{\parallel}$ σ^* band. The s-p σ^* band is approximately 16eV wide in $Ba(Pb,Bi)O_3$ due to the closeness in energy of the Pb 6s and O 2p levels. The bare energies of the 5s orbitals, are significantly higher than the 6s orbitals. Therefore the primary difference in band structures between the 5s and 6s cases might be a smaller 5s-2p overlap, leading to a narrower sp σ^* band and a gap between this band and the oxygen $p_{\perp}$ band. Because the gap between the $p_{\perp}$ and $sp_{\parallel}$ σ^* bands is wide, it is likely that the carriers that are introduced by Sb substitution never get into the σ^* band. The difference between the Sn and Sb orbital energies suggests that the carriers are found in a narrow impurity-like mid-gap band. Thus we propose that carrier doping in $BaSnO_3$ by Sb substitution never leads to the presence of carriers in the hybridized $sp_{\parallel}$ σ^* band, even though it does lead to conductivity. It is the nearness of that band in energy to the $p_{\perp}$ oxygen bands in the 6s bonding systems that makes it accessible to doping.

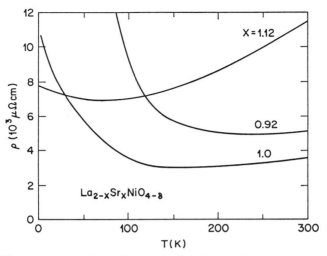

Fig. 4:   Temperature dependent resistivities for various $(La, Sr)_2 NiO_4$ compositions, as measured on polycrystalline pellets.

These results clearly point to the special situation in $Ba(Pb, Bi)O_3$ which leads to superconductivity, and which is obviously rooted in the unique combination of metal and oxygen energy levels. The change from 6s to 5s metal states destroys the delicate balance. The same rationale holds true for the cuprate superconductors, as can be seen e.g. by comparing $(La, Sr)_2 CuO_4$ with $(La, Sr)_2 NiO_4$, where the slightly higher Ni 3d energy levels apparently also change the properties profoundly.

$La_{2-x}Sr_xNiO_4$

Superconductivity in the $(La, Sr)_2 CuO_4$ solid solution with the $K_2 NiF_4$ structure type has been studied extensively in recent years. In particular, the evolution of the system from antiferromagnetism to superconductivity is at the center of the debate on the nature of the microscopic pairing mechanism responsible for high T_c superconductivity. Studies of the analogous $(La, Sr)_2 NiO_4$ solid solution have been reported by many groups before the advent of the high T_c copper oxides[7-11], and also since then.[12] The similarity in the energies of the Ni and Cu 3d states, the well known ability of Ni to occur in a low spin state in oxides, and the observation that both Ni and Cu display Jahn-Teller distortions, suggests that the comparison of the Cu and Ni based $K_2 NiF_4$ type compounds is of considerable interest.

The synthesis of the polycrystalline materials is described elsewhere.[13] The resistivities in the $La_{2-x}Sr_xNiO_4$ solid solution change several orders of magnitude with changes in composition. Our results are summarized in figure 3, which shows the measured resistivities at 100 and 300K. The data show that a dramatic decrease in resistivity begins near $x = 0.5$, the same composition where the c axis

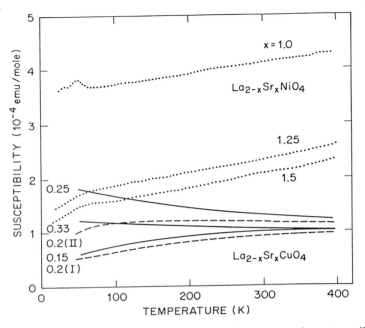

Fig. 5:   Comparison of the temperature dependent magnetic susceptibilities of various $La_{2-x}Sr_xNiO_4$ and $La_{2-x}Sr_xCuO_4$ compositions. A small extraneous Curie term was subtracted from the nickelate curves.

length begins to decrease.[13] A minimum measured resistivity occurs near $x = 1$, and is relatively sharply defined in composition. Although the temperature dependent resistivities are weakly monotonically decreasing for $x > 1.12$, the absolute magnitudes increase. The increase in measured resistivity for $x > 1$ is no doubt due to the lower densities of the ceramic pellets for $x > 1$: the intrinsic resistivities may actually continue to decrease. The temperature dependencies of the resistivities for our materials are similar to those reported in reference 12, except that our absolute resistivities are lower than theirs for $x < 1.2$ and their data show a minimum resistivity at $x = 1.2$ instead of $x = 1.0$. Representative temperature dependent resistivities for samples with compositions near $LaSrNiO_4$ are shown in figure 4. The best resistivities observed are similar to those for $(La, Sr)_2CuO_4$ polycrystalline pellets. For the materials with compositions $La_{.75}Sr_{1.25}NiO_4$ and $La_{.6}Sr_{1.4}NiO_4$ (not shown), the resistivity is metallic over the whole temperature range.

Magnetic susceptibilities were measured between 300 and 5K in a dc field of 10kOe in a commercial magnetometer. The values of $\chi(T)$ for selected $La_{2-x}Sr_xNiO_4$ compositions are compared with various $La_{2-x}Sr_xCuO_4$ compositions[14,15] in figure 5. Shown is $\chi(T)$ after subtracting a very small extrinsic Curie-Weiss term. If Ni were in a conventional low spin state, no paramagnetic contribution would be expected. Since a significant paramagnetism is observed, we suggest that it cannot arise solely from the small expected Pauli contribution of the conduction electrons. When viewed on the scale of figure 5, the temperature dependence of χ is quite pronounced, and is of opposite sign to that of highly metallic $La_{2-x}Sr_xCuO_4$ ($x > 0.2$), but is similar to that in the cuprates at low Sr content.

Magnetic measurements to check for the presence of superconductivity in the conductive samples near $x = 1$ were performed in a dc magnetometer in a field of 20 Oe, to a temperature of 1.5K. No superconductivity was found. Further, the most conductive samples were cooled to 30mK in a dilution refrigerator and checked for superconductivity with a 1kHz ac susceptibility technique, also without the observation of a superconducting transition. We must therefore conclude that the $(La,Sr)_2NiO_4$ solid solutions, although isostructural with superconducting $(LaSr)_2CuO_4$ solid solutions, do not become superconductive to very low temperatures.

The properties of $La_{2-x}Sr_xNiO_4$ are somewhat enigmatic. At compositions where x is between 1 and 1.4 a conductive phase exists with no sign of magnetic ordering, a phase which does not become superconducting down to very low temperatures. An explanation of the evolution of the observed properties by a "high spin" to "low spin" transition of the Ni d electrons alone as a function of Sr content appears to be inadequate. We suggest that a microscopic picture which considers the hole distribution among the Ni and the surrounding oxygens, in close analogy to $(La,Sr)_2CuO_4$ and also $(Ni,Li)O^{16}$, is necessary for complete understanding. The lattice parameters for the conductive compositions of the $La_{2-x}Sr_xNiO_4$ solid solution indicate that the relative charge distribution among the d_{z^2} and $d_{x^2-y^2}$ orbitals is different from that in the conductive $La_{2-x}Sr_xCuO_4$ compositions.

More likely to be critical, however, are the relative energies of the nickel d and oxygen p states, which are considerably further separated in energy (by an additional 1-2eV) than are those of Cu d and oxygen p. This difference in energy of a few eV results in a larger charge transfer gap in La_2NiO_4 than in La_2CuO_4,[17] and is responsible for the necessity for higher Sr concentrations to induce metallic conductivity. This weakening of the metal-oxygen bond hybridization, may be sufficient to prevent the appearance of superconductivity in nickel based oxides.

$La_{2-x}Sr_xCaCu_2O_6$

In the three years since the publication of the pioneering work of Bednorz and Muller, many new copper-oxide based superconductors have been discovered. All contain layers of copper-oxygen squares, pyramids or octahedra so their electronically active structural components. One structure type, first reported[18] for $La_2CaCu_2O_6$ and $La_2SrCu_2O_6$, has stood since the beginning of high T_c research as an enigma in the field. The crystal structure, shown in figure 6, is the least complex of all the structures with the double copper oxide pyramidal layers

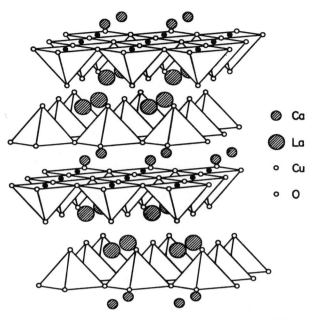

Fig. 6:   The   crystal   structure   of   the   new   60K   superconductor $La_{1.6}Sr_{0.4}CaCu_2O_6$.

common to the highest T_c superconductors. It is derived from that of La_2CuO_4 by insertion of an additional CuO_2 plane, with the doubled plane now separated by a layer of Ca or Sr in eight-fold coordination. The crystal structure of $Ba_2YCu_3O_7$ is derived from $La_2CaCu_2O_6$ by the insertion of a CuO layer (the chains) between the apices of the pyramids, shifted so that they are aligned. Despite considerable effort, both published and unpublished, it had not been made superconducting.

This has been particularly puzzling as the bond lengths of the pyramids in $La_{1.9}Ca_{1.1}Cu_2O_6$, for example, are virtually identical to those of $Ba_2YCu_3O_6$.

We have very recently shown[19] that the compound can infact be made superconducting, through the proper selection of host structure and doping ion, and oxygen annealing at 20 atmospheres pressure. Details of the synthesis and characterization are found in reference 19. A typical superconducting transition, measured on cooling in a dc field of 1.8Oe, is shown in figure 7. The highest superconducting transition temperature of 60K is obtained at the composition $La_{1.6}Sr_{0.4}CaCu_2O_6$. This 60K T_c is what we suggest is intrinsic for a double pyramidal copper oxide based superconductor without the intervention of a charge reservoir layer. Our preliminary determination of the optimal Sr content for $(La,Sr)_2CaCu_2O_6$ results in a hole concentration of 0.20/Cu, only slightly higher than the 0.175/Cu optimal for the single layer $(La,Sr)_2CuO_4$. If this holds up on further study, it implies either a very large change in T_c with a small increase in hole concentration, or it implies that the ubiquitous universal curves of T_c vs. hole

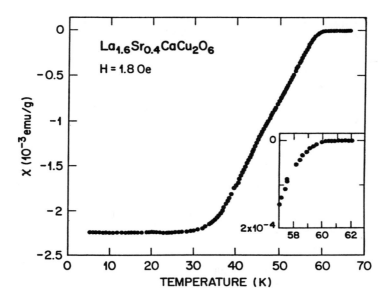

Fig. 7: Representative superconducting transition measured magnetically (dc, 1.8Oe) on field cooling of $La_{1.6}Sr_{0.4}CaCu_2O_6$.

concentration, which ignore factors such as the number of copper planes or copper-oxygen coordination geometries, cannot tell the whole story. This is a uniquely simple double layer superconductor, which like $(La,Sr)_2CuO_4$, becomes superconducting through the introduction of carriers in an unambiguous manner: by straightforward atomic substitution without the intervention of charge reservoir layers with flexable valence states. Detailed study of its composition dependent properties should lead to a more unambiguous understanding of double-layer superconducting copper oxides.

REFERENCES

1. J. G. Bednorz and K. A. Müller, Z. Phys. *B64*, 189 (1986).

2. A. W. Sleight, Science *242*, 1519 (1988).

3. R. J. Cava, Science *247*, 656 (1990).

4. A. W. Sleight, J. L. Gillson and P. E. Bierstedt, Sol. St. Comm. *17*, 27 (1975).

5. See, for instance, L. F. Mattheiss and D. R. Hamann, Phys. Rev. *B26*, 2686 (1982).

6. R. J. Cava, P. Gammel, B. Batlogg, J. J. Krajewski, W. F. Peck, Jr., L. W. Rupp, Jr., R. Felder and R. B. van Dover, submitted to Phys. Rev. B.

7. Gerard Demazeau, Michel Pouchard and Paul Hagenmüller, J. Sol. St. Chem. *18*, 159-162 (1976).

8. J. Gopalkrishnan, G. Colsmann and B. Reuter, J. Sol. St. Chem. *22*, 145-149 (1977).

9. I. F. Kononyuk, N. G. Surmach and L. V. Makhnach, Isvestiya Akademii Nauk SSSR, Inorganic Materials, *18*, 1222-1225 (1982).

10. J. B. Goodenough, Mat. Res. Bull. *8*, 423-432 (1973).

11. K. K. Singh, P. Ganguly and J. B. Goodenough, J. Sol. St. Chem., *52*, 259-273 (1984).

12. Y. Takeda, R. Kanno, M. Sakano, O. Yamamoto, M. Takano, Y. Bando, H. Akinaga, H. Takita and J. B. Goodenough, Mat. Res. Bull. *25*, 293 (1990).

13. R. J. Cava, B. Batlogg, T. T. Palstra, J. J. Krajewski, W. F. Peck, Jr., A. P. Ramirez and L. W. Rupp, Jr., submitted to Phys. Rev. B.

14. J. B. Torrance, A. Bezinge, A. I. Nazzal, T. C. Huang, S. S. P. Parkin, D. T. Keane, S. J. LaPlaca, P. M. Horn and G. A. Held, Phys. Rev. *B40*, 8872 (1989).

15. H. Takagi, Y. Tokura, and S. Uchida in Springer Series in Matl. Sci. Vol. 11, Mechanisms of High Temperature Superconductivity, H. Kamimura and A. Oshiyama, Eds., Berlin (1989) p. 238.

16. P. Kuiper, G. Kruizinga, J. Ghijsen, G. A. Sawatzky and H. Verweij, Phys. Rev. Lett. *62*, 221 (1989).

17. G. Sawatzky, private communication.

18. N. Nguyen, L. Er-Rakho, C. Michel, J. Choisnet and B. Raveau, Mat. Res. Bull. *15*, 891 (1980).

19. R. J. Cava, B. Batlogg, R. B. van Dover, J. J. Krajewski, J. V. Waszczak, R. M. Fleming, W. F. Peck, Jr., L. W. Rupp, Jr., P. Marsh, A. C. W. P. James, and L. Schneemeyer, Nature *345*, 602 (1990).

Electron Transfer Between Layers
in High-Temperature Superconductivity

Edward Teller

Lawrence Livermore National Laboratory

Livermore, CA

I'm afraid that I am unable to do the one thing I want to do. I can not really explain superconductivity. If I had any such temptation, the news this morning that vanadium is as good as copper makes anybody who tries to explain it think and think again.

Instead, I will try to give you a point of view which, I believe, establishes a number of connections between the phenomena we associate with superconductivity. I start by telling you that in all the new superconductors, the cations form something that is an approximation to a body-centered cubic lattice.

We have a layer cake, with the cations lying in alternate layers in the corners and in the centers of squares -- this is, of course, the body-centered cubic lattice. The anions are oxygens. We have three kinds of layers: layers without oxygen; layers that are monoxides with the oxygen usually, but not always, in the center of the square (the exception is the copper chain layer in the 123 compound); and finally we have the perovskite layers, the CuO_2 layers, in which there is an oxygen in the center of each line connecting two neighboring coppers.

Three of us, Art Broyles, Brian Wilson and I, (referred to as BTW) published a short paper. This paper needs a very essential modification about which I am going to talk. The paper is available; it is in the Journal of Superconductivity.

The essential thing (not sufficiently specified in our paper) is that all participating electrons have two nodes crossing on every cation at 45° in this square lattice that I described to you. It may be f rather than d because an extra node in the plane of barium cations may make little difference.

I will start now by showing you the electron configuration in terms of Bloch waves. Incidentally, what I'm going to tell you will be, in essence, the BCS theory with a couple of modifications.

In Fig. 1, I plot the energy as a function of k in the x-y plane. Actually, I show only the dependence on k_x near $k_x=k_y=0$ and $k_x=k_y=1$ (in units of π/a) (similar signs and opposite signs on neighboring cations). In a ceramic, it is reasonable that all the states in one band should be filled and all the states in the next band should be empty.

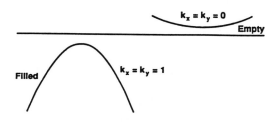

Fig. 1a. Filled and empty states for a ceramic model.

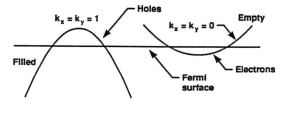

Fig. 1b. Filled and empty states with holes and electrons for a metallic model.

What we did wrongly in our paper was to assume that one of the bands in question was associated with oxygen rather than with copper and that the filled and empty states had the same wave number. I am, on purpose, not drawing it that way. What we did rightly in our paper was to associate the filled band with the perovskite layer and the empty band with a neighboring oxide layer. I furthermore believe that this particular explanation holds only for the bismuth and thallium compounds while, for the other compounds, mixing of wave functions in the the perovskite layer alone maybe relevant. (I am indebted to Broyles and Wilson for this remark.) In the following I shall concentrate on the bismuth case.

The remarkable thing, I think, is that the maximum and the minimum shown in the figure have almost the same energy. In fact, I would like to draw it in such a way that the minimum is a little bit below the maximum as in Fig. 1b. The holes and electrons forming shallow Fermi "puddles" will be equal in number if the distances of the crossing points of the parabolae with the Fermi surface are equal. But as long as the curvature of the lower parabola is greater, the holes will have a smaller effective mass and a higher velocity than the electrons. We have a positive Hall effect since the hole-conductivity will predominate. The energy difference between the maximum and minimum of the two parabolae may be in the neighborhood of one tenth of a volt.[*]

To describe the essential wave function in the perovskite plane, we show the ions in this plane in Fig. 2 writing Cu and O at the appropriate locations. The dotted lines at 45° show the intersections of the nodes of the last (9th) d-electron on the copper ions or, for that matter, on the strontium or calcium ion above or below this plane. The highest filled electron, however, is not a p_z electron on the oxygen as stated by BTW. Rather, it is, as has been originally stated, the ninth electron on the copper, the unpaired electron, which through the interaction in the oxygen can be antiferromagnetic.

---

[*] The very nice article by Pickett (Reviews of Modern Physics, 1989, p. 433) discusses calculations that indicate unfilled levels from the monoxide layers descend into the neighborhood of the Fermi surface. My discussion assumes a simple model which differs from this calculation.

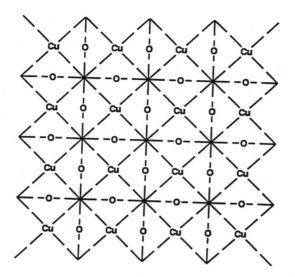

Fig. 2. The perovskite plane showing intersections of nodes with the plane.

Now, this electron on the copper is not purely on the copper. The wave function, has an angular dependence proportional to $x^2 - y^2$. That gives 45° node lines with maximum extensions in the direction of the oxygens. These wavefunctions will then interact with p electrons on the copper, but not the p_z electrons, but rather the p_x electrons and the p_y electrons. If I put the main amplitude on the oxygen and add the wave function on the copper with a smaller amplitude, then I get completely filled states. If instead, I take the main wave function on the copper and subtract a little of the oxygen wave function so as to establish an extra node between oxygen and copper, then I have the ninth electron on the copper which is then, in fact, half filled. The model of distributing one electron per copper should be valid because otherwise the repulsion of two electrons on the same copper becomes too strong.

That the p_x and p_y electrons participate in the wave function causes vertical and horizontal nodes to appear in Fig. 2. These cause the wave functions to change sign when we move from one copper to its neighbor. (There are additional nodes due to our having used the repulsive wave functions between Cu and O, but these do not influence the changes of sign we are discussing and have not been entered.)

I wanted to say that this wave function in the perovskite plane mixes with a wave function in a neighboring strontium oxide plane. This is not possible because, in the perovskite layer, we have at the top of the Bloch parabola the state shown in Fig. 2 which changes sign if we move from one copper to the next. But in the SrO plane or the Ca-plane, there is at the bottom of the Bloch parabola no such change of sign. Therefore, we have in one case $k_x=k_y=1$, in the other $k_x=k_y=0$.

Now comes the essential point. A group in Karlsruhe, Pintschovius, Reichardt and others, have found a vibration shown in Fig. 3 where alternate coppers are approached by their four neighboring oxygens, whereas the other coppers see the oxygens go away from them. In half a period, the role of these two kinds of oxygen will

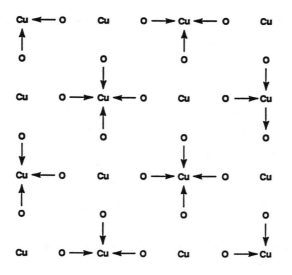

Fig. 3. Breathing vibration with symmetry $k_x = k_y = 1$

be reversed. The symmetry of this vibration is precisely the symmetry difference between the maximum in one parabola and the minimum in the other. In other words, the two wave functions don't mix, but one wave function can mix with the other if one vibrational quantum is added.

First, taking this situation as I described it, I want to explain to you some of the properties of these new superconductors above the transition point. You know that the resistance in the planes has a linear dependence on the temperature. The conductivity is not as good as in good metals, but this linear dependence is the same as that of metals at high temperature. At the low temperatures of liquid nitrogen, 100° K or thereabouts, most metals have a much lower resistivity, and the resistivity, instead of being linear, goes with a power of the temperature.

Actually, I should not be talking about the wave functions in the precise maximum or the precise minimum. Nor, should I talk about the vibration as shown in the figure. Instead, in all three cases, I should add a small wave number vector. The wave number vector must be small otherwise I go too low on the lower parabola or too high on the higher parabola. I have talked to you about processes that amount to the creation or annihilation of electron-hole pairs. If that should occur, they can not have high wave numbers and the vibration will have to be modified by the difference of their wave numbers. When you scatter them, which gives rise to resistance, you are concerned only with acoustic vibrations of a low energy $\hbar\omega$ because the wave length is long. Therefore the linear dependence of the resistance becomes obvious.

On the other hand, if you look at the resistance along the z axis, there is some evidence that it is rising rapidly between room temperature and the transition point. It may rise by a factor five or a factor thirty. The reason may be that the electrons are confined to their planes except when the breathing vibration shown in Fig. 3 is participating.

This vibration has been found at approximately seventy millivolts. Conductivity is helped by this vibration in the z direction, but hindered by the acoustic vibration in the x-y plane. Actually, the latter conclusion appears firm. The former needs more experimental study and theoretical discussion.

I now only have to say "BCS" because from here on I will proceed as Bardeen, Cooper, and Schrieffer do. I talk about the creation or annihilation of an electron-hole pair together at the same time with the excitation of this vibration. That is one step. The second step is that I create a second electron-hole pair with opposite spins and k-values and at the same time de-excite the breathing vibration. That leads to a coupling between electrons. These elementary steps are not precisely, as described by Bardeen, Cooper, and Schrieffer, to excite from just below the Fermi surface to just above the Fermi surface, but they are to excite from one region of the k space into another with approximately one-tenth of a volt for the magnitude of the energy difference. The total act consists either in the creation of an electron pair with opposite spins together with a hole pair with opposite spins or else the annihilation of an electron pair with opposite spins and a hole pair with opposite spins. All these should be included in any calculation of this kind. The difference, therefore, from Bardeen, Cooper, and Schrieffer is that we move between two regions of the single electron spectrum and that the lattice displacement that we invoke -- that is the cause for the coupling attraction of the electrons -- is not an acoustic vibration, but a particular optic vibration. I must add that, according to discussions with Broyles and Wilson, in the lanthanum and 123 compounds the transitions may occur in the same electronic band but will still be due to the breathing (optical) vibration.

I want to go a little beyond this, and I want to give you now a more detailed description of the wave functions in the superconducting state. From the wave functions of different k-values on the copper plane and on the neighboring plane or planes, I can construct wave packets. As shown in our paper (BTW) the interaction for a two dimensional case is proportional to $1/r^2$ where r is the linear dimension of the wave-packet in the x-y plane. To sketch the procedure in a formal way, we may start from the simple case corresponding to $k_x=k_y=1$ in the perovskite plane and $k_x=k_y=0$ in the strontium oxide plane with the superposition,

$$\bar{\phi}_p\psi_0 + \alpha^2\bar{\phi}_{Sr}\psi_1 = (\bar{\phi}_p + \alpha\bar{\phi}_{Sr})(\psi_0 + \alpha\psi_1) + (\bar{\phi}_p - \alpha\bar{\phi}_{Sr})(\psi_0 - \alpha\psi_1)$$

where ϕ_p and ϕ_{Sr} are the electron wave functions in the perovskite and strontium planes and Ψ_0 and Ψ_1 are the lowest and first excited wave functions of the breathing vibration. The number $\alpha<1$ depends on the strength of the coupling. Thus Ψ_0 is a gaussian. Ψ_1 is its first derivative and $(\Psi_0 + \alpha\Psi_1)$ as well as $(\Psi_0 - \alpha\Psi_1)$ are approximately gaussians displaced forward one set and the other set of the copper sites as shown in Fig. 3. Since ϕ_p changes its sign on a displacement from one copper to the next and ϕ_{Sr} does not, ϕ_p and ϕ_{Sr} will reinforce each other for $\phi_p + \phi_{Sr}$ on one set of copper ions and tend to cancel on the other set. For $\phi_p - \phi_{Sr}$, the roles are interchanged. Thus the superposition can be so arranged that electrons escape from one set of sites to which the oxygen ions are approaching and are retained on the other sites.

The next step is to take wave functions at k-values $k_x = 1 + \epsilon$ for ϕ_p and $k_x=\eta$ for ϕ_{Sr} (and similarly for k_y). The modulated breathing vibrations should have k-values that satisfy the conservation law in k-space. For small ϵ and η-values distributed on an appropriate gaussium, one can obtain a wave packet extending a few Cu-Cu distances in the x-y plane where the vibration occurs together with the electron transfers from a copper-site to the Sr-plane.

If all of this is done for an electron of a given spin and also for an electron of the opposite spin, then two electrons will be transferred to the neighboring plane. The displacements on the breathing vibration will be doubled and the resulting energy multiplied by 4.

By changing the phases with which the $k_x = 1 + \varepsilon$ (for ϕ_p) and $k_x = \eta$ (for ϕ_{Sr}) wave functions are superposed, one can change the locations of the wave packets.

The locations of the electrons pairs and the corresponding displacements of the O^{--} ions in a breathing vibration may form a close packed hexagonal lattice on the x-y plane which is incommensurate with the crystal lattice.

To obtain the ground state of the superconducting phase, one more step is needed. We must take the configuration described above and displace it by integral multiples of the lattice distance and take the sum. This gives the state with zero current. Superposition of the same states with appropriate phase shifts gives the current carrying states.

There is an interesting possible consequence of all this. The question: can one see these actual displacements in interference patterns of x-ray, electron, or neutron-scattering? A displacement will give, in the diffraction pattern, changes in k values which are in the x direction and the y direction one half the normal changes because the periodicity of the breathing vibration is double that of the usual. So, in the reciprocal lattice, we obtain half of the deflection.

But, in addition, you will have a modulation due to the close-packed superlattice I have described. Thus I am predicting the appearance of half values in the reciprocal lattice, but these, in turn, should not be single maxima, but spread out to account for the superlattice that I have described. The small amplitudes of the displacements should give rise to low intensities for these maxima. This may not be easy to see.

The group in Karlsruhe have seen something similar. What they have done is to study vibrations by inelastic neutron scattering where the energy has changed by one quantum of the vibration. In the superconducting state, one should observe the change in wave number in the inelastic scattering. This will not be strongly altered at the transition point. But if one goes to low temperatures, maybe 20 - 30° K, then there should be, with a low probability, a momentum change in the neutrons that corresponds to a displacement characteristic of the vibration which they have already seen. Unlike the earlier experiment, this should not be accompanied by an energy change of the neutrons, but it should be only a momentum change in an otherwise elastic scattering.

One last question: what happens to the energy gap? We have a transition from the perovskite electrons into the monoxide region. There is in every case more than one monoxide layer. If transitions occur into different layers or into states that are linear combinations between several layers, this should give rise not to a single gap, but to a multiplicity of gaps. This corresponds to the general experience.

I started by telling you that I don't understand superconductivity. Indeed I don't. I would like to understand not only why perovskite layers -- in combination with other layers where d or f electrons can occur -- why this gives superconductivity? The approach of the oxygen to the copper, as in the breathing vibration, is strongly coupled with the electron state -- much more strongly than the coupling of acoustic vibrations on which BCS relies. But I think that a similar analysis will be required to show why mercury has superconductivity and sodium does not. Why the A15's have it? Why the C1 crystals, a sodium-chloride-like structure of carbides and nitrides, why these have superconductivity?

I want to conclude with an obviously unfounded prediction. By pursuing superconductivity along the lines that we are now working, I don't think we'll get to too much higher temperatures. I do believe, however, that the Schrödinger equation as applied to 10^{20} electrons will produce many more surprises. It may well be that we shall find still other types of lattices, maybe organic, maybe something entirely different, that will give superconductivity at really high temperatures.

Work performed under the auspices of the U.S. Department of Energy by Lawrence Livermore National Laboratory under contract W-7405-ENG-48.

Fundamental Properties

NORMAL-STATE RESISTIVITY AND THE ORIGIN OF HIGH-TEMPERATURE SUPERCONDUCTIVITY IN Cu OXIDES

C.C. Tsuei

IBM Research Division
Thomas J. Watson Research Center
Yorktown Heights, NY 10598

ABSTRACT

The in-plane resistivity data, $\rho_\parallel(T)$, as a function of temperature for various Cu-oxide superconductors have provided strong evidence for a Fermi-liquid normal state and also important clues for understanding the mechanism responsible for high-temperature superconductivity. A dominant quadratic temperature dependence of $\rho_\parallel(T)$ above T_c is observed in the electron- doped NdCeCuO system and several relatively low-T_c hole- doped cuprates. On the other hand, high T_c Cu oxides ($T_c \gtrsim 80K$) are always characterized by a linear temperature dependence of $\rho_\parallel(T)$. Within the framework of the BCS phonon-mediated pairing, the closeness of a 2D van Hove singularity in the density of states to the Fermi level provides a basis for understanding the origin of high-temperature superconductivity, the correlation between T_c and $\rho_\parallel(T)$, anomalous isotope effects and several other normal-state and superconducting properties.

INTRODUCTION

More than four years have elapsed since high-temperature superconductivity was discovered in copper oxides. Its origin, however, remains elusive in spite of the fact that numerous novel theoretical models have been proposed. Clues to the mechanism responsible for high-temperature superconductivity in these oxides presumably can be found in their unusual normal-state and superconducting properties. Early experimental findings such as the near-zero oxygen isotope effect and the linear temperature dependence of the in-plane normal-state resistivity tend to favor the theoretical models based on non-Fermi liquid and non-phonon pairing mechanisms. Recent results, however, have changed this viewpoint significantly. Experimental data accumulated so far have shown that the new superconducting cuprates are BCS-like superconductors. Furthermore, evidence mostly from experiments such as NMR and photoelectron spectroscopy suggests a Fermi-liquid like picture is viable for the description of the normal state.

In this brief review, in-plane resistivity data in various Cu-oxide superconductors will be presented and its implications for the origin of high-
temperature superconductivity will be discussed.

IN-PLANE RESISTIVITY

The fact that the hole-doped high-T_c (say $T_c \gtrsim 80K$) cuprates are generally
characterized by a linearly temperature-dependent in-plane resistivity, $\rho_{\parallel}(T)$,
is well-established in the literature.   A recent resistivity measurement[1] on
epitaxial   films   of   the   electron-doped   superconducting   oxide
$Nd_{1.85}Ce_{0.15}CuO_{4-y}(y\sim0.04)$ has revealed a dominant quadratic temperature dependence of $\rho_{\parallel}(T)$ for $T_c < T < 200K$ ($T_c = 20K$) and a logarithmic
correction to the T^2 dependence for $T > 200K$. (Fig.1) A quadratic temperature dependence of $\rho_{\parallel}(T)$ is also observed by others in single crystals[2] and
thin epitaxial films.[3]   Such a non-linear temperature dependence is apparently not a unique normal-state property of the electron-doped Cu oxides.
A similar T^2 term in $\rho_{\parallel}(T)$ is also found in several hole-doped cuprate
superconductors.[4] Examples are:

1.   The "60K" phase of the YBaCuO system (e.g. $YBa_2Cu_3O_x$, x = 6.6).

2.   The "248" phase of the YBaCuO system (i.e. $Y_2Ba_4Cu_8O_{16-\delta}$).

3.   The $Tl_2Ba_2CuO_{6-y}$ system.

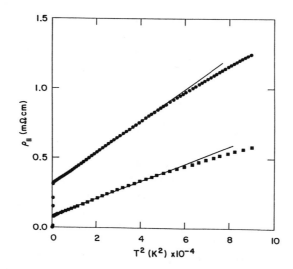

Fig.1: In-plane resistivity $\rho_{\parallel}(T)$ of the electron- doped NdCeCuO superconductors as a function of T^2 for an epitaxial film, 2500 Å thick (• data
taken from Ref.1) and a single crystal (■ data taken from Ref.2).

In particular, the resistivity data of the single CuO_2-layer system $Tl_2Ba_2CuO_{6-y}$ by Kubo et al.[5] (Fig. 2) shows a systematic change of the temperature dependence of $\rho_{\parallel}(T)$ with T_c which varies from 0K to 80K as the oxygen deficiency y changes only from 0 to 0.1. As shown in Fig. 2, the resistivity of the sample A ($T_c = 80$K) is definitively linear in T. As the oxygen deficiency decreases, both T_c and resistivity decrease and the temperature dependence becomes increasingly non-linear. In fact, for the sample D ($T_c \sim 15$K) a T^2 term dominates the $\rho(T)$ for 20K < T < 150K. The significance of these results can be briefly stated as follows:

1. Electron-phonon interaction is probably not important in determining the temperature dependence of $\rho(T)$, because the small change in oxygen deficiency ($\triangle y < 0.1$ for $y \sim 6$) involved is too small to change the crystal structure.

2. A dominant quadratic temperature dependence of $\rho_{\parallel}(T)$ is consistent the Fermi-liquid model for electron-electron interaction which is greatly enhanced by the quasi-2D nature of the CuO_2 - layered systems. In terms of the Luttinger theorem, if the T^2 dependence (sample B,C and D) is a manifestation of Fermi-liquid normal state then the linear T of $\rho(T)$ for the high-T_c sample (sample A) is also consistent with a Fermi-liquid model. Basically, this is because there is no structural or any other phase transition as a consequence of small change in the oxygen content. And the Fermi-liquid model should prevail through out the whole composition range (ie. $y=0$ to 0.1). In other words, if the linear temperature dependence of $\rho_{\parallel}(T)$ can not be attributed to electron-phonon interaction, it does not necessarily imply a non-Fermi-liquid normal state.

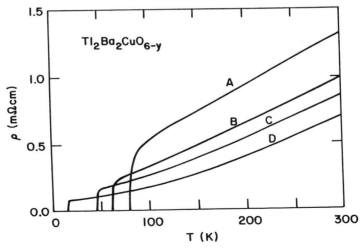

Fig.2: In-plane resistivity $\rho_{\parallel}(T)$ as a function of temperature for $Tl_2Ba_2CuO_{6-y}$ ($y=0$ to 0.1, after Kubo et al.[5].)

3. Together with the in-plane resistivity data described earlier in this article, these resistivity results clearly demonstrate the fact that the linear temperature dependence of $\rho_\parallel(T)$ is not a universal normal-state property of the cuprate superconductors independent of the type of charge carries (electrons or holes). Furthermore, there exists a correlation between T_c and the degree of linearity of $\rho_\parallel(T)$. The high-T_c Cu-oxide superconductors are invariably characterized by a linearly temperature-dependent $\rho_\parallel(T)$ and a very small residual resistivity. On the other hand, the relatively low-T_c (say $T_c < 40K$) Cu oxides show a dominant quadratic temperature dependence in $\rho_\parallel(T)$.

THE VAN HOVE SCENARIO IN HIGH-TEMPERATURE SUPERCONDUCTIVITY

The in-plane resistivity data, $\rho_\parallel(T)$, as a function of temperature for various Cu-oxide superconductors have provided strong evidence for a Fermi-liquid normal state and also important clues for understanding the mechanism responsible for high-temperature superconductivity. The basic scenario is that the quasi-2D nature of the CuO_2-layered cuprate superconductors gives rise to a logarithmic van Hove singularity (vHs) in the density of states, $N(E)$, for charge carriers:

$$N(E) = N_0 \, \ell n \left| \frac{E_F}{E - E_F} \right|, \tag{1}$$

where E_F is the Fermi energy and N_0 is the density of states normalized to a flat band with a bandwidth of $2E_F$. And the close proximity of E_F to the van Hove singularity in $N(E)$ can lead to enhanced T_c and strong low energy electron-electron scattering, resulting in the correct quasiparticle lifetime which is consistent with many normal-state and superconducting properties. A van-Hove-like energy-dependent $N(E)$ near E_F has been proposed in the past as a T_c-enhancement mechanism for the conventional A15 superconductors[6] and more recently for the high-T_c Cu oxides.[7] This T_c enhancement mechanism has not attracted much attention. Recently, the vHs mechanism is resurrected on the basis of experimental results of resistivity and other properties. It is suggested that the closeness of the vHs in $N(E)$ to E_F determine temperature dependence of $\rho_\parallel(T)$ and the magnitude of T_c. The effect of vHs on T_c can be seen from the standard BCS gap equation:

$$\frac{2}{V} = \int_{E_F - \hbar\omega_c}^{E_F + \hbar\omega_c} \tanh\left(\frac{E - E_F}{2k_BT_c}\right) N(E) \left(\frac{dE}{E - E_F}\right) \qquad (2)$$

where ω_c is the cut-off frequency ($\hbar\omega_c$ can be identified with the energy of certain phonon mode or the Debye energy, ie. $\hbar\omega_c \sim k_BT_D$), and N_0V is a measure of the electron-phonon interaction strength. An approximate solution to Eq (2) is:

$$T_c = 1.36\, T_F \exp\left[-\sqrt{\frac{2}{N_0V} + \left(\ell n\, \frac{k_BT_F}{\hbar\omega_c}\right)^2} - 1\right] \qquad (3)$$

If $N(E)$ is energy-independent, as assumed in the standard BCS treatment, Eq. (2) will lead to the well-known BCS formula for T_c in the weak-coupling limit:

$$T_c = 1.14\, T_D \exp\left[-\frac{1}{N_0V}\right] \qquad \text{for } N_0V << 1 \qquad (4)$$

By substituting realistic parameters (such as $NoV = 0.3$, $T_D = 360K$, and $T_F = 5800K$), the BCS formalism with vHs (Eq. (3)) yields a T_c of 200K. The standard BCS T_c-formula predicts a T_c of ~15K.

In spite of the apparent success of the vHs high- T_c mechanism in producing a T_c of the order of 100K, various objections to this mechanism have been raised. These issues are fully addressed and shown that they do not invalidate the sharp-peak-in-N(E) as a viable mechanism for high-temperature superconductivity in Cu oxides.[4,8]

Furthermore, a recent slave boson mean-field realistic band structure calculations[9] for CuO_2-layered systems shows that E_F is generally pinned very close to the van Hove singularity in $N(E)$. The vHs is not located at the half-filling point, as in the Hubbard model, but corresponds to a doping level of $x = 0.15$ to 0.35 for the quasi-2D Cu-O systems, depending on the long-range interaction in the model. Based on the same model, the lifetime due to electron-electron scattering is calculated.[10] The calculated, parameter-free lifetime, $1/\tau(E)$, is linearly dependent on E when E_F is close to the vHs and has the right magnitude compared with the angle-resolved photo-emission and infrared conductivity data. The conventional $1/\tau(E) \propto E^2$ in E dependence, typical of 2D systems is expected when E_F is far away from the vHs. These results are also consistent with the idea that close proximity of E_F to vHs is essential to high-T_c Cu-oxide superconductors which are all characterized by $\rho_\parallel(T) \propto T$. The observation of $\rho_\parallel(T) \propto T^2$ or $T^2 \ell nT$ in relatively low-T_c cuprates also finds a natural explanation in the vHs model.

Another important piece of supporting evidence for the vHs high-T_c mechanism comes from a recent development in understanding the anomalous oxygen isotope effect in various Cu-oxide superconducting systems.[11] Recent experimental work by Crawford et al.[12] has shown that a near-zero isotope effect (i.e. $\alpha_0 \sim 0$, where $T_c \sim M^{-\alpha_0}$ and M is the atomic mass of oxygen) is not a general feature of superconducting cuprates. The isotope exponent, α_0, in $La_{2-x}Sr_xCuO_4$ is found to depend strongly on the doping concentration (x) of Sr, and can exceed significantly the standard BCS limit for α (i.e. $\alpha^{BCS} = 0.5$). A recent preprint by Franck et al.[13] indicates the δ_0 increases in the $Y_{1-x}Pr_xBa_2Cu_3O_7$ system as T_c decreases as a result of increasing Pr content. Within the framework of the BCS phonon-mediated pairing, when the effect of vHs or a more general energy-dependent peak in N(E) is taken into account, all the above-mentioned anomalous isotope effects and the T_c-variation with α_0 can be understood.

SUMMARY AND CONCLUSIONS

In summary, there exists a correlation between T_c and the degree of linearity of the temperature dependent in-plane resistivity $\rho_\parallel(T)$. The high-T_c copper oxides are always characterized by a linear temperature dependence of $\rho_\parallel(T)$. On the other hand, a T^2 term usually dominates the $\rho_\parallel(T)$ over a wide range of temperature above T_c for relatively low-T_c cuprates. These resistivity data clearly demonstrate that the linear temperature dependence of $\rho_\parallel(T)$ is not a universal normal-state property and is probably not due to electron-phonon interaction. In terms of electron-electron interaction, these $\rho_\parallel(T)$ results have provided strong supporting evidence for a Fermi-liquid normal state and also important clues for understanding the mechanism responsible for high-temperature superconductivity.

Within the framework of the BCS phonon-mediated pairing, the proximity of the Fermi level to the 2D van Hove singularity in N(E) determines the magnitude of T_c and its correlation with the temperature dependence of $\rho_\parallel(T)$. Even with a modest electron-phonon interaction (i.e. the electron-phonon coupling constant $\lambda < 1$), the vHs T_c-enhancement mechanism can lead to a T_c of the order of 100K. This high T_c mechanism is consistent with many other observed normal-state and superconducting properties.

REFERENCES

1. C.C. Tsuei, A. Gupta and G. Koren, Physica C 161, 415 (1989).

2. Y. Hidaka and M. Suzuki, Nature 338, 635 (1989).

3. T. Terashima et al. Appl. Phys. Lett. 56, 677 (1990).

4. C.C. Tsuei, "Origin of High Temperature Superconductivity in Cooper Oxides -- Clues from the Normal - State Resistivity", and the references therein, Proceedings of the 3rd. Bar - Ilan Conference on Frontiers in Condensed Matter Physics (Israel, 1990), Physica A, 168, 238 (1990).

5. Y. Kubo, Y. Shimakansa, T. Manako, T. Satoh, S. Iijima, T. Ichihashi and H. Igarashi, Physica C 162-164, 991 (1990).

6. J. Labbe, S. Barisic, and J. Freidel, Phys. Rev. Lett. 19, 1039 (1967); G. Kieselmann and H. Rietschel, J. of Low Temp. Phys. 46, 27 (1982); J.E. Hirsch and D.J. Scalapino, Phys. Rev. Lett. 56, 2732 (1986) and the references therein.

7. P.A. Lee and N. Read, Phys. Rev. Lett. 58, 2691; A. Virosziek and J. Ruvalds, Phys. Rev. B 42, 4064, (1990); J.E. Dzyaloshinskii, Soviet Phys. JETP Lett. 46, 118 (1987); Soviet Phys. JETP 66, 848 (1989); J. Labbe and J. Bok, Europhys. Lett. 3, 1225 (1987); J. Friedel, J. Physique 48, 1787 (1987); ibid. 49, 1435 (1988); J. Phys. Condens. Matter 1, 7757 (1989).

8. R.S. Markiewicz, J. Phys. Condens. Matt. 2, 665 (1990); R.S. Markiewicz and B.C. Giessen, Physica C 160, 497 (1989).

9. D.M. Newns, P.C. Pattnaik and C.C. Tsuei, "Role of van Hove Singularity in High Temperature Superconductors I: Mean Field (to be published in Phys. Rev. B).

10. P.C. Pattnaik, D.M. Newns and C.C. Tsuei "Evidence for van Hove Scenario in High Temperature Superconductivity" (preprint).

11. C.C. Tsuei, D.M. Newns, C.C. Chi and P.C. Pattnaik, "Anomalous Isotope Effect and van Hove Singularity in Superconducting Cu Oxides", (preprint).

12. M.K. Crawford, M.N. Kunchur, W.E. Farneth, E.M. McCarron III and S.J. Poon, Phys. Rev. B41, 282 (1990).

13. J.P. Franck, J. Jung, M.A.K. Mohamed, S. Gygax, and I.G. Sproule (preprint).

NEW DEVELOPMENTS IN FLUX CREEP
OF HIGH-TEMPERATURE SUPERCONDUCTORS: EVIDENCE FOR A VORTEX GLASS AND QUANTUM TUNNELING

A. P. Malozemoff *

IBM Research, Yorktown Heights NY 10598-0218 USA

ABSTRACT

Four different temperature regimes can be distinguished in flux-creep relaxation-rate data on many high-temperature superconductors: 1) the ultra-low temperature region where initial results show a temperature-independent relaxation suggesting vortex tunneling, 2) a low-temperature region where the relaxation increases linearly with temperature in agreement with conventional flux creep theory, 3) an intermediate region where the normalized relaxation rate becomes constant, as suggested by a vortex glass or collective pinning model, and finally 4) a high-temperature region near the irreversibility line, where experiments are still inconclusive.

INTRODUCTION

The literature on magnetic relaxation of high-temperature superconductors[1-3] is extensive. Here some recent developments are reviewed, particularly the new evidence in favor of the vortex glass or collective pinning models, and also the new data on ultra-low-temperature relaxation, indicating quantum tunneling of vortices.

Fig. 1 illustrates typical features of magnetic relaxation in the critical state of YBaCuO materials, as observed in many recent studies.[4-14] This particular study[9] was on a fully-oxygenated YBaCuO crystal with a field of 1 T applied parallel to the c-axis. The crystal was studied before and after irradiation with 3 MeV protons, which increased the critical current density by more than an order of magnitude at 77 K. The field was cycled in such a way as to achieve a fully-penetrated vortex critical state, and the magnetization was measured as a function of time in a SQUID magnetometer, with due experimental precautions,[4,6] the neglect of which has created considerable uncertainty about much early SQUID data in the literature. As is well known, the magnetization M relaxes in an approximately logarithmic manner with time t, and the logarithmic relaxation rate $dM/d\ln(t)$ can be studied as a function of temperature T.

For reasons to be explained below, it is useful to normalize this relaxation rate by the initial irreversible magnetization M_i. It is important to use the irreversible magnetization (half the difference of field-increasing and field-decreasing magnetizations at the same field) rather than the magnetization itself because reversible diamagnetism can cause a spurious offset particularly at high temperature. If the change in magnetization during the measurement time is small, $(1/M_i)d(M)/d\ln(t)$ is close to the dimensionless quantity

*Work performed in part at Université Paris-Sud, Physique des Solides, Bât. 510, 91405 Orsay-Cedex, France,

$$S \equiv (1/M)dM/d\ln(t) = d\ln(M)/d\ln(t) , \qquad (1)$$

which will henceforth be referred to as the normalized relaxation rate. This quantity is plotted in Fig. 1 versus temperature T.

Figure 1 indicates four separate temperature regimes labeled as Regions I to IV. The data in the region below about 20 K appear to extrapolate linearly to a finite intercept at $T = 0$. Since thermal processes are frozen out at $T = 0$, finite relaxation here would suggest some kind of quantum tunneling process for vortex motion. This ultra-low-T region is called Region I. With increasing temperature, S(T) appears to increase linearly up to about 20 K; this is called Region II. Above 20 K S(T) saturates, forming region III. In many studies,[5,6] this region appears as a broad peak rather than as a plateau, although S is usually[14] in the range 0.02 to 0.035, whether at the peak or on the plateau, in YBaCuO samples ranging from powders[5,6] to flux-grown or melt-processed crystals[7 – 9] to thin films.[4,10] Finally, above about 60 K, the data of different groups conflicts, with some showing increasing, others showing decreasing S(T). This is called Region IV. Since signals are so small, a reliable picture in this region remains to be established, and it will not be considered further here.

REGION II - CONVENTIONAL FLUX CREEP

The linear increase in S(T) in Region II is most simply explained in the conventional Anderson-Kim theory of thermally activated flux creep.[15 – 17] In the standard theory, the net potential barrier to vortex motion $U_J = U_0[1 - (J/J_{c0})]$ is assumed to be linearly dependent on the applied current density J, which exerts a Lorentz force on the vortices. Combined with the standard Arrhenius relation for the hopping time $t = t_{eff}\exp(U_J/kT)$, where t_{eff} is an effective[18] attempt time, this leads to the classic formula[17]

$$J(T,t) = J_{c0}[1 - (kT/U_0)\ln(t/t_{eff})] . \qquad (2)$$

Here J_{c0} is the critical current density in the absence of thermal activation. Since in the critical state,[17] the magnetization M is proportional to J, the logarithmic derivative of Eq. 1 can be evaluated, givingsup18

$$S = -kT/[U_0 - kT\ln(t/t_{eff})] , \qquad (3)$$

which in the low-T limit is just kT/U_0. The advantage of considering S rather than dM/d$\ln$(t) is apparent here: by normalizing, one eliminates J_{c0} from Eq. 2 and makes possible the direct determination of U_0. Since U_0 is expected, like J_{c0} and most fundamental BCS parameters, to be independent of T at low temperatures and to drop to zero at the critical temperature T_c, S should be linear in T at low temperatures and to curve **upwards**. The linear prediction can be compared to experiment to extract the $J = 0$ barrier height U_0 which lies in the range 20 to 50 meV for the data of Fig. 1, for the unirradiated and irradiated YBaCuO crystals.

REGION III - EVIDENCE FOR THE VORTEX GLASS MODEL

The real questions begin, however, with the experimental observation[19] that at higher temperatures, in Region III, S curves **downwards** rather than upwards, indicating a limit to the applicability of the conventional theory. The recent literature is full of proposed explanations[4,6,10,18,19] for this effect. Some involve distributions of pinning barriers[18,19] or distributions of critical current densities.[10] Perhaps the most widely considered approach[5,6,20] is based on the nonlinear dependence of U_J on J, in contrast to the simple linear dependence assumed above. Beasley, Labusch and Webb[16] first pointed out that the linear U_J dependence is only an approximation; in fact it corresponds in a simple one-dimensional model to a triangular or saw-tooth potential with unphysical slope discontinuities. Any smooth potential as a function of position will give an upward curvature, of the type illustrated in Fig. 2. As J decreases with increasing temperature, U_0, which is just the extrapolation to J = 0 of the tangent to the U_J curve at the given J, shifts to higher values, thus causing S to curve down according to Eq. 3.

Fig. 2 shows several recently-discussed upward-curving forms for U_J, including two with another feature which at first glance might appear to be unphysical, namely a divergence at J = 0. Yet a divergence of the power-law form $U_0[(J_{c0}/J)^\mu - 1]$, where μ is an exponent, is just what emerges from treatments based either on the vortex glass and scaling theory models[21−2] or on collective pinning theory.[22] The divergence reflects the collective nature of the pinning process and implies a phase transition, in the limit of zero current density, to a true superconducting state with zero ohmic resistivity. Another proposal, by Zeldov et al.,[20] is of the logarithmic form $U_0 \ln(J_{c0}/J)$, arising from the logarithmic forces between vortices.

Substituting the power-law form leads to the so-called interpolation formula

$$J(T,t) = J_{c0}/[1 + (\mu kT/U_0) \ln(t/t_{eff})]^{1/\mu} , \qquad (4)$$

so named because it interpolates between the conventional low-T Anderson-Kim behavior and the new high-T vortex-glass behavior. S deduced from this formula is

$$S = -1/[(U_0/T) + \mu \ln(t/t_{eff})] , \qquad (5)$$

which reduces in the high-T limit to the simple result $S = 1/\mu \ln(t/t_{eff})$. Since μ is expected to be a universal exponent in the vortex glass model,[21] and since the dependence on the measurement time t and the effective attempt time t_{eff} is only logarithmic, this formula predicts an almost universal behavior, with S falling close to 0.03, if one takes $\mu = 1$, t = 1000 sec (as in a typical relaxation measurement) and $t_{eff} = 10^{-10}$ sec. As pointed out recently by Malozemoff and Fisher,[14] the prevalence of just such values in experiments on a wide variety of materials provides strong evidence for the vortex glass model (or the phenomenologically equivalent collective pinning theory) and against the distribution models, for which such universality is hard to rationalize.

Further evidence for this picture comes from measurements[11 – 13] determining the direct time-dependence of S. The curvature predicted by Eq. 5 has been observed in these studies and the exponent μ is found to be around 1, consistent with the vortex glass theory prediction.   Eq. 4 can also explain[24] the well-known quasi – exponential[25] T-dependence of $J_c(T)$.

REGION I - EVIDENCE FOR QUANTUM TUNNELING OF VORTICES

There have been persistent reports[26 – 29] of ultra-low-T magnetic relaxation in a variety of high and low-T superconductors, usually based on linear extrapolations of S(T) to T = 0. Indeed, such an extrapolation is tempting in Fig. 1. However such extrapolations are suspect because of the following theoretical argument, presented recently by Fruchter et al. and Griessen.[30,31]

While the simplest thermal activation model predicts a linear-T dependence of S(T), this model corresponds to the unphysical case of a sawtooth potential, as we have already remarked above. In a one-dimensional model with a smooth potential barrier U(x) as a function of position x, it can be shown quite generally[16,30,31] that in the low-T limit where $J \to J_{c0}$, the net barrier goes as

$$U(J) = U_0[1 - (J/J_{c0})]^{3/2}, \tag{6}$$

where the 3/2 power comes from the fact that the distance from the minimum to the maximum of the tilted barrier does not stay constant, as in the sawtooth potential, but tends to zero as $J \to J_{c0}$. Eq. 6 is plotted in Fig. 2, where U(J) can be seen to have zero slope as $J \to J_{c0}$. Using the Arrhenius relation as before, one can now solve for J to obtain

$$J = J_{c0}\{1 - [\frac{T}{U_0} \ln(\frac{t}{t_{eff}})]^{2/3}\}. \tag{7}$$

This result implies a normalized time-logarithmic relaxation rate $S = d\ln(M)/d\ln(t)$ which scales as $T^{2/3}$! Thus a conventional thermally activated flux creep model which takes into proper account the expected J-dependence of the barrier predicts a **downward-curving** temperature dependence of the normalized relaxation rate near T = 0. It is **not** linear as has generally been assumed in the past.

Another important phenomenological result[30] is that if some kind of quantum tunneling process is assumed for vortex motion, with an exponential dependence on the tunneling barrier width and height as in the WKB approximation, it gives rise to an essentially T-independent relaxation even in the presence of thermal activation. This persists until a crossover, at higher temperature, to T-dependent relaxation. In other words, the highest-rate relaxation process dominates the magnetic relaxation; this is in contrast to transport measurements for which the voltage contributions of the two processes add linearly.

Thus direct measurements of magnetic relaxation, and in particular the determination of the T-dependence, are of great importance in investigating the interesting possibility of quantum vortex tunneling. Just such measurements have recently been carried out by Fruchter et al.[30] on an YBaCuO crystal in the temperature range 0.1 to 1 K. The results are shown in Fig. 3; S is essentially T-independent, confirming the prediction of the phenomenological theory for quantum tunneling of vortices. Alternative models based on distributions of energy barriers are conceivable but unlikely, since they require such low energies and such a special distribution to give a T-independent S.

Quantum tunneling of vortices in high-temperature superconductors is plausible in view of their small coherence lengths. This follows from the fact that in the WKB approximation for quantum tunneling, the barrier width enters exponentially, and it is reasonable to expect that the minimum barrier width will scale as the coherence length (or perhaps as the unit-cell dimension perpendicular to the CuO_2 planes). In this context, an interesting aspect of the results of Fig. 3 is the lower relaxation rate at 0.2 T as compared to 1.7 T. A key difference between these two regions is that self-fields curve the vortices towards the plane at low applied fields, while they are almost straight at high applied fields. This suggests that quantum tunneling of vortices lying in the CuO_2 planes is lower than that for vortices lying perpendicular to them. One might understand this qualitative difference by assuming that vortex motion is mediated by the motion of kinks[32] in the vortex core. For slightly bent vortices along the c-axis, the kinks lie in the CuO_2 planes, and they move in the direction of the c-axis; hence the distance between potential wells should be the greater of either the c-coherence length ξ_c or the separation between CuO_2 planes. For vortices along the ab-planes, the kinks lie along c and move along ab; so the longer coherence length ξ_{ab} represents the nearest distance between potential wells. An alternative explanation ignores the kinks but focuses on the larger electronic mass for k-vector perpendicular to the CuO_2 planes, and possibly a higher barrier, both of which would suppress tunneling in the WKB approximation. In any case, since significant numbers of electrons are involved in the vortex core, this phenomenon may represent a new kind of macroscopic quantum tunneling,[33] though little is known about the possible microscopic mechanism at this point.

In summary, there has been significant progress recently in clarifying long-standing questions about the nature of flux-creep in high-temperature superconductors. Most of this progress has been of an experimental and phenomenological nature, but a great deal remains to be done to clarify microscopic aspects of these phenomena.

The author thanks I. A. Campbell and the Université de Paris-Sud for their hospitality during a summer at the Laboratoire de Physique des Solides, during which time many of the ideas as well as data described in this paper were developed. The author also thanks J. R. Thompson of the Oak Ridge National Laboratory for discussion of his latest relaxation data on YBaCuO crystals. He also thanks R. Griessen, L. Fruchter, M. Konczykowski, M. P. A. Fisher, Y. Yeshurun, L. Civale and L. Conner for many useful discussions.

REFERENCES

1. K. A. Müller, M. Takashige and J. G. Bednorz, Phys. Rev. Lett. **58**, 1143 (1987).

2. Y. Yeshurun and A. P. Malozemoff, Phys. Rev. Lett. **60**, 2202 (1988).

3. A. P. Malozemoff, in **Physical Properties of High Temperature Superconductors**, ed. D. Ginsberg (World Scientific, Singapore, 1989), p. 71.

4. G. M. Stollman, B. Dam, J. H. P. M. Emmen and J. Pankert, Physica C **159**, 854 (1989); also Physica C **161**, 444 (1989).

5. Y. Xu, M. Suenaga, A. R. Moodenbaugh and D. G. Welch, Phys. Rev. B **40**, 10882 (1989).

6. I. A. Campbell, L. Fruchter and R. Cabanel, Phys. Rev. Lett. **64**, 1561 (1990).

7. C. Keller, H. Kuepfer, R. Meier-Hirmer, U. Weich, V. Selvamanickam and K. Salama, Cryogenics, May 1990.

8. Y. Y. Xue, Z. J. Huang, L. Gao, P. H. Hor, R. L. Meng, Y. K. Tao and C. W. Chu, to be published.

9. L. Civale, A. D. Marwick, M. W. McElfresh, T. K. Worthington, A. P. Malozemoff, F. Holtzberg, J. R. Thompson and M. A. Kirk, Phys. Rev. Lett. **65**, 1164 (1990).

10. J. Z. Sun, C. B. Eom, B. Lairson, J. C. Bravman and T. H. Geballe, to be published.

11. P. Svedlindh, K. Niskansen, P. Norling, P. Nordblad, L. Lundgren, C. Rossel, M. Sergent, R. Chevrel and M. Potel, Phys. Rev. B, submitted.

12. K. Niskanen, P. Norling, P. Svedlindh, P. Nordblad, L. Lundgren, C. Rossel and G. V. Chandrashekhar, Proc. International Meeting on High Temperature Superconductivity, Cambridge, England, Aug. 1990, to be published.

13. J. R. Thompson, Y. Sun and F. Holtzberg, to be published.

14. A. P. Malozemoff and M. P. A. Fisher, Phys. Rev. B, Oct. 1990, to be published.

15. P. W. Anderson and Y. B. Kim, Rev. Mod. Phys. **36**, 39 (1964).

16. M. R. Beasley, R. Labusch and W. W. Webb, Phys. Rev. **181**, 682 (1969).

17. A. M. Campbell and J. E. Evetts, Adv. Phys. **21**, 199 (1972).

18. C. W. Hagen and R. Griessen, in **Studies of High Temperature Super-conductors**, Vol. 3, ed. A. V. Narlikar (Nova Science Publishers Inc., Commack NY, 1990), p. 159; also Phys. Rev. Lett. **62**, 2857 (1989).

19. A. P. Malozemoff, T. K. Worthington, R. M. Yandrofski and Y. Yeshurun, Intl. J. Mod. Phys. B **1**, 1293 (1988).

20. E. Zeldov, N. M. Amer, G. Koren, A. Gupta, R.J. Gambino, and M.W. McElfresh, Phys. Rev. Lett. **62**, pp. 3093-3096 (1989); E. Zeldov, Physica A, in press.

21. M. P. A. Fisher, Phys. Rev. Lett. **62**, 1415 (1989); D. S. Fisher, M. P. A. Fisher and D. A. Huse, Phys. Rev. B, to be published.

22. T. Nattermann, Phys. Rev. Lett. **64**, 2454 (1990).

23. M. V. Feigel'man, V. B. Geshkenbein, A. I. Larkin and V. M. Vinokur, Phys. Rev. Lett. **63**, 2303 (1989); M. V. Feigel'man and V. M. Vinokur, Phys. Rev. B **41**, 8986 (1990).

24. A. P. Malozemoff and J. R. Thompson, unpublished.

25. S. Senoussi, M. Oussena, G. Collin and I. A. Campbell, Phys. Rev. B **37**, 548 (1988).

26. A. V. Mitin, Sov. Phys. JETP **66**, 335 (1988).

27. A. C. Mota, A. Pollini, P. Visani, K. A. Müller and J. G. Bednorz, Phys. Rev. B **36**, 4011 (1987), and

28. A. Hamzic, L. Fruchter and I. A. Campbell, Nature **345**, 515 (1990).

29. L. Fruchter, A. Hamzic, R. Hergt and I. A. Campbell, Proc. VI Int. Conf. on Valence Fluctuations, Rio de Janeiro, 1990, to be published.

30. L. Fruchter, A. P. Malozemoff, I. A. Campbell, J. Sanchez, M. Konczykowski, R. Griessen and F. Holtzberg, Phys. Rev. Lett., submitted.

31. R. Griessen, Physica C, submitted.

32. D. Feinberg and C. Villard, Mod. Phys. Lett. **B4**, 9 (1990), and Phys. Rev. Lett. **65**, 919 (1990).

33. A. O. Caldeira and A. J. Leggett, Phys. Rev. Lett. **46**, 211 (1981).

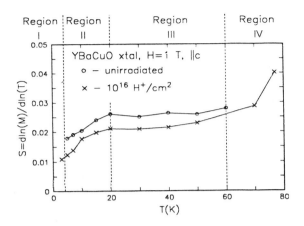

Figure 1. Normalized relaxation rate S (see Eq. 1) vs. temperature T for a fully oxygenated YBaCuO crystal, unirradiated and irradiated, prepared by hysteresis-loop cycling in the critical state at an applied field of 1 T parallel to the crystal c-axis. Dashed lines indicate schematically different regions of relaxation behavior discussed in the text. (After Civale et al., Ref. 9.)

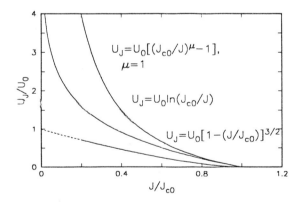

Figure 2. Depedence of the thermal activation barrier U_J on current density J in several models: The power-law dependence represents the effective barrier of the "interpolation formula" of the vortex glass[21,22] or collective pinning[23] models, the natural-logarithm dependence is that proposed by Zeldov et al.,[20] and the 2/3-power dependence represents the prediction[16,30,31] for a spatially-smooth one-dimensional barrier in the limit $J \rightarrow J_{c0}$, where J_{c0} is the maximum critical current density in the absence of thermal activation.

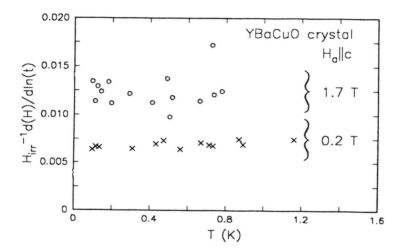

Figure 3. Normalized low-temperature relaxation rate, determined with a Hall-probe technique, of an YBaCuO crystal versus temperature T in two applied fields parallel to the crystal c-axis. Here H is the field detected by the Hall probe and H_{irr} is one-half the width of the Hall-probe-field hysteresis loop. (After Fruchter et al., Ref. 30.) The strong relaxation extending to the lowest temperatures points to the existence of a vortex quantum tunneling process.

Harmonic Generation in the AC Magnetic Response of High-Tc Superconductors - Applications

R.N. Bhargava, S.P. Herko and A. Shaulov
Philips Laboratories
North American Philips Corporation
Briarcliff Manor, New York

ABSTRACT

Nonlinear magnetic behavior in high-Tc superconductors as reflected by the appearance of harmonics in ac magnetic response, yields valuable information about transition temperature, critical fields and the dynamics of fluxons. In this paper we discuss the specific applications of the third harmonic measurements in developing (i) a high resolution spectrometer for determining Tc's and (ii) a microscope which reveals local variation of Tc and the other superconducting properties.

INTRODUCTION

AC magnetization experiments using a coupled primary and a pair of balanced secondary coils, has been extensively used to characterize linear magnetic behavior of high-Tc superconductors. The resulting off-balance voltage induced across the secondary coil pair is measured using a two-phase lock-in amplifier. However, at certain temperature and external steady dc magnetic field conditions, the high-Tc superconductors exhibit a non-linear magnetic behavior. Thus, a nonsinusoidal voltage is inducted across the sample coil and a voltage signal appears at harmonics of the driving frequency. Harmonic generation in the alternating magnetic response of high-Tc superconductors has been the topic of many recent studies [1-8]. These harmonic measurements of the magnetic response, have been recently successfully used by Shaulov et al. to understand the contribution of weak links and grains to remanent nonlinear magnetic response [7,9]. In this paper we emphasize the non destructive measurements, based on third harmonic generation, which provide information regarding transition temperature and local inhomogeneities.

MEASUREMENTS

The block diagram of the experimental set up used in measuring harmonics of the ac magnetization is shown in figure 1. The balanced coil system is the same as in the conventional ac susceptometer. The output of the balanced pick-up coils goes to a lock-in detector that measures the phase contents of the difference signal at the driving frequency. Another output goes to a spectrum analyzer to allow measurement of the harmonics. A scanner and a computer are included in the system to permit simultaneous measurements of the in-phase and out-of-phase signals as well as the amplitude of the harmonic components as a function of temperature and external bias fields. The system is distinguished from the conventional systems by its ability to measure the spectrum of the magnetic response of materials to alternating fields. An ac field of 0.04 Oe at a frequence of 20kHz was normally used as a driving signal while the third harmonic signal at 60 kHz was primarily used to characterize the non-linear magnetization properties. The steady state field H_{dc} varied from 0 to 1.4 kOe.

Determination of Superconducting Transitions:

For small applied ac signals, type II superconductors exhibit a nonlinear magnetic behavior in a close vicinity of the transition temperature, T_c. Above the transition temperature, and far below this point, the magnetic behavior is linear. Thus, the superconducting transition is accompanied by the generation of harmonic components in the spectrum of the material's response to a sinusoidal field[10]. Measurement of the amplitude of the third harmonic component as a function of temperature in figure 2 shows a peak near the transition temperature for a sintered YBaCuO sample. The transition temperature is more accurately determined by the onset of the third harmonic signal as the material transforms into the superconducting phase. Figure 3 describes the magnetic behavior of a 1000A multiphase YBaCuO thin film sample prepared by laser ablation. This data reveals at least four transitions at about 88K, 89K, 89.5K and 90K. It is important to note that resistivity measurements in this sample indicated a single transition at 90K. As demonstrated in figures 2 and 3 , the harmonic measurements offer an accurate method to determine transition temperatures of single and multiphase samples. Thus the method can be effectively considered as a high resolution spectrometer for Tc determination. In order to determine whether several transitions observed in the thin film shown in figure 3, are indeed associated with different superconducting phases, we studied the variation of the peak position as a function of the applied dc magnetic field H_{dc}. The peak shift as a function of H_{dc} is shown as an insert in figure 3. The shift of the peak position as a function of H_{dc} is similar to that observed for a single phase material. From this initial data it seems that four peaks between 88 - 91K can be associated with four distinct phases of the superconductor.

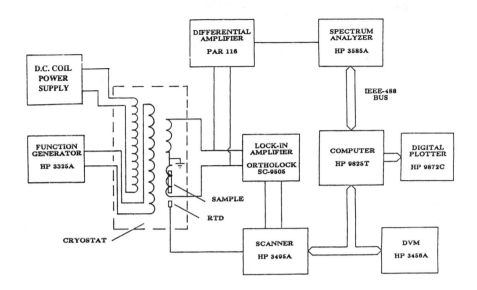

Figure 1   Block diagram of the experimental set up to measure the magnetic susceptibility and harmonics under ac magnetization conditions.

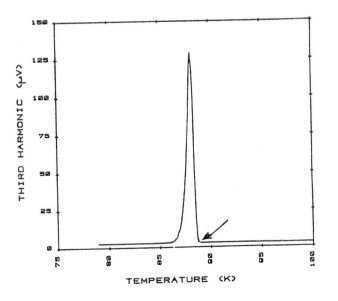

Figure 2   Third harmonic signal vs. temperature for a sintered YBaCuO sample. The arrow depicts Tc.

Recent studies on magnetic susceptibility performed on single crystals of YBa-CuO showed several superconducting transitions in ac susceptibility measurements. These results on single crystals were interpreted as due to varying degrees of oxygen deficiency in a staircase like structure[11,12]. The discrete nature of several transitions observed in films by harmonic response as shown in figure 3 gave well resolved peaks separated by ~1K. It may be that a staircase like super-lattice structure indeed is responsible for slight variations of oxygen stoichiometry in a discrete steps and results in several sharp Tc's.

Based on the detection of the third harmonic in ac magentic field signal, we have designed and built a simple device to screen samples for possible superconductivity as demonstrated by nonlinear magnetic response. In this simplified susceptometer, we measure the strength of the 60 kHz harmonic signal across the secondary coil as the temperature of sample is varied. The appearance of the signal at 60 kHz at various temperatures reflects the presence of single or multiphase superconducting materials which could be in the form of bulk, thin film or powder. Such a quick tester is valuable to material scientists who are preparing superconducting materials.

Local Measurements of Tc:

Since the harmonic measurements offer rather high resolving power in distinguishing the various values of Tc's, we have expanded the measurements to extract local values of Tc's. The principle behind such a measurmeent is to have a coupled magnetic head, consisting of a write and a read head. The write head is driven at the fundamental frequency, while the read head is tuned to pick up the third harmonic. When the superconductor is just below Tc, the strength of the third harmonic signal can be measured by the read head. By moving the coupled head over the superconductor, a map showing the local variation in the amplitude of the third harmonic signal at a given Tc is generated.

The principle of the technique is illustrated in figure 4. The superconductor is held at a constant temperature just below its Tc where the third harmonic signal peaks. The combined read/write magnetic head is held in close proximity to the surface of the superconductor. A local sinusoidal magnetic field at 200 kHz is produced at the gap of the write head by driving its coil with a sine wave signal generator. This induces small oscillations in the magnetization of the sample. Due to nonlinear magnetic behavior near Tc, a nonsinusoidal voltage is induced in the read head producing odd harmonics of the driving frequency. The presence of the third harmonic can be detected. The read/write magnetic head is scanned over the surface of the superconductor, resulting in a map of the surface and near surface inhomogeneities as indicated by the amplitude variation of the third harmonic signal.

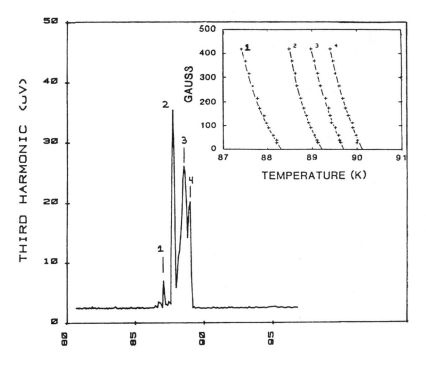

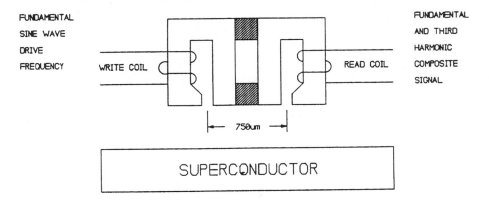

Figure 3   Third harmonic signal from a multiphase thin film YBaCuO sample. The insert depicts the shifts of the various peaks as a function of applied dc magnetic field.

Figure 4   Schematic representation of the principal of magnetic head mapping using read/write head.

The magnetic head is moved over the sample in the x and y directions and the amplitude of the third harmonic signal is stored. An additional temperature sensor is utilized to measure the magnetic head temperature to ensure that it has remained stable during the measurement. In our experiments the magnetic head was of a coaxial design where the gap of write head is at a right angle to the gap of the read head with the central axis perpendicular to the sample surface. This configuration allows detection of the third harmonic component in the same area of the sample that is exposed to the fundamental component of the magnetic field. The read/write gaps and the distance of the head above the sample determines the resolution.

The computer software includes programs to generate surface plots and contour maps. Figure 5a is a typical contour map of a YBaCuO thin film sample. The scanned area is 6mm by 8mm and contained thirty thousand data points spaced at 40 micrometer intervals. Figure 5b is the surface map of the same sample. A clear interpretation of these maps requires more analysis of the third harmonic signal in this configuration.

This concept of a microscope for superconductivity with a resolution of ~50μm has been demonstrated. Further increase in the resolution of such a microscope could yield important information regarding the role of defects, grain boundaries and microstructures as they specifically relate to local transition temperature.

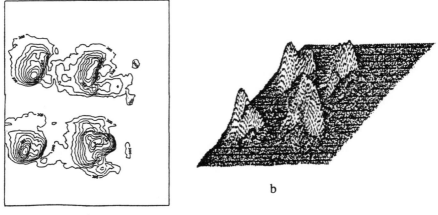

a   b

Figure 5   The contour map of a thin film of YBaCuO sample. The surface was scribed in four sections and the data was taken by a coaxial read/write magnetic head. Both the contour map (a) and the topological map (b) clearly show the scratched line demarkation as well as strength of the third harmonic signal.

SUMMARY

The harmonic response of ac magnetization offer a simple and high resolution tool for studying properties of high-Tc superconductors susceptometer. Extension of such measurements to study the relationship between the microstructure and superconductivity continues to be a challenging area of research. The local inhomogeneities as measured by a coupled read/write head can be very useful in determining device performance. Additionally, if the head design could be reduced to 1-5μm gaps, it may allow us to study the local properties as well as possibly isolate localized areas showing higher Tc's than normally observed.

ACKNOWLEDGEMENT

We thank Don Dorman for the assistance in taking some of the measurements.

REFERENCES

1. C. Jeffries, Q.H. Lam, Y.Kim, L.C. Bourne, and A. Zettl, Phys. Rev. **B37,** 9840 (1988); C.D. Jeffries, Q.H. Lam, Y. Kim, C.M. Kim, A. Zettl and M. Klein, Phys. Rev. **B39,** 11526 (1989).

2. A. Shaulov and D. Dorman, Appl. Phys. Lett. **53,** 2680 (1988).

3. K. H. Muller, J.C. McFarlane, and R. Driver, Physica C 158, 366 (1989).

4. L. Ji, R.H. Sohn, G.C. Spalding, C.J. Lobb, and M. Tinkham, Phys. Rev. **B40,** 10936 (1989)

5. T. Ishida and R.B. Goldfarb, Phys. Rev. **B41,** 8937 (1990)

6. D.G. Xenikos and T.R. Lemberger, Phys. Rev. **B41,** 869 (1990)

7. A. Shaulov, R. Bhargava and D. Dorman, Proceedings of the MRS Meeting, Boston, Vol. 169, p. 943, (1989); A. Shaulov, D. Dorman, R. Bhargava and Y. Yeshurun, Appl. Phys. Lett. 57, 724 (1990)

8. M.W. Johnson, D.H. Douglass and M.F. Bocko, Phys. Rev. B (to be published)

9. A. Shaulov, Y. Yeshurun, S. Shatz, R. Hareureni, Y. Wolfus and S. Reich (to be published)

10. C.P. Bean, Rev. Mod. Phys. 36, 31 (1964)

11. M. Conach, A.F. Khoder, F. Monnier, B. Barbara and J.Y. Henry, Phys. Rev. **B38,** 748 (1988)

12. M. Touminen, A. M. Goldman and M.L. Mecartne, Physica **C153,** 324 (1988)

Transport Properties In Magnetic Fields and Microstructural Studies of the High-T_c Superconducting Composites

M. K. Wu, M. J. Wang, J. L. Lin, D. C. Ling[*] and K. R. Ma[+]

Department of Physics and Materials Center
National Tsing-Hua University
Hsinchu, Taiwan, Republic of China

and

F. Z. Chien

Department of Physics
Tamkang University
Tamsui, Taiwan, Republic of China

ABSTRACT

A systematic study of the $YBa_2Cu_3O_y$ (YBCO)/Ag_2O composite shows that the presence of Ag at high temperature is to stabilize the superconducting YBCO phase. Ag is believed to serve as catalyst which causes partial melting of $YBa_2Cu_3O_y$ phase at $985^\circ C$ and aided grain growth. Some of Ag particles were trapped in the grain or at grain boundaries during rapidly grain growth process and resulted as pinning centers. For the strong pinning samples, two distinguished resistive transitions attributed to the intra-grain and the inter-grain region are observed. Thermally activated flux creep is insignificant in these samples.

INTRODUCTION

After the first observation of the suspension effect [1] in the silver added $YBa_2Cu_3O_y$ superconducting composite, we have demonstrated the flux pinning and mechanical workability enhancement in these composites [2-5]. Thermal studies indicated that the addition of silver enhances the phase stability of the high T_c superconducting oxide and limits the oxygen out-diffusion [6]. The results also suggested that larger grains growth and precipitation of silver at the grain boundaries might be responsible for the observed enhancements. However, the detailed knowledge of the large grain growth mechanism is not available. Consequently, we still do not have any guideline for the process to synthesize materials with designed characteristics. On the other hand, the detailed pinning mechanism for this class of materials has not been understood.

In order to gain more insight into these problems, we have carried out a systematic study of the crystal growth processes and the electrical properties of the Ag-added composites. In this paper, we report the results of: (1) the thermal and microscopic studies and (2) the electrical and magnetic studies of the strong flux pinning YBCO/Ag$_2$O composites. We have developed a process to grow large size and high quality single crystal by using these silver-added composite without the introduction of any unwanted flux materials. The mechanism of such a large grain growth process is believed due to the presence of low melting Ag-Ba-Cu alloy which serves as an intermediate medium to cause the liquid phase crystal growth of the high T_c YBCO compound.

EXPERIMENTALS

YBa$_2$Cu$_3$O$_y$ (YBCO) powder was prepared by solid state reaction. Ag$_2$O powder was added with the weight ratio of 1 to 3 to YBCO. The powder was mixed and pressed in the form of disc (~ 10 mm in diameter) under 3 ton/cm^2 pressure. Two sintering processes were used to prepare the final crystals. 1) Directly increases the temperature to 980°C and holds at this temperature for 2, 4, 6, 8, 16 hours and then furnace cools to room temperature with continuous oxygen flow (process 1). 2) Increases the temperature to 950°C and holds at this temperature for 3 hours, followed by a slow heating to 985°C and holds there for 2, 4, 6, 8, 16 hours, and then decreases temperature slowly to 950°C and keeps for 3 hours, finally furnace cools to room temperature under oxygen flow (process 2). Microstructure analysis were determined using a PHILIPS SEM with EDX attachment. Magnetization was measured with the Oxford Faraday Balance. A conventional four-probe method was used to determine the resistivity. Thermal studies were carried out using ULVAC Multi-TAS-7000 thermal analyzer with DTA and TGA capabilities.

RESULTS AND DISCUSSIONS

A. Microstructure and Crystal Growth

Typical microstructures of the samples prepared by process 1 with 2 hours sintering are shown in Figure 1. Although majority phase is YBCO, the sample uniformity is poor where BaCuO$_2$, CuO, Y$_2$BaCuO$_5$ phases and small amount of unreacted BaO particles exist. As the sintering time increases, the samples become more uniform and larger irregular shape grains form. Grains size (typical grain size is

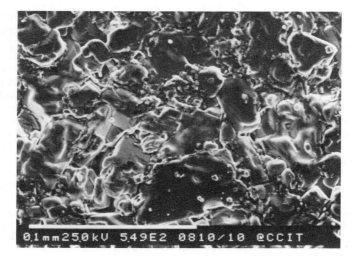

Figure 1.   Microstructure of YBCO/Ag$_2$O prepared using
process 1.   Some of the precipitations are Ba-Cu-O and Ag
particles.

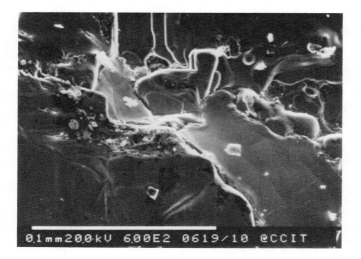

Figure 2.   Microstructure of YBCO/Ag$_2$O prepared using
process 2, showing the formation of plate-like grain.

about 80 micron) increases with sintering time but not conspicuous even for 16 hours. Energy Dispersive of X-ray
(EDX) Analysis showed that Ag has reacted and dissolved with
Ba-Cu-O and forms a glassy phase. X-ray diffraction shows
that the superconducting phase remains identical to the
undoped Y123. The change in lattice parameters is negligible within our experimental resolution (.03° 2O). It also
shows that slight alignment of the crystal in a preferred
orientation along c-axis.

The sample prepared using process 2 shows large grain
growth and becomes slightly flat with random orientation
after sintering at 980°C for 2 hours. EDX analysis of the
whole sample shows that small amount of Y_2BaCuO_5 phase and
Ag rich particles dispersed in the grain or at grain boundaries. Ba-Cu-O flux resulted from partial melting of YBCO
was extruded out of the grain and scattered at voids. Most
of Ag particles embed in the flux and react with Ba-Cu-O and
form complex network structure.

As the sintering time at 980°C increased to 4 hours,
the sample distortion was obvious. Microstructure observations (Fig. 2) showed that large and stacked plate-like
grains appear and the amount of Y_2BaCuO_5 (211) and Ag rich
particle increases. We have also observed the presence of
volcano-like convex precipitates on the sample surface, as
shown in Fig. 3. EDX analysis indicates that these convex
precipitates are Y-rich phase or Ag-rich phase and there
appears to have composition gradient around the precipitates. After sintering for 16 hours, sample microstructure
is more uniform. Well stacked plate-like grains arrange in
a-b direction and intend to increase in size (typical size
of 0.6 mm) by further coalescence and growth process. In
fact, single crystals of $3x1.5x1$ mm^3 in size can be grown
using this process with sintering time larger than 8 hours.

From the microstructural and chemical composition analysis, we propose the following grains growth mechanism.
During the sintering process at 950°C, it prefers the Cu-
Ba-O low melting phase formation to the YBCO facet-like
grain growth. This is consistent with the differential
thermal analysis (DTA) of the unreacted YBCO/Ag$_2$O where a
reaction at ~ 945°C is observed. The Ag particles result
from the dissociation of Ag$_2$O scatter primarily in the void,
and the Ba-Cu-O liquid phase distributes at grain boundaries
and grain corners. Ag melted and caused peritectic partial
melting of YBCO phase in the following sintering process at
980°C. Rearrangement of grains occurs if enough liquid
forms due to the partial melting of YBCO. Consequently,

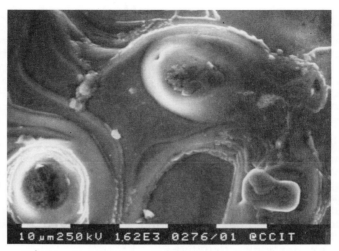

Figure 3.   Microstructure of YBCO/Ag$_2$O prepared using process 2.   The volcano like precipitates are Y-rich or Ag-rich phase.   Composition gradient exists around the precipitates.

stacked plate-like grains form by solution and reprecipitation process due to different interfacial energy in various crystal orientations and grains contact together by liquid capillary force.   Some of the Ag rich small liquid particles and 211 precipitates were trapped in the grain or in the facet grain boundaries because they were unable to diffuse out in time while plate-like grains rapidly contact and coalescence.

Most of Ag particles embed in Cu-Ba-O liquid phase were extruded and jammed into pores or on the sample surface during the process when stacked plates continue to grow in a-b direction.   During the second step of sintering at 950°C (cooling from 980°C), stacked plate like grain continues to grow in c axis by solid diffusion, and small YBCO crystals precipitate at grain adjoining regions or voids.   Finally, except a small amount of Cu-Ba-O liquid phase remains in void, most of the Cu-Ba-O liquid were jammed to the sample surface and reacted with 211 phase and Ag rich particles which act as nucleation and growth centers, resulted in convex shape microstructure (Fig.3) with composition gradient around the convex. Increasing the sintering time at 985°C, in addition to the more densification and uniformity of the microstructure, the larger and well stacked plate-like grains continue to grow. Meanwhile, the amount of 211 and Ag-rich precipitations increase and disperse evenly in the grain or at grain boundaries, which may contribute to flux pinning resulting in the enhancement in critical current density J_c.

Based on the above description, it is apparent that the 950°C sintering step is essential in the crystal growth process. This is why the grain growth phenomena is not conspicuous when the samples are prepared using process 1. It only results in the formation of irregular shape and inhomogeneous composition for not being through 950°C sintering, which hindered the rearrangement and further growth of the grain. Extending the sintering time at both 950°C and 980°C of process 2 has resulted in the formation of large size single crystals. Preliminary microscopy study of these crystals indicates that their defect densities are as high as those of the single crystal thin films with high current density [7].

B. Electrical and Magnetic Characterization

Figures 4a and 4b are the magnetizations as a function of temperature for YBCO and YBCO/Ag$_2$O prepared with process 2. The large difference in magnetization between the zero field and field cool measurements at low temperature indicates that the sample prepared with process 2 indeed has large pinning effect [8]. However, the temperature dependence of resistance the sample sintered for 8 hours indicates that the superconducting transition remains almost identical to that of undoped YBCO. Figure 5a, and 5b show the resistive transition in magnetic fields of the YBCO/Ag$_2$O prepared using process 1 and process 2, respectively. It is clear that the resistive transition is narrower for sample prepared with process 2. The magnetic field effect on the transition width is also smaller for samples prepared with process 2. Although the pinning effect is weaker in the samples prepared with process 1, its superconducting characteristics (magnetic and resistive transition) is almost the same as that of the single crystal sample [8]. By fitting the transition width as function of the applied field, it is found that for process 1 sample T_c H$^{2/3}$. This result is similar to that observed in a single crystal sample [9] and can be described in terms of the phase slippage model proposed by Tinkham [10]. For the process 2 sample the transitions in magnetic fields clearly show two different regions. One can be attributed to the characteristic of the superconducting grains, the other to the inter-grain regions. The magnetic field effect on resistive transition is much smaller than that of the process 1 sample, indicating strong pinning in these superconducting grains.

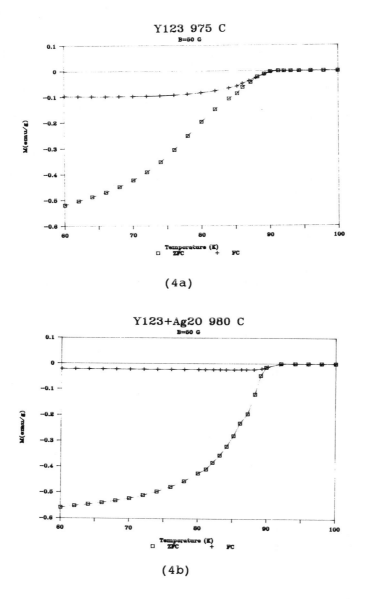

(4a)

(4b)

Figure 4.   Magnetization as a function of temperature for YBCO (4a) and YBCO/Ag$_2$O prepared with process 2 (4b).

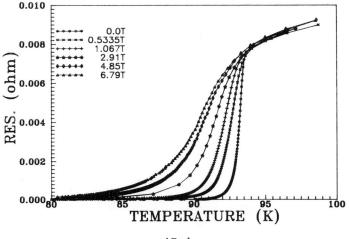

(5a)

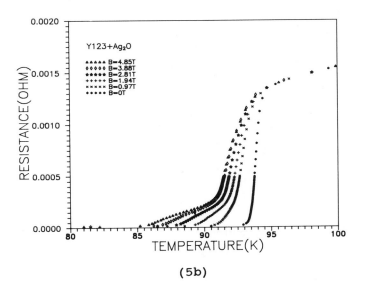

(5b)

Figure 5. Resistive transition under magnetic fields for process 1 YBCO/Ag$_2$O (5a) and process 2 YBCO/Ag$_2$O (5b).

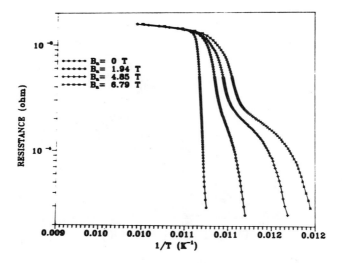

Figure 6. Log R vs. 1/T for process 2 YBCO/Ag$_2$O at different fields.

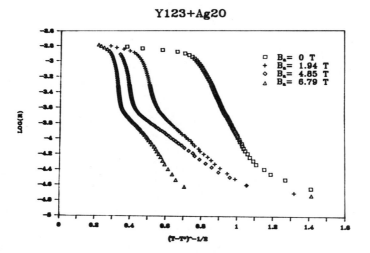

Figure 7. Log R vs. $(T-T^*)^{-0.5}$ for process 2 YBCO/Ag$_2$O at
different fields, where T^* is the zero resistance temperature.

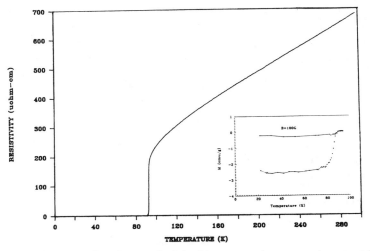

Figure 8. Electrical resistivity vs. temperature of a single crystal grown by process 2. Inset is the temperature dependence of the magnetization

It should be noted that true zero resistance state still exists even at 82K and under a magnetic field of 5 T, suggesting that the thermally activated flux creep is insignificant in this sample. The data at the transition region can not be fitted using an Arrhenius plot, as shown in figure 6. The behavior in the inter-grain regions for sample 2 is similar to that of the array of weak links which intergrain coupling is due to proximity effect [11]. As shown in Figure 7, the data at high fields can be described with a linear dependence of the resistance on $Aexp[-B(T-T^*)^{-0.5}]$, where T^* is the zero resistance temperature and A, B are constants. This result suggests that at high magnetic fields the intergrain coupling is weakening and the thermally activated unbinding of the vortex pairs appears. The exact mechanism for the observed strong pinning is likely due to the high defect density present in this material. Preliminary transmission electron microscopy of the superconducting grains indeed indicates that these grains have extremely high twinning densities. More detailed correlation of the microstructure of these grains with their transport and magnetic properties are currently under investigation.

Figure 8 is the temperature dependence of the resistance and the magnetization of a single crystal prepared by process 2. The crystal size is about 0.5x0.8x1 mm^3. The excellent quality of the crystal is evident from the narrow transition width (< 1 K) and the linearly dependent of resistivity. The Meissner signal displayed shows the large pinning of the crystal. In fact, magnetic measurement of the crystal suggests that the critical current density is at least as high as that of the melt grown crystals. Details of these results will be reported elsewhere [12].

ACKNOWLEDGEMENTS

The authors like to acknowledge C. H. Kao, W. L. Chen and at Tsing-Hua University for their contributions. We also thank Drs. H. C. Ku, C. C. Chi, D. H. Lee, C. C. Tsuei and Profs. C. S. Ting and C. R. Hu for their support and fritful discussions. The work at Tsing-Hua was supported by the ROC National Science Council grants NSC79-0208-M007-95.

REFERENCES

*Present address, Department of Physics, Wayne State University, Detroit, Michigan 48202; + Also at Chung-Cheng Institute of Techonology, Taoyen, Taiwan.

1. P. N. Peter, R. C. Sisk, E. W. Urban, C. Y. Huang and M. K. Wu, Appl. Phys. Lett. 52, 2066, (1988).
2. C. Y. Huang, P. N. Peter, B. B. schwartz, Y. Shapira and M. K. Wu, Mod. Phys. Lett. B2, 1027, (1988).
3. C. Y. Huang, H. H. Tai and M. K. Wu, Mod. Phys. Lett. B3, 525, (1989).
4. C. Y. Huang, T. J. Li, Y. D. Yao, L. Gao, Z. J. Huang, Y. Shapira, E. J. McNiff,Jr., and M. K. Wu, Mod. Phys. Lett. B3, 1251, (1989).
5. J. K. Tien, J. C. Borofka, B. C. Hendrix, T. Caulfield, and S. H. Reichman, Mat. Trans. 19A, 1841, (1988).
6. M. K. Wu and D. C. Ling, Chinese J. Phys. April, (1990).
7. C. H. Chen, T. L. Kuo, M. J. Wang, M. K. Wu and Y. Huang, to be published.
8. Y. Yeshurun, and A. P. Malozemoff, Phys. Rev. Lett. 60, 2202, (1988).
9. Y. Iye, T. Tamegai, T. Sakakibara, T. Goto, N. Miura, H. Takeya and H. Takei, Physica C, 153-155, 26, (1988).
10. M. K. Tinkham, Phys. Rev. Lett., 61, 1658 (1988).
11. D. H. Sanchez and J. L. Berchier, J. Low. Temp. Phys. 43, 65, (1981).
12. M. J. Wang, C. H. Kao, J. L. Lin, M. K. Wu and C. C. Chi, to be published.

LORENTZ FORCE INDUCED DISSIPATION AND PINNING IN TRANSPORT PROPERTIES OF $YBa_2Cu_3O_{7-\delta}$ SINGLE CRYSTALS

W. K. Kwok[+], U. Welp[#,+], S. Fleshler[+$], K. G. Vandervoort[+*], H. Claus[*], Y. Fang[+], and G. W. Crabtree[+]

[#]Science & Technology Center for Superconductivity and [+]Materials Science Division, Argonne National Laboratory, Argonne, IL   60439

J. Z. Liu
Department of Physics, University of California-Davis, Davis, CA   95616

ABSTRACT

We present direct observation of twin boundary pinning and Lorentz force induced flux motion in twinned, untwinned, and oxygen deficient crystals of $YBa_2Cu_3O_{7-\delta}$. We find that twin boundaries act as strong pinning centers only when the vortex lines are pinned within the twin boundary planes. In addition, we find anomalous broadening in the superconducting resistive transition in magnetic fields in an oxygen deficient crystal where $T_c < 30$ K.

INTRODUCTION

The anomalous broadening of the resistive transition in magnetic fields of the the high T_c superconductors still remains to be understood. This broadening is unique, in that the onset of the superconductive resistive transition remains independent of field while the zero resistive temperature decreases rapidly with increasing magnetic field giving a 'fan shape' appearance to the resistive transitions. In our earlier work,[1] we showed an angular dependent Lorentz force induced flux flow component of the resistive broadening superimposed over an angular independent part. In this manuscript we extend our work to higher magnetic fields and to several single crystals of $YBa_2Cu_3O_{7-\delta}$, including untwinned, and oxygen deficient samples. In addition, we investigate the effect of twin boundary pinning for both field orientations, H ‖ c and H ‖ ab. We find that twin boundaries act as strong pinning sites only when entire vortex lines are pinned within the twin boundary planes.

RESULTS AND DISCUSSION

Twinned single crystals of $YBa_2Cu_3O_{7-\delta}$ were grown by a flux method and had T_c's in excess of 92 K with a width of less than 0.5 K in a field of 1 Gauss as determined from DC magnetization measurements. Typical dimensions of the crystals were $1.0 \times 0.6 \times 0.04$ mm^3 platelets. Three crystals were used in this experiment. One crystal possessed twins only along one direction and has been the subject of an earlier paper[1]. An untwinned crystal was obtained by annealing a fully oxygenated $T_c > 92$ K twinned crystal under uniaxial pressure in flowing oxygen as described

$Also at Dept. of Physics, Purdue University, West Lafayette, IN
*Also at Dept. of Physics, University of Illinois at Chicago, Chicago, IL

elsewhere[2]. Finally, an oxygen deficient single crystal was obtained by removing oxygen by annealing in 1.5 Torr of oxygen for seven days at 520 °C. The crystal then had a diamagnetic onset of 29 K and a transition width of ~5 K in 1 Gauss, with zero resistivity occurring at 23.5 K.

AC resistance data were obtained at 17 Hz as a function of field and temperature with a four probe technique. DC magnetization measurements were performed in a commercial SQUID magnetometer. Figure 1 shows the resistive transition of a single crystal with twin boundaries only along the <110> direction (top panel) and an untwinned crystal (center panel) for magnetic fields up to 8 Tesla along the crystallographic c-axis and measuring current in the ab plane of the crystal (current is along one of the two 'a/b' axes for the twinned crystal and along the a axis for the untwinned crystal). The twinned and untwinned crystals show almost identical T_c in zero field with ΔT_c(10-90%) <200 mK and zero resistance at T=92.2K and 91.5K respectively. In this current and field orientation, a maximum Lorentz force perpendicular to the current and field direction is exerted on the vortices. For the twinned crystal, the vortex motion induced by the Lorentz force is such that the vortices move across the twin boundary planes. The effect of pinning due to twin boundary planes in this current-field geometry is shown by comparing the resistive transition of the twinned and untwinned crystal in the bottom panel in Figure 1, for a 4 Tesla magnetic field. The shape of the resistive transitions for the two crystals normalized at T=95K is remarkably similar except that at low temperatures the untwinned crystal shows a longer tail. The zero resistance temperature of the two curves differ by about 3 K. We attribute this difference to pinning by twin boundaries in the twinned crystal. This pinning reduces the dissipation induced by Lorentz force more rapidly at lower temperature as compared to the untwinned crystal devoid of twin boundaries.

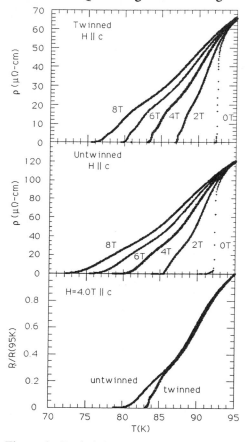

Figure 1. Resistivity versus temperature for twinned and untwinned crystals in magnetic fields up to 8 Tesla for H ‖ c.

Figure 2 shows the resistive transition of both crystals again normalized at 95K in a magnetic field of 4 Tesla in the ab plane with the measuring current also in the ab plane. The top panel shows the maximum Lorentz force configuration (H ⊥ I), and the lower panel shows the nominal zero Lorentz force case (H ‖ I) for both the twinned and untwinned crystals. For H ⊥ I, the Lorentz force is directed parallel to the crystallographic c-axis and the vortices intersect twin planes at a finite angle. We see very little effect due to twin boundaries as witnessed by the similar shape of the transition curves throughout their entirety. The kink in the resistive transition for the twinned crystal at lower temperatures is also seen in the untwinned crystal, but not as pronounced. The resistive transition of the untwinned crystal is shifted by about 0.3 K down in temperature relative to the twinned crystal. This is mainly due to the slight difference in the zero field T_c 's of the two crystals and not due to twin boundary effects.

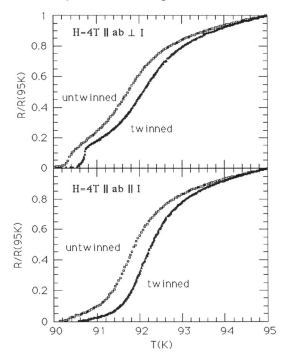

The difference in the pinning effect due to twin boundaries between the current-field orientation in Figure 1 and Figure 2 is seen clearly in the Arrhenius plot in Figure 3. In this plot, the effective pinning potential U_{eff}/k is proportional to the slope of the curve. For H ∥ c and I ∥ ab, the relative difference in the effective pinning potential is about a factor of four larger in the twinned crystal compared to the untwinned crystal. On the other hand, the effective pinning potential in the H ∥ ab ⊥ I configuration is almost identical for the two crystals. The different behavior seen in Figure 3 is due to the interaction of the twin

Figure 2. Normalized resistance versus temperature for twinned and untwinned crystals in a magnetic field of 4 Tesla for H ∥ ab ⊥ I and H ∥ ab ∥ I.

boundaries with the vortices. In the first case H ∥ c, the Lorentz force is directed along the ab plane perpendicular to the magnetic field and measuring current and causes vortex lines which are parallel to the twin planes to cross them. In the latter case (H ∥ ab), the vortices intersect the twin planes at a finite angle and the Lorentz force which induces vortex motion along the c-axis does not increase or decrease the number of point intersections of the vortices with the twin planes. However, in the latter case, if the magnetic field were rotated with respect to the crystal such that it were aligned with the twin planes, then entire vortex lines may be pinned by the twin planes similar to the case for H ∥ c. This effect is best seen in the angular scan shown in Figure 4. Here, the temperature is held constant while the angle between the magnetic field and the measuring current in the ab plane is varied. A very sharp drop in resistivity is observed whenever the magnetic field is along the <110> twin boundary direction (at -55 and -125 degrees in Figure 4). Such effect is of course not observed in the untwinned crystal. This sharp drop in resistivity is observed every 180 degrees due to the existence of twin boundaries only along one direction in this crystal. The relative effective pinning potential for twin boundary pinning in this orientation (H ∥ twin boundaries) is about a factor of five greater than the unpinned case and is independent of field up to 7 Tesla[3]. Thus twin boundaries act as strong pinning centers only when the vortex lines are aligned parallel to them.

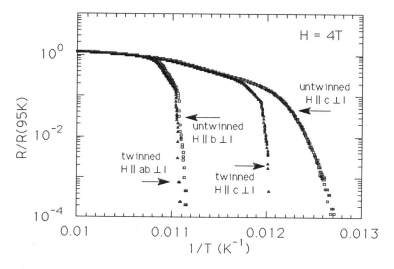

Figure 3. Arrhenius plot of normalized resistance for both twinned and untwinned crystal for the maximum Lorentz force configuration H ∥ c ⊥ I and H ∥ ab ⊥ I.

The sin$^2(\theta)$ modulation of the resistivity in Figure 4 is a direct result of flux flow dissipation at this temperature range near T$_c$. A minimum and maximum is observed for H ∥ I (nominal zero Lorentz force) and H ⊥ I (maximum Lorentz force) respectively. However, there exists a large angular independent contribution to the dissipation which cannot be simply explained by Lorentz force. This is seen in the shape difference between the curves of the same crystal for H ∥ I and H ⊥ I in Figure 2 and more clearly in Figure 5 where we show the resistive field broadening up to 8 Tesla for the two current-field orientations. The maximum of the angular dependent portion of the dissipation, ie. the difference in resistivity between the two current-field orientations, is shown in the bottom panel. Above T$_c$ flux flow resistivity goes to zero, and at lower temperatures it decreases to zero due to the onset of pinning. Thus a maximum is observed between the two temperature limits. Even in the nominal zero Lorentz force configuration (H ∥ I), there is a large degree of broadening. The magnitude of this angular independent dissipation seems to be independent of crystal quality as shown previously in measurements of several imperfect crystals,[4] thus indicating the intrinsic nature of this anomaly.

One explanation that has been proposed for the angular independent background is vortex line bending and entanglement[5] in the case of high critical temperatures which may lead to substantial Lorentz force dissipation even for H ∥ I. These effects could conceivably be minimized in oxygen deficient single crystals of YBa$_2$Cu$_3$O$_{7-\delta}$ with much lower T_c's and large oxygen vacancies which will contribute to increased number of pinning sites.

Figure 6 shows the resistive transition of an oxygen deficient single crystal for H ‖ c and H ‖ ab. The measuring current is in the ab plane of the crystal. The difference in zero field T_c is due to oxygen vacancy short range ordering in the sample.[6] This sample originally displayed a zero resistance temperature of 23.5 K. In order to change the field orientation, the sample was removed from the cryostat and then cooled again from room temperature to T_c. Subsequent zero field measurements yielded a zero resistance temperature of 26.5 K. Concurrently, the normal state resistivity decreased by about 20 percent. The aging effect has been attributed to the ordering of oxygen vacancies in the chains which occurs even at room temperature[8].

We find a substantial degree of broadening $\Delta T \sim 20$ K in a magnetic field of 8 Tesla for H ‖ c, much larger than in the fully oxygenated samples. DC magnetization measurement of the mean field T_c at 1 Tesla yields $T_c = 24$ K, about 10 K higher than the zero resistance temperature of 13.5K. In addition, the

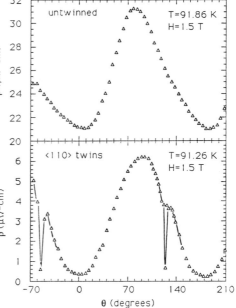

Figure 4. Resistivity versus angle in a magnetic field of 1.5T (H ‖ ab) at a constant temperature for an untwinned and twinned single crystal.

shape of the resistive broadening undergoes a transition at high fields. For fields less than 3 T, a 'fan shaped' broadening is observed, similar to the fully oxygenated samples shown in Figure 1. However, at fields greater than 3T, the entire transition curve appears to shift in parallel to lower temperatures. The difference in the shape induced by the magnetic field is easily seen in the comparison between the 8 Tesla and 1 Tesla curve. Measurements on the recently discovered layered organic superconductor κ-(BEDT-TTF)$_2$Cu[N(CN)$_2$]Br also show similar behavior[7].

The shape of the resistive transition for H ‖ ab ‖ I is very similar to the fully oxygenated samples, except that here again, the broadening is greatly enhanced (Figure 6, lower panel). This is surprising, considering the decrease in thermal energies and increase in defects due to oxygen vacancies compared to the fully oxygenated crystals. In addition, pinning by twin boundaries and an angular dependent $\sin^2(\theta)$ Lorentz force dissipation which is again superimposed on a large angular dependent component has also been observed[8] in the oxygen deficient single crystals. These results suggest that the large angular independent resistive broadening in these layered materials is an intrinsic property not related to Lorentz force induced dissipation. A more microscopic theory based on the dynamics of the order parameter in magnetic field may be required.

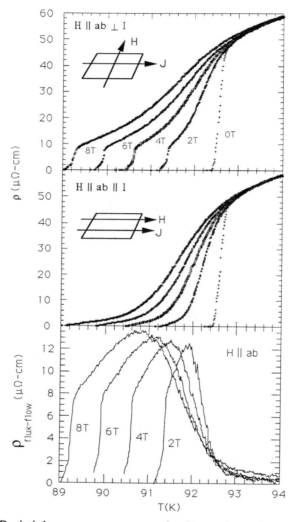

Figure 5. Resistivity versus temperature for the maximum Lorentz force configuration H ⊥ I (top panel) and nominal zero Lorentz force configuration H ‖ I (center panel) for magnetic fields in the ab plane of the crystal. The difference in resistivity between the two configurations is the flux flow resistivity (bottom panel).

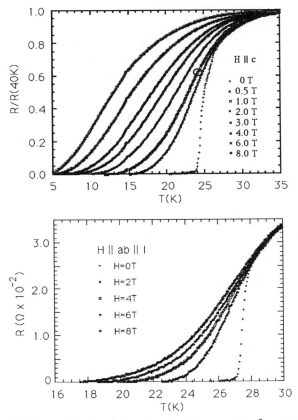

Figure 6. Normalized resistance versus temperature for an oxygen deficient crystal in magnetic fields up to 8 Tesla for H ∥ c (top panel). The circle denotes the mean field T_c at 1 Tesla determined from DC magnetization measurements. Resistance versus temperature for H ∥ ab (bottom panel).

SUMMARY

In summary, we have shown from comparison of twinned and untwinned single crystals of $YBa_2Cu_3O_{7-\delta}$ that twin boundaries act as strong pinning centers when entire vortex lines lie within the twin boundaries. The Lorentz force induced flux dissipation due to vortex motion manifests itself as a $\sin^2(\theta)$ angular dependent resistivity superimposed over a substantially large angular independent component. Our investigation on oxygen deficient crystals where critical temperatures are lower and defects from oxygen vacancies are greater show markedly enhanced broadening, suggesting that the angular independent broadening may be an intrinsic property not related to Lorentz force induced dissipation.

ACKNOWLEDGEMENTS

This work was supported by the U. S. Department of Energy, BES--Materials Science under contract #W-31-109-ENG-38 (WKK, SF, YF, JZL, GWC) and the NSF--Office of Science and Technology Centers under contract #STC8809854 (UW, KGV) Science and Technology Center for Superconductivity). SF and KV acknowledge partial support from the Division of Educational Programs, Argonne National Laboratory.

REFERENCES

1. W. K. Kwok, U. Welp, G. W. Crabtree, K. G. Vandervoort, R. Hulscher, J. Z. Liu, Phys. Rev. Lett. 64, 966, (1990).
2. U. Welp, M. Grimsditch, H. You, W. K. Kwok, M. M. Fang, G. W. Crabtree, J. Z. Liu, Physica C 161, 1 (1989).
3. W. K. Kwok, U. Welp, S. Fleshler, K. G. Vandervoort, G. W. Crabtree, J. Z. Liu, J. Brooks, J. Hettinger, S. T. Hannahs, Proceedings of the LT-19 Satellite Conference on High T_c Superconductors, Cambridge U. K., Aug 13-15, 1990.
4. W. K. Kwok, U. Welp, K. G. Vandervoort, Y. Fang, G. W. Crabtree, J. Z Liu, Appl. Phys. Lett. 57, 1 (1990).
5. D. Nelson, Phys. Rev. Lett. 60, 1973 (1988).
6. B. W. Veal, A. P. Paulikas, H. You, H. Shi, Y. Fang, J. W. Downey, Phys. Rev. B (in press).
7. W. K. Kwok, U. Welp, K. D. Carlson, G. W. Crabtree, K. G. Vandervoort, H. H. Wang, A. M. Kini, J. M. Williams, D. L. Stupka, L. K. Montgomery, J. E. Thompson, Phys. Rev.B (inpress).
8. G. W. Crabtree, W. K. Kwok, U. Welp, H. Claus, K. G. Vandervoort, S. Fleshler, Y. Fang, Proceedings of the LT-19 Satellite Conference on High T_c Superconductors, Cambridge U. K., Aug 13-15, 1990.

CRITICAL CURRENTS IN ALIGNED YBCO AND BSCCO SUPERCONDUCTORS

K. W. Lay
GE Corporate Research & Development, Schenectady, NY 12301

ABSTRACT

Low Jc values are typical for polycrystalline high temperature superconductors, especially in magnetic fields. On the other hand, recent results on some polycrystalline samples are showing encouraging Jc values. Grain boundary defects which limit current will be considered. The intrinsic properties of grain boundaries in the high Tc materials will be emphasized. These properties such as the atomic arrangements at grain boundaries are inherent to a given material. It will be shown that a plausible explanation for the low Jc values in c-axis aligned $YBa_2Cu_3O_7$ compared to $Bi_2Sr_2Ca_{1-x}Cu_xO_y$ is the ability to transfer supercurrents across basal plane boundaries in the latter material.

INTRODUCTION

Fairly quickly after the initial discovery of the high Tc cuprate superconductors by Bednorz and Muller[1], three different classes of materials were discovered which all can have Tc values above 77K[2-4]. The first and still most studied compound is $YBa_2Cu_3O_7$, referred to in this paper as YBCO. Silver-clad tapes of the bismuth-containing compounds have been extensively studied, especially in Japan. These compounds will be called BSCCO throughout the paper using the notation 2212, for example, to denote the ratio of Bi:Sr:Ca:Cu layers in the crystal structure. In the case of the 2223 compound Pb is generally substituted for some of the Bi but for simplicity the material will still be called BSCCO. The thallium containing compounds (TBCCO) will be mentioned occasionally but the bulk of the discussion in this paper will be focused on a comparison of the YBCO and BSCCO systems.

High power applications of these materials for magnetic coils or transmission lines will require the ability to carry large currents through long conductors. The expected high costs of very long single crystals requires the wires to be polycrystalline. Unfortunately polycrystalline cuprate superconductors often exhibit very low Jc values[5]. In this paper the focus will be on a comparison the experimental Jc values in c-axis aligned YBCO and BSCCO systems. The question to be answered is: Why are much better Jc's obtained in BSCCO samples compared to YBCO?

It will be shown that a plausible explanation for the higher Jc values in BSCCO may be the ability to carry appreciable currents across basal plane grain boundaries. It will be argued that YBCO basal plane grain boundaries do not carry such currents.

Jc IN SINGLE- AND POLY-CRYSTALS OF YBCO

Some critical currents reported in the literature for various forms of YBCO are shown in Figure 1. The highest values are obtained for single crystal films. The best aligned polycrystalline films are about 2 orders of magnitude lower. Melt processed samples show intermediate Jc values. Unfortunately, single crystal epitaxial films, single crystals, and even the melt grown samples are not made by processes amenable to the fabrication of long, inexpensive wires. The Jc values of c-axis aligned polycrystalline samples are representative of processes which might be used to make wires. These current density values are not high enough for most applications. Considering the extensive effort world-wide on this problem, it seems that the prospects for high Jc wires are not promising -- at least for YBCO.

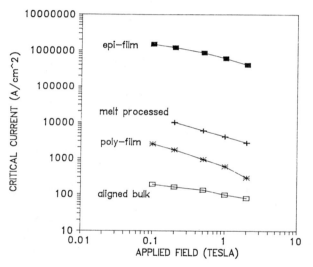

Figure 1   Critical current of YBCO at 77K. Epi-film & poly-film = films laser ablated onto yttria stabilized zirconia (100) single crystal or polycrystal substrates[6]. Aligned bulk = bulk sample sintered from magnetically aligned powder compact[7]. Melt processed = bulk sample heated above melting point during processing[8].

EXTRINSIC & INTRINSIC GRAIN BOUNDARY DEFECTS

The poor current carrying capacity of polycrystalline samples is due to disruptions of the superconducting current at grain boundaries. This disruption can be due to either extrinsic defects, defects not inherent to the superconductor, or to intrinsic defects.

Examples of extrinsic defects are cracks, non-superconducting impurity phases, or possibly impurities in the superconducting grain which are segregated at the grain boundaries. Cracks at or adjacent to (001) grain boundaries in YBCO are particularly present in non-aligned samples due to the very large contraction in the [001] direction during oxidation. Aligned samples of YBCO which did not contain cracks or impurity phases at the grain boundaries still showed low critical currents[9]. In fact, a higher critical current was obtained in a sample which was intentionally doped to contain a small amount of second phase material. The extent of detrimental effect of the second phase is very dependent on the location of the phase with respect to the current carrying path. Isolated second phase particles at 3- and 4-grain intersections are much, much less detrimental than thin layers coating the 2-grain faces which mist carry most of the current. It should also be pointed out here that low energy grain boundaries will be less likely to be coated by second phase particles than random grain boundaries[10]. All these extrinsic defects can be eliminated by the correct material processing techniques.

Intrinsic grain boundary disorder could be expected to be important in the oxide superconductors. The coherence length is of the same magnitude as the thickness of the grain boundary disordered region which is about one or two atomic spacings thick. In addition, the anistotropy of the crystals results in different grain boundary structures depending on the orientation of the boundaries. All the high temperature superconductors are pseudo-tetragonal; they show a large anisotropy in properties in the [001] direction compared to those in the perpendicular (001) basal plane. This anisotropy is also seen in the interfacial energy of surfaces and grain boundaries and in the kinetics of crystal growth. A striking manifestation of the anisotropy is the characteristic platy grain shape with larger dimensions in the a- and b-directions compared to the c-direction. The result is a preponderance of grain boundaries parallel to the (001) basal plane in most materials.

GRAIN ORIENTATION & SHAPES FOR ALIGNED MATERIALS

Grain alignment has been used to obtain favorable orientations of the grains and/or obtain favorable grain boundaries. Two different special grain structures are commonly obtained depending on the alignment technique used.

In one case the c-axes of the grains are aligned, or nearly aligned, in a common direction. I will call such a material a c-axis aligned sample. This type of alignment is obtained by a variety of techniques such as the application of a uniaxial pressure during processing, magnetic alignment of powder particles, or the tendency of the platy grains to grow parallel to a substrate. This alignment results in a "poor" intra-granular conduction direction perpendicular to a "good" conduction plane. An example of such a material is shown in Figure 2. The disc-shaped grains are stacked with (001) grain boundaries on the large faces of the grains perpendicular to the short coherence length poor conduction c-axes of the grains. The relative orientation of the two grain at these basal plane boundaries involves a rotation about the c-axis -- a twist boundary. The grain boundaries around the edges of the disc-shaped grains will be tilt boundaries with the [100] directions (and/or the [010] since the materials are nearly tetragonal) of the two grains misaligned by some angle up to $45°$. Conduction across these boundaries would be within the ab-plane which has a larger coherence length.

Figure 2   Microstructure of magnetically aligned sintered sample sectioned in two orthogonal directions. C-axes of grains are perpendicular to plane of photograph for top micrograph and perpendicular to the direction of the micron marker in the lower micrograph.

The second common alignment is obtained by directional solidification -- and presumably also by direction grain growth as well. The movement of a temperature gradient through the sample results in the alignment of the a-axes parallel to a common direction. The grains will be elongated "needles". The grain boundaries at the ends of the grains will be parallel to (100) planes in the grains. The grain boundaries on the long sides of the grains will be parallel to (001) and (100) planes. In the limit of very slow growth rates, all the grain boundaries can be eliminated and single crystals will result. In light of the slow grow rates required for directional solidification, it is unlikely that this process will be economically feasible for the manufacture of the hundreds of meter lengths necessary for the fabrication of magnetic coils. Therefore the a-axis aligned samples will not be discussed further in this paper.

SPECIAL GRAIN BOUNDARIES

Several special grain boundaries can arise in c-axis aligned materials. A very prevalent boundary type is the (001) basal plane grain boundary. These are pure twist boundaries if the c-axes of the adjacent grains are exactly aligned. In the rest of the paper these boundaries will be call basal plane twist boundaries or c-axis twist boundaries. For low misorientations of the [100] directions in the two grains, the boundary is made up of an array of dislocations. These boundaries will be considered in much more detail for both YBCO and BSCCO.

The other major grain boundary type found in c-axis aligned materials is the tilt boundary found at the grain edges. These grain boundaries are perpendicular to the ab-planes. For grains with perfectly aligned c-axes, the a-axes of the adjacent grains are misaligned by a tilt angle. These boundaries will be called a-axis tilt boundaries. Dimos et al.[11,12] studied bi-crystal films of YBCO containing these boundaries. They found the critical current carried across these boundaries was a function of the misalignment between the two grains. For misalignments up to about 5° there was little loss in current carrying capacity of the grain boundary compared to the intragranular value. At higher misalignments, the critical current was rapidly reduced to as little as 1/50 the intragranular value for misalignments between 15 and 45°.

If the c-axes of the grains in an aligned material are not perfectly aligned, the grain boundaries will not be pure twist or tilt boundaries, but will contain both twist and tilt components. At small total misalignments, the misalignment can be accommodated by dislocation arrays separating relatively unchanged regions at the grain boundaries. At larger misalignments, the entire grain boundary area will be disordered.

COMPARISON OF YBCO VS BSCCO & TBCCO

It has been shown that polycrystalline YBCO shows much lower critical currents than single crystals, even for c-axis aligned materials. It might be expected that the situation for BSCCO and TBCCO materials would be the same or worse due to their even greater anisotropy and shorter c-axis coherence lengths. Just the opposite is seen. Some polycrystalline samples of the latter two materials show quite respectable Jc values. The remainder of this paper will be devoted to examining possible reasons for the critical current

differences in the two classes of materials. Since much more
information is available on the BSCCO materials, the TBCCO polytypes
will not be discussed further. It might be assumed that the similar
crystal structures will result in similar grain boundary properties.
The details, however, may not be the same. Better bonding between the
thallium double layers is seen in comparison to the bismuth double
layers. The layer structure at basal plane grain boundaries in double
(or single) layer TBCCO has not been studied.

Figure 3 shows a comparison of some representative Jc values
for aligned polycrystalline BSCCO and YBCO at 4K. (At this
temperature flux creep is minimal in the BSCCO samples.) After a drop
at low magnetic field values, the critical current is nearly field-
independent up to very high field values. The superiority of BSCCO is
obvious. The grain boundaries in BSCCO are able to carry large
currents, in contrast to YBCO where a small fraction of the
boundaries are carrying the current. We will now examine possible
reasons for the differences in the two materials.

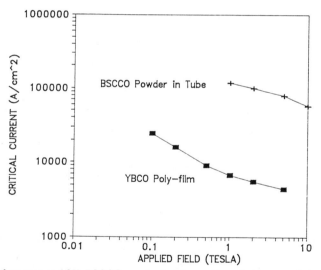

Figure 3   Critical current at 4.2K. BSCCO powder in tube = silver clad tape made by deformation of powder-filled tube[13]. YBCO poly-film = laser ablated film deposited onto polycrystalline yttria stabilized zirconia[6].

GRAIN ASPECT RATIOS

One quite striking difference between the microstructures of
YBCO and BSCCO is the much larger aspect ratio of the grains in BSCCO
materials. The grains are very thin platelets many microns long with
grain diameter to thickness ratios of 10 or more. The much larger
aspect ratios of BSCCO are not surprising in light of the greater
anisotropy in this material. The large anisotropy in growth rates of
the different crystallographic directions results in much more rapid

growth in the ab-plane compared to the c-direction where new basal planes must be continually nucleated.

Malozemoff[15] pointed out the importance of grain aspect ratio on critical current in polycrystalline materials. Current transfer from grain-to grain need not be in a direction parallel to the macroscopic current flow. This lateral transfer of current across grain boundaries parallel to the macroscopic current flow can result in a macroscopic Jc (amps/cm^2) in a polycrystalline sample which could be much greater than the critical current per unit area of the individual grain boundaries. The ideal situation would be long needle-shaped grains pointing in the direction of the macroscopic current flow. This general concept, however, must also take into account the different types of grain boundaries in an aligned sample of an anisotropic material. Not all grain boundaries are the same, and the particular grain shapes and grain boundary types must be considered as well as the grain aspect ratios.

The grains in a c-axis aligned material can be modeled as mono-sized, flat hexagonal plates with six faces around the hexagon and two hexagonal faces. Each grain is surrounded by 12 neighboring grains. The grain boundaries around the six faces of the grain are a-axis tilt boundaries. At the top or bottom hexagonal faces of the grain are 3 neighboring grains on each face. These grain boundaries will be c-axis twist boundaries. The amount of supercurrent which can be carried across each of these 12 boundaries will be a function of the boundary type and the misorientation angle between the two adjacent grains at the boundary. It should be noted that the percolation problem requires consideration of: (1) the two different types of grain boundaries, (2) the misalignment between the two grains at a boundary and (3) the aspect ratio of the grains (which will determine the relative areas of the two types of grain boundaries).

No bicrystal measurements have been made on a-axis tilt boundaries in BSCCO. It seems reasonable that the dependence of Jc on misorientation angle for BSCCO might be similar to that in YBCO. The in-plane coherence lengths in the two materials are similar. The structure of current carrying copper oxide planes and the neighboring BaO or SrO planes are also similar. Therefore, in the absence of conflicting data it will be assumed here that the Jc dependence of the a-axis tilt boundaries in BSCCO is the same as that in YBCO. At misorientation angles of 0 to 5° the critical current at the grain boundary is nearly the same as the intragrain value. At larger misorientation angles the intergranular Jc will be rapidly reduced.

The grain aspect ratio determines the relative fraction of grain boundary area which is a-axis tilt boundaries compared to that which is c-axis twist boundaries in c-axis aligned material. If the grains are approximately disc-shaped with diameter d and thickness t, the ratio of the area of c-axis twist boundaries per grain to the area of a-axis tilt boundaries d/2t. The grain aspect ratio in one study of c-axis aligned YBCO[9] was about d/t=2/1. With this aspect ratio, the area of c-axis twist boundaries is equal to the area of a-axis tilt boundaries. However, if the aspect ratio is 100/1, as might be seen for a BSCCO sample, the area of c-axis twist boundaries per grain is 50 times that of the area of tilt boundaries per grain. It is quite apparent that the importance of the basal plane c-axis twist boundaries becomes much more important as the grain diameter to thickness ratio increases.

(001) GRAIN BOUNDARIES IN YBCO

In light of the importance of the basal plane grain boundaries, the structures of these boundaries will be considered in more detail. The character of these boundaries will undoubtedly be quite different in YBCO compared to BSCCO due to the very different cations in the layers in the respective crystal structures. Therefore the two materials will be considered separately.

High resolution transmission electron microscopy has been done of YBCO basal plane grain boundaries[16,17]. The layers adjacent to a grain boundary are shown schematically in Figure 4. The BaO_x designation for the layer at the grain boundary indicates the possibility that excess oxygen is present here compared to the BaO composition of the corresponding layers in the perfect crystal structure. It would not be surprising for excess oxygen to be present since barium peroxide is stable at the temperatures typically used to oxidize YBCO to form the required orthorhombic high-Tc material. As a result of the modification in the structure it is quite probable that the layers next to a basal plane boundary in YBCO are not superconducting.

CuO_2

BaO CuO_2

CuO   SrO

BaO_x BiO_x

————   ————

BaO_x BiO_x

CuO   SrO

BaO CuO_2

CuO_2

Figure 4   Schematics of atomic layers adjacent to a basal plane twist boundary in YBCO (left) & BSCCO (right).

The presumed non-superconducting layers adjacent to (001) grain boundaries in YBCO result in very important ramifications on the ability to carry appreciable critical currents through polycrystalline YBCO. If these grain boundaries cannot pass large currents, a large fraction of the grain boundary area do not participate in current transport. Even more important is the geometry of the c-axis aligned material. The removal of the basal plane grain boundaries from consideration as current carrying surfaces converts the percolation path from a 3-dimensional problem to essentially a 2-dimensional problem. The allowed current paths from grain to grain are only through the a-axis tilt boundaries, and only a fraction of these boundaries will be low angle boundaries which will carry large currents. (The percolation path is not exactly 2-dimensional since in current can be carried in the c-direction by transfer from one grain across an a-axis tilt boundary to a neighboring grain which extends further in the c-direction.) We will assume that only 0 to 5° mismatched tilt boundaries carry current, and assume the misorientations are randomly distributed between 0 and 45°. Then the probability that current can be transferred from one grain to another is about 5/9 since the current must enter through one low angle grain boundary and exit through one of the remaining 5 tilt grain boundaries with a 1/9 probability that a given boundary is low angle. It is expected that this percolation problem will result in continuous current paths through only a fraction of the sample. This would explain the low Jc values seen in well aligned YBCO samples.

(001) GRAIN BOUNDARIES IN BSCCO

The basal plane grain boundaries in BSCCO are not as well studied as those in YBCO. In fact, the crystal structure of the BSCCO polytypes are not yet completely determined[18]. In particular the location and number of the oxygen ions in the BiO_x layers are not known. The structure is "micaceous" with a tendency to cleave between the bismuth-containing layers. The bonding in the crystal structure is weakest between the two adjacent BiO_x layers[19].

The high critical currents now being obtained in silver-clad BSCCO tapes requires that appreciable supercurrents be carried across basal plane boundaries. The critical current in these samples have been shown to depend on the degree of alignment of the c-axes. As the alignment improves, the critical current is increased. A consideration of the microstructure of these c-axis aligned materials almost demands that the macroscopic current path must traverse the basal plane boundaries. These materials show platy grains with very high aspect ratios. The ratio of basal plane boundaries to non-basal plane boundaries is very large. If current can be carried across both c-axis twist boundaries as well as the a-axis tilt boundaries the number of adjacent grains available for transport is doubled. Also the percolation path for the current will be 3-dimensional, in contrast to the nearly 2-dimensional path proposed for YBCO.

Eibl has studied some special grain boundaries in BSCCO 2212[19]. He found that basal plane twist boundaries often showed special orientations. One commonly seen was a $90°$ twin where [100] and [010] were parallel in the adjacent grains. He also saw $45°$ misorientations. This would be a low sigma coincidence site grain boundary for this pseudotetragonal material similar to the $45°$ boundaries seen in YBCO[20]. One could expect a series of special, low-energy c-axis twist boundaries to be present in aligned BSCCO. The $90°$ twins should carry nearly the same current as the intragranular material. Whether the low sigma CSL boundaries show higher Jc values than random twist boundaries is unknown.

In contrast to the case with YBCO the layer structure at BSCCO basal plane grain boundaries is unchanged from that in the grain interior. The grain terminates in a BiO_x layer[19]. The layer structure across a basal plane boundary is shown schematically in Figure 4. The contrast with the YBCO case is obvious. In BSCCO the basal plane twist grain boundaries involve only a rotation about the common c-axes in the two grains. The layer sequence continues across the grain boundary exactly as it would within a grain. In YBCO there is a change in the layer sequence in crossing a basal plane twist boundary. An extra CuO and an extra BaO layer have been inserted. there may be extra oxygen in the barium oxide layers to form a barium peroxide. It is suggested that this difference in the intrinsic grain boundary structures of the two materials is crucial to the apparent ability carry high supercurrents across basal plane boundaries in BSCCO while similar boundaries in YBCO do not seem to carry such currents.

SUMMARY

It is known that well aligned BSCCO polycrystalline samples show less weak link behavior than YBCO samples. It has been argued here that the difference in these materials is the ability to carry currents across basal plane grain boundaries in BSCCO where the atomic layer structure adjacent to these grain boundaries is the same as is present within the grain. The YBCO basal plane grain boundaries exhibit a different layer structure than within the grain. It is hypothesized that this modified layer structure at basal plane grain boundaries is non-superconducting and leads to an inability to carry supercurrents across these boundaries.

ACKNOWLEDGMENTS

Funding for this study was provided by DARPA/ONR contract N00014-88-C-0681. The author wishes to acknowledge many helpful discussions with Eric Tkaczyk.

REFERENCES

1. J.G.Bednorz & K.A.Muller, Z. Phys. 193,B64(1986)

2. M.K.Wu, J.R.Ashburn, C.J.Torng, P.H.Hor, R.L.Meng, L.Gao, Z.J.Huang, Y.Q.Wang & C.W.Chu, Phys. Rev. Lett.58,908(1987)

3. H.Maeda, Y.Tanaka, M.Fukutomi & T.Asano, Jpn. J. Appl. Phys. 27, L209(1988)

4. Z.Z.Sheng, A.M.Hermann, A.El Ali, C.Almasan, J.Estrada, T.Datta & R.J.Matson, Phys. Rev. Lett. 60, 937(1988)

5. A.P.Malozemoff, in Physical Properties of High Temperature Superconductors, ed. D. Ginsberg (World Publishing, Singapore, 1989), pp.,71-150

6. D.P.Norton, D.H.Lowndes, J.D.Budai, D.K.Christen, E.C.Jones, K.W.Lay & J.E.Tckaczyk, Appl. Phys. Lett. 57, 1164(1990)

7. J.W.Ekin, H.R.Hart,Jr. & A.R.Gaddipati, J. Appl Phys. 68, 2285(1990)

8. H.Kupfer, C.Keller, K.Salama & V.Selvamanickam, Appl. Phys. Lett. 55, 1903(1989)

9. J.E.Tkaczyk & K.W.Lay, J. Mater. Res. 5, 1368(1990)

10. L.T.Romano, P.R.Wilshaw, N.J.Long & C.R.M.Grovenor, Supercond. Sci. Technol. 1, 285(1989)

11. D.Dimos, P.Chaudhari, J.Mannhart & F.K.LeGoues, Phys. Rev. Lett. 61, 219(1989)

12. D.Dimos, P.Chaudhari, J.Mannhart, Phys. Rev. B 41, 4038(1990)

13. N.Enomoto, H.Kikuchi, N.Uno, H.Kumakure, K.Togano &K.Watanabe, Jap. J. Appl. Phys. 29, L447(1990)

15. A.P.Malozemoff, IBM Research Report RC15621 3/29/90

16. H.W.Zandbergen, R.Gronsky & G.Thomas, Physica C 153-155, 1002(1988)

17. H.W.Zandbergen, R.Gronsky & G.vanTendeloo, J. Supercond. (To Be Published)

18. O.Eibl, Physica C, 168, 215(1990)

19. O.Eibl, Physica C, 168, 239(1990)

20. S.E.Babcock & D.C.Larbalestier, J. Mater. Res. 5, 919(1990)

EFFECTS OF Ag DOPING IN THE N-TYPE SUPERCONDUCTOR Nd-Ce-Cu-O

Y. D. Yao[a], J. W. Chen[b], Y. Y. Chen[a], C. S. Fang[a],
C. C. Wu[b], C. H. Cheng[b] and J. Y. Chen[c]
a Institute of Physics, Academia Sinica, Taipei, Taiwan 11529, ROC
b National Taiwan University, Taipei, Taiwan 10764, ROC
C Physics Department, Fu Jen University, Taipei, Taiwan 24205, ROC

ABSTRACT

Changes in various physical properties due to both addition
of Ag and substitution of Ag for Cu in the N-type superconductor
system Nd-Ce-Cu-O were investigated by measurements of electrical
resistivity, magnetization, X-ray powder diffraction, X-ray
photoelectron spectroscopy (XPS) and UV photoelectron spectroscopy
(UPS). There is only a small decrease in the superconducting
transition temperature with Ag dopants up to 40% (compared with
Cu). The normal-state resistivity decreases with the addition of
Ag; however, it increases with the substitution of Ag for Cu. The
M-H hysteresis loops at 5 K were enhanced for the 20% Ag-doped
samples. The superconductivity in the Ag-doped Nd-Ce-Cu-O samples
depends greatly on the reduction processing conditions.

INTRODUCTION

Recently, the electron-doped copper oxide superconductor
$Nd_{2-x}Ce_xCuO_y$ (NCCO) has been discovered by tokura et al. [1].
It has a tetragonal $T'-Nd_2CuO_4$ structure with no apical oxygen
around the Cu ions; this means that the copper atoms in the "T'-
phase" are in square planar coordination with oxygen forming the
Cu-O sheets [1-3]. As sintered samples in air or oxygen show
semiconducting behavior, while superconductivity with T_c = 24 K
appears only when they are annealed in a reducing atomosphere.
   The variation of the physical properties of the hole-doped
copper oxide superconductors with Ag doping and addition have
recently been studied [4-8]. However, relatively little effort
has been devoted to the effect of Ag dopant in the electron-doped
superconductor [9]. Therefore, it is very interesting to compare
the effects of Ag doping and addition in both the electron-doped
and the hole-doped copper oxide superconductors.
   The purpose of the present investigation was to understand
the effect of both addition of Ag and substitution of Ag for Cu
in NCCO system on the superconductivity with various experimental
techniques; such as electrical resistivity, magnetization, X-ray
powder diffraction, XPS and UPS.

EXPERIMENTAL

Samples used for the present study were synthesized by solid state reaction from stoichiometric mixtures of high purity CeO_2, Ne_2O_3, CuO and Ag_2O. Three different conditions were taken for the solid state reaction method to prepare the samples. For series A samples, the powders were mixed thoroughly and calcined at 925 C in air for 24 hours. After grounding and mixing, samples were pressed into pellets and sintered at 1150 C in flowing oxygen for 12 hours. The pellets were then annealed at 1000 C in N_2 for10 hours following annealing at 1050 C in a vacuum of 10^{-3} Torr for 5 to 10 hours. For series B samples, the powders were mixed thoroughly and calcined at 925 C in air for 24 hours. After pulverizing and mixing, samples were pressed into pellets and sintered at 1150 C in flowing oxygen for 12 hours, then quenched into liquid nitrogen. The pellets were then annealed at 900 C in vacuum of 10^{-3} Torr for 6 to 8 hours. For series C samples, the powders were mixed thoroughly and calcined at 900 C in air for 22 hours and removed from the furnace and quenched in air to room temperature, reground again and fired in air at 900 C for another 22 hours and quenched to room temperature. The resulting powders were then pressed into pellets and fired in air at 1100 C for 20 hours and quenched to room temperature. The pellets were then sintered in flowing helium at 990 C for 20 hours and furnace cooled to room temperature in helium flow.

A standard four-probe technique and a SQUID magnetrometers were used to measure the electrical resistivity and magnetic moment as a function of temperature. X-ray powder diffraction data were obtained from a Rigaku powder diffractometer with filtered copper radiation.

For surface analysis, samples were introduced into a VG ESCA MK II system. Clean surfaces were obtained by low-energy ion sputtering until no carbon could be detected by X-ray photoelectron spectroscopy (XPS). XPS with Mg anode emission was used for Ag 3d, Cu $2p_{3/2}$ and O 1s core-level spectra. A helium discharge lamp provided a HeI light source of 21.2 ev for valence band spectra. Energy distribution curves (EDC's) of UV photoelectron spectroscopy (UPS) and XPS were taken in angle integrated mode.

RESULTS AND DISCUSSION

For series A samples, we have shown that superconductivity occurs only for NCCO samples with Ag-dopant less than 10%. [9] It becomes non-superconducting when the Ag content is increased to 20%, and Ag clusters were observed in the 20% Ag-doped samples. However, for series B and C samples, there is only a slightly decrease in the superconducting transition temperature with Ag dopants up to 40% (compared with Cu). As an example, Fig. 1

shows the electrical resistivity as a function of temperature for the series B samples with additional Ag ($Nd_{1.85}Ce_{0.15}CuAg_xO_y$, $0 < x < 0.4$). Fig. 2 presents the electrical resistivity as a function of temperature for the series C samples with substitution of Ag for Cu ($Nd_{1.85}Ce_{0.15}Cu_{1-x}Ag_xO_y$, $0 < x < 0.4$). The room temperature electrical resistivity decreases with the addition of Ag; however, generally speaking, it increases for the samples with the substitution of Ag for Cu. Details will be reported later.

   The powder X-ray diffraction patterns indicate that there exists a T'-phase in all the samples which show superconductivity. As an example, Fig. 3 plots the X-ray diffraction patterns for the series A samples (Curves a and b) and series B samples (Curve c and d). Curves a and c are NCCO samples with 20% Ag addition, and Curves b and d are NCCO samples with 20% substitution of Cu

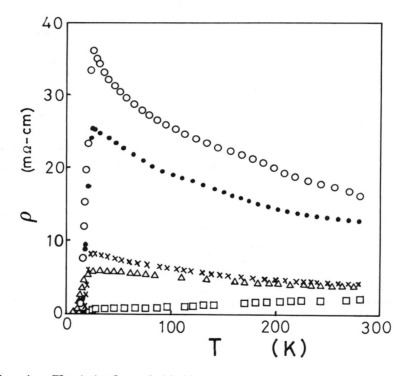

Fig. 1. Electrical resistivity vs. temperature of series B samples with various addition of Ag. (o: $Nd_{1.85}Ce_{0.15}CuO_y$, ●: $Nd_{1.85}Ce_{0.15}CuAg_{0.1}O_y$, x: $Nd_{1.85}Ce_{0.15}CuAg_{0.2}O_y$ △: $Nd_{1.85}Ce_{0.15}CuAg_{0.3}O_y$, □: $Nd_{1.85}Ce_{0.15}CuAg_{0.4}O_y$)

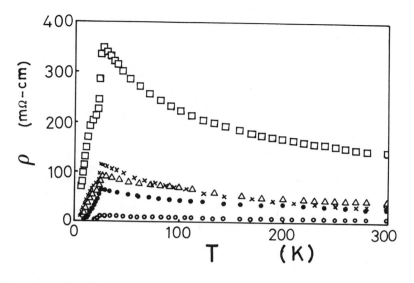

Fig. 2. Electrical resistivity vs. temperature of series C samples with various substitution of Ag for Cu.
(o: $Nd_{1.85}Ce_{0.15}CuO_y$, ●: $Nd_{1.85}Ce_{0.15}Cu_{0.9}Ag_{0.1}O_y$,
x: $Nd_{1.85}Ce_{0.15}Cu_{0.8}Ag_{0.2}O_y$, △: $Nd_{1.85}Ce_{0.15}Cu_{0.7}Ag_{0.3}O_y$,
□: $Nd_{1.85}Ce_{0.15}Cu_{0.6}Ag_{0.4}O_y$)

by Ag. The main peaks of the diffraction patterns of curves c and d are essentially identical to that of the undoped T'-phase NCCO samples.

From the magnetization study for the series B samples, both Meissner and shielding effects show maximun values for the samples with 20% Ag. The M vs H curves for samples with Ag-dopants less than 20% are plotted in Fig. 4. It is manifest that the area of the magnetic loop is enhanced for the Ag-doped samples. This enhancement suggests that the superconducting critical current is enhanced by the Ag-doping. Similar effect has also been observed in the Ag-doped YBaCuO samples. Studying by the polarized optical microscope, we found that the grain size of the doped samples is larger than that of the undoped sample.

By comparing the Ag $3d_{5/2}$ XPS of high-purity Ag foil with the XPS of NCCO samples with Ag dopants, addition of Ag dopant results in greater Ag XPS peak than substitution of Ag for Cu. The Cu $2p_{3/2}$ XPS spectra of undoped and Ag-doped series B samples with Ag-dopants up to 20% are presented in Fig. 5. By comparing this result with that of the series A samples [9], we can say that the difference in Cu $2p_{3/2}$ B. E. was quite small between the undoped sample (curve a) and the Ag-doped samples (curve b to e). All these samples show superconductivity behavior. The Cu $2p_{3/2}$

B. E. of the non-superconducting samples shifts to a lower-energy
value [9]. From the oxygen 1s XPS spectra for samples of series
A and B, a shoulder in the O 1s spectrum observed at a higher
binding energy than that of the main peak. This has been assigned
to traces of contamination [10] and the tendency of Cu^{+1} formation
[11]. However, this shoulder is also likely due to the sputtering
procedure [12]. Sputtering is known to cause outdiffusion species
in the bulk. The lighter oxygen atoms may segregate towards
the surface and form a different phase, giving rise to a shifted
oxygen 1s core level. Nevertheless, effects of silver doping in
NCCO samples still can be concluded from the oxygen 1s main peak.
The oxygen 1s main peak seems to shift to an lower-energy value
for samples that either the main peak of the Ag $3d_{5/2}$ XPS spectra

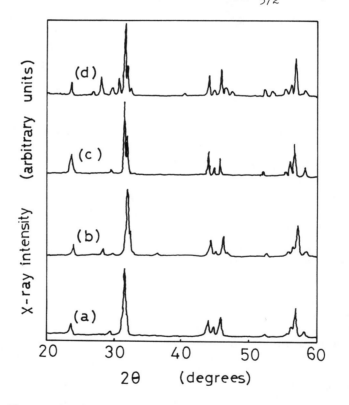

Fig. 3. The powder X-ray diffraction pattern of series A
samples (a & b) and series B samples (c & d).
(a & c: $Nd_{1.85}Ce_{0.15}CuAg_{0.2}O_y$,
b & c: $Nd_{1.85}Ce_{0.15}Cu_{0.8}Ag_{0.2}O_y$)

shifts to a higher-energy value or the main peak of the Cu $2p_{3/2}$
XPS spectra shifts to a lower-energy value. This suggests that
either adding Ag dopants or reducing Cu atom in the NCCO system
may weaken the Cu-O bonds.

Analysing the result of the energy distribution curves of
both series A and series B samples, we found that as the doped
Ag concentration is increased there is an increase in the
intensity of Cu 3d - O 2p hybridized states [13]. The increase
in the photoemission intensity of the Cu-O hybrids would be
expected from a decrease in the loss of oxygen involved in the
hybridization with copper. The broad feature extending from
8.5 eV to 13 eV below the Fermi edge is probably assigned to
Cu $3d^8$ sattellite [14], indicating the strongly correlated nature
of the 3d electrons. Note that a peak at -12.5 eV owing to Cu
3p -3d resonance photoemission appears only on Cu^{2+} materials[15].
Since ion sputtering usually causes a conversion of Cu^{2+} to Cu^{1+}

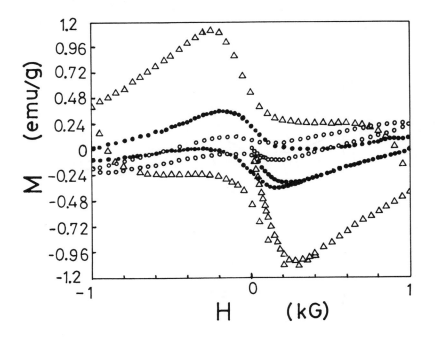

Fig. 4. Magnetization vs magnetic field for series B
samples with Ag dopants less then 20%. (o: $Nd_{1.85}Ce_{0.15}CuO_y$,
•: $Nd_{1.85}Ce_{0.15}CuAg_{0.1}O_y$, Δ: $Nd_{1.85}Ce_{0.15}CuAg_{0.2}O_y$)

[12], the Cu^{2+} resonance peak shall not be observed on the Ar^+ sputter-cleaned surface in this study. It can be seen that the photoemission intensity of Cu $3d^8$ sattellite for samples which

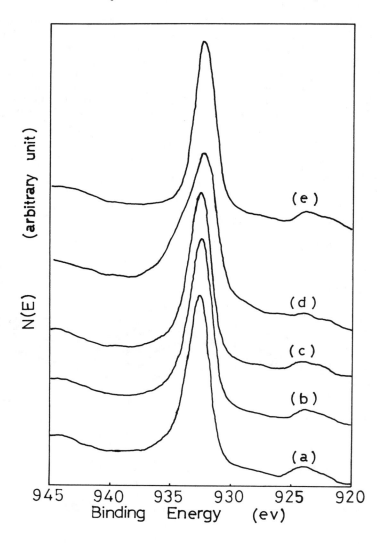

Fig. 5.   The Cu $2p_{3/2}$ XPS spectra of series B samples with Ag dopants less than 20%. [(a) $Nd_{1.85}Ce_{0.15}CuO_y$, (b) $Nd_{1.85}Ce_{0.15}Cu_{0.9}Ag_{0.1}O_y$, (c) $Nd_{1.85}Ce_{0.15}CuAg_{0.1}$, (d) $Nd_{1.85}Ce_{0.15}Cu_{0.8}Ag_{0.2}O_y$, (e) $Nd_{1.85}Ce_{0.15}CuAg_{0.2}O_y$]

do not show superconductivity is greatly reduced. This indicates the great decrease in the correlated nature of the Cu 3d electrons for the non-superconducting Ag-doped NCCO samples.

In conclusion, we have briefly reported the experimental study on changes in various physical properties due to both addition of Ag and substitution of Ag for Cu in the electron-doped NCCO superconductors. There is only a small decrease in the superconducting transition temperature with Ag dopants up to 40% (compared with Cu). The room temperature electrical resistivity decreases with the addition of Ag; however, it increases with the substitution of Ag for Cu. The M - H hysteresis loops at 5 K were enhanced for the 20% Ag-doped samples. Finally, the superconductivity in the Ag-doped NdCeCuO samples depends greatly on the reduction processing conditions.

ACKNOWLEDGEMENT

This research is supported by the National Science Council, Taiwan, ROC, under the Grant No. NSC79-0208-M001-60.

REFERENCES

1. Y. Tokura, H. Takagi and S. Uchida, Nature $\underline{337}$, 345 (1989).
2. H. Takagi, S. Uchida and Y. Tokura, Phys. Rev. Lett. $\underline{62}$, 1197 (1989).
3. C. L. Seaman, N. Y. Ayoub, T. Bornholm, E. A. Early, S. Gharnaty, B. W. Lee, J. T. Markert, J. J. Neumeier, T. S. Tsai and M. B. Maple, Physica C $\underline{159}$, 391 (1989).
4. Y. H. Kao, Y. D. Yao and L. W. Song, Int. J. Mod. Phys. B $\underline{3}$, 573 (1989).
5. C. Y. Huang, Y. Shapira, E. J. McNiff, Jr., P. N. Peters, B. B. Schwartz, M. K. Wu, R. D. Shull and C. K. Chiang, Mod. Phys. Lett. B$\underline{2}$, 869 (1988).
6. C. Y. Huang, T. J. Li, Y. D. Yao, L. Gao, Z. J. Huang, Y. Shapira, E. J. McNiff, Jr. and M. K. Wu, Mod. Phys. Lett. B$\underline{3}$, 1251 (1989).
7. Y. H. Kao, Y. D. Yao, L. Y. Jang, F. Xu, A. Krol, L. W. Song, C. J. Sher, A. Darovsky, J. C. Phillips, J. J. Simmins and R. L. Snyder, J. Appl. Phys. $\underline{67}$, 353 (1990).
8. J. J. Lin, T. M. Chen, Y. D. Yao, J. W. Chen and Y. S. Gou, Jap. J. Appl. Phys. $\underline{29}$, 497 (1990).
9. C. S. A. Fang, Y. D. Yao and C. C. Wu, Physica C $\underline{167}$, 89 (1990).
10. S. Uh, M. Shimoda and H. Aoki, Jap. J. Appl. Phys. $\underline{28}$, L804 (1989).
11. M. K. Rajumon, D. D. Sarma, R. Vijayaraghavan and C. N. Rao, Sol. St. Comm. $\underline{70}$, 875 (1989).

12. P. A. P. Lindberg, Z. X. Shen, I. Lindau, W. E. Spicer, C. B. Eom and T. H. Geballe, Appl. Phys. Lett. $\underline{53}$, 529 (1988).

13. M. S. Hybertsen and L. F. Mattheiss, Phys. Rev. Lett. $\underline{60}$, 1661 (1988).

14. Z. X. Shen, P. A. P. Lindberg, B. O. Wells, D. B. Mitzi, I. Lindau, W. E. Spicer and A. Kapitulnik, Phys. Rev. B$\underline{38}$, 11820 (1988).

15. R. Zanoni, Y. Chang, M. Tang, Y. Hwu, M. Onellion, G. Margaritondo, P. A. Morris, W. A. Bonner, J. M. Tarascon and N. G. Stoffel, Phys. Rev. B$\underline{38}$, 11832 (1988).

ANISOTROPY OF THE NORMAL STATE RESISTIVITY AND SUSCEPTIBILITY OF UNTWINNED YBa$_2$Cu$_3$O$_{7-\delta}$ SINGLE CRYSTALS

U. Welp[#+], W. K. Kwok[+], S. Fleshler[+$], K. G. Vandervoort[+*], J. Downey[+], Y. Fang[+] and G. W. Crabtree[+]

[#] Science & Technology Center for Superconductivity and [+] Materials Science Division, Argonne National Laboratory, Argonne, IL 60439

J. Z. Liu
Department of Physics, University of California-Davis, Davis, CA 95616

ABSTRACT

We present measurements of the **a-b** anisotropy of the normal state resistivity ρ_a/ρ_b and measurements of the normal state magnetization along the three crystal axes. ρ_a/ρ_b increases with increasing temperature and reaches a (sample dependent) value of 1.2 to 1.85 near room temperature. The results show that the CuO chains are metallic and that they can have a conductivity comparable to that one of the CuO$_2$ planes. A model of a parallel circuit of planes and chains is used to analyse the data. Superconducting fluctuations are found to be much more pronounced in the planes than in the chains. The normal state susceptibility shows the onset of strong diamagnetic fluctuations at temperatures below 200 K. At higher temperatures the anisotropy $\chi_c > \chi_a > \chi_b$ is observed and compared to predictions based on a model of local Cu-moments in the planes and chains.

INTRODUCTION

The orthorhombic structure of YBa$_2$Cu$_3$O$_{7-\delta}$ is characterized by the corrugated CuO$_2$ planes sandwiching the Y-sites and the one dimensional CuO chains along the **b**-direction in the basal plane. YBa$_2$Cu$_3$O$_{7-\delta}$ has the remarkable property that in spite of the large structural anisotropy in the **a-b** plane the superconducting properties (upper critical field[1], lower critical field[2], gap[3], magnetic torque[4]) are found to be almost isotropic in the **a-b** plane. This leads to the conclusion that superconductivity in this material is dominated by the CuO$_2$ planes. However, in the normal state electronic and magnetic properties are expected to show additional anisotropies due to the presence of the chains. These properties can be accessed with measurements of the resistivity and magnetization on untwinned single crystals. Measurements of the anisotropy of the resistivity in the basal plane will allow a determination of the chain and plane contributions of the conductivity and a comparison to models of the Fermi surface. Their temperature dependence shows how superconducting fluctuations develop in the planes and in the chains while approaching T$_c$. Magnetization measurements give additional information on diamagnetic fluctuations and the behavior of local Cu-moments in the different surroundings of the planes and chains.

$Also at Dept. of Physics, Purdue University, West Lafayette, IN 47907
*Also at Dept. of Physics, University of Illinois at Chicago, Chicago, IL 60680

RESULTS AND DISCUSSION

The crystals studied here were grown by a self-flux method[5] and detwinned[1] by annealing under uniaxial strain at 400 $^{\circ}$C. The anisotropy ρ_a/ρ_b of the normal state resistivity in the **a-b** plane was measured on six detwinned single crystals using the Montgomery method. In order to avoid large geometrical anisotropies, samples of almost square shape were selected. Contacts were made with silver paste and had less than 1 Ω resistance[6]. In order to obtain absolute values of the resistivities, the contacts on two crystals were extended to the conventional linear four probe geometry after finishing the measurements of ρ_a/ρ_b. The normal state magnetization was measured in a commercial SQUID magnetometer in a field of 1 T along the three crystallographic axes in the temperature range from 90 K to 350 K. A special sample holder which is compensated for thermal expansion was used.

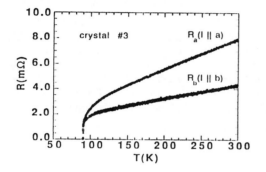

Figure 1. Temperature dependence of the resistances in the two Montgomery configurations for crystal #3.

Figure 1 shows the temperature dependence of the measured resistances R_a (I ‖ **a**) and R_b (I ‖ **b**) of crystal #3. At high temperatures a linear temperature dependence is observed whereas at temperatures below 150 K there is a downward curvature due to superconducting fluctuations. The results indicate that fluctuations are more pronounced in the a- than in the b-direction. This different behavior leads to the temperature dependence of ρ_a/ρ_b shown in Fig. 2 for six detwinned crystals. ρ_a/ρ_b increases with increasing temperature and reaches sample dependent saturation values of 1.20 to 1.85 near room temperature. All crystals show a higher resistivity along **a** than along **b** implying that the chains are metallic rather than localized. This result is in

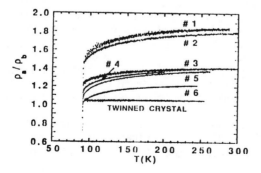

Figure 2. Temperature dependence of the anisotropy of the resistivity for six detwinned crystals and one twinned crystal.

agreement with results of IR-measurements[3] and band structure calculations[7] which indicate enhanced conductivity along **b** due to the presence of the CuO-chains. Also included in Fig. 2 are the results for a twinned crystal. The temperature independent value of 1.05 for ρ_a/ρ_b confirms that the Montgomery method gives reliable results. The difference between the measured value and the expected value of 1.00 is due to uncertainties in the distance between the contacts. The large observed anisotropies imply the interesting result that the

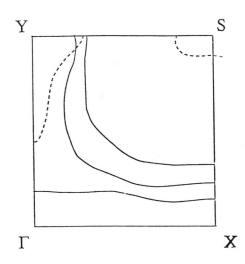

Y   S

Γ   X

Figure 3.   Fermi surface sheets in the central plane as calcualted in Ref. 7.   The one dimensional electron sheet between Γ and X and the dotted hole surfaces around Y and S are due to the chains.

chains in $YBa_2Cu_3O_{7-\delta}$ are as good an electrical conductor as the CuO_2-planes are. Fig. 3 shows the Fermi surface sheets in the central plane as determined from band structure calculations[7]. The one dimensional electron sheet along Γ-X is the probable origin of the high chain conductivity, since the chain surfaces around S and Y are small and have high effective masses. Our results imply then that this Fermi surface sheet exists, even though 2D-ACAR and ARPES have so far failed to detect it.[8]

Fig. 4. shows the temperature dependence of ρ_a and ρ_b for crystal #2 (ρ_a/ρ_b = 1.8) and crystal #4 (ρ_a/ρ_b = 1.35). ρ_b has been measured after extending the contacts and ρ_a has been calculated with the data in Fig. 2. The data are analyzed in a simple model of a parallel circuit of planes and chains. Whereas along the **a**-axis only the plane contribution is measured, a measurement along **b** yields the parallel resistivity of planes and chains. With the additional assumption that the plane contribution is isotropic, the chain contribution ρ_{ch} can be calculated from

$$\rho_{ch} = \rho_b/(1-\rho_b/\rho_a)$$

which is also shown in Fig. 4. Besides the planes and chains there is a third conduction path via the bridging O(4) sites, which allows charge to be transferred from the chains to the planes. This path might become important when the chains are interrupted. In this situation ρ_{ch} will be given by the contribution from the complete chain fragments plus the contribution due to the 'detours' through the O(4) sites into the planes and back into the chains. Since motion of the charge carriers along **c** involves large effective masses (i. e. large resistivities) the interruptions will increase the macroscopically measured chain resistivity ρ_{ch} and decrease the anisotropy ρ_a/ρ_b. We propose this mechanism as explanation for the wide variation of the observed values of ρ_a/ρ_b: the chains have a sample dependent amount of interruptions which might be due to inhomogeneities in oxygen content or oxygen disorder. This might also account for differences between our results and values of $\rho_a/\rho_b \approx 2.2$ reported recently[9].

For crystal #2 (ρ_a/ρ_b = 1.8) ρ_{ch} shows metallic behavior with a downward curvature only at temperatures very close to T_c. The magnitude of ρ_{ch} is similar to the plane contribution. In contrast, ρ_{ch} of crystal #4 (ρ_a/ρ_b = 1.35) is about four times as large as ρ_a. After an initial linear decrease of ρ_{ch} with decreasing temperature, there is a tendency of saturation at temperatures below 120 K. No downturn is observed near T$_c$. The strong suppression of the downturn in ρ_{ch} with respect to the behavior of ρ_a and ρ_b for both crystals indicates that superconducting fluctuations are predominantly confined to the CuO$_2$ planes. This conclusion is supported by IR data showing normal state behavior in the chains below 90 K. However NMR data[10] and ARPES data[11] suggest the opening of a gap in the chains. This situation is not resolved at present and may be related to the oxygen content in the chains.

The temperature dependence of the anisotropy ρ_a/ρ_b can in a qualitative way be understood by assuming superconducting fluctuations only in the planes. Using the parallel resistor formula and a linear temperature dependence for the normal resistivity of the chains ($\rho_{ch} = c_0 + c_1 T$) and planes ($\rho_{pl} = a_0 + a_1 T$), we obtain

$$\frac{\rho_a}{\rho_b} = 1 + \frac{1/(c_0 + c_1 T)}{1/(a_0 + a_1 T) + \sigma_{fl}}$$

where σ_{fl} is the fluctuation conductivity. For large T, ρ_a/ρ_b will saturate at $1 + a_1/c_1$ (which for crystals #2 and #4 is equal to 2 and 1.6 respectively, using linear fits to the high temperature values of ρ_a and ρ_{ch} in Fig. 4) while at T$_c$ the anisotropy ρ_a/ρ_b will drop sharply to 1 since σ_{fl} diverges. This model also explains why the downturn due to fluctuations in ρ_b is less pronounced in crystal #2 with large ρ_a/ρ_b than in crystal #4 with a small ρ_a/ρ_b: the low chain resistivity more effectively shorts out the changes in the plane resistivity due to fluctuations.

Fig. 5 shows the temperature dependence of the susceptibility χ in a field of 1 T along the three crystal axes. We note the following points:

1) For H ∥ c χ is essentially temperature independent for temperatures above 200 K, whereas at lower temperatures there is a downward curvature due to superconducting fluctuations.

2) For H ∥ a,b χ shows a weak Curie-Weiss like increase with decreasing temperature with a Curie

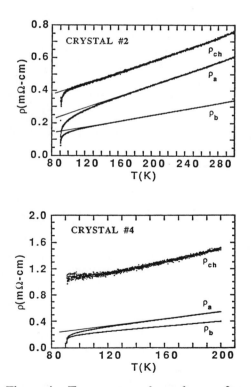

Figure 4. Temperature dependence of ρ_a, ρ_b and ρ_{ch} for crystal #2 (top panel) and crystal #4 (bottom panel). The solid lines are extrapolations of the normal state background.

constant of 6×10^{-6} cm^{3}K/g. Similar behavior has been interpreted[13] as originating from paramagnetic impurities, possibly Cu^{2+} ions in imperfect surroundings. The observed value of the Curie constant would correspond to Cu^{2+} ions with their theoretical effective moment of 1.7 μ_B in about 1.2 % of the unit cells.

3) There is a clear anisotropy $\chi_c > \chi_a > \chi_b$. At high temperatures $\chi_c - \chi_b = 2.3 \times 10^{-7}$ cm^3/g and $\chi_a - \chi_b = 0.5 \times 10^{-7}$ cm^3/g. $\chi_a - \chi_b$ is temperature independent as shown in Fig. 6.

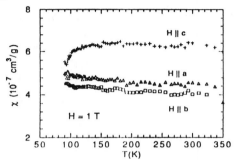

Figure 5. Temperature dependence of the susceptibility in 1 T along the three crystal axes.

Besides the fluctuation and impurity contribution the susceptibility contains also the van Vleck and spin contributions of the local Cu^{2+} moments and the Pauli paramagnetic and Landau diamagnetic contributions of the itinerant carriers. All the latter contributions can be considered[13] temperature independent at temperatures below 300 K. The impurity contribution appears isotropic in the **a-b** plane. Here we make the additional assumption that this is also true for the **c**-direction. Then the temperature dependence of $\chi_c - \chi_a$ is given by the fluctuation diamagnetism. The results are shown in Fig. 7. Shown are also fits to the expressions for the fluctuation diamagnetism χ_{fl} in the low field limit for the 2D and 3D regimes with H $\parallel$ **c**. For the 2D regime χ_{fl} is given by [13]

$$\chi_{fl} = -\frac{2 \pi k_B T \xi_{ab}^2(0)}{3 \Phi_0^2 d} \frac{T_c}{T-T_c}$$

where $\xi_{ab}(0)$ is the zero temperature coherence length in the **a-b** plane and d the CuO$_2$-plane repeat distance. The 3D regime of an anisotropic superconductor can be described with anisotropic Ginzburg-Landau theory. Diamagnetic fluctuation in this regime are given by [14]

$$\chi_{fl} = -\frac{\pi k_B T \xi_{ab}(0)}{6 \Phi_0^2} \sqrt{\frac{M}{m}} \sqrt{\frac{T_c}{T-T_c}}$$

M/m is the effective mass ratio for pair motion perpendicular and in the planes. The fits to these expressions are shown in the Figure. A constant $\chi_{fl,o} = 2.1 \times 10^{-7}$ cm^3/g has been included into the 2D expression in order to match the high temperature data: $\chi_c - \chi_a = \chi_{fl} + \chi_{fl,o}$. The effective mass ratio[4] M/m $\approx$ 50 and the coherence length[14] $\xi_{ab}(0) = 15$ Å have been taken from independent experiments. The data can be described in a qualitative way, a quantitative fit however and a determination of the point of dimensional cross-over are not possible with the present resolution.

Fig. 8 shows χ_a as function of 1/T. A Curie law is followed down to temperatures very close to T_c; no indications of superconducting fluctuations can be detected in this field orientation. This result might be expected[12] since for the **a(b)** orientation fluctuations are suppressed by the mass ratio M/m .

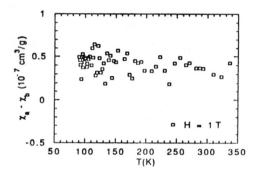

Figure 6. Temperature dependence of the anisotropy χ_a - χ_b.

The anisotropy of the susceptibility at high temperatures is caused by the local Cu-moments and the Landau diamagnetism. In the absence of strong spin-orbit coupling the Pauli paramagnetism is isotropic. Since the plane Cu (Cu(2)) has to first approximation uniaxial symmetry around **c** and the chain Cu (Cu(1)) has to first approximation uniaxial symmetry around **a**, the anisotropy χ_c - χ_b will contain mostly effects due to Cu(2), whereas χ_a - χ_b reflects the anisotropy of Cu(1). The Landau diamagnetism is at this point not well characterized. Since it is an orbital effect and scales like the inverse of the single particle effective mass, it reflects the anisotropy of the Fermi surface. Using an ionic model the anisotropy of the van Vleck and spin contributions of Cu(1) and Cu(2) have been estimated[15] (not including Landau diamagnetism) on the basis of NMR data[10] to χ_c - χ_b $\approx$ 2x10^{-7} cm^3/g and χ_a - χ_b $\approx$ 0.9x10^{-7} cm^3/g. The agreement for Cu(2) is reasonably good, indicating that the ionic model is a valid description for Cu(2). Whether the discrepancy in χ_a - χ_b is due the fact that Cu(1) can not be treated as completely local moment or due to anisotropies of the Landau diamagnetism is open yet. For this field geometry the anisotropy of the Landau diamagnetism for the chains is expected to be negative.

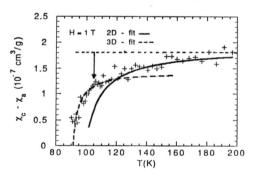

Figure 7. Temperature dependence of the fluctuation diamagnetism in 1 T ∥ **c**. The solid line is a fit to the 2D and the broken line a fit to the 3D expressions for the fluctuation diamagnetism. The dotted line indicates an extrapolation of the normal state from high temperatures. The arrow corresponds to fluctuation diamagnetism.

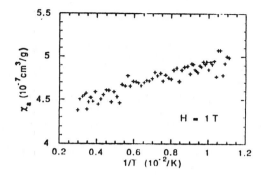

Figure 8. χ_a as function of $1/T$

CONCLUSIONS

Signatures of the CuO chains are observed in the normal state resistivity and magnetization of detwinned $YBa_2Cu_3O_{7-\delta}$. The resistivity along **a** is found larger than along **b** with a sample dependent anisotropy ranging from 1.2 to 1.85. The results show that the CuO chains are metallic and that they can have a conductivity which is comparable to that one of the planes. This indicates that the Fermi surface contains the one dimensional chain electron sheet which band structure calculations predict but which spectroscopic techniques could not detect so far. In $YBa_2Cu_4O_8$ which has a smilar structure but contains twice as many chains than $YBa_2Cu_3O_{7-\delta}$ a resistivity anisotropy of 3 has been reported[16] suggesting that the intrinsic conductivity of CuO chains in this class of materials is high. The large sample dependence of the observed anisotropy of the resistivity is attributed to a high sensitivity of this conduction path along the chains to oxygen vacancies in the O(1) sites. Superconducting fluctuations are much stronger developed in the plane resistivity than in the chain resistivity which is consistent with the picture that superconductivity in $YBa_2Cu_3O_{7-\delta}$ is dominated by the planes. In the normal state susceptibility the anisotropy $\chi_c > \chi_a > \chi_b$ is observed. This anisotropy can be interpreted as arising from the different surroundings of a local Cu-moment in the chain and in the plane sites.

ACKOWLEDGEMENTS

This work was supported by the U. S. Department of Energy, BES--Materials Science under contract #W-31-109-ENG-38 (WKK, SF, JD, YF, GWC, JZL) and the NSF--Office of Science and Technology Centers under contract #STC8809854 (UW, KGV) Science and Technology Center for Superconductivity). SF and KGV acknowlegde partial support from the Division of Educational Programs, Argonne National Laboratory.

REFERENCES

1. U. Welp, M. Grimsditch, H. You, W. K. Kwok, M. M. Fang, G. W. Crabtree, J Z. Liu, Physica C **161**, 1 (1989)
2. A. Umezawa, G. W. Crabtree, U. Welp, W. K. Kwok, K. G. Vandervoort, Phys.   Rev. **B** (in press)
3. Z. Schlesinger, R. T. Collins, F. Holtzberg, C. Feild, S. H. Blanton, U.Welp, G. W. Crabtree, Y. Fang, Phys. Rev. Letters **65**, 801 (1990)
4. D. E. Farrell, J. P. Rice, D. M. Ginsberg, J. Z. Liu, Phys. Rev. Letters **64**,   573 (1990)
5. D. L. Kaiser, F. Holtzberg, M. F. Chisholm, T. K. Worthington, J. Chryst. Growth **85**, 593 (1987)
6. U. Welp, S. Fleshler, W. K. Kwok, J. Downey, Y. Fang, G. W. Crabtree, J. Z. Liu, Phys. Rev **B** (in press)
7. J. Yu, S. Massida, A. J. Freeman, D. D. Koelling, Physics Letters A 122, 203 (1987); S. Massida, J. Yu, A. J. Freeman, D. D. Koelling, ibid. 198 H. Krakauer, W. E. Pickett, R. E. Cohen, J. Supercond. 1, 111 (1988)
8. J. C. Campuzano, L. C. Smdskjaer, R. Benedek, G. Jennings, (preprint, August 1990)
9. T. A. Friedman, M. W. Rabin, J. Giapintzakis, J. P. Rice, D. M. Ginsberg, Phys. Rev. **B** (in press)
10. S. E. Barrett, D. J. Durand, C. H. Pennington, C. P. Slichter, T. A. Friedman, J. P. Rice, D. M. Ginsberg, Phys. Rev. **B 41**, 6283 (1990)
11. J. C. Campuzano, G. Jennings, N. Rivier, A. J. Arko, R. S. List, B. W. Veal, A. P. Paulikas, (preprint, Sept. 1990)
12. R. A. Klemm, Phys. Rev. **B 41**, 2073 (1990)
13. D. C. Johnston, S. K. Sinha, A. J. Jacobson, J. M. Newsman, Physica C **153-155**, 572 (1988) K. Westerholt, H. Bach, Phys. Rev. **B 39**, 858 (1989) W. C. Lee, D. C. Johnston, Phys. Rev. **B 41**, 1904 (1990)
14. R. R. Gerhardts, Phys. Rev **B 9**, 2945 (1974); and references there in
15. D. C. Johnston, J. H. Cho, Phys. Rev. **B** (submitted, July 1990)
16. J. Schoenes, J. Karpinski, E. Kaldis, J. Keller, P. de la Mora, Physica C **166**, 145 (1990)

THE EFFECT OF PINNING ON FLUX LATTICE MELTING

R.S. Markiewicz

Northeastern University, Boston, MA 02115

ABSTRACT

In order to interpret the 'irreversibility line' found in most high-T_c superconductors, it is necessary to understand how pinning modifies the flux lattice melting transition. Here, the effect of pinning is incorporated in a quasi-two-dimensional formulation, which should be appropriate for these highly anisotropic materials. Comparing the results to experiment, it is found that the flux lattice appears to be very different in $YBa_2Cu_3O_{7-\delta}$ than in the Bi- and Tl- layered materials.

THE IRREVERSIBILITY LINE

Unlike conventional superconductors, the critical currents in the new high-T_c superconductors appear to vanish above a 'critical field'[1] B_{irr} which is considerably less than H_{c2}. This irreversibility line can be measured by many different techniques, but all seem to reproduce the same function[2] $B_{irr}(T)$. This line can be identified with the point at which vortices can move freely in the material. However, a number of questions remain to be answered about this line. For instance, does it represent a true phase transition or merely an activated vortex mobility? Is it associated with flux lattice melting[3-8] or with thermal depinning of vortices[9-11]?

The answer to these questions depends on the nature of pinning in these materials, and in particular, on the pinning center density. If the pins are so dense that the flux lattice correlation length R is less than the vortex separation a_0, then no lattice exists and each vortex must be individually depinned. In the other limit, $R >> a_0$, there should be a well defined flux lattice, and the melting transition should significantly enhance the vortex mobility.

It is important to recognize that, since $a_0 \propto B^{-1/2}$, the strong pinning limit will *always* prevail in sufficiently low fields. The question then arises as to whether the magnetic field can be made sufficiently large to cross over into the weak pinning regime. As will be demonstrated below, the answer seems to be that $YBa_2Cu_3O_{7-\delta}$ (YBCO) can easily be brought into this weak pinning limit, while the layered, Bi- and Tl- based superconductors are much closer to the strong pinning limit.

THERMALLY ACTIVATED FLUX FLOW

Recently, a theory of thermally activated flux flow (TAFF) has been developed[11] to describe the physics of flux motion in the vicinity of the irreversibility line. The basic assumption is that the resistivity produced by flux-line motion is thermally activated

$$\rho_{TAFF} \propto exp(-U/k_B T). \tag{1}$$

While this model has been extremely successful in correlating a large body of experimental data, the interpretation of the activation energy U has been open to question. This was originally considered to be the energy associated with depinning a single vortex or flux bundle, but the early calculations neglected the large energy contribution associated with compressing the flux lattice – a depinned vortex line cannot move unless it pushes other vortices out of the way. (This contribution will remain large even in the disordered or fluid phases, since the compressional modulus does not change suddenly at a melting transition.)

When this compressional energy is properly accounted for, a very different picture arises[12,13]. Eq. 1 no longer holds. The effective U becomes dependent on current, $U_{eff} \propto j^{-\alpha}$, so that ρ vanishes in the limit $j \to 0$. To recover a TAFF-like resistivity, it is necessary to incorporate defects (e.g., dislocations)[14] into the flux lattice. In this case, U becomes the energy to nucleate a dislocation or to activate its motion. In either case, U will involve the sliding of vortices by one another. Hence, $U \propto c_{66}$, and so $U \to 0$ at a melting transition.

However, Feigel'man and Vinokur[15] have postulated a 'collective depinning' transition, which could arise prior to the melting transition. They suggest that, as soon as the average thermal displacement of a vortex is equal to ξ, the range of the pinning potential, the vortex will depin, even if $U << k_B T$. I have analyzed this situation, and find that even in the collective pinning regime, the criterion for depinning remains $U \simeq k_B T$, and therefore that collective depinning does not occur[16].

In this case, the irreversibility line should be associated with the melting transition (always assuming that the sample is in the weak pinning regime). However, the irreversibility line need not be the field at which the flux lattice melts. Dislocations could become mobile along grain boundaries (of the flux lattice) at fields significantly below the melting field[17]. One might, indeed, identify B_{irr} with grain boundary melting.

THE EFFECT OF PINNING ON FLUX LATTICE MELTING

Experimentally, pinning has been found to have a very different effect on the irreversibility line in YBCO and in the Bi- layered compounds. In the Bi compounds, introducing pins via irradiation damage shifts the line to higher fields (or higher temperatures, at fixed field)[18]. This is the behavior expected in the strong pinning limit, where each vortex is individually pinned, and increasing the pinning strength should reduce vortex mobility.

However, in YBCO, irradiation increases the strength of the pinning (as measured by j_c), *but leaves the irreversibility line virtually unchanged*[19]. Furthermore, increasing the twin plane density (by substituting Fe for Cu)[20] *decreases* j_c and causes the irreversibility line to shift to substantially *lower* fields (Fig. 1). Both of these features can be understood if the flux lattice is in the weak pinning limit.

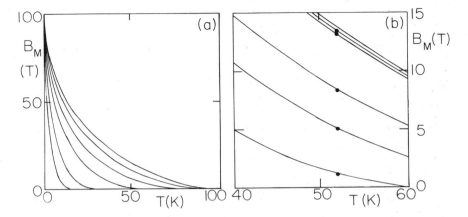

$Fig.$1 (a) Melting phase diagram of YBCO, assuming $H_{c2}(T = 0) = 100T$. Curve with largest $B_M(T)$ corresponds to the pinning-free case; increasing the pinning strength monotonically reduces B_M. Details of the calculation, and the precise pinning levels, are given in Ref. 25. (b) Blow-up of region near T=52K, comparing theoretical curves to data of Ref. 20 (filled circles).

The effect of pinning has been incorporated[21] into the theory of flux-lattice melting in a quasi-two-dimensional limit[22], which should be appropriate for these highly anisotropic materials. It was shown that, in the limit of weak electronic interlayer coupling, the anisotropic Lindemann criterion reduces to a Kosterlitz-Thouless-like melting transition, in a layer of effective thickness $d_{eff} \simeq a_0$, caused by magnetic interlayer coupling. In this limit, the effects of impurities can be incorporated[21,23,24], and they are found to have a most surprising effect. In the weak impurity limit, it is the unpinned impurities which can become mobile (this is why $U \propto c_{66}$). Impurities introduce dislocations into the flux lattice, and these dislocations reduce c_{66}. Hence, the addition of impurities should *reduce* the pinning strength, and *shift the irreversibility line to lower fields*, just as experimentally observed in YBCO.

A detailed analysis[25] of the data of Ref 20 shows that this picture is quantitatively correct. From the doping dependence of B_{irr}, it is possible to calculate how the flux lattice correlation length R varies with doping (Fig. 2). It is found that at all Fe doping levels, R is exactly ten times the average twin spacing (which was independently measured for each of these samples). Moreover, in the Fe-free sample, R is found to be $\sim 1\mu m$, comparable to the value directly measured in flux decoration experiments[26], which were carried out at considerably lower fields.

Finally, this can explain the results of the irradiation experiments. From the Fe doping experiments, it is found that a very large pin center density is required to significantly reduce B_{irr}. Moreover, pinning by twin boundaries is much more effective in reducing R than it is in enhancing j_c. This is because the twins provide only a force opposed to flux lines *crossing* the twin planes, but little resistance for motion parallel to the planes[27]. Hence, the addition of even a few radiation-damage centers can greatly enhance j_c, since the twin planes limit the

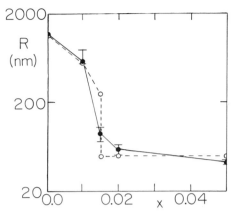

Fig.2 Doping dependence of correlation length in YBCO. Solid circles = values of R derived from analysis of Fig. 1b; open circles = 10 × *D*, where *D* is the average twin spacing, from Ref. 20. Lines are guides to the eye.

ability of unpinned vortices to slide by those that are already pinned. At the same time, it will take a very large pinning center density to reduce R from the value determined by twin planes alone. This would also explain why a large variety of experiments, on many different kinds of samples (films, single crystals, polycrystalline samples) finds virtually the same irreversibility line[2]. (Note that the irreversibility line, Fig. 1, is only weakly affected by the twin density, until the microtwinning regime is entered.)

Why are the Bi-based layered compounds different? A similar analysis[21] of the effect of doping on B_{irr} suggests that these materials may have much smaller values of R, close to the strong pinning limit. In the absence of pinning, the irreversibility line must pass through $B_{irr} = 0$ only at T=T_c. Pinning shifts this transition to a $T^* < T_c$. To make $T^* \simeq 30K$, as found experimentally[28], requires $R \simeq 30nm$ (Fig. 3). Since $a_0 = 48nm/\sqrt{B}$, with B in Tesla, the flux lattice will be in or close to the strong pinning limit for all available magnetic fields. Evidence for small R values for positional disorder seems also to follow from decoration experiments, although there is still long-range orientational order[29].

The difference between the two materials may be a consequence of the greater anisotropy of the Bi-based compounds, but it could also be due to differences in doping. Both of these compounds are modulation doped, with holes introduced into the CuO_2-planes from ionic sites in other layers. In YBCO, these sites are oxygen vacancies in the CuO chain layers, which are relatively remote from the conducting planes. In the Bi-based layered compounds, however, the holes are associated with a variety of site substitutions, including the possibility of Bi on Ca sites. Since these layers are adjacent to the CuO_2 planes, the Bi sites may provide much more effective pinning centers. Note that for such microscopic pins, when the average pin spacing is $\ll \xi$, the vortices will be pinned on fluctuations in the pinning site density[21].

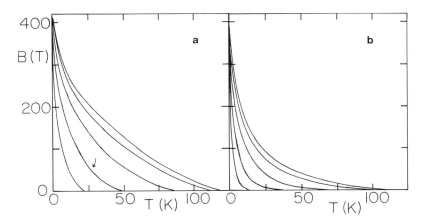

*Fig.*3 Melting phase diagram of Tl-1234, from Ref. 21. Curves are similar to those in Fig. 1a, with (a) and (b) referring to two possible choices of a renormalization parameter. The arrow in Fig. 3a points out the curve which most closely matches the experimental situation, with the flux lattice melting near 50K as $B \rightarrow 0$; this corresponds to $R \simeq 30nm$. (In this respect, Tl-1234 is similar to Bi-2212).

REFERENCES

1. K.A. Müller, M. Takashige, and J.G. Bednorz, Phys. Rev. Lett. **58**, 408 (1987).

2. P. Esquinazi, A. Gupta, and H.F. Braun, to be published, Proc. LT19 (XIX Int. Conf. on Low T. Physics, Sussex, Eng., Aug., 1990).

3. D.R. Nelson and H.S. Seung, Phys. Rev. B**39**, 9153 (1989).

4. M.A. Moore, Phys. Rev. B**39**, 136 (1989).

5. E.H. Brandt, Phys. Rev. Lett. **63**, 1106 (1989).

6. A. Houghton, R.A. Pelcovits, and A. Sudbo, Phys. Rev. B**40**, 6763 (1989).

7. R.S. Markiewicz, J. Phys. C: Solid State Phys. **21**, L1173 (1988).

8. R.S. Markiewicz, J. Phys. Condens. Matt. **2**, 1197 (1990).

9. Y. Yeshurun and A.P. Malozemoff, Phys. Rev. Lett. **60**, 2202 (1988).

10. M. Tinkham, Phys. Rev. Lett. **61**, 1658(1988).

11. P.H. Kes, J. Aarts, J. van den Berg, C.J. van den Beek, and J.A. Mydosh, Supercond. Sci. Tech. **1**, 242 (1989).

12. M.V. Feigel'man, V.B. Geshkenbein, A.I. Larkin, and V.M. Vinokur, Phys. Rev. Lett. **63**, 2303 (1989).

13. M.P.A. Fisher, Phys. Rev. Lett. **62**, 1415 (1989).

14. M.V. Feigel'man, V.B. Geshkenbein, and A.I. Larkin, Physica C**167**, 177 (1990).

15. M.V. Feigel'man and V.M. Vinokur, Phys. Rev. B**41**, 8986 (1990).

16. R.S. Markiewicz, unpublished.

17. H.J. Jensen, A. Brass, A.-C. Shi, and A.J. Berlinsky, Phys. Rev. B**41**, 6394 (1990).

18. A. Gupta, P. Esquinazi, H.F. Braun, H.-W. Neumüller, G. Ries, W. Schmidt, and W. Gerhäuser, Physica C, to be published.

19. L. Civale, A.D. Marwick, M.W. McElfresh, T.K. Worthington, A.P. Malozemoff, F.H. Holzberg, J.R. Thompson, and M.A. Kirk, Phys. Rev. Lett. **65**, 1164 (1990).

20. R. Wördenweber, K. Heinemann, G.V.S. Sastry, and H.C. Freyhardt, Cryogenics **29**, 459 (1990).

21. R.S. Markiewicz, to be published, Proc. LT 19.

22. R.S. Markiewicz, J. Phys. Condens. Matt., **2**, 4005 (1990).

23. S.J. Mullock and J.E. Evetts, J. Appl. Phys **57**, 2588 (1985).

24. D.S. Fisher, Phys.Rev. B**22**, 1190 (1980).

25. R.S. Markiewicz, unpublished.

26. G.J. Dolan, G.V. Chandrasekhar, T.R. Dinger, C. Feild, and F. Holtzberg, Phys. Rev. Lett. **62**, 827 (1989).

27. R. Wördenweber, K. Heinemann, G.V.S. Sastry, and H.C. Freyhardt, Supercond. Sci. Technol. **2**, 207 (1989).

28. P.L. Gammel, L.F. Schneemeyer, J.V. Waszczak, and D.J. Bishop, Phys. Rev. Lett. **61**, 1666 (1988); R.S. Markiewicz, K. Chen, B. Maheswaran, A.G. Swanson, and J.S. Brooks, J. Phys. Condens. Matt. **1**, 8945 (1989).

29. C.A. Murray, P.L. Gammel, D.J. Bishop, D.B. Mitzi, and A. Kapitulnik, Phys. Rev. Lett. **64**, 2312 (1990).

ON ELECTRODYNAMICS OF THE CRITICAL STATE IN SUPERCONDUCTORS

V.V.Borisovskii, L.M.Fisher, N.V.Il'in, I.F.Voloshin

All-Union Lenin Electrical Engineering Institute,
Krasnokazarmennaya st., 12, 111250, Moscow, USSR,
273-32-63, 361-53-84

N.M.Makarov, V.A.Yampol'skii

The Institute For Radiophysics and Electronics, Academy of
Science of Ukr.SSR, Acad. Proskura st.,12, Kharkov, USSR

This report bases on our papers devoted to study of the electrodynamics of high-T_c ceramics in the critical state. It is well known now that the electrodynamics of high-T_c ceramics is described well by the critical state model in a wide region of magnetic fields at low frequencies. This model was proposed by professor Bean about thirty years ago to describe the magnetization curves of hard superconductors. Derch and Blatter /1/ have done the theoretical foundation for the applicability of this equation for high-T_c ceramics. This equation has a form:

$$\mathrm{rot}\mathbf{B} = (4\pi/c)\cdot \mu \cdot j_c(B)\cdot E/E.$$

Here **B** is the magnetic induction, **E** is the electric field, μ is the magnetic permeability of ceramics without intergranular links where intragranular currents are taken into account, $j_c(B)$ is the critical current density.

In the first part I shall consider some results concerning the main characteristic of high-T_c superconductor – the dependence of critical current density on the magnetic field. Unfortunately we have not now the theory described this dependence. So, the information about the function $j_c(B)$ may be obtained from the experimental data only.

The commonly used method of determination of the dependence $j_c(B)$ in ceramics based, as a rule, on measurement of a critical value of the full transport current I_c as a function of an external magnetic field. However, such method has serious fails. Firstly, it is impossible to neglect an inhomogeneous self- magnetic field of the current. So the full current I_c is defined not only

by external magnetic field but by a self-field too. In
addition this method is not convenient because of the
necessity to use the electric contacts.

We have theoretically proposed and have experimentally
realized the new contactless method of determination of
critical current dependence on magnetic field. The method
bases on a simple connection between the critical current
density and the surface impedance of ceramics. It is well
known that the surface impedance is strictly proportional to
the ac field penetration depth δ. On the other hand this
depth is inversely proportional to the critical current
density j_c.

Let us consider a cylindrical sample inserted in the
magnetic field having the form $H + h \cos \omega t$. Here H is a
constant magnetic field and h is a small amplitude of probe
signal. If the amplitude of the probe signal is so small
that we can neglect the dependence j_c on h, $j_c(H+h) \approx j_c(H)$
the surface impedance is connected with j_c by relation

$$\mathcal{Z} = (2/3\pi) \cdot \{ \mu\omega h / [c j_c(\mu H)] \} \cdot (1 - 3\pi i/4).$$

Thus, the measurement of impedance vs H gives the full
information about the function $j_c(B)$.

We realized this method to measure the critical current
density $j_c(B)$ of cylindrical YBaCuO ceramic samples with
different size of grains. The fine-grained sample with
radius $R = 0.25$ cm had average grain size $a \approx 1\mu m$ and the
coarse-grained one with $R = 0.45$ cm had $a \approx 10\mu m$. The real
and imaginary parts of impedance were measured by usual
method. The peak-up coil was wounded directly on a sample by
copper wire 30 μm in diameter. The sample was immersed in
liquid nitrogen. The peaked up signal which is proportional
to the surface impedance was detected by the lock-in
amplifier PAR-124A. The amplitude h and the frequency $\omega/2\pi$
of ac magnetic field were 1 Oe and 1 kHz correspondingly.

The dependence of impedance vs H for the fine-grained
sample is presented on this fig.1 The results were obtained
at frequency $\omega/2\pi = 1$ kHz. The values $\mathcal{R}$ and $\mathcal{X}$ were measured
in units Y_n, where $\mathcal{X}_n$ is reactance of the sample in the

normal state $\chi_n = 2\pi R\omega/c^2$. Using the results of surface impedance measurements and the relation between the impedance and the critical current density it is not difficult to define the dependence $j_c(\mu H)$. These dependencies are plotted on the Fig.2 for both samples. The parameter μ is taken 0.5.

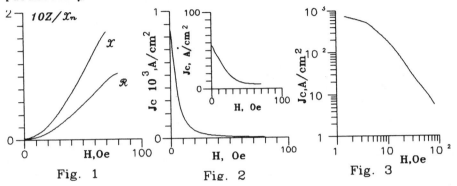

Fig. 1   Fig. 2   Fig. 3

It is found that the function $j_c(H)$ for all of samples has a good approximation by the power function in a wide range of magnetic field. To our surprise the power exponent proves to be the same and equals $-3/2$ for different samples. In the case of fine-grained sample

$$j_c(H) = 5000/H^{3/2} \qquad \text{at } H \geq 5 \text{ Oe}$$

for coarse-grained sample

$$j_c(H) = 2050/H^{3/2} \qquad \text{at } H \geq 15 \text{ Oe}$$

where H and j_c are measured in Oe and A/cm^2 accordingly. In the region of low magnetic fields these relations become invalid. It is clear seen the deviation of the graph from the straight line in the region $H < 5$ Oe on Fig.3 demonstrating the dependence $j_c(H)$ for the fine-grained sample in double logarithmic scale. In order to describe the function $j_c(H)$ taking into account this circumstance we present other, more complicated fit function $j_c(H)$, which is correct for the all magnitudes of H:

$$j_c = 865/[1+(H/H^*)^{7/4}] \qquad H^* = 4.5 \text{ Oe}$$

for the fine-grained and

$$j_c = 55 / [1 + (H/H^*)^{7/4}] \qquad\qquad H^* = 16.5 \quad Oe$$

for the coarse-grained samples.

Let us pay attention to exponents in these relations. In spite of different structure of our sample and great difference in critical current density the exponent value is the same for both types of samples. May be this circumstance is not accidental and it reflects the some unknown objective regularity.

Using these formulae we calculated the dependence of full transport critical current I_c on a sample radius R: $I_c \sim R^{7/5}$. This relation is in a good agreement with our previous data of direct measurements.

Several words about the obtained dependence $j_c(B)$. Independently of ceramics structure the critical current density in the field higher than H^* decreases with field as $B^{-3/2}$ that is much stronger than in Kim-Anderson model. This fact is in contradiction with widely disseminated conception that the behaviour of function $j_c(B)$ must coincide with average dependence of critical current I_{cr} of single Josephson junction on magnetic field. At high fields this averaging has to produce $j_c \sim B^{-1}$. This contradiction may be understood if we take into account that at low magnetic field the first oscillations of the function $I_{cr}(B)$ are essential while the dependence $j_c \sim B^{-1}$ would be observed at higher magnetic fields. Unfortunately, we cannot identify the form of function $j_c(B)$ by described above method at higher fields because from the first critical field of the grains $H_{c1} \approx 50$ Oe magnetic field penetrates into the grains. As a result, the parameter μ becomes a function of magnetic field and increases with H. In these conditions the dependence $\mathcal{Z}(H)$ stops to reproduce the function $j_c^{-1}(H)$.

In conclusion I shall say about physical situation where the critical state model equation leads to the new interesting phenomenon.

Let us consider the situation when a fixed direct transport current I flows along z-axis of the cylindrical sample with radius R. Besides, let there is a magnetic field

parallel to the transport current. This is a typical
forceless configuration and one would think that nothing
interesting phenomenon can exist. However let us consider
this situation in detail.

If the current I is less than critical one the
distribution of current density has a form shown
schematically on a Fig.4a . Let a magnetic field parallel to
z-axis is switched on. In such case azimuthal electric field
E_φ arises in the surface layer according to Faradey law.
This electric field produces current j_φ that will screen an
external field H. It is extremely important that there is
φ-component of electric field in a sample only. The field E_z
equals to zero as long as a transport current density does
not exceed its critical value j_c anywhere. The azimuthal
current j_φ exists in the whole region where external
magnetic field has penetrated while the current component j_z
is absent here. This means that the part of transport
current is displaced to more deep layers (see Fig.4b).

The collapse of a transport current will be continuing
until the external magnetic field achieves its threshold
value H_t when the current density j_z becomes nonzero on axis
of cylinder (Fig.4c). At $H(t)>H_t$ the transport current
density becomes higher then critical one and sample transits
to resistive state. The homogeneous electric field E_z
appeared in a sample leads to transport current flow through
the whole cross section of sample. Thus the critical state
model equation leads to unexpected phenomenon of collapse of
the transport current.

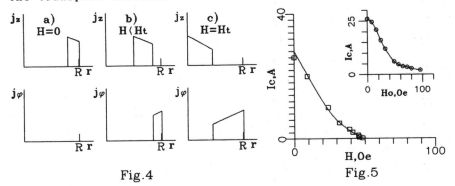

Fig.4   Fig.5

In the framework of the critical state model one can calculate a dependence of magnetic field H_t on a transport current. Using Kim-Anderson relation which takes into account the dependence of a critical current density on a magnetic induction in the form

$$j_c(B) = j_0 / (1 + |B|/B^*),$$

we get following relations

$$H_t = H_p \begin{cases} 1 - 0.5(I/I_c)^{2/3}, & I \ll I_c \\ 3^{-1/2}(1 - I/I_c), & I_c - I \ll I_c \end{cases}.$$

Here H_p is value of an external magnetic field when a magnetic flux achieves cylinder axis in the absence of transport current. The field H_p is equal to

$$H_p = (8\pi j_0 B^* R / c\mu)^{1/2}$$

and critical current value is described by formula

$$I_c = (2\pi j_0 B^* c R^3 / 3\mu)^{1/2}$$

From above consideration it follows that the space distribution of transport current in presence of longitudinal magnetic field H_z has to depend on sequence of their switching on. If one applies the magnetic field to zero field cooled sample before transport current, the current distribution will correspond to Fig.4 where the parameters j_z and j_φ will exchange.

In spite of all our expectations, the results of measurements of z-component magnetic moment by the vibration magnetometer appeared to be independent from current and field order of switching on. The critical value of transport current defined by four-probe method at a fixed magnetic field does not depends on pre-history also. We established that the reason of such distinct contradiction between the experiment and the theory is connected with the small fluctuations of electric field E_z, which exist always in nonequilibrium state of superconductors. These fluctuations lead to the re-distribution of currents in a short time. So, the distributions shown on Fig.4 does not realized. In accord to this consideration, it takes rather quick changing of $H(t)$ to observe the scenario of transport current collapse stimulated by magnetic field. In other words, the induced electric field responsible for the current I displacement deep into a cylinder. has to exceed considerably the fluctuation electric fields. Really, the collapse of transport current is experimentally observed when ac magnetic field $H(t)$ oscillates with frequency about several Hertz.

We studied the critical state experimentally on a cylindrical samples of yttrium HTSC-ceramics. The samples had 0.5 cm in diameter, and the length of 4-5 cm and were cooled by liquid nitrogen. For every fixed I we determined the value of h_o which corresponds to resistive state transition. On the Fig.5 this dependence is shown by points. It is practically insensitive to field frequency in a range $10 - 10^4$ Hz. On the same Fig. the theoretical dependence I vs h_o is presented also. In order to plot the theoretical curve we took the parameter H_p = 50 Oe from the independent experiment. The value of I_c in is picked up in such a manner that the calculated curve should pass near the experimental points.

The experimental dependence I vs H in static conditions in a wide range of magnetic field is presented on insert of the Fig.5. The dependence $I(H)$ is similar to behaviour of critical current in ac magnetic field. The main differences are follows: 1) the values of $I(H)$ exceed the corresponding values of $I(h_o)$ elsewhere in the region $0 < H, h_o < H_p$; 2) the main difference is observed at the condition $H, h_o > H_p$. Here, due to collapse in ac field, transport current equals to zero, while in a static condition superconducting transport current exists even in a field higher than 10 kOe.

The collapse may be observed not only for cylindrical samples but for superconductors rings too. In this case the collapse leads to the damping of induced current which we have observed in experiment.

So, the observed suppression of the critical current by ac magnetic field one can understand in the framework of critical state model. The ac magnetic field displaces the transport current deep into a sample. As a result, current cord has formed, which radius decreases with increase of amplitude h. The sample transits to a resistive state, when a critical current density in a cord achieves its critical value at $h = H_t$.

Refference

1. Derch H., Blatter G. Phys. Rev. B, 38, (1988) 11391.

ANGLE RESOLVED PHOTOEMISSION SPECTRA OF HIGH–T$_c$ SUPERCONDUCTORS: A FERMI-LIQUID BASED ANALYSIS

Hayoung Kim and Peter S. Riseborough
Polytechnic University, Brooklyn, New York 11201

ABSTRACT

Recent high resolution angle resolved photoemission spectra for the high temperature superconductors are analyzed on the basis of a fermi-liquid theory of highly correlated metals. Due to the presence of large manybody effects, the region over which typical Landau fermi-liquid properties hold is small. The spectrum is dominated by the self-energy variation over an intermediate range of energies, and in this region the spectrum bears a close resemblance to that found in the phenomenological marginal fermi-liquid theory of Varma et. al.

INTRODUCTION

The angle resolved photoemission spectra of doped high T$_c$ superconductors has recently been measured with high resolution, by Olson et. al.[1]. These measurements yield an experimental determination of the bands and the <u>k</u> points at which they cross the fermi-surface. The fermi-surface volume derived from these experiments is in good agreement with that obtained from Local Density Function Approximation (LDA) calculations of the electronic structure[2], even though the corresponding experimentally determined band masses are found to be enhanced by a factor of roughly 4. These measurements, when considered together with the recent high field measurements of the de Haas – van Alphen oscillations[3], do indicate that the high temperature superconductors might be described by a Fermi-Liquid based theory. In this manuscript we shall show that the gross features of the Angle Resolved Photoemission Spectra (ARPES) are described by the off fermi-surface properties of a highly enhanced fermi-liquid, and that these features are also quite consistent with the Marginal Fermi-Liquid recently proposed by Varma and co-workers[4].

CALCULATION AND RESULTS

We consider an N fold (spin-orbit) degenerate Hubbard model, characterized by a screened coulomb interaction U between the electrons. The bare electronic density of states is modeled by a square band of width

2W, the discontinuous band edges are characteristics of a two dimensional system. The dimensionless coupling constant is given by the parameter $u = U(N-1)/W$. The electronic spectrum $A(\underline{k}:\omega)$ is given in terms of the one-electron Green's function by $A(\underline{k}:\omega) = - 1/\pi$ Im $G(\underline{k},\underline{k}:\omega)$. The irreducible self-energy is calculated[5] by expanding in powers of $1/N$.   The lowest non-trivial order is equivalent to the Random Phase Approximation[6,7] (RPA). The self-energy $\Sigma(\underline{k}:\omega)$ can be written as

$$\Sigma(\underline{k}:\omega) = u^2/(N-1) \int_0^\infty dx/\pi \sum_{\underline{q}} \text{Im } \Pi(\underline{q}:x)$$

$$\{ [1 - f(\underline{k}-\underline{q}) + N(x)].[\omega - x - e(\underline{k}-\underline{q}) + \mu]^{-1}$$

$$+ [f(\underline{k}-\underline{q}) + N(x)].[\omega + x - e(\underline{k}-\underline{q}) + \mu]^{-1} \}$$

$$(1),$$

where the spectral density for the coupled charge and spin fluctuations Im $\Pi(\underline{q},x)$ is given by

$$\text{Im}\Pi(\underline{q}:x) = \text{Im}\Pi^0(\underline{q}:x) [1 + 1/(N-1) | u \Pi^0(\underline{q}:x) |^2]$$

$$.| 1 + u \Pi^0(\underline{q}:x)|^{-2}.| 1 - u \Pi^0(\underline{q}:x)/(N-1)|^{-2}$$

$$(2),$$

and the non-interacting polarizability $\Pi^0(\underline{q},x)$ is expressed as

$$\Pi^0(\underline{q}:x) = \sum_{\underline{k}'} [f(\underline{k}') - f(\underline{k}'+\underline{q})].[x - e(\underline{k}'+\underline{q}) + e(\underline{k}')]^{-1}$$

$$(3).$$

The summation over wave-vectors that appear in the RPA self-energy is simplified by utilizing the local approximation[5], which is reasonable for the heavily doped superconductors where the fermi-surface does not nest.

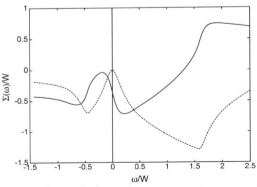

Fig. 1. The real and imaginary parts of the self energy are shown by the solid and broken lines. They are plotted in units of W, versus the dimensionless energy ω/W. The chemical potential is set at $\mu/W = -4/7$, the other parameters are chosen as $N = 14$ and $u/u_c = 0.85$.

The system undergoes a magnetic instability at the critical value of u, $u_c = 2(N-1)$. The real and imaginary parts of the self-energy, for a value of u close to u_c, are shown in figure 1. The characteristic energy scale over which the self-energy shows behavior typical of an enhanced Fermi-Liquid is given by $(1-u/u_c)$ $(W \pm \mu)$. At larger excitation energies the imaginary part of the self-energy shows an approximate linear dependence on ω. The angle integrated photoemission spectrum shows a satellite above and below the upper and lower band edges due to rapid charge fluctuations, in addition to a low excitation energy fermi-liquid peak which is only observable at low temperatures. In figure 2, the angle resolved spectra are shown for various $\underline{k}$ values.

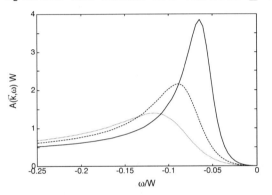

Fig. 2. The spectral density $A(\underline{k}:\omega)$ in units of N/W versus ω/W. The solid, dashed and dotted curves correspond to different $\underline{k}$ values given by $e(\underline{k})/W = -0.5$, -0.6 and -0.7, respectively.

The spectra exhibit an almost constant intensity at higher energies and an asymmetrical peak close to the fermi-energy, the width of the peak scales linearly with ω. For $\underline{k}$ vectors such that $e(\underline{k}) > e(\underline{k}_f)$, the occupied portion of the spectrum shows the tail of the asymmetrical peak located in the unoccupied portion of the spectrum.

DISCUSSION

The ARPES measurements have been shown to be consistent with the off fermi-energy behavior of a highly enhanced Fermi-Liquid. The range of excitation energies over which Fermi-Liquid theory prevails becomes increasingly narrower as the mass enhancement is increased, also the region of linear variation of Im $\Sigma(\omega)$ concomitantly increases. The experimentally observed spectra[1] does show evidence for this linear ω variation and is equally consistent with the description either as an highly enhanced fermi-liquid[5,6,7] or as a marginal Fermi-Liquid[4]. Furthermore, the experimental data[1] does also show evidence of a large constant intensity contribution to the spectrum, which persists for $\underline{k}$ values above the fermi-surface, in accord with the results of this work.

ACKNOWLEDGEMENTS

This work was supported by the U.S. Department of Energy, Office of Basic Energy Science through grant DE-FG01-84ER45127. The authors would like to thank C.G. Olson and A.J. Arko for making reference (1) available to us prior to publication, and we would like to thank I. Dzyaloshinskii, A.P. Kampf and J.R. Schrieffer for enlightening conversations.

REFERENCES

1. C.G. Olson, R. Liu, A.B. Yang, D.W. Lynch, A.J. Arko, R.S. List, B.W. Veal,, Y.C. Chang, P.Z. Jiang and A.P. Paulikas, preprint submitted to Phys. Rev. B and Science 245 (1989)
2. S. Massida, J. Yu, A.J. Freeman and D.D. Koelling, Phys. Lett. A 122, 198 (1987)
3. F.M. Mueller, C.M. Fowler, B.L. Freeman, J.C. King, A.R. Martinez, N.N. Montoya, J.L. Smith, W.L. Hults, A. Migliori and Z. Fisk, Bull. Am. Phys. Soc. 35, 550 (1990)
4. C.M. Varma, P.B. Littlewood, S. Schmitt-Rink, E. Abrahams, and A.E. Ruckenstein, Phys. Rev. Lett. 63, 1996 (1989)
5. P.S. Riseborough, Phys. Rev. B (in press).
6. A.P. Kampf and J.R. Schrieffer, Phys. Rev. B 41 6399 (1990)
7. A.P. Kampf and J.R. Schrieffer, preprint.

WHAT DO MADELUNG POTENTIALS IMPLY FOR HIGH T_c

A.J.Bourdillon and N.X.Tan

Department of Materials Science and Engineering, SUNY,
Stony Brook, N.Y. 11794-2275

ABSTRACT

Some high temperature superconductors require
atmospheres of oxygen for the formation of the
superconducting phase, and others do not. The former
corrode more easily and includes the doped compound
$(La_{1.5}Ba_{0.5})CuO_4$. In the latter class, Cu ions inhabit sites
with large ionic potential, which apparently allows the
formation of high charge states such as Cu(d8) or CuO^+. An
elementary code was used to examine potentials on a number
of defect sites. Thus a high charge state on polyvalent Cu
is a feature common to both types of superconductor. The
use of Madelung potential calculations is reviewed for the
construction of band energy diagrams, and for determining
cohesion energies and stability of defects.

INTRODUCTION

Since the discovery of high temperature superconduction
in doped La_2CuO_4 [1] the chemical states and charge
distribution in this and in other systems have demanded
considerable attention. The many variations in chemical
conditions obtained in the various crystal structures, and
the variety of processing routes required to produce
optimized material with high transition temperatures (T_c)
and critical current densities (J_c), form clues to the
nature of the superconducting mechanism, about which
consensus is as remote as ever.

The most fundamental feature in the bonding of ceramic
materials is site Madelung potential. These potentials
determine a variety of features including cohesion energies
[2], band gaps [3], elastic constants etc. Since the
superconductivity in these materials depends on holes [4],
the nature of the top of the valence band is of critical
importance. This could depend on many features including
defect states, charge distributions and layer reservoirs
etc. The features can be raised above speculation and
conjecture by the application of Madelung potential
calculations, which are simple enough for evaluation to the
accuracy required, about one per cent, on personal
computers.

Band gaps in fluoride perovskites are firstly
considered, to establish methods for understanding the high
T_c superconductors. Roles played by monovalent anions can

be interpreted with more certainty than corresponding roles in oxides. The band gap end energy band considerations lead to conclusions about charge distributions inside various systems, and to speculation about the nature of the superconductivity.

CALCULATIONS

The Madelung potential, V_i, at any site, i, can be calculated conventionally by summing potentials due to point charges, q_j, from within one unit cell and from shells of surrounding cells, separated from the first by integral numbers of lattice constants, $h\underline{a}, k\underline{b}, l\underline{c}$:

$$V_i = \sum_{h,k,l=0}^{n} \sum_{i,j}{}' \frac{q_j}{\underline{r}_j - \underline{r}_i} \qquad\qquad 1$$

where the prime indicates a null value when $\underline{r}_j = \underline{r}_i$, $\underline{r}_i$ represents the position vector of the ith site with coordinates u_i, v_i, w_i in the unit cell, and

$$\underline{r}_j = ((u_j - ha)^2 + (v_j - kb)^2 + (w_j - lc)^2)^{1/2} \qquad\qquad 2$$

is the position vector of the q_jth charge in the unit cell at $h\underline{a} + k\underline{b} + l\underline{c}$.

The chief problem lies in convergence. As n increases the potential due to each new atom decreases as r^{-2} but the number of contributing atoms increases as r^2. However the potentials due to dipoles and multi-poles fall off faster than r^{-2} and the rate at which convergence is reached depends on the sizes of unit cells. Then there can be systematic distortions due to surface charges, which without precautions can represent enormous potentials generated across the material. In centrosymmetric cells, the surface charges can be compensated by the use of Evjen's method of fractional charges [5], but in non-centrosymmetric systems, careful checks must be used to monitor any changes in potential calculated at equivalent sites. The simplicity of these techniques leads to unambiguous results and they are easily adapted to examinations of defects and surfaces. The main uncertainties then lie in the ionicities of the materials, in small repulsive hard-sphere energies and finally in the effects of itinerant charge. In later work we shall address these problems, but will meanwhile consider what can be learnt from idealized potentials firstly as applied to fluoride perovskites and secondly to oxides.

a) bandgaps in fluoride perovskites.
The following calculations demonstrate the applicability of Madelung potentials to the determination of band gaps by comparison of optical spectra from fundamental edges of

some fluoride perovskites, KMgF$_3$, KNiF$_3$ and RbMnF$_3$.   These
are ionic insulators.   Term diagrams for the core, valence
and conduction bands of these solids can be constructed by
adjusting known respective free atom levels by Madelung
potentials (Figure 1).

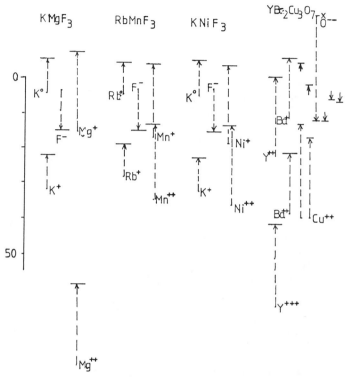

Figure 1. Term diagrams for KMgF$_3$, RbMnF$_3$ KNiF$_3$ and YBa$_2$Cu$_3$O$_7$
(table II, column 2).   Upper levels are broadened by band
structure effects.   Terms (full lines) result from
ionization levels shifted by Madelung potentials (broken
lines).

In the solid, valence and conduction terms are broadened
by crystal field effects. On such diagrams the fundamental
band gaps seem at first sight anomalously large but they
are in fact modified by many body effects, as the
calculated energy for a charge transfer exciton shows: The
energy, E_g, of the exciton at the fundamental edge is the
energy required to transfer an electron from the anion to
a neighboring cation.   This depends on the Madelung
potentials at the two sites, V_c and V_a, on the ionization
energy of the cation, I_c, the electron affinity of the
anion, I_a, and finally on the local change in energy due to
the charge equivalent of local Schottky disorder:

$$E_g = |V_c| + |V_a| - I_c + I_a - e/r_{ac}, \qquad\qquad 3$$

where e is the electronic charge in atomic units and r_{ac} is the distance separating neighboring anions and cations. In table I calculated and measured [6] values for the charge transfer excitons are compared.

TABLE I

| compound | transition | exciton energy | | ratio |
		calculated	observed	calc/ob
KMgF$_3$	F-K	15.29	11.83	1.29
	F-Mg$^+$	15.58	-	
KNiF$_3$	F-K	15.41	11.75	1.31
	F-Ni$^+$	11.89	10.10	1.17
KCoF$_3$	F-K	15.19	10.93	
	F-Co$^+$	12.64	10.30	1.23
KMnF$_3$	F-K	14.47	10.42	1.29
	F-Mn$^+$	13.32	11.62	1.25
RbMnF$_3$	F-K	14.67	11.47	1.28
	F-Mn$^+$	13.00	11.11	1.17

The calculated values are over-valued by 25% on average, due to neglect of hard core repulsion and other approximations mentioned above. However **comparative** changes in low lying conduction band states from alkali metal to transition metal states are faithfully reflected in both calculated and observed [6] values. For example KMgF$_3$ has a K-like conduction band, while KNiF$_3$ has a conduction band based on Ni. Further confirmation of this is obtained from core exciton spectra [6]. Thus within defined limits, elementary band structure models can be constructed, though these depend on a correct determination of charge distribution which is not always clear cut, as in some examples below.

b) Oxide perovskites

Whereas the F$^-$ ion is stable in vacuo, with a known electron affinity, O^{--} is not. I_a in equation 3 becomes a negative repulsive energy, which can be estimated from known data on CuO. Using formula 3 and treating I_a as unknown, the equation solves if I_a =-17.0 eV, though the true value may be only 80% of this value. (For SrTiO$_3$ the equivalent calculated value for I_a is 19.4 eV). Using a semi-empirical value of -17.0 eV it is possible to study cohesive energies and charge distributions in various high temperature superconductor and associated systems.

In YBa$_2$Cu$_3$O$_{7-x}$ (YBCO) The sum of cationic charges, as normally found, is odd while the anionic charges are all even. In such chemically complex systems, knowledge is first required about the most stable charge distribution, so that Madelung potentials are optimized with respect to maximum cohesive energies. These energies are the sum of Madelung energies, ionization energies and electron affinities. It is known that in ionic materials cohesive energies are well represented by Madelung potentials, since hard-core repulsive energies are an order of magnitude smaller [2].

Calculated Madelung potentials of various superconductor and associated systems are shown in tables II, III and IV. Also shown are the mean Madelung energies per ionic charge calculated for each system and cohesive energies calculated from summing Madelung energies, ionization energies and electron affinities, but neglecting core repulsion.

TABLE II

Calculated Madelung Potentials/V in YBCO with various charge states. Shown also are mean energies/eV per ionic charge (MEPIC) and cohesive energies (CE). sites as in ref. 7.

atom site	Orthorhombic with				Tetragonal with	
	Cu^{3+} in chains	$Cu^{2.5+}$ in planes	O^- in chains	$O^{1.5-}$ in planes	$Cu^{1.5+}$ in planes	Cu^+ in chains
Y	-31.4	-22.1	-30.6	-19.9	-39.0	-29.6
Ba	-13.8	-16.4	-12.6	-15.7	- 8.6	-11.3
Cu1 plane	-28.7	-25.8	-27.8	-19.6	-29.8	-26.9
Cu2 chain	-21.6	-22.3	-10.5	-21.7	1.4	6.3
O1 plane	19.7	28.3	20.6	26.4	13.9	21.4
O2 plane	19.6	28.1	20.5	26.4	13.9	21.4
O3 Ba	25.0	21.9	25.1	22.2	13.9	25.4
O4 chain	33.7	22.8	24.4	23.4		
MEPIC	-11.7	-11.9	-10.9	-10.9	-9.5	-10.7
CE	-246.0	-251.6	-221.3	-221.3	-170.7	-199.5

Several conclusions can be drawn regarding the most stable charge configuration:
1. From those configurations calculated, the one with the greatest Madelung energy occurs with high charge states on Cu planes (adjacent to the layer of Yttrium ions). This is the plane commonly thought to be most responsible for superconducting properties because of similarities with $(La,Ba)_2CuO_4$ (LBCO) and with the $Bi_2Sr_2Ca_nCu_{n+1}O_{6+2n}$, n=0,1,2 system (BSCCO). This configuration also has the highest cohesive energy compared with free atom states, if core repulsion is neglected. The core repulsion will be higher in the first two columns of table II owing to the greater density of electrons in these configurations.
2. Cohesion energies are calculated to be less when holes

are placed on O⁻ ions though the difference will be less if core repulsion is reduced.

TABLE III

Calculated Madelung Potentials/V in BSCCO and mean energies/eV per ionic charge (MEPIC). Sites as in ref. [7]

atomic site		2201 phase (n=0)	2212 phase (n=1)	2223 phase (n=2)
Bi		-14.3	- 7.3	- 4.0
Sr		-30.9	-24.2	-20.9
Cu1		-51.6	-47.2	-43.9
Ca1			-44.5	-41.3
Cu2				-43.9
O1	Bi plane	31.5	38.5	41.7
O2	Sr "	9.2	15.9	19.2
O3	Cu "	-1.71	.6	3.9
O4				3.8
MEPIC		-11.7	-10.0	-10.3

3. In BSCCO charge distribution is at first sight comparatively unambiguous. However when the band structure is constructed from the data in table III, and modeled on figure 1, anomalies appear: the Bi conduction band occurs well below the oxygen and copper valence bands and the oxygen valence bands are too widely scattered in energy, especially those based on the O1 site. The anomalies can be greatly reduced by introducing Bi vacancies and increasing cationic charge on Cu planes. This procedure increases the potentials on the Bi sites and is consistent with electron microscope reports of superlattices [8] and vacancies [9]. Such vacancies are in turn consistent with weak Madelung potentials on the Bi site.
4. The Madelung potentials for the Tl-Ba-Ca-Cu-O system are similar to those for BSCCO, though in these cases, singly valent Tl⁺ can take the place of Bi vacancies. This would explain why superlattices in this system are much weaker.
5. Some high temperature superconductors require oxygen environments during processing, such as YBCO [10] and LBCO [11]; while others, such as BSCCO, require low oxygen pressures [12,13]. This implies that oxygen vacancies poison superconductivity in the former systems, but are beneficial in the latter. Increases in oxygen vacancy concentration will increase cationic vacancies through the operation of Schottky defects. The Bi plane could then work as a charge reservoir to induce holes on the CuO_2 plane, but the mechanism remains unclear.

TABLE IV

Calculated Madelung Potentials/V in LBCO with various charge distributions and mean energies/eV per ionic charge (MEPIC). Sites as in ref.7

atomic site	La_2CuO_4	$(La_{1.5}Ba_{.5})^{2.75+}CuO_4$ with $Cu^{2.5+}$, O^- in		
			La plane	Cu Plane
$(La_{1.5}Ba_{.5})$	-24.9	-23.8	-22.6	-23.6
Cu	-29.5	-33.0	-29.4	-27.7
O1 La plane	23.1	22.0	21.0	21.7
O2 Cu "	20.5	22.1	20.2	19.0
MEPIC	-12.0	-12.2	-10.6	-10.6

6.   Systematic trends which relate the T_cs of these materials to crystal potentials and which contain explanatory force, are difficult to identify. In all of the superconducting systems discussed above the Cu ion is contained in a significantly deeper potential well than the other cations. This provides an inducement to high ionicity and the creation of holes on Cu.

7. These calculations exclude some models already proposed for superconductivity, such as the peroxiton $(Cu^+O^-_2)$ model [14] in YBCO, unless core repulsion has here been wrongly estimated. As calculated, the cohesion energy implied by this model is unstable.

In conclusion, Madelung potentials provide explanations at an elementary and unambiguous level for several of the phenomena observed in high temperature superconductors, including states of maximum cohesion. Further work is needed to make the results more precise through the measurement of core repulsion from elastic constants.

REFERENCES

1. J.G.Bednorz and K.A.Muller, Z. Phys. B-Cond. Matter, 64, 189 (1986)

2.   M.P.Tosi Solid State Physics 16 1 (1964)

3. N.F.Mott and R.W Gurney, Electronic Processes in Ionic Crystals (Clarendon, Oxford, 1948), p.98

4. W.A.Groen, D.M.de Leeuw and L.F.Feiner, Physica C, 165, 55 (1990)

5. H.M.Evjen, Phys. Rev. 39, 675, (1932)

6. J.H.Beaumont, A.J.Bourdillon and J.Bordas, J. Phys. C: Solid State Phys., 10, 333 (1977)

7. N.X.Tan and A.J.Bourdillon, Int. J. Mod. Phys. B4, 517 (1990)

8.   N.X.Tan,   A.J.Bourdillon,   S.X.Dou,   H.K.Liu   and C.C.Sorell, Phil. Mag. Lett., 58, 149 (1988)

9. Y.Matsui, H.Meada, Y.Tanaka and S.Horiuchi, Jpn. J. Appl. Phys., 27, L361 (1988)

10. K.E.Easterling, C.C.Sorrell, a.J.Bourdillon, S.X.Dou, G.J.Sloggett and J.C.MacFarlane, Materials Forum, 11 30 (1988)

11. T.M.Tarascon, C.H.Greene, B.G.Bagley, W.R.McKinnon, P.Barbour and G.W.Hull, Novel Superconductivity, eds. S.A.Wolf and V.Z.Krezin, (Plenum 1988) p.705

12. S.X.Dou, H.K.Liu, A.J.Bourdillon, M.Kviz, N.X.Tan and C.C.Sorrell, Phys. Rev. Lett. B40, 5266 (1989)

13. M.Mimura, H.Kumakura, K.Togano and H.Maeda, Appl. Phys. Lett., 54, 1582 (1989)

14. B.K.Chakraverty, D.D.Sarma and C.N.R Rao, Physica C, 156, 413, (1988)

Nonlinear interaction between high-T_c superconductors and weak external fields

Mark Jeffery and Robert Gilmore

Department of Physics and Atmospheric Science
Drexel University, Philadelphia, PA 19104

The magnetic properties of crystalline superconductors in an externally applied field can be modeled by a system of interacting weak link supercurrent loops. The quantum state of the interacting system is described by a set of integers n_i. In each quantum state $\mathbf{n}$ the magnetic flux through each independent loop is quantized. The magnetic field dependence of the energy in each quantum state, $E(\mathbf{n}, \mathbf{B}_{ext})$, is computed in terms of classical parameters (mutual inductance matrix), quantum numbers $\mathbf{n}$ and the external magnetic field $\mathbf{B}_{ext}$. Transitions between states are driven by absorption and emission of microwave quanta. Back reaction of the network on the field eliminates 'frustration.' The model is applied to bulk and single crystal high T_c superconductors and the results are in qualitative agreement with microwave absorption experiments. The theory predicts the possibility of energy bands in superconductors.

I. INTRODUCTION

There exists a large amount of microwave absorption experimental data on single crystal and bulk high T_c superconductors[1-9]. These data demonstrate the remarkable reproducible magnetic properties of these materials at and well below the critical temperature. Traditionally these experimental results are explained by solving the Ginzburg-Landau phenomenological equations. Unfortunately the derivation of these equations is only valid at temperatures close to T_c. In this paper we take a different point of view.

The pairing theory put forward by Bardeen, Cooper, and Schrieffer for low temperature superconductors does not work at high temperatures because the phonon mechanism is too weak. However, the theory works for many materials at low temperatures and we may assume the basic idea is correct although the mechanism is not known. If we assume that paired electrons form a Bose condensate then at temperatures well below the transition temperature we can write the wave function of the superconducting state as[10]

$$\Psi = \sqrt{\rho}\, e^{i\phi} \tag{1}$$

where the wave function is normalized to the charge density, ρ, in the superconductor. This macroscopic wave function contains the long range phase coherence information that is responsible for the superconductor's properties in an externally applied magnetic field.

Microwave absorption experiments measure magnetization and susceptibility as a function of applied microwave power. If the power is increased at a constant rate the measured quantities are proportional to derivatives of the energy of the sample with respect to the external field. The fundamental problem is to calculate the energy of the wavefunction (1) as a function of applied magnetic field subject to the multiply connected topology of the superconductor. In general the externally applied field will induce macroscopic screening currents, and these screening currents will induce magnetic fields which then induced more currents. The calculation of energy is therefore a difficult self consistent problem.

In Sec. II we review the properties of a single loop of superconductor containing a weak link and calculate the energy of the loop as a function of externally applied magnetic field. The calculation takes into account the back reaction of the sample on the field. In Sec. III we extend the calculation to a system of N-interacting loops connected together by weak links, and calculate the ground state energy of a typical network. The theory is then applied to a 2-loop model of a single crystal of high temperature superconductor and compared to microwave absorption experiments in Sec. IV.

II. THE ENERGY OF A SUPERCONDUCTING RING AS A FUNCTION OF FIELD

If the wavefunction (1) is required to be single valued we are lead directly to the quantization of flux in a solid ring of superconductor. The phase change of the wavefunction in one cycle around the ring must equal an integer times 2π, but the change in phase of the wavefunction in one cycle is the magnetic action integrated around the loop

$$\frac{2e}{\hbar} \oint \mathbf{A} \cdot ds = \Phi \frac{2e}{\hbar} = 2\pi n \tag{2}$$

so the flux is quantized in units of $\Phi_0 = h/2e$. Once the flux is trapped in the ring the only way to change it is to heat the ring above the transition temperature.

A more interesting case exists when the ring contains a "weak link" schematically shown in Fig. 1. Flux can then slip through the link and the total entrapped flux can change as a function of applied external magnetic field. We define a weak link as a region where flux can penetrate and yet macroscopic supercurrents can still flow[11]. The weak link is therefore not necessarily a true Josephson junction.

The ground state energy of the weak link ring can be calculated if we require the wavefunction to be single valued. The condition of single valuedness is equivalent to requiring the quantization of flux threading the ring. The quantized flux is equal to the flux from the external field, $\mathbf{B}_{ext} \cdot \mathbf{A}$, less the flux due to the macroscopic screening current or

$$n\Phi_0 = \mathbf{B}_{ext} \cdot \mathbf{A} - Li \tag{3}$$

Here i is the induced current and L is the geometric self inductance of the ring. The ground state energy of the ring can be calculated since

$$E = \frac{1}{2} Li^2 \tag{4}$$

or on inverting (3) for the current

$$E = \frac{1}{2} L^{-1} (\mathbf{B}_{ext} \cdot \mathbf{A} - n\Phi_0)^2 \tag{5}$$

FIG. 1. A schematic diagram of a single superconducting loop containing a weak link. The external magnetic field is applied perpendicular to the loop area.

The energy is therefore a quadratic form in the externally applied field and in the quantum number n. Fig. 2 shows a plot of solutions to (5) given different flux quantum numbers. In this simulation $\Phi_0 = 20.7 G\mu m^2$, $A = 10\mu m^2$ directed perpendicular to B_{ext}, and L is 10 in the scale used. The ground state energy can be obtained by connecting the bottoms of the parabolas. Note that where the parabolas cross corresponds to emission or absorption of a microwave photon. The microwave photon just changes the trapped flux by one quantum.

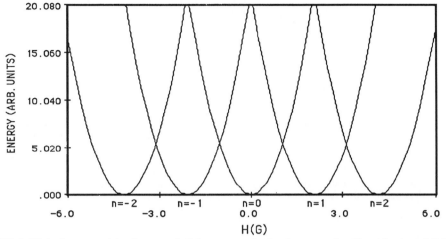

FIG. 2. Plot of energy states for a weak-link superconducting loop as a function of external field. The integer n labels the number of fluxons trapped in the ring. For a given state n the parabola defines the energy of the ring due to the external field and induced shielding current. The parabola crossings correspond to external field values where the loop can emit or absorb a microwave photon.

The general expression for the energy of the ring is a little more complicated than (5) since the weak link has a capacitance. The coupled inductor and capacitance can be excited by photons with energy $\hbar\omega_c$, where $\omega_c = 1/\sqrt{LC}$. If we neglect the zero point energy of the oscillator the Hamiltonian matrix for the weak link ring can be written as

$$<n',m'| H_o | n,m> = (1/2)L^{-1}(B_{ext} \cdot A - n\Phi_0)^2\delta_{nn'}\delta_{mm'} + m\hbar\omega_c\delta_{nn'}\delta_{mm'} \qquad (6)$$

so the total energy of the ring depends upon the two quantum numbers n for trapped fluxons and m for ring-oscillator modes. If we include in (6) the nearest-neighbor coupling of quantum states differing by $n=\pm 1$ quantum of flux then we are led to a total Hamiltonian matrix of the form

$$H = H_o + H_{int} \qquad (7)$$

where the matrix elements of the interaction hamiltonian are given by

$$<n',m'| H_{int} | n,m> = \varepsilon C_{mm'}\delta_{n'n\pm 1} \qquad (8)$$

Here $C_{mm'}$ is the coupling coefficient between ring-oscillator modes and ε is the interaction energy between flux states differing by one quantum of flux. If the matrix (7) is diagonalized as a function of magnetic field the resulting plot (Fig. 3) shows energy bands. In this calculation A and L are the same as those used for Fig. 2, ε is

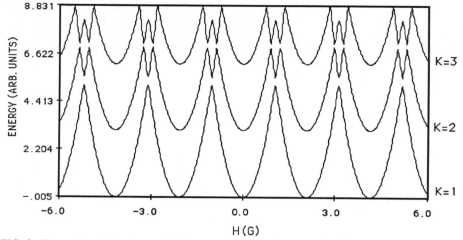

FIG. 3. Energy bands for the weak-link superconducting loop as a function of external field. The ground state and the first two excited states are shown. Energy bands are generated when the interaction between flux states different by ±1 quantum is included in the Hamiltonian.

$0.2\hbar\omega_c$, and the coupling coefficients are calculated from[11]

$$C_{mm'} = \sum_{r=0}^{\min\{m,m'\}} (m!m'!)^{\frac{1}{2}}[r!(m-r)!(m'-r)!]^{-1}\lambda^{(m+m'-2r)} \tag{9}$$

where λ is a dimensionless parameter set to 0.5, and the combinatorial factors arise because photons are identical bosons. In this calculation n and m are clearly not good quantum numbers and can be replaced by a band quantum number K.

T. D. Clark has measured energy bands like those shown in Fig. 3 for a weak link ring of niobium in the superconducting state. This result is a remarkable verification of quantum mechanics operating on a macroscopic scale[11].

III. A SYSTEM OF N-INDEPENDENT LOOPS

We can generalize the calculation of the preceding section to N-independent loops since the magnetic flux in each loop must be quantized. The quantized flux in the r-th loop consists of two parts, the flux due to the uniform external field and the flux due to current flow in all superconducting loops. This flux depends linearly on the currents in each loop through the classical mutual inductance matrix so we can write

$$n_r\Phi_o = B_{ext}\cdot A_r + \sum_s \Lambda_{rs} j_s \tag{10}$$

where n_r is the quantum number of flux in the r-th loop, A_r is the directed area of the r-th loop and j_s is the supercurrent in the s-th loop. We can invert this equation for the current in each loop

$$j_s = \sum_r \Lambda_{sr}^{-1}[n_r\Phi_o - B_{ext}\cdot A_r] \tag{11}$$

The energy associated with the current flow is

$$E = \frac{1}{2}\sum_{r,s} j_r\Lambda_{rs}j_s \tag{12}$$

or substituting (11)

$$E(\mathbf{n},\mathbf{B}) = \frac{1}{2}\sum_{r,s}[n_s\Phi_o - A_s\cdot B_{ext}]\,\Lambda_{sr}^{-1}[n_r\Phi_o - A_r\cdot B_{ext}] \tag{13}$$

In each quantum state $\mathbf{n} = (n_1, n_2, \ldots, n_k)$ the network energy is quadratic in the external magnetic field B_{ext}.

A typical graph of $E(\mathbf{n}, B_{ext})$ vs. B_{ext} is shown in Fig. 4. A total of 15 loops of average area $57.5\mu m^2$ and standard deviation $27\mu m^2$ were distributed at random with random orientation throughout a volume $10^6 \mu m^2$. The mutual inductance matrix for this geometry was then estimated. The network energy was minimized to find the parabolas closest to the lowest energy. These parabolas were then plotted. The minima over all energy parabolas is the ground state of the system.

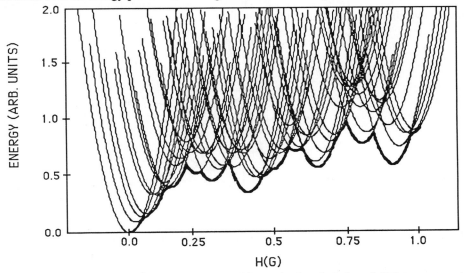

FIG. 4. Energy states for a system of 15 independent randomly oriented weak-link supercurrent loops. The ground state of the system can be obtained by connecting the bottoms of the parabolas.

If we assume that each loop has identical ring-oscillator modes then the general Hamiltonian of the system will have the form

$$\langle\mathbf{n'},m'|\,H\,|\mathbf{n},m\rangle = E(\mathbf{n},\mathbf{B})\delta_{nn'}\delta_{mm'} + m\hbar\omega_c\delta_{nn'}\delta_{mm'} + \varepsilon C_{mm'}\delta_{n'n\pm1} \tag{14}$$

where the interaction is between quantum states $\mathbf{n} = (n_1, n_2, \ldots, n_k)$ and $\mathbf{n'} = (n_1', n_2', \ldots, n_k')$, where all quantum numbers $n_i' = n_i$ except for one, with $n_r' = n_r \pm 1$.

Diagonalization of (14) as a function of external field will produce complicated energy bands.

The theory developed in this section can be applied to microwave absorption experiments on single crystal and bulk high temperature superconductors. In a bulk superconductor, experiments suggest that large supercurrent loops exist in the sample. The current loops are probably connected together by weak links at the grain boundaries of the crystals. In the low temperature limit the behavior of the superconductor as a function of magnetic field must be defined by the Hamiltonian (14). The groundstate energy of the system should then look very similar to Fig. 4 with the microwave absorption showing "random" peaks corresponding to the transitions between lowest energy parabolas.

IV. APPLICATION TO A SINGLE CRYSTAL OF YBaCuO

A simple model of a single crystal microwave absorption experiment is schematically shown in Fig. 5. We assume that two shielding supercurrent loops exist on adjacent faces of the crystal. Experiment suggests that the loops are connected by a weak link on the edge of the crystal[12]. Two perpendicular fields, H_{dc} and $H_{\mu w}$, are applied to the crystal. The fields are along the x and z axes, and the crystal is rotated through an angle Ø about the y-axis. The DC field is modulated by a small AC field for phase sensitive detection. Typically the amplitude of the microwave field is held constant and the DC field is scanned. The resulting spectrum is therefore modulated microwave absorption as a function of DC field.

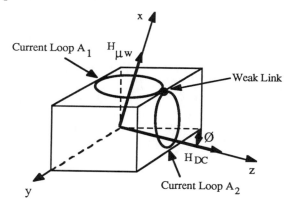

FIG. 5. A schematic diagram of a single crystal microwave absorption experiment. The $H_{\mu w}$ and H_{dc} fields are oriented along the x and z axes respectively. The AC modulation field (not shown) is in the direction of H_{dc}. The crystal is rotated at an angle Ø about the y axis. Two supercurrent loops are connected by a weak link on the edge of the crystal.

We assume that the microwave frequency is sufficiently low that flux slip will occur in a small fraction of a cycle. The microwave field is therefore assumed to have a constant RMS value set to 0.8 Gauss in all calculations. We have carried out a large number of simulations on two loop networks. Fig. 6 shows a typical graph of Energy vs H_{dc} for Ø=7° from 0 to 5 Gauss. In this calculation the areas of the two loops were taken as $A_1 = 50\mu m^2$ and $A_2 = 30\mu m^2$ and the mutual inductance matrix was approximated by

$$\Lambda = \begin{bmatrix} 10 & -0.01 \\ -0.01 & 5 \end{bmatrix} \tag{15}$$

The energy $E(\mathbf{n}, \mathbf{B}_{ext})$ is calculated from (13) with $\mathbf{B}_{ext} = (H_{\mu w}, 0, H_{dc})$. Two quantum numbers, $\mathbf{n} = (n_1, n_2)$, label each parabola in the ground state of the system. The ground state energy can be obtained by connecting the bottoms of the parabolas.

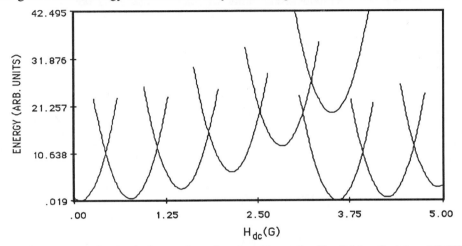

FIG. 6. Energy states for the single crystal as a function of increasing H_{dc} field on the interval [0,5]G. The $H_{\mu w}$ field is assumed constant at 0.8G. The crystal is rotated at an angle $\emptyset = 7°$ to the perpendicular $H_{\mu w}$ and H_{dc} fields. The parabolic increase of the energy states as a function of field will generate quadratically spaced microwave absorption peaks.

Fig. 6 demonstrates a number of interesting properties of the model. The energy of each parabola increases quadratically as the DC field is increased until a state with different flux quantum numbers is energetically more favorable. This quadratic increase is a function of the angle $\emptyset$ which couples the fields to the loops. If $\emptyset$ is zero then the scan looks similar to Fig. 2 for a single loop. In Fig. 2 the parabola crossings are evenly spaced, and therefore the microwave absorption peaks are periodic. In general the H_{dc} and $H_{\mu w}$ fields are not aligned with the crystal and the energy of the parabolas increases as shown in Fig. 6. Therfore, the microwave absorption peaks grow further apart quadratically. This result explains recent experiments on single crystal YBaCuO which have shown the quadratic increase of absorption peak spacing[13].

If the simulation is carried out by first fixing the H_{dc} field and then scanning the amplitude of the $H_{\mu w}$ field we obtain a similar graph to Fig. 6. Experiments have also shown that similar microwave absorption traces are obtained if the DC field is made constant and the amplitude of microwave field is scanned[14].

The measured quantities in microwave absorption experiments are proportional to derivatives of the energy with respect to the increasing field. If the H field is scanned at a constant rate the field depends linearly upon time

$$H = H_o + \frac{\Delta H}{T_o} t \tag{16}$$

where T_0 is the period of the field scan. The microwave power absorbed is then given by

$$P = \frac{dE}{dt} = \frac{\partial E}{\partial H}\frac{dH}{dt} = \frac{\Delta H}{T_0}\frac{\partial E}{\partial H} \tag{17}$$

Modulating the absorption is equivalent to taking the derivative of the power with respect to the field. The modulated absorption, χ, can be calculated from

$$\chi = \frac{\partial P}{\partial H} = \frac{\Delta H}{T_0}\frac{\partial^2 E}{\partial H^2} \tag{18}$$

The modulated absorption is therefore proportional to the second derivative of the energy with respect to the field.

   If we include in the single crystal model the ring oscillator modes and the interaction between flux states $\mathbf{n}=(n_1,n_2)$ different by ± 1 quantum we can calculate the energy bands in the system. The Hamiltonian (14) was diagonalized as a function of the H_{dc} field on the interval [-2,2] Gauss. The parameters used in this calculation are identical to those used for Fig. 6. with the interaction energy, ε, between flux states set to $0.8\hbar\omega_c$. The resulting plot of the first three energy bands is shown in Fig. 7, and the second derivative of the energy bands is shown in Fig. 8.

   The possibility exists that energy bands similar to those shown in Fig. 8 can be observed using microwave absorption experiments.

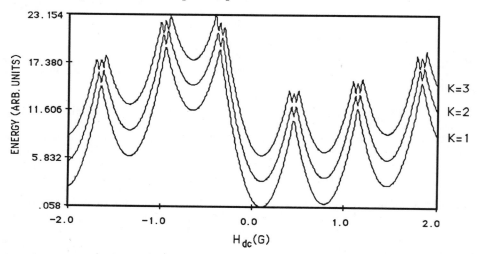

FIG. 7. Energy bands for the single crystal as a function of increasing H_{dc} field on the interval [-2,2]G. The ground state and the first two excited states are shown.

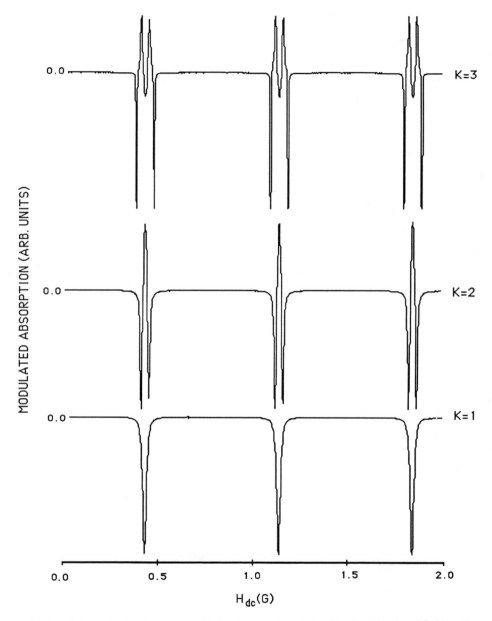

FIG. 8. The second derivate of the Energy bands shown in Fig. 7 as a function of external field on the interval [0,2]G. The second derivate of the energy is proportional to the modulated microwave absorption measured experimentally.

V. CONCLUSION

We have derived a general formula for the calculation of the energy of n-independent weak link current loops interacting with an external magnetic field. The ground state energy of the system depends upon flux quantum numbers, $\mathbf{n} = (n_1, n_2, ... , n_k)$, and the classical mutual inductance matrix. Transitions between flux states correspond to the emission or absorption of microwave quanta. If the interaction of flux states different by ±1 quantum is included then energy bands result as a function of the external field.

The ground state energy of a large random loop network was calculated and found to be in qualitative agreement with microwave absorption experiments on bulk high T_c superconductors. The general energy calculation was also applied to a simple model of a single crystal of high temperature superconductor. The results explain experimentally observed microwave absorption peak spacing, and show that the same type of data will be obtained independent of the field, $\mathbf{H}_{dc}$ or $\mathbf{H}_{\mu w}$, scanned. The model also predicts the possibility of observing energy bands in superconductors. These are a consequence of induced supercurrents and their interaction with the electromagnetic field.

The calculations are only valid in the low temperature limit when the superconducting state can be represented by a single macroscopic wave function. The quantization of flux in all closed loops is equivalent to requiring single valuedness of the macroscopic wavefunction. Therefore, the concept of "frustration" is eliminated by this model.

ACKNOWLEDGMENTS

We thank Prof. S. Tyagi for useful discussions. This work was supported in part by NSF Grant PHY 85-20634.

REFERENCES

1. K. W. Blazey, A. M. Portis, K. A. Muller, and F. H. Holtzberg, Europhys. Lett.. **6** (1988).
2. A. Dulcic, R. H. Crepeau, and J. H. Freed, Physica **C160** (1989).
3. H. Vichery, F. Beuneau, and P. Lejay, Physica **C159** (1989).
4. K. W. Blazey, K. A. Muller, J. G. Bednorz, W. Berlinger, G. Amoretti, E. Bulluggiu, A. Vera, and F. C. Matacotta, Phys. Rev. **B36**, 7241 (1987).
5. J. Stankowski, P. K. Kahol, N. S. Dalal, and J. S. Moodera, Phys. Rev. **B36**, 7126 (1987).
6. S. Tyagi and M. Barsoum, Supercond. Sci. and Tech. **1**, 20 (1988) .
7. S. Tyagi, M. Barsoum and K. V. Rao, J. Phys. C: Solid State Phys. **21**, L827 (1988).
8. S. Tyagi, M. Barsoum, K. V. Rao, V. Skumryev, Z. Yu and J. L. Costa, Physica **C156**, 73 (1988).
9. R. S. Steinman, P. Lejay, J. Chaussy, and B. Pannetier, Physica **C153-155**, 1487 (1988).
10. R. P. Feynman, *Statistical Mechanics*, Benjamin. Reading, Mass.. 1972, p304.
11. T. D. Clark, *Quantum Implications* Essays In Honor Of David Bohm, B. J. Hiley & F. D. Peat, Eds. Roatledge Kegan Paul. 1987. pp121-150.
12. K. W. Blazey , A. M. Portis, and F. H. Holtzberg. Physica **C157** (1989).
13. S. Tyagi, M. Jeffery, K. Kish, and R. Gilmore. *To be published.*
14. S. Tyagi and K. Kish. *To be published.*

STUDIES ON THE ROLE OF ALIOVALENT SUBSTITUTIONS AND THEIR COOPERATIVE EFFECT IN $Y_{1-x-y}Pr_xCa_yBa_2Cu_3O_{7-\delta}$ SYSTEM

G.K.Bichile,Smita Deshmukh,D.G.Kuberkar and S.S.Shah
Department of Physics,Marathwada University
Aurangabad 431 004 INDIA.

R.G.Kulkarni
Dept.of Physics,Saurashtra University
Rajkot 360 005 INDIA.

ABSTRACT

With a view to monitor the formal valence of copper in YBaCuO system, substitutions of divalent (Ca) and tetravalent (Pr) impurities at Y-site have been carried out. In addition to this the cooperative effect of these substitutions at Y-site on the structural and superconducting behaviour of YBaCuO system has been investigated using X-ray and susceptibility measurements. The changes in the superconductining transition temperature have been discussed in the context of oxygen content and changes in the formal copper valence in the YBaCuO system.

INTRODUCTION

Studies on the role of Praseodymium substituting at Y-site in $YBa_2Cu_3O_7$ (90 K) superconductor has attracted considerable attention over the past few months, owing to its exceptional behaviour in the $PrBa_2Cu_3O_{7-\delta}$ stoichiometry showing a structural similarity but non superconductive[1].

It has been established that the superconducting properties of the so called 1-2-3 superconductors depend markedly on the changes in the effective copper valence[2]. The effective valence can be monitored by the changes in the oxygen stoichiometry or the chemical substitutions at copper and non copper sites[3,4]. The total charge Q 'on the non copper cations can be altered by the substitution of the aliovalent elements at non copper sites which in turn modifies the formal valence of the copper[5].

The aim of the present work is to undertake a detailed investigation of Pr and Ca substitution at Y - site of $YBa_2Cu_3O_{7-\delta}$ superconductor. Praseodymium when substituted for Y, do not alter the crystal structure but

T_c decreases rapidly with Pr concentration up to $x = 0.5$ in $Y_{1-x}Pr_xBa_2Cu_3O_{7-\delta}$ system[6].Pr exists in nature with 3^+ and 4^+ valence states,but when valency of Pr changes from 3^+ to 4^+,superconducting property disappears[7].It would be interesting to study how this fluctuating valency of Pr,could affect the formal copper valence.In order to probe Y-site by Pr,which has a fluctuating valency,we have simultaneously substituted Ca having definite 2^+ valency at Y -site.These studies,especially, on the simultaneous substitutions of Pr and Ca at Y - site would help in resolving the effect of valence fluctuation and its effect on the superconducting and magnetic behaviour of 1-2-3 superconductor.

EXPERIMENTAL

A series of compounds having the stoichiometric compositions $Y_{1-x}Pr_xBa_2Cu_3O_{7-\delta}$ ($x = 0.05,0.1,0.15,0.2,$ 0.25,0.3), $Y_{1-x}Ca_xBa_2Cu_3O_{7-\delta}$ ($x = 0.05,0.1,0.15,0.2,$ 0.25,0.3), $Y_{1-x-y}Pr_xCa_yBa_2Cu_3O_{7-\delta}$ ($x = 0.05$ y = 0.15, $x = 0.1$ y = 0.1,$x = 0.15$ y = 0.05) have been synthesised using the standard solid state reaction method as discussed in our earlier paper[8].Based on our literature study on the limit of solid solubilities of Pr and Ca at Y-site in YBaCuO system, proper stoichiometric compositions of the above systems were selected[9,10] In this paper we are reporting the part of the work on $Y_{1-x}Pr_xBa_2Cu_3O_{7-\delta}$ ($x = 0.1,0.2$) and $Y_{1-x-y}Pr_xCa_yBa_2Cu_3O_{7-\delta}$ ($x = 0.05$,y = 0.15) samples. Samples were characterised by X-ray diffraction technique.The oxygen content analysis was carried out using the standard iodometric titration method[11]. The superconducting transition temperatures were recorded using the a.c.susceptibility method employing the mutual inductance technique with a frequency of 313 Hz.at 50 mv.D.C.magnetization measurements were performed using a EG & G PAR Model 4500 Vibrating Sample Magnetometer (VSM) with 10 KOe electromagnet.

RESULTS AND DISCUSSIONS

The X-ray analysis revealed that all the samples are single phased and orthorhombic with few additional peaks as shown in the X-ray pattern of one of the samples (fig.1).
The lattice parameters were obtained using standard least squares fitting programme.It is observed that the orthorhombocity (b/a) decreases as the Pr concentration

increases suggesting a tendancy of structural transformation towards tetragonality

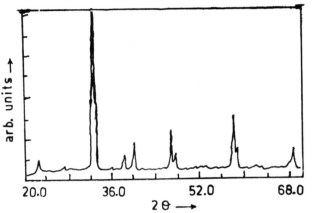

Fig.1. XRD pattern of Sample - C

The values of the unit cell parameters, orthorhombocity and the oxygen content for the samples are listed in Table - I.

TABLE - I

Values of lattice parameters, orthorhombocity & oxygen content for $Y_{1-x-y}Pr_xCa_yBa_2Cu_3O_{7-\delta}$ system.

| Sample | Lattice parameters (A) | | | | oxygen |
	a	b	c	b/a	content
x=0.1 y=0 (A)	3.8202	3.8862	11.6751	1.017	6.64
x=0.2 y=0 (B)	3.8301	3.8902	11.6851	1.015	6.76
x=0.15 y=0.05(C)	3.8255	3.8879	11.6616	1.016	6.79

Figures 2 & 3 show the temperature dependence of a.c.susceptibility and the magnetic susceptibility measurments derived from magnetisation studies resp.The values of the superconducting transition temperatures obtained from these measurements are listed in Table-II

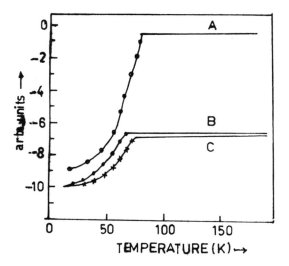

Fig.2. Temperature dependence of a.c.
susceptibility of samples A,B & C.

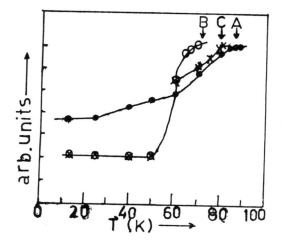

Fig.3. Temperature dependence of magnetic
susceptibility for samples A,B & C.

It is evident from the Table –II, that $T_c^{on}(\propto ac)$ and $T_c^{on}(\chi_{mag})$ agree fairly well with each other. This shows the consistancy of the measurements and the high quality of the samples.

TABLE - II

Values of T_C obtained from the a.c.susceptibility and d.c.magnetisation measurements for the samples A,B & C.

Sample	$T_C{}^{on}$ (K)		Q
	(χ_{ac})	($\chi_{d.c.mag}$)	
A	82.8	84	7.1
B	67.1	70	7.2
C	75.7	78	7.1

It is apparent from the above table that the T_C decreases for x = 0.1 (A) to x = 0.2 (B) while the simultaneous addition of 0.05 Calcium alongwith Pr (sample C) at Y - site in YBaCuO system has enhanced the T_C value appreciably as compared to that of for the partial substitution of Pr or Ca at Y - site in this system.

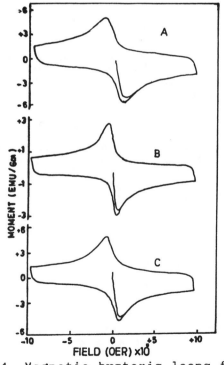

Fig.4. Magnetic hysteris loops for samples A,B & C at 13 K.

Figure 4 shows the typical magnetisation (H) scans of the three samples obtained at 13 K. We have estimated the intragrain critical current density J_c from the relation[12]

$$J_c = \frac{15\ (\ M^+ - M^-\ )}{R}$$

where M^+ and M^- are the moments observed from the hysteresis scan.Using R = 10 μ m and ζ = 5 gm/cm^3 , J_c at 13 K is found to be 2.13×10^5 amp/cm^2 for sample A , 0.9×10^5 amp/cm^2 for sample B and 1.81×10^5 amp/cm^2 for sample C. This suggest that the addition of calcium has not only enhanced T_c but also has helped to increase J_c.

In summary,it appeares from the present findings that the superconducting properties vary markedly as the charge Q on the non copper cations changes which in turn also is reflected in the changes in the formal valence of copper.Further indepth investigations on simultaneous substitutions of Ca alongwith Pr and Hf at Y - site are under progress.

ACKNOWLEDGMENTS

Authors (DGK & SD) thankfully acknowledge Prof. P.Boolchand & Prof.V.V.Itagi for the encouragement and Dr.L.J.Raibagkar , Mr.K.M.Jadhav and Mr.Sunil Patil for the helpful discussions during the course of this work.

REFERENCES

1. L.Soderholm,K.Zhang,D.G.Hinks,M.A.Beno,J.D.Jorgensen, C.U.Segre and Ivan K.Schuller, Nature,38,604 (1987).
2. J.M.Tarascon,W.R.McKinnon,L.H.Greene,G.W.Hull and E.M.Vogel, Phys. Rev. B36, 226 (1987).
3. Y.Maeno,T.Tomita,M.Kyogoku,S.Awaji,Y.Aoki,K.Hosino, A.Minami and T.Fujita, Nature, 328, 512 (1987).
4. M.Kikuchi,Y.Syomo,A.Tokiwa,K.Oh-ishi,H.Arai,K.Hiraga, N.Kobayashi,T.Sasaoka and Y.Muto, Jpn.J.Appl.Phys., 26, L1066 (1987).
5. Y.Tokura,J.B.Torrance,T.C.Huang and A.I.Nazzal, Phys. Rev. B38, 7156 (1988).
6. Kyum Nahm,Bo Young Cha and chul Koo Kim, Sol.St.Comm. 72, 6, 7 (1988).
7. J.K.Liang,X.T.Xu,S.S.Xie,G.H.Rao,X.Y.Shao and   Z.G. Duan, Z.Phys.B - Cond.Mat., 69, 137 (1987).
8. G.K.Bichile,D.G.Kuberkar and S.S.Shah,   Sol.St.Comm. 74, 7, 629 (1990).

9. A.Kebede,Chan Soo Jee,D.Nichols, M.V.Kuric, J.E.Crow, R.P.Guertin,T.Mihalisin,G.H.Myer,I.Perez, R.E.Salomon and P.Schlottmann J.Mag.& Mag.Mat.,76&77,619 (1988).
10. A.Manthiram, S.J.Lee and J.B.Goodenough, J.Sol.St. Chem. 73, 278 (1988).
11. A.I.Nazzal,V.Y.Lee,E.M.Engler, R.D.Jacowitz, Y.Tokura and J.B.Torrance, Physica C, 153, 1367 (1988).
12. J.O.Willis,J.R.Cost,R.D.Brown,J.B.Thompson and D.E. Peterson, IEEE Trans.Magn.MAG - 25 (1989).

EVIDENCE OF ELECTRON-PHONON INTERACTION IN THE INFRARED SPECTRA OF OXIDE SUPERCONDUCTORS

T. Timusk

McMaster University, Hamilton, Ontario, Canada L8S 4M1

D.B. Tanner

University of Florida, Gainesville, Florida 32611

ABSTRACT

We show that the structure in ab-plane infrared spectra of the oxide super-conductors can be understood in terms of the electron-phonon interaction. The minima in the optical conductivity seen in the oxygen breathing mode region translate to peaks in the frequency dependent scattering rate. The position and the width of the peaks are in reasonable agreement with structure seen in time-of-flight neutron spectroscopy. We interpret the appearance of phonon modes in the highly metallic ab-plane in terms of a charged phonon model borrowed from the theory of organic conductors.

INTRODUCTION

The role of phonons in the high T_c superconductors is subtle. On the one hand there is evidence that the charge carriers responsible for superconductivity are weakly coupled to the lattice. For example, the isotope shift in in $YBa_2Cu_3O_7$ is small if not zero and calculations based on the electron-phonon mechanism, while quite good at predicting T_c's in conventional superconductors, do not yield the high transition temperatures actually observed in the oxides.[1] The temperature dependence of the resistivity is not what one expects of phonons. The oxygen breathing modes at 500 cm^{-1} (62 meV) should give rise to a positive curvature to the resistivity at room temperature instead o the straight line that is actually observed. The magnitude of the scattering rate, $1/\tau \approx (1-2)k_B T$, suggests weak scattering of the carriers from som low frequency excitation, below typical phonon frequencies. Thus transpor measurements point to a weak interaction between the dc current carriers an phonons.

However spectroscopic measurements on the phonons themselves show tha certain phonons are in strong contact with the electrons. Neutron time-of-fligh (TOF) spectra,[2] show pronounced softening in the 50 meV region as the mater als are changed from insulating parent compound to conductor/superconduct by oxygen doping. Fig. 1a and and 1b, from Renker $et\ al.$[2] illustrate the effects in $YBa_2Cu_3O_{7-\delta}$. Phonon branches—related to Cu-O stretches—sho anomalies near the zone boundary.[3] Raman spectroscopy too shows that certa optic phonon frequencies shift as the materials become superconducting[4-7] a the lines show asymmetric shapes that suggest interaction of the Raman-acti modes with an underlying electronic continuum.[8] This continuum gives rise t broad band of scattering at low temperature that peaks in the 350 cm^{-1} regi in A_{1g} symmetry.

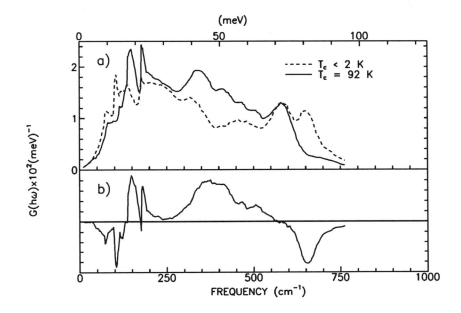

Fig. 1 a) Neutron density of states for phonons from Renker *et al.*[2] for YBa$_2$Cu$_3$O$_x$. The solid curve is for a superconducting sample ($x = 7$) the dashed one for a non-superconducting one ($x = 6$). There is a strong enhancement of density of states in the 20 and the 50 meV regions. b) A difference spectrum between the two curves in a).

Here we are concerned with the influence of phonons on the infrared properties of the high T_c materials. Early work on ceramic samples exhibited a rich spectrum dominated by phonons characteristic of perovskites.[9] These phonon lines were shown by Collins *et al.* to originate from the c-axis response of the crystallites.[10] There are no obvious signatures of phonons in the crystal or thin film spectra for the ab-plane reflectance. This is not surprising since the materials are good metals and the charge carriers, not the phonons, are expected to dominate the electromagnetic response.

It is therefore surprising to see in YBa$_2$Cu$_3$O$_7$, at low temperature, ab-plane reflectance features in the form of two "knees", one at 155 cm^{-1} (20 meV) and the other, more prominent one, at 420 cm^{-1} (52 meV) in the phonon region of the spectrum. These features exhibit characteristics of phonons: their frequency does not depend on changes in the electronic structure associated with oxygen doping[11] and they do not undergo dramatic changes when the materials become superconducting, and they are present in the normal state.[12] We have suggested that these features can be understood in terms of a model where the phonons acquire electronic oscillator strength by the "charged" phonon mechanism, a phenomenon seen in organic conductors, where charge density waves can give internal molecular vibrations electronic-strength infrared intensities.[13,14]

PHONONS IN METALS

Phonons in metals can absorb electromagnetic energy in several different ways. The simplest is a direct coupling of the ionic charges to the external field, the familiar reststrahlen absorption. It is expected to be weak in the oxides. Tajima *et al.*[15] have recently measured the oscillator strength of the *ab*-plane infrared active phonon transitions for a variety of insulating parent compounds of the high T_c superconductors. They find rather small effective charges for the oxygens, with Z_O ranging between 0.87 and 1.45 electronic charges. Oxygen, being the light ions of the structure, would dominate the absorption.

It is easy to estimate the strength of phonon lines expected in the superconductor. Tajima *et al.* find that in La_2CuO_4 a phonon has for example $\omega_{TO} = 140$ cm^{-1}, $\omega_{LO} = 320$ cm^{-1}, $\gamma = 42$ cm^{-1}. With parameters that fit typical of good crystals in the *ab*-plane,[12] a phonon of this strength at 155 cm^{-1} would only give a 0.01 % feature in reflectance. This kind of weak structure is well below what can be seen experimentally. Clearly optical phonons are difficult to observe in superconductors, even with an electronic background of the type seen in the high T_c materials. This simple direct mechanism cannot be responsible for the observed phonon features at 155 cm^{-1} and at 360 cm^{-1}.

In ordinary metals with strong electron-phonon interaction the Holstein[16] mechanism provides observable phonon structure in the infrared.[17] Each phonon branch gives rise to a threshold in absorption, above this threshold the incoming photon energy is divided between the phonon and electron-hole pair in the final state. In the superconducting state this threshold becomes a step at a frequency that is the sum of the phonon frequency and the energy necessary to break the Cooper pair 2Δ. This predicted 2Δ shift at the superconducting transition of the absorption threshold of the phonons has not been observed for any feature in the spectra of high $T_c$superconductors. Model calculations[18−20] show that for a strong coupling Eliashberg model with a gap of $4k_BT_c$, coupled to oxygen breathing modes, a strong threshold appears at approximately $8k_BT_c$ in the normal state, shifting to $10k_BT_c$ as superconductivity sets in. The absence of this 2Δ shift rules out the Holstein mechanism as the cause of the phonon features.

A mechanism that enhances the oscillator strength of optic phonons and breaks down selection rules making totally symmetrical phonons optically active is the "charged phonon" mechanism of Rice,[13,21,22] known to occur in organic conductors. Associated with low-lying electronic states the symmetric lattice vibrations pump charge over large distances giving rise to absorption at phonon frequencies that is of electronic strength. The characteristic signature of this process is the appearance of sharp peaks where the background electronic absorption has gaps and sharp notches or antiresonances where the electronic background has non-zero oscillator strength.

THE OPTICAL CONDUCTIVITY AND SCATTERING RATE

The best evidence for the charged phonon mechanism in the high T_c superconductors comes from an analysis of the optical conductivity in the superconducting state where the Drude conductivity has collapsed to form a delta function at zero frequency, better revealing the electronic background. Fig. 2a shows the conductivity from the work of Kamarás et al. for $YBa_2Cu_3O_{7-\delta}$. The overall amplitude of the conductivity varies between 1000 and 2500 $\Omega^{-1}cm^{-1}$. The conductivity is dominated by a broad continuum peaking at 900 cm^{-1} with a pronounced minimum in the 500 cm^{-1} region. There is a a gap-like onset in the 130 cm^{-1} region. Other investigators generally find similar results[23,24,11,10] although there is some variation in the details of the spectra. The gap in conductivity results from the region of unity reflectance below 130 cm^{-1}. While accurate measurements of absolute reflectance at low frequency are difficult several groups have reported, to within an experimental accuracy of 1 % or so, unity reflectance below this frequency. Structure is always observed at 155 cm^{-1} but not all experiments show a gap. For example recent bolometric experiments by Pham et al.[25] show a a small peak at this frequency superimposed on an absorption with a quadratic frequency dependence suggestive of a finite conductivity to the lowest frequencies. A gap has also been reported by Reedyk et al.[26] in $Bi_2Sr_2CaCu_2O_8$ at 300 cm^{-1} and like the gap in $YBa_2Cu_3O_7$ this feature is present in the normal state and likely associated with phonons too.

We next calculate the frequency dependent scattering rate $\Gamma(\omega)$ associated with the conductivity in Fig. 2a. We assume that the dielectric response is made up of three parts, the Drude part ϵ_D, a midinfrared electronic part ϵ_e and the high frequency contribution ϵ_∞:

$$\epsilon(\omega) = \epsilon_D + \epsilon_e + \epsilon_\infty. \tag{1}$$

Since ϵ_e has all the structure at low frequency[27] it is helpful to write ϵ_e in terms of the frequency dependent scattering rate:[28,29]

$$\epsilon_e = \frac{\omega_{pe}^2}{\omega I(\omega) - \omega^2 - i\omega\Gamma(\omega)}, \tag{2}$$

where ω_{pe} is the plasma frequency of the electronic midinfrared band and $\Gamma(\omega)$ and $I(\omega)$ describe the real and imaginary parts of the frequency dependent scattering rate. With $\epsilon_e = \epsilon_{e1} + i\epsilon_{e2}$, we can get ϵ_{e2} by Kramers-Kronig analysis from reflectance experiments and ϵ_{e1} from a second Kramers-Kronig transformation of ϵ_{e2}. Thus, the frequency dependent scattering rate,

$$\Gamma(\omega) = \frac{\omega_{pe}^2 \epsilon_{e2}}{\omega(\epsilon_{e1}^2 + \epsilon_{e2}^2)} \tag{3}$$

can be obtained from the experimentally determined $\sigma(\omega)$ at low temperature. The only fitted parameter used is the plasma frequency of the midinfrared band,

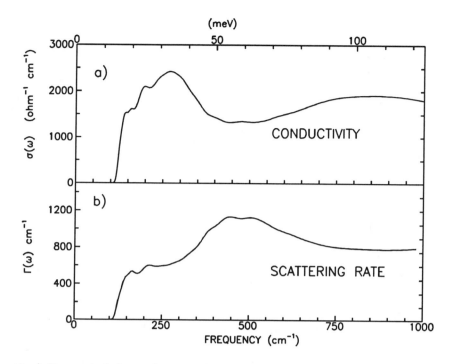

Fig. 2 a) Frequency dependent conductivity of YBa$_2$Cu$_3$O$_7$ from Kamarás *et al.*. The spectrum is characterized by a notch-like minimum in the 450 cm^{-1} region and a threshold at 130 cm^{-1}. We associate both these features with phonons. b) The frequency dependent scattering rate $\Gamma(\omega)$ obtained from the conductivity be Kramers Kronig analysis. There is a peak at the frequency of the notch in the conductivity. We attribute this peak to the same phonon branch that causes the anomalous neutron scattering seen in Fig. 1b.

ω_{pe} (taken from oscillator fits by Kamarás *et al.*[12]), which sets the overall scale of $\Gamma(\omega)$.

Fig. 2b shows $\Gamma(\omega)$ obtained from the low temperature conductivity spectra of Kamarás *et al.* for a laser ablated film of YBa$_2$Cu$_3$O$_7$. The spectrum has an onset in the 130 cm^{-1} (16 meV) region, a broad peak centered at 450 cm^{-1} (55 meV), followed by a fairly constant scattering rate of the order of 1000 cm^{-1}. We will focus on the structure in the phonon region, 100–700 cm^{-1}.

The most striking feature of the $\Gamma(\omega)$ spectrum is a peak in the scattering rate just below the oxygen breathing mode region. The observation of a peak rather than a threshold is clear evidence that the electron-phonon interaction here is not of the Holstein type which yields onsets but of a type where the final state only has a phonon, *i.e.* a charged phonon process described by Rice[13]. The center frequency of the peak is good agreement with the phonon band found by Renker *et al.* (Fig. 1b) in YBa$_2$Cu$_3$O$_7$ that appeared when to doping level was altered to change the material from the non-superconducting O_6 to the superconducting concentration O_7.

The strength of the band can be described in terms of λ the electron-phonon coupling constant. We can make a rough approximation by setting $\lambda \approx \pi\Gamma(\omega_0)d\Gamma/2\omega_0^2$, where $d\Gamma$ is the width of the peak in $\Gamma(\omega_0)$ and ω_0 its center frequency. With a value of $\Gamma(\omega_0) = 1000$ cm^{-1} a λ of approximately 2 is found for the 450 cm^{-1} band. Thus the broad phonon band centered at 450 cm^{-1} is *strongly* coupled to the electronic continuum.

The onset of scattering in the 130 cm^{-1} (16 meV) region in Fig. 1c is a reflection of the gap-like depression in conductivity below this frequency in the spectra of Kamarás *et al.*. At low frequency the accuracy of the experimental reflectance that has been used as input to the Kramers Kronig analysis becomes worse. We have therefore also used a model calculation to analyze the data.[14] Approximating the frequency dependent scattering rate by Lorentz oscillators plus a constant term to describe the background damping, the model describes the experimental reflectance quite well with only two oscillators chosen match the neutron TOF difference peaks of Fig. 1b. In this model the lower knee is produced by the sharp 150 cm^{-1} double feature in the TOF spectrum and the upper one by the broader peak at 400 cm^{-1} (50 meV). The conclusion that we reach from our model calculation is that the onset and the shoulder in $\Gamma(\omega)$ at 150 cm^{-1} that we obtain by Kramers Kronig analysis of the existing experiments may not be correct. Instead $\Gamma(\omega)$ may have a peak at 150 cm^{-1} superimposed on a background. More accurate experiments are needed to settle the nature of the scattering at low frequency.

Fig. 2 demonstrates that the conductivity and scattering rate in the superconducting state of $YBa_2Cu_3O_7$ has the experimental signatures of charged phonons. In reflectance optic phonons generally give rise to reflectance *minima* when there is a free carrier background whereas charged phonons in the presence of an electronic continuum give rise to *peaks*. The opposite holds for the optical conductivity. We find from our model that when the charged phonon coupling gets to be strong which is the case here, particularly in the superconducting state, the reflectance peaks turn into thresholds.

DISCUSSION

The charged phonon mechanism is well studied in organic conductors where A_{1g} symmetry internal vibrations of the molecules are coupled by symmetry breaking charge density waves to the external electromagnetic field. In the case of the high-T_c oxides, a potential will also be needed to allow IR activity of modes that are not at $q = 0$, in particular the branch at the zone boundary modes seen by neutron spectroscopy to couple strongly to the electrons. Since it appears from the temperature dependent spectra of Kamarás *et al.* that the depth of the notch in the conductivity increases as the temperature is lowered, that is the coupling between the electrons and the zone boundary branch grows at low temperature, an incipient structural instability may generate the symmetry breaking potential. One cannot rule out the role of defects in the activation of the phonons since there is quite a large amount of sample to sample variation in the infrared spectra in the phonon region.[30] The high degree of variability of the 155 cm^{-1} mode in particular suggests that its activation may be associated with defects.

In our picture of the frequency dependent conductivity, the phonons interact with the electronic continuum producing, by a Fano interference between the discrete and continuous states, the characteristic v shaped minima. This electronic absorption band, we argue, should be distinguished from the Drude free carrier band as a separate parallel channel of conductivity. Our analysis of the frequency dependent scattering of the part of the conductivity should be distinguished from models where the entire infrared conductivity, Drude plus midinfrared, is analyzed according to Eqn. 2.[23,10]

The idea of two parallel contributions to the low frequency $\sigma(\omega)$ was proposed by Schützmann et al.[24] The two contributions come from the two ab-plane conduction bands that occur in the electronic band structure of YBa$_2$Cu$_3$O$_7$. The two bands arise because there are two copper oxide layers per unit cell and these are coupled by a transverse hopping integral $t_\perp$. This coupling splits the degeneracy of the plane orbitals by $\pm t_\perp$. In the picture proposed by Schützmann et al. these two bands have different effective masses and scattering rates: both contribute to the dc conductivity. In contrast, we view one band (the Drude carriers) as being weakly coupled or uncoupled to the phonons and determining the dc conductivity. The other band couples strongly as described above. This coupling reduces the dc conductivity of the carriers in this band to zero, gives in only modest temperature dependence, and leads to absorption throughout the phonon region of the spectrum. The reason for the different behavior of these two bands we are unable to specify, but it must in some way be related to the different symmetry of the two bands (one even and the other odd under inversion, for example).

The experiments of of Kamarás et al.[12] that our analysis leading to Fig. 2b is based on were done on high quality twinned films and thus represent a superposition of a- and b-axis conductivities. There is evidence that the chains make a separate contribution to the conductivity in the form of a peak at 0.5 eV[31−35] and recent work by Schlesinger et al.[35] shows that the low frequency tail of this absorption extends down into the phonon region analyzed in this work. To account for the chains one should subtract this contribution from the midinfrared conductivity. This is difficult since the ab-plane itself is quite anisotropic, both because of anisotropic bond-lengths and interactions with the chain layer below.

Raman spectroscopy provides an independent confirmation of the anisotropy of the electronic continuum in the phonon region. Slakey et al.[8] working with untwinned crystals find that the electronic Raman scattering is stronger in the chain direction and the coupling to phonons is stronger at 116 cm^{-1} in the chain direction than normal to it. Nevertheless electronic continuum is present in both directions in the ab-plane along with the Fano type coupling to the optic phonons at $q = 0$.

It is tempting to identify the 350 cm^{-1} peak in the Raman continuum with the peak in $\Gamma(\omega)$ since this quantity is according to Eq. 3 proportional to $Im(1/\epsilon)$ which in turn enters the expression for the electronic Raman crossection. There is a problem with this simple picture however since the dielectric function in the Raman crossection is the total one including the contribution from the condensate whereas the $\Gamma(\omega)$ in Fig. 2 only refers to the midinfrared electronic band.

As in the case of the 350 cm^{-1} peak in the Raman spectrum the infrared feature at 450 cm^{-1} as well as the onset at 130 cm^{-1} have been interpreted as the superconducting gap.[23,24,10] The difficulties with this interpretation have been pointed out by several groups.[11,12] In this review we have outlined the evidence for an alternate interpretation of these low lying structures in the infrared spectra of the high T_c superconductors in terms of charged phonons interaction with an electronic background.

We thank J.P. Carbotte, H.D. Drew, C. Kallin, M. Reedyk and M.J. Rice for valuable discussions. The research at McMaster is supported by the Natural Science and Engineering Research Council of Canada (NSERC) and the Canadian Institute for Advanced Research (CIAR). Work at Florida is supported by DARPA through contract MDA972-88-J-1006.

REFERENCES

1. W. Weber and L.F. Mattheiss, *Phys. Rev. B* **37**, 599 (1988).

2. B. Renker, F. Gompf, E. Gering, D. Ewert, H. Rietschek, and A. Dianoux, *Z. Phys. B – Condensed Matter* **73**, 309 (1988).

3. W. Reichard, N. Pyka, L. Pintschovius, B. Hennion, and G. Collin, Int. Conf. on Superconductivity, Stanford, 1989, *Physica C.* **162 – 164**, 464, (1989).

4. S.L. Cooper, M.V. Klein, B.G. Pazol, J.P. Rice, and D.M. Ginsberg, *Phys. Rev. B* **37**, 5920 (1988).

5. R.M. Mcfarlane, H.J. Rosen, and H. Seki *Solid State Comm.* **63**, 831 (1987).

6. C. Thomsen, M. Cardona, B. Gegenheimer, R. Liu, A. Simon, *Phys. Rev. B* **37**, 9860 (1988).

7. T. Ruf, C. Thomsen, R. Liu, M. Cardona, *Phys. Rev. B* **3**, 11985 (1988).

8. F. Slakey, S.L. Cooper, M.V. Klein, J.P. Price, E.D. Bukowsky, and D.M. Ginsberg, *Phys. Rev. B* **38**, 11934 (1989).

9. For a review of the early work on the infrared properties of the high T_c materials see T. Timusk, D.B. Tanner, in *Physical Properties of High Temperature Superconductors I,* Donald M. Ginsberg, editor, (World Scientific, Singapore, 1989) p. 339.

10. R.T. Collins, Z. Schlesinger, F. Holtzberg, and C. Field, *Phys. Rev. Lett.* **63**, 422 (1989).

11. S.L. Cooper, G.A. Thomas, J. Orenstein, D.H. Rapkine, M. Capizzi, T. Timusk, A.J. Millis, L.F. Schneemeyer, and J.V. Waszczak, *Phys. Rev. B* **40**, 11358 (1989).

12. K. Kamarás, S.L. Herr, C.D. Porter, N. Tache D.B. Tanner, S. Etemad, T. Venkatesan, E. Chase, A. Inham, X.D. Wu, M.S. Hegde, and B. Dutta, *Phys. Rev. Lett.* **64**, 84 (1990).

13. M.J. Rice, *Phys. Rev. Lett.* **37**, 36 (1976).

14. T.Timusk and D.B. Tanner, *Physica C* **169**, 425 (1990).

15. S. Tajima, S. Uchida, S. Ishibashi, T. Ido, H. Takagi, T. Arima, Y. Tokura, *Physica C*, **168**, 117 (1990).

16. T. Holstein, *Phys. Rev* **96**, 539, (1954); *Ann. Phys. (N.Y.)* **29**, 410, (1964).

17. R.R. Joyce and P.L. Richards, *Phys. Rev. Lett.* **24**, 1007 (1970).

18. W. Lee, D. Rainer, and W. Zimmerman, *Physica C* **159**, 535 (1989).

19. N.E. Bickers, D.J Scalapino, R.T Collins,and Z. Schlesinger, *Phys. Rev. B* **42**, 67 (1990).

20. E. Nicol, J.C. Carbotte, and T. Timusk, *Phys. Rev. B* (to be published).

21. M.J. Rice, V.M. Yartsev,and C.S. Jacobsen, *Phys. Rev. B* **21**, 3437 (1980).

22. M.J. Rice and Y.R. Wang, *Phys. Rev. B* **36**, 8794 (1987).

23. G.A. Thomas, J. Orenstein, D.H. Rapkine, M. Capizzi, A.J. Millis, R.N. Bhatt, L.F. Schneemeyer, and J.V. Waszczak, *Phys. Rev. Lett.* **61**, 1313 (1988).

24. J. Schützmann, W. Ose, J. Keller, K.F. Renk, B. Roas, L. Schultz, and G. Saemann-Ischenko, *Europhys. Lett.*, **8**, 679 (1989).

25. T. Pham, H.D. Drew, S.H. Mosley, and J.Z. Liu, *Phys. Rev. B* **41**, 11681 (1990).

26. M. Reedyk, D.A. Bonn, J.D. Garrett, J.E. Greedan, C.V. Stager, T. Timusk, K. Kamarás, and D.B. Tanner, *Phys. Rev. B* **38**, 11981 (1988).

27. A detailed analysis done in ref. 12 of the optical conductivity above and below the T_c shows that the portion of the conductivity that condenses at T_c is smooth and Drude-like. The frequency dependent structure is associated with temperature independent midinfrared band. This analysis also demonstrates that the Drude absorption does not develop a gap at T_c.

28. W. Götze and P. Wölfle, *Phys. Rev. B* **6**, 1266 (1972).

29. J.W. Allen and J.C. Mikkelsen, *Phys. Rev. B* **15**, 2952 (1977).

30. G.A. Thomas, M. Capizzi, T. Timusk, S.L. Cooper, J. Orenstein, D. Rapkine, S. Martin, L.F. Schneemeyer, and J.V. Waszczak, *J. Opt. Soc. Am.* **6**, 415, 1989

31. K. Kamarás, C.D. Porter, M.G. Doss, S.L. Herr, D.B. Tanner, D.A. Bonn, J.E. Greedan, A.H. O'Reilly, C.V. Stager, and T. Timusk, *Phys. Rev. Lett.* **59**, 919 (1987).

32. J. Tanaka, K. Kamya, M. Shimizu, M. Simada, C. Tanaka, H. Ozeki, K. Adachi, K. Iwahashi, F. Sato, A. Sawada, S. Iwata, H. Sakuma, and S Uchiyama, *Physica C* **153-155**, 1752 (1988).

33. M.P. Petrov *et al. Pis'ma Zh. Eksp. Teor. Fiz.*, **50**, 25, (1989).

34. J. Orenstein, G.A. Thomas, A.J. Millis, S.L. Cooper, D. Rapkine, T. Timusk, L.F. Schneemeyer, and J.V. Waszszak, *Phys. Rev. B* (to be published).

35. Z. Schlesinger, R.T. Collins, F. Holtzberg, C. Feild, S.H. Blanton, U. Welp, G.W. Crabtree, Y. Fang, and J.Z. Liu, *Phys. Rev. Lett.* **65**, 801 (1990).

THERMODYNAMICS OF THE La-Sr-Cu-O HIGH Tc SUPERCONDUCTORS: HEAT CAPACITIES OF $SrCuO_2$, Sr_2CuO_3, $Sr_{14}Cu_{24}O_{41}$, and $(La_{1-x}Sr_x)CuO_{4-x}$ ($x = 0.00, 0.05, 0.30, 0.5$)

Roey Shaviv* and Edgar F. Westrum Jr.
Department of Chemistry, University of Michigan, Ann Arbor, MI 48109

Charles B. Alcock, and Baozhen Li
Center for Materials, University of Notre Dame, Notre Dame, IN 46556

ABSTRACT

The subambient heat capacities of seven La-Sr-Cu-O samples were measured by adiabatic calorimetry over the temperature range 5 to 350 K. The heat capacity of $Sr_{14}Cu_{24}O_{41}$ was derived from that of the mixed phase of nominal composition $SrO:CuO = 1:2$. No anomalous contribution was identified in the heat capacities of $SrCuO_2$, Sr_2CuO_3, $Sr_{14}Cu_{24}O_{41}$, $LaSrCuO_{3.5}$ and La_2CuO_3. A small anomaly associated with the orthorhombic to tetragonal transformation was identified at the vicinity of 300 K in the heat capacity of $La_{1.90}Sr_{0.10}CuO_{3.90}$. Onset of Meisner effects in the magnetic susceptibilities of $La_{1.90}Sr_{0.10}CuO_{3.90}$ and $La_{1.4}Sr_{0.6}CuO_{3.7}$ were identified at about 32 and 36 K respectively. The onset of superconductivity in these samples is not accompanied by a break in the heat capacity at the critical temperatures. A series of thermal history dependent anomalies were identified in the heat capacity, magnetic susceptibility and electrical conductivity of $La_{1.4}Sr_{0.6}CuO_{3.7}$ between 234 and 290 K. Annealing at 234 K causes the anomalies to shift to higher temperatures. Cooling below the transition temperature restores the anomalies. The anomalies tend to disappear after prolonged annealing at 380 K in air.

INTRODUCTION

The La-Sr-Cu-O (LSCO) chemical system which shows an unusual array of electric and magnetic properties is in the focus of much scientific activity since the discovery of superconductivity above 30 K in this, and in other chemically related, systems. The thermodynamic properties of this system have been the center of a broad investigation in which the subambient heat capacities, superambient electrochemical properties, magnetic susceptibilities, and electrical resistivities of seven samples have been studied. This thorough investigation has been the subject of several recent communications.[1-5] The samples used in the studies and their compositions are listed in Table I. For clear identification and easier communication the compounds studied are designated with Roman numerals I - VII. This designation is especially useful when the composition of the sample used in the experiment is different than that of a stable phase. This is the case with the sample whose nominal composition is $SrCu_2O_{3+x}$ which is believed to consist of the two stable compounds CuO and $Sr_{14}Cu_{24}O_{41}$. All the other samples are believed to be of stable single-phase compositions.

---

* Present address: Department of Chemistry, University of Illinois at Chicago, Box 4348, Chicago, IL 60680

Table I   Compounds and designations in this study

Formula	Designation	M/g·mol^{-1}	Ref.
$SrCuO_2$,	I	183.166	1
Sr_2CuO_3,	II	286.785	1
$Sr_{14}Cu_{24}O_{41}$,	III	3407.764	1
$La_{1.9}Sr_{0.1}CuO_{3.95}$	IV	399.437	2
$La_{1.4}Sr_{0.6}CuO_{3.70}$	V	369.792	2
$LaSrCuO_{3.50}$	VI	346.076	3
La_2CuO_4	VII	405.366	3

This report makes no attempt to repeat the material that we have already published, submitted, or committed to other channels of communications. The major goal of the authors here is to supplement their previous communications and to attempt to summarize and unify our research efforts concerning this chemical system under one title. The readers are referred to references 1 through 5 for further details regarding experimental procedures, sample preparation, tabulated data, thermophysical properties, and further discussion.

EXPERIMENTAL

The oxide samples were prepared by a conventional ceramic mixing technique using high purity $SrCO_3$, CuO, and La_2O_3. All samples were pressed into pellets (under a pressure of 10,000 Kg·Cm^2) and sintered in air. X-ray analysis was used to confirm the existence of a single-phase product. The sample whose Sr/Cu composition ratio was 1/2 was found to consist of at least two phases of which one is the oxygen rich phase $Sr_{14}Cu_{24}O_{41}$ and the other is excess CuO. Chemical analysis of this sample revealed the pressure of carbonate ion (about 0.03 % by mass of carbon).

The subambient heat capacities of the seven samples were measured in an adiabatic calorimetric cryostat (laboratory designated Mark XIII[6, 7]) from 5.3 to 350 K. The same gold plated, oxygen-free, high-conductivity, (OFHC) copper calorimeter (laboratory designation W-99) was used for all samples. Samples massing between 20 and 35 g were used in these experiments.

The magnetic susceptibilities at 50 Gauss of about 1 g samples of $La_{1.9}Sr_{0.1}CuO_{3.95}$ (between 4 and 60 K) and $La_{1.4}Sr_{0.6}CuO_{3.70}$ (between 2 and 300 K) and the electrical resistivity of the former (between 2.4 and 333 K) were measured using a SQUID magnetometer (Quantum Design). Powder samples were used for the susceptibility measurements and two different pelleted (of different thermal history) samples were used for the measurements of electrical resistivity.

RESULTS AND DISCUSSION

$SrCuO_2$, Sr_2CuO_3, and $Sr_{14}Cu_{24}O_{41}$

As indicated above the tabulated experimental results and their graphical representations are presented elsewhere.[1-3] The first three compounds in the series

(I-III) show no anomalous contribution apart from the lattice and electronic heat capacities. While the thermodynamic properties of I can be well approximated as a sum of the properties of CuO and SrO this is not the case for compounds II and III. All three compound exhibit Debye-like behavior at low temperatures with a small but finite Sommerfeld coefficient.

The actual stoichiometry of the third sample was $Sr_{14}Cu_{28}O_{45}$ and it is believed to contain two phases, $Sr_{14}Cu_{24}O_{41}$ and CuO. The mass ratio between the phases was calculated as 91.54/8.46. The existence of the oxygen rich compound $Sr_{14}Cu_{24}O_{41}$ was reported by several authors[8,9] and was supported by thermogravimetric analysis of our sample. However, more recently it was suggested that the apparent excess oxygen in the sample may be due to a small residue of carbonate ions which remain in the sample after synthesis.[10] Consequently our sample was analyzed for any excess carbonate and small traces of carbonate ions (0.178 % carbon by mass) were identified. It is not yet clear whether this amount of carbonate ions can explain the oxygen stoichiometry and the physical properties of the sample. It should be noted in this context that the magnetic transition characteristic of CuO between 210 and 240 K cannot be found in the heat capacity of this sample although well within the precision of the experiments. The absence of these transitions may suggest that the sample is more complex than a simple mixture of $Sr_{14}Cu_{24}O_{41}$ and CuO.

$La_{1.9}Sr_{0.1}CuO_{3.95}$ and $La_{1.4}Sr_{0.6}CuO_{3.70}$

Compounds IV and V are high Tc superconductors with critical temperatures of 32 and 36 K respectively. However, no break in the heat capacity curve of either compound is observed at the critical temperature for superconductivity. Heat capacity breaks in the heat capacity of IV ($La_{1.9}Sr_{0.1}CuO_{3.95}$) or samples of very close composition were observed by several authors.[11] However, the existence and magnitude of the break depends on the uniformity of the sample, sample preparation methods, and exact stoichiometry. There is not yet an agreement in the literature concerning the magnitude of the break and its dependence on stoichiometry in this chemical system. In contrast to classical superconductors, the transformation from normal state to the superconducting state in this system is broad, as evident from the magnetic susceptibilities of these samples.[2] This broadness may mask the break in the heat capacity. It is also evident that the transformation to the superconducting state in V is not complete and even at 4 K only about 1 % of the material is superconducting. The magnitude of the Meisner effect in IV is about an order of magnitude larger than that of V under the same conditions. During the course of the heat-capacity measurement two series of runs through the critical temperature region of IV were conducted. Three series of measurement through the critical temperature region were conducted for sample V.

A small hump in the heat capacity of IV was observed at about 300 K. This excess contribution is associated with the orthorhombic to tetragonal transformation at that temperature. The agreement of this transition temperature with literature values strongly supports the sample composition reported here and provides further evidence for its being a single phase.

Anomalies Between 230 and 290 K

Anomalous behavior is observed in the heat capacity, magnetic susceptibility, and electrical resistivity of V ($La_{1.4}Sr_{0.6}CuO_{3.70}$) between 234 and 290 K.[2] These anomalies are strongly dependent on the thermal history of the sample. Annealing the sample at or slightly above the temperatures in which an anomaly is observed tends to suppress the phenomena. Consequently, repeating a measurement does not necessarily reproduce the same results since the thermal history of the specimen was affected by the previous measurement. However, measurements taken under the same condition and the same thermal history tend to be reproducible.

During the course of the heat-capacity experiment, three series of measurements were conducted. Prior to the first series the sample (which was quenched from 1050 K to room temperature during synthesis and was maintained at that temperature) was cooled to 4.9 K and was heated slowly during the course of the measurement. The sample was heated and cooled through the critical temperature (36 K) three times during that experiment before being heated through the region in which the anomalous contributions were observed. Three peaks, each larger then the experimental uncertainty at that temperature, are observed in the heat-capacity curve between 234 and 290 K. These anomalies were not reproduced in the next two series of heat-capacity measurements. The sample was cooled slowly to 180 K and to 190 K, respectively, prior to these two series of measurements. No anomalous behavior was observed in either series.

Two sets of susceptibility measurements over the temperature region 200 to 300 K were conducted using a 1 g powdered sample. For the first set the sample was cooled from 303 to 200 K during the course of the measurement. For the second set the sample was cooled to liquid helium temperature, maintained at that temperature for some time then heated to 200 K and the measurement repeated. The two sets of measurements clearly disagree although performed on the same sample, (which was not removed from the magnetometer) during experiments on a given sample.

An approximate 1 g portion of the sample used for the heat capacity measurement was pelleted and sintered at 1050 K for about 48 hours. The pellet was then quenched to room temperature and cut into several rods. The electrical resistivity of two of these rods was measured between 2.4 and 330 K. At low temperatures the sample is a semiconductor with a small decrease in the resistivity at the critical temperature for superconductivity (ref. 2 figure 6). The resistivity between the critical temperature and 234 K is almost constant. At 234 K an increase of two orders of magnitude in the resistivity is observed for the first sample rod. The magnitude of the jump varies with thermal history and heating rate. Maintaining the sample at 234 K causes the resistivity at that temperature to return to its pre 234 K level and the anomaly is shifted to a higher temperature. The same phenomena reoccurs upon annealing the sample at the temperature to which the anomaly is shifted. However, the annealing time required to suppress the anomaly increases with temperature. The anomaly is restored only upon cooling below the critical temperature, although the exact temperature to which the sample has to be cooled was not identified. Both the resistivity anomaly and its dependence on thermal history are reproducible. The second sample rod was annealed at 380 K in air prior to the resistivity measurement. No anomalous behavior is observed in the resistivity of this sample. It is believed that the different thermal history of these samples accounts for the difference in properties.

Anomalous behavior of thermophysical properties of high Tc superconductors, between 200 and 300 K, have been previously reported.[11] However these anomalies tend to be irreproducible and were often described as the "now you see it now you don't" phenomena. The anomalies reported here do not fit the above terminology; they are all reproducible and their dependence on thermal history is characterized. The dependence of these anomalies on thermal history and their tendency to shift to higher temperatures upon annealing explains the oddly shaped heat-capacity curve between 230 and 290 K. It is likely that these series of anomalies observed over that temperature region are a single anomaly which is propagating to higher temperature as the sample temperature increases. Various methods of heat-capacity measurements (such as scanning calorimetry, AC, or continuous heating methods) may reveal a variety of apparently different anomalies as a function of heating rates and thermal histories.

$LaSrCuO_{3.50}$ and La_2CuO_4

The major motivation for measuring the heat capacities of these two compounds[3] is to assist in the development of an interpolation method between compounds of different stoichiometry within the same chemical system. Substitution of La in La_2CuO_4 with Sr allows for the compositions which produce high Tc superconductivity. Thus by simple chemical substitution one may move almost freely from La_2CuO_4 (VII) through $LaSrCuO_{3.50}$ (VI) to Sr_2CuO_3 (II). Thus knowing the thermophysical properties of these three phases, an interpolation method may be developed and the properties of related compounds may then be calculated. A similar interpolation method has been previously developed by the authors for other chemical systems such as the lanthanide sesquisulfides.[6,12] These methods take advantage of developments in macroscopic properties such as the molar volumes,[13] masses, or combinations of the two.[14] Once such a method is obtained a "generic" heat capacity for compounds in this chemical system may be generated and the lattice contribution to the heat capacity of compounds of any given composition within the system may readily be calculated. The development of such an interpolation system along with the analysis of the heat capacities of VI and VII are the subject of reference 3.

ACKNOWLEDGMENTS

The authors wish to acknowledge John Shewchun for his assistance in the magnetic susceptibility and electrical resistivity measurements. The preparation of this manuscript at the University of Illinois at Chicago was supported by the Solid State Chemistry Program of the Division of Materials Research of the National Science Foundation under grant DMR-8815798.

REFERENCES

1. R. Shaviv, E. F. Westrum Jr., T. L. Yang, C. B. Alcock, B. Li, J. Chem. Thermodyn. (in press).
2. R. Shaviv, E. F. Westrum Jr., C. B. Alcock, B. Li, J. Chem. Thermodyn. (to be submitted).
3. B. Li, C. B. Alcock, R. Shaviv, E. F. Westrum Jr., J. Chem. Thermodyn. (to be submitted).
4. B. Li, Ph.D. Dissertation, University of Notre Dame, Notre Dame, IN, USA, 1990.

5. C. B. Alcock, B. Li, J. Am. Ceram. Soc. (in press).
6. R. Shaviv, Ph. D. Dissertation, The University of Michigan, Ann Arbor, MI, USA, 1988.
7. E. Fuente, C. K. Kenesey, R. Shaviv, E. F. Westrum Jr., H. Fjelvag, A. Kjekshus, J. Solid State Chem. (to be submitted).
8. H. Kitaguchi, M. Ohno, M. Kiach, J. Takada, A. Osaka, Y. Mitura, Y. Ikeda, M. Takano, Y. Bando, Y. Takeda, R. Kanno, O. Yamamoto, J. Ceram. Soc. Jpn. Intl. Ed. 96, 388 (1988).
9. R. S. Roth, C. J. Rawn, J. D. Whitler, C. K. Chiang, W. K. Wong-Ng, J. Am. Ceram. Soc. 72(3), 395 (1989).
10. P. Ferloni, private communication.
11. For reviews on the subject see: R. A. Fisher, J. E. Gordon, N. E. Phillips, Superconductivity, 1, 231 (1988); H. E. Fischer, S. K. Watson, D. G. Cahill, Comm. on Condensed Matter Phys. 14, 65 (1988); E. Gmelin, in High Temperature Superconductors, A. V. Naritkar, Ed. (Nova Science, New York, 1989); A. Junod, in Physical Properties of High Temperature Superconductors II, D. M. Ginsberg, Ed. (World Scientific, Singapore, 1989), and references within.
12. R. Shaviv, E. F. Westrum Jr., J. B. Gruber, B. J. Beaudry, P. E. Palmer, J. Chem Phys. (to be submitted).
13. R. D. Chirico, E. F. Westrum Jr., J. Chem Thermodyn, 12, 71 (1980); 12, 311 (1980); 13, 519 (1981); 13, 1087 (1981).
14. N. Komada, Ph.D. Dissertation, University of Michigan, Ann Arbor, MI, USA, 1985.

O 1s x-ray absorption spectroscopy of $Tl_2Ba_2Ca_2Cu_3O_{10}$ high-T_C superconductors

A. Krol, C. S. Lin, Y. L. Soo, Z. H. Ming, and Y. H. Kao

Department of Physics and New York State Institute on Superconductivity,

State University of New York at Buffalo, New York 14260.

Y. Ma, C. T. Chen, and F. Sette,

AT&T Bell Laboratories, Murray Hill, New Jersey 07974

Jui H. Wang and Min Qi,

Department of Chemistry, State University of New York at Buffalo, New York

14260.

G. C. Smith

Brookhaven National Laboratory, Upton, New York 11973

Abstract

X-ray absorption around oxygen K-edge of $Tl_2Ba_2Ca_2Cu_3O_{10}$ high-T_C superconductors was measured by means of a bulk sensitive x-ray fluorescence yield detection method. Three distinct pre-edge peaks are revealed. They are ascribed to core-level excitations of oxygen 1s electrons to empty states at Fermi level which have predominantly oxygen 2p character. These oxygen holes are located in CuO_2, BaO and TlO planes. The strong dependence of T_C on oxygen holes concentration on O(1) site in CuO_2 layer is found.

We have recently demonstrated that fluorescence yield is a bulk-sensitive method in x-ray absorption spectroscopy (XAS) and gives rise to results independent of the superconductor surface condition.[1] This method was employed in the XAS studies of single phase polycrystalline $Tl_2Ba_2Ca_2Cu_3O_{10+\delta}$ high-T_C superconductors.

The samples were grown by the method described elsewhere.[2] The oxygen content in the investigated samples was controlled by reduction in flowing H_2 in the 200-280 °C range.[2] The characteristic x-ray diffraction (XRD) pattern was not altered by this process, and was very similar to that of the single phase $Tl_2Ba_2Ca_2Cu_3O_{10}$ superconductor.[3] The oxygen content δ was estimated by weight loss measurements and titration method. These two independent methods gave rise to the establishment of the relation between T_C and δ in this superconductor system.[2] T_C and XRD were measured before and after XAS experiment thus providing us with information on the oxygen content in the investigated samples.

The O 1s normalized (in the range 532-560 eV) absorption spectra of $Tl_2Ba_2Ca_3Cu_3O_{10+\delta}$ with T_C equal to 114 and 92 K are shown in **Fig. 1 and 2**, respectively.

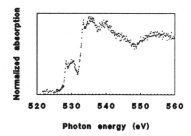

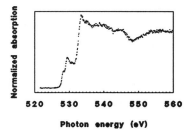

Fig. 1. O 1s normalized absorption spectrum of $Tl_2Ba_2Ca_2Cu_3O_{10+\delta}$ with $T_C = 114$ K.

Fig. 2. O 1s normalized absorption spectrum of $Tl_2Ba_2Ca_2Cu_3O_{10+\delta}$ with $T_C = 92$ K.

The XAS measurements were carried out at the AT&T Bell Laboratories soft-x-ray Dragon beamline at the National Synchrotron Light Source. A low-pressure parallel-plate chamber was used to detect the fluorescence emission in single-photon counting.[4] The obtained spectra can be roughly divided into two parts: below and above photon energy $h\nu \approx 532$ eV. The low energy part, which consists of so called pre-peaks, is attributed to transitions from the oxygen 1s core states to holes, with 2p symmetry, on the oxygen sites. The high energy part, which appears as the very steep rise in the spectral weight, is mainly due to transitions to Tl 6p and Ba 5d, Ca 3d and Ba 4f empty states hybridized with O 2p states.[5] Contrary to the high energy part, the low energy portion of the XAS spectrum exhibits strong dependence on T_C, i.e. on the oxygen content. In the crystal structure of $Tl_2Ba_2Ca_2Cu_3O_{10+\delta}$ one should expect 4 nonequivalent oxygen sites.[3] See Fig. 3.

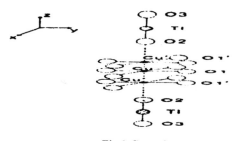

Fig.3 Crystal structure of $Tl_2Ba_2Ca_2Cu_3O_{10+\delta}$.

O(1) and O(1)′ in the CuO_2 layers, O(2) in the Ba and O(3) in the Tl planes. Mayer et al.[5] decomposed the broad XPS line into three components as follows: the lowest line is due to O(2) atoms, second line due to O(1) and O(1)′ atoms, and the third line with the highest-binding-energy is due to O(3) atoms. However, from the local density approximation (LDA) the calculated core levels of O(1) and O(1)′ (which are expected to be very close to each other) are predicted to be the highest, 1.2 eV below is O(2) level, which in turn is 1 eV above O(3) level which is the deepest.[6]   Romberg at al.[7] have arrived at the same assignment based on EELS investigations of $Tl_2Ba_2CaCu_2O_{8+\delta}$ single crystals.   Utilizing the two latter results one can readily ascribe the first absorption edge at 527.8 eV to the core-level excitations of oxygen 1s electron to O 2p holes located on O(1) and O(1)′ atoms, the second edge at 529 eV to O 2p holes on O(2) atoms in Ba planes, and the third edge at 530.5 eV to the O 2p on the O(3) atoms in Tl planes. The position of these absorption edges are very close to those observed by EELS.[7]   From the band structure calculations formation of O(2)-Tl-O(3) bridges in the direction of c is expected.[6,8,9]   Those bridges consist of Tl 6s and Tl $5d_{3z^2-r^2}$ orbitals hybridized with O(2) and O(3) $2p_z$ orbitals.   Hybridization is very strong due to relatively short distances of about 2 Å between Tl and O atoms.   One can estimate the relative concentration of O 2p holes on different lattice sites by comparing intensity of the relevant pre-peaks in the absorption spectra obtained for different oxygen contents.   From our experiment we obtained $I[O(1),O(1)′]_{114}/I[O(1),O(1)′]_{92} = 4.0$, $I[O(2)]_{114}/I[O(2)]_{92} = 1.0$, and $I[O(3)]_{114}/I[O(3)]_{92} = 1.2$, where $I[O(j)]_T$ denotes intensity of the absorption line related to O(j) lattice site at the temperature T. From these results one can conclude that concentration of holes on oxygen in CuO_2 planes strongly decreases with decreasing oxygen content, and only very moderately on oxygen in the O(2)-Tl-O(3) bridges. This confirms the active role of the dpσ^* band created in the CuO$_2$ planes in Tl- based cuprates similar to that found in other cuprate superconductors.

This research is supported at the State University of New York at Buffalo by AFSOR Grant No. AFSOR-88-0095 and DOE Grant No. DE-FG02-87ER45283 and at Brookhaven by DOE Grant No. DE-AC02-76CH00016.

References

1. A. Krol, C. S. Lin, C. J. Sher, Z. H. Ming, and Y. H. Kao, C. T. Chen, and F. Sette, Y. Ma, D. C. Smith, Y. Z. Zhu, and D. T. Shaw, Phys. Rev. B **42**, 2635 (1990).

2. J. H. Wang, M. Qi, X. Liu, and A. Petru, Physica C **166**, 399 (1990).

3. C. C. Toradi et al. Science, **240**, 631 (1988).

4. G. C. Smith, A. Krol, and Y. H. Kao, Nucl. Instr. Meth. A **291**, 135 (1990).

5. H. M. Meyer III, T. J. Wagener, and J. H. Weaver, and D. S. Ginley, Phys. Rev. B **39**, 7345 (1989).

6. P. Marksteiner, J. Yu, S. Massidda, A. J. Freeman, J. Reinger, and P. Weinberger, Phys. Rev. B **39**, 2894 (1989). J. Yu, S. Massidda, and A. J. Freeman, Physica C **152**, 273 (1990).

7. H. Romberg, N. Nücker, M. Alexander, J. Fink, D. Hahn, T. Zetterer, H. H. Otto, and K. F. Renk, preprint.

8. R. V. Kasowski, W. Y. Hsu, F. Herman, Phys. Rev. B **38**, 6470 (1988).

9. D. R. Hamann and L. F. Mattheiss, Phys. Rev. B **38**, 5138 (1988).

EMPIRICAL SEARCH FOR NEW AND HIGHER-TEMPERATURE PEROVSKITE-TYPE OXIDE SUPERCONDUCTORS

Z. Z. Sheng

Department of Physics, University of Arkansas,
Fayetteville, AR 72701

ABSTRACT

Although high Tc oxide superconductors have different chemical compositions, they are all perovskite-type. In this paper, we define high Tc oxide superconductor-forming elements, and show distribution of these elements in the periodic table. We present a novel expression for the high Tc perovskite-type oxide superconductors, which is as simple as a chemical formula, but contains important structural information. Using this expression, we list most of known high Tc superconductors. We found that a perovskite-type superconductor with certain layers of metallic atoms may reach a highest Tc, and the highest Tc is a function of total metallic atom layers: Tc(highest) = f x L (K) when L $\leq$ 3 x (E-1), where E is the number of metallic elements, L is the number of metallic atom layers, and f is a coefficient. Therefore, higher Tc perovskite-type superconductors should be sought in more complex systems.

INTRODUCTION

The initial discovery of 30 K superconductivity in the LaBaCuO system by Bednorz and Muller[1] has stimulated a race for even higher Tc superconductors. The 90 K YBaCuO system[2] increased Tc to above liquid nitrogen temperature for the first time, the 110 K BiSrCaCuO system[3] first showed a superconducting onset above 100 K, and the 120 K TlBaCaCuO system[4,5] reached the highest Tc of 125 K to date. In addition to these milestone discoveries, various other oxide

210   © 1991 American Institute of Physics

superconductors have also been discovered. Some of them are very different from each other, for example, the rare earth-free 90 K $Tl_2Ba_2CuO_6$[6-8], the alkaline earth-free 40 K La_2CuO_4[9], the Cu-free 30 K $(Ba,K)BiO_3$[10,11], the n-type 24 K $(Nd,Ce)_2CuO_4$[12], and the six-metal 105 K $(Tl,Pb)Sr_2(Ca,Y)Cu_2O_x$[13]. Despite this difference, all high Tc oxide superconductors reported so far are perovskite-type. In this paper, we do not describe crystallographical details, rather give an overview of the perovskite-type superconductors and summarize some general guidelines of empirical search for new and even higher Tc superconductors. We define high Tc oxide superconductor-forming elements, and show distribution of these elements in the periodic table. We present a novel expression for the perovskite-type oxide superconductors, which is as simple as a chemical formula, but contains important structural information. Using this expression, we list most of known high Tc oxide superconductors. We found that the highest Tc that a perovskite-type superconductor with certain layers of metallic atoms may reach is a function of total metallic atom layers: Tc(highest) = f x L (K) when L ≤ 3 x (E-1), where E is the number of metallic elements, L is the number of metallic atom layers, and f is a coefficient. We conclude that higher Tc perovskite-type superconductors should be sought in more complex systems.

HIGH Tc OXIDE SUPERCONDUCTOR-FORMING ELEMENTS

In this paper, a high Tc oxide superconductor is sometimes called a high Tc superconductor, or an oxide superconductor, or simply, a superconductor. We would like to define the term "superconductor-forming elements". A superconductor-forming element is an element which is required for a superconducting crystal. For example, Y, Ba, and Cu are superconductor-forming elements because they all are required for the superconducting crystal $YBa_2Cu_3O_x$[2,14]. A superconductor-forming element may also be an element whose occurrence is required for superconductivity of a superconducting crystal. For example, K is a superconductor-forming element since K

is required for the 30 K superconductivity of the $(Ba,K)BiO_3$[10,11].
Figure 1 shows the periodic table of the elements in which the
superconductor-forming elements are marked by dark color. It can be
seen that the superconductor-forming elements are distributed in the

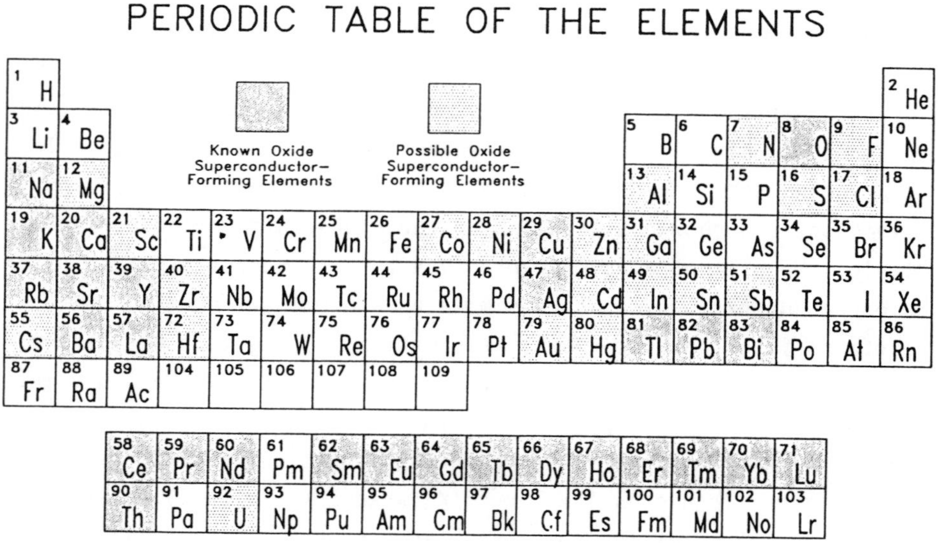

PERIODIC TABLE OF THE ELEMENTS

Figure 1   Distribution of known and possible oxide
superconductor-forming elements in the periodic
table of the elements.

left and right parts of the periodic table. We call the
superconductor-forming elements in the left part as L elements, and
those in the right part as R elements. So far, Na, K, Rb, Ca, Sr,
Ba, RE (rare earths), and Th are known L elements, whereas Cu, Ag,
Tl, Pb, and Bi are known R elements. In figure 1, elements Cs, Mg,
Sc, U, Zr, and Hf are marked, by light grey color, as L elements,
and elements Co, Ni, Au, Zn, Cd, Hg, Al, Ga, In, Sn, and Sb as R
elements. These elements are called as possible superconductor-
forming elements. They are selected according to their chemical
properties and ionic radii, and the preliminary experimental results
in our own lab and other labs. Although this paper does not deal
with non-metallic elements, for completion, oxygen is marked as a

superconductor-forming element, and N, S, F, and Cl as possible
superconductor-forming elements. Other non-metallic elements and
the elements between L and R elements have very few data, and remain
to be investigated. It is noteworthy that some radioactive
elements, such as Pm and Ra, are certainly superconductor-forming
elements, but are not marked in the periodic table because of their
high radioactivity.

NOVEL EXPRESSION FOR PEROVSKITE-TYPE OXIDE SUPERCONDUCTORS

The chemical formula of a perovskite-type superconductor, for
example, $YBa_2Cu_3O_x$, does not give any structural information.
However, structural information is extremely important to the study
of superconductors. Here we present a novel expression, which is as
simple as a chemical formula, but gives basic structural information
of a superconductor. Figure 2 shows schematically the arrangements
of metallic atoms in a unit cell of $YBa_2Cu_3O_x$[14] and in half unit
cell of $Tl_2Ba_2CaCu_2O_x$[8,15]. It can be seen that in $YBa_2Cu_3O_x$, Ba and
Y atoms of L elements are in a line Ba-Y-Ba, and Cu atoms of R
element are in a line Cu-Cu-Cu. Similarly, in $Tl_2Ba_2CaCu_2O_x$, Ba-Ca-
Ba are in a line, and Tl-Cu-Cu-Tl in a line. Therefore, if oxygen

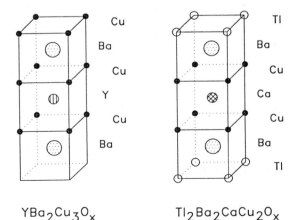

$YBa_2Cu_3O_x$ $Tl_2Ba_2CaCu_2O_x$

Figure 2   Arrangements of metallic atoms in a unit cell of
$YBa_2Cu_3O_x$ and in half unit cell of $Tl_2Ba_2CaCu_2O_x$.

atoms are omitted, $YBa_2Cu_3O_x$ may be written as BaYBa/CuCuCu, or YBa_2/Cu_3, or Ba_2Y/Cu_3, and $Tl_2Ba_2CaCu_2O_x$ as BaCaBa/TlCuCuTl, or BaCaBa/TlCu$_2$Tl. Crystallographically, the atoms to the left of the slash are A atoms, and the atoms to the right of the slash are B atoms. Using this expression, various crystalline structures of perovskite-type oxide superconductors can be presented by a manner as simple as chemical formulas. We found that all known high Tc oxide superconductors can be classified into three groups:

(a) A_{n+1}/B_n,

(b) A_n/B_n, and

(c) A_n/B_{n+1},

where n is the number of metallic atoms in a line. Using this expression, we list most of known high Tc oxide superconductors in Table I for further discussion. Note that for A_{n+1}/B_n type superconducting crystals, arrangements of metallic atoms in a unit cell are

$$
\begin{vmatrix} A_{n+1} & B_n \\ B_n & A_{n+1} \end{vmatrix}
$$

for example, for $(La,Ba)_2/Cu$,

$$
\begin{vmatrix} La,Ba & La,Ba & Cu & \\ & Cu & La,Ba & La,Ba \end{vmatrix}
$$

Similarly, for A_n/B_{n+1} type superconducting crystals, arrangements of metallic atoms in a unit cell are

$$
\begin{vmatrix} A_n & B_{n+1} \\ B_{n+1} & A_n \end{vmatrix}
$$

for example, for BaCaBa/TlCu$_2$Tl,

$$
\begin{vmatrix} & Ba & Ca & Ba & Tl & Cu & Cu & Tl \\ Tl & Cu & Cu & Tl & Ba & Ca & Ba & \end{vmatrix}
$$

However, for A_n/B_n type superconducting crystals, arrangements of metallic atoms are simply

$$
\begin{vmatrix} A_n \\ B_n \end{vmatrix}
$$

for example, in BaYBa/Cu$_3$,

$$
\begin{vmatrix} Ba & Y & Ba \\ Cu & Cu & Cu \end{vmatrix}
$$

Table I   Perovskite-type High Tc Oxide Superconductors

Structure	Tc (K)	Reference
(1) A_{n+1}/B_n type (half unit cell)		
La_2 /Cu	40	9
$(LaCa)_2/Cu$*	20	17
$(LaSr)_2/Cu$	40	17
$(LaBa)_2/Cu$	30	1,17
$(LaNa)_2/Cu$	40	27
$(PrCe)_2/Cu$	24	28
$(NdCe)_2/Cu$	24	12,28,29,30
$(SmCe)_2/Cu$	24	28
$(EuCe)_2/Cu$	24	28
$(PrTh)_2/Cu$	24	28
$(NdTh)_2/Cu$	24	28,30
(2) L_n/R_n type (unit cell)		
Ba_2RE/Cu_3	90	2,14
Sr_2Y /Cu_3	80	31
Ba_2Y /Ag_3	50	16
Ba /(BiPb)	13	32
(BaK)/ Bi	30	10,11
(BaRb)/ Bi	30	11
Ba_2 /TlCu	13	24,33
Ba_2Ca /$TlCu_2$	91	24,33
$Ba_2Ca_2/TlCu_3$	116	24,33
$Ba_2Ca_3/TlCu_4$	122	24
$Ba_2Ca_4/TlCu_5$	117	24
Sr_2Ca /$TlCu_2$	70	18,34
$Sr_2Ca_2/TlCu_3$	100	34
$Sr_2Ca/(TlBi)Cu_2$	75	35
$(SrRE)_2$ / Tl Cu	46	36
$Sr_2(CaRE)/$ Tl Cu_2	85	19
$Sr_2(CaRE)/(TlPb)Cu_2$	105	13
Sr_2Ca /(TlPb)Cu_2	85	20

Table I continued.

Structure	Tc (K)	Reference
$Sr_2Ca_2/(TlPb)Cu_3$	115	20
$SrCuSr(CaRE)/Pb_2Cu_2$	70	21,22,23
(3) L_n/R_{n+1} type (half unit cell)		
Ba Ba/TlCu Tl	90	6,7,8,37
BaCa Ba/TlCu$_2$Tl	110	4,8,15,38,39,40
BaCa$_2$Ba/TlCu$_3$Tl	125	4,8,39,40
BaCa$_3$Ba/TlCu$_4$Tl	115	24
Sr Sr/BiCu Bi	20	41
SrCa Sr/BiCu$_2$Bi	80	3,42
SrCa$_2$Sr/BiCu$_3$Bi	110	3,42
SrCa Sr/(BiPb)Cu$_2$(BiPb)	80	43
SrCa$_2$Sr/(BiPb)Cu$_3$(BiPb)	110	43

* ( ) denotes the atoms which occupy the same site in a crystalline lattice.

DISCUSSIONS

(1) New High Tc Superconductors

From Table I, it can be seen that in the existing perovskite-type superconductors, A atoms consist of L elements, and B atoms consist of R elements. New oxide superconductors may also consist of L elements and R elements. In other words, appropriate combinations of known and possible L and R elements may lead to new and even higher Tc superconductors. Note that L elements have low electronegativity, and most of them are valence-fixed, whereas R elements have high electronegativity, and all of them are valence-variable. The planes of A atoms and B atoms arrange one by one at either c-axis or a-and b-axis. We believe that this configuration is one of the pre-requirements for a perovskite-type oxide being a high Tc superconductor. It is noteworthy that according to the above point of view, Cu is not required for high Tc oxide superconductivity. In fact, Cu-free Ba_2Y/Ag_3 is superconducting at 50 K[16], and in particular, Cu-free (Ba,K)/Bi is superconducting at 30 K[10,11]. Therefore, the search for new high Tc superconductors should not be restricted to cuprates, although cuprate have been and will be the major object in the search for new superconductors.

(2) Elemental Substitutions May Lead to Higher Tc

In a unit cell (or half unit cell) of a superconducting crystal, A atoms and B atoms are in their respective line. A atoms may be substituted fully or partially by other A atoms, while the crystal structure remains unchanged. The same is true for B atoms. This kind of substitutions may lead to higher Tc superconductors, for example, 30 K $(La,Ba)_2/Cu^1$ -> 40 K $(La,Sr)_2/Cu^{17}$. The following is another example:

(1) Sr_2 Ca   / Tl Cu_2 (70 K)[18]
(2) $Sr_2(CaRE)/$ Tl Cu_2 (85 K)[19]
(3) Sr_2 Ca   /(TlPb)Cu_2 (85 K)[20]
(4) $Sr_2(CaRE)/(TlPb)Cu_2$ (105 K)[13]

It is interesting to point out that in $SrCuSr(Ca,RE)/Pb_2Cu_2$[21-23], one of Cu atoms plays the role of an A atom. This is due to the fact that this Cu atom is in its low oxidized state (Cu^{1+}). It seems that the atom of L element going to the A site is not helpful to increase, rather decrease the Tc. Partial substitution of Tl (L element) for Ca (A site) in $BaCaBa/TlCu_2$ also decrease Tc[44].

(3) The Crystals with More Metallic Atom Layers May Reach
    Higher Tc

It is believed that Cu-O planes in oxide superconductors are responsible for the high Tc superconductivity. It is also believed that Tc's of Tl-based superconductors are a function of the number of Cu-O layers in a unit cell (or half unit cell) (see, for example, Ref. 24). However, Tc's of $BaCa_{n-1}Ba/TlCu_nTl$ with double Tl-O layers are not the same as, but higher than those of the corresponding $BaCa_{n-1}Ba/TlCu_n$ with single Tl-O layer. We found that Tc's of the Tl-based superconductors are, more reasonably, a function of the number of total layers of metallic atoms in a unit cell (or half unit cell), as shown in Figure 3 (data taken from Ref. 24). This means that even if only Cu-O layers are responsible for high Tc oxide superconductivity, the B atom Tl and the A atoms Ba and Ca in the Tl-based superconductors also influence Tc in a similar manner. In figure 3, we also plotted Tc of the two-layer superconductor (Ba,K)/Bi and Tc of the three-layer superconductor La$_2$/Cu. These two data points fall fairly well into the same trend,

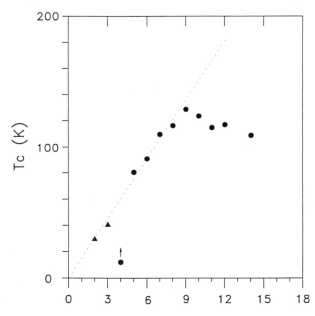

Figure 3
Tc's as a function of the number of total layers of metallic atoms in a unit cell (or half unit cell) of Tl-based superconductors (data taken from Ref. 24). The triangles are Tc of 30 K (Ba,K)/Bi (two layers) and 40 K La$_2$/Cu (three layers).

The number of layers of metallic atoms

suggesting that this relationship between Tc and the number of metallic atom layers is generally valid to all high Tc superconductors. The Tc of the four-layer Ba$_2$/TlCu reported in Ref. 24 is low (13 K). We believe that the highest Tc of four-layer superconductors would be about 60 K, and that Tc of Ba$_2$/TlCu could increase up to about 60 K by improved preparation procedures, or a new 60 K four-layer superconductor could be discovered in the future. In fact, four-layer superconductor (Sr,RE)$_2$/TlCu with Tc of about 46 K has been reported[45]. It is noteworthy that the nine-layer BaCa$_2$Ba/TlCu$_2$Tl reaches a Tc of 127 K. More than nine layers,

the Tc's decrease with increasing number of metallic atom layers.
There are two possibilities: (i) the Tc of 127 K (or slightly
higher) is the highest possible Tc in the four-metal systems, or
(ii) TlBaCaCuO superconductors with more than nine layers can reach
higher Tc, because the Tc of a superconducting crystal depends
strongly on its preparation procedure. We note that 162 K
superconductivity in the TlBaCaCuO system has been reported[25].
However, according to the analyses in the next paragraph, we are
inclined to reject the possibility (ii).

(4) <u>The Systems with More Metallic Elements May Reach a Higher
    Highest Tc</u>

The existing data show that a system with a certain number of
metallic elements has a certain highest Tc. For example,

one-metal   (non-perovskite-type)
  NbO   1.25 K[26]

two-metal   (three layers, A_{n+1}/B_n type)
  La_2/Cu 40 K[9]

three-metal (six layers, A_n/B_n type)
  Ba_2Y/Cu_3 90 K[2,14]

four-metal (nine layers, A_n/B_{n+1} type)
  $BaCa_2Ba/TlCu_3Tl$ 127 K[4,8,24]

It can be seen that a system reaches the highest Tc in a structure
with the number of metallic atom layers L = 3 x (E-1), where E is
the number of metallic elements in the system. Assuming that the
six-layer Ba_2Y/Cu_3 has reached the highest Tc of the three-metal
systems (90 K), we can estimate an increase of 45 K for each
metallic element and 15 K for each layer. Thus we can write an
empirical formula of the possible highest Tc for a superconductor
with L layers: when L ≤ 3 x (E-1),

Tc (highest) = 15 x L (K)--------------------------(1)

According to this equation, a four-metal system could reach a
highest Tc of 135 K. The Tc of 127 K of the $BaCa_2Ba/TlCu_3Tl$ has
been close to this highest Tc. If this equation can be extend to
more complicated system, one may surmise that a twelve-layer
superconductor in the five-metal systems could reach a highest Tc of

180 K. The dash line in Fig. 3 represents such a suggested trend.
Equation (1) may be written more generally as

Tc (highest) = f x L (K)----------------------------(2)

where f is a coefficient. This relationship of Tc and metallic atom
layers would be explained by any successful theory of high Tc
superconductivity. We would like to point out that the number of
metallic atom layers (L) of a perovskite-type crystal is roughly
proportional to its c-axis (or volume), and that the carrier
concentration in the normal state of a superconducting crystal is
related to its chemical formula.

(5) <u>Search for Higher Tc Superconductors in More Complex</u>
     <u>Systems</u>

A trend that the Tc which a superconducting system may reach
increases with complexity of the system has already been noted in
our previous paper[19]. Now we have had a much better understanding
of this trend. First, a more complex system containing more
elements may adjust more easily a crystal to reach higher Tc;
secondly, a more complex system may form crystals with more layers
which may reach higher Tc; third, a more complex system may reach a
higher highest Tc. At any rate, higher Tc perovskite-type oxide
superconductors should be sought from more complex systems. It is
noteworthy that the history of the discoveries of high Tc oxide
superconductors shows that preparation conditions are vital to
successful experiments. In the empirical search for higher Tc
perovskite-type superconductors, in addition to studying potential
higher Tc superconducting systems, one should also study appropriate
experimental conditions for the preparation of higher Tc
superconducting phases in these systems.

Acknowledgment The author thanks L.Sheng for his assistance in
preparing the figures in the manuscript, and Mrs. M.Megginson for
her assistance in the preparation of this manuscript. The author
would like to express thanks to Y.Xin and J.M.Meason for their
helpful discussion and assistance. This work was supported by the
Arkansas Energy Office grant AR/DNRG/AEO-UAF-88-005/S.

REFERENCES

1. J.G Bednorz, and A.K. Muller, Z.Phys.B 64, 189 (1986).
2. M.K.Wu, J.R.Ashburn, C.T.Torng, P.H.Hor, R.L.Meng, L.Gao, Z.J.Huang, Y.Q.Wang, and C.W.Chu, Phys.Rev.Lett. 58, 908 (1987).
3. H.Maeda, Y.Tanaka, M.Fukutomi, and T.Asano, Jpn.J.Appl.Phys.Lett. 27, L207 (1988).
4. Z.Z.Sheng and A.M.Hermann, Nature 332, 138 (1988).
5. Z.Z.Sheng, W.Kiehl, J.Bennett, A.El Ali, D.Marsh, G.D.Mooney, F.Arammash, J.Smith, D.Viar, and A.M.Hermann, Appl.Phys.Lett. 52, 1738 (1988).
6. Z.Z.Sheng and A.M.Hermann, Nature 332, 55 (1988).
7. Z.Z.Sheng, A.M.Hermann, A.El Ali, C.Almason, J.Estrada, T.Datta, and R.J.Matson, Phys.Rev.Lett. 60, 937 (1988).
8. R.M.Hazen, L.W.Finger, R.J.Angel, C.T.Prewitt, N.L.Ross, C.G.Hadidiacos, P.J.Heaney, D.R.Veblen, Z.Z.Sheng, A.El Ali, and A.M.Hermann, Phys.Rev.Lett. 60, 1657 (1988).
9. P.M.Grant, S.S.P.Parkin, V.Y.Lee, E.M.Engler, M.L.Ramirez, J.E.Vazquez, G.Lim, and R.D.Jacowitz, Phys.Rev.Lett. 58, 2482 (1987).
10. L.F.Mattheiss, E.M.Gyorgy, and D.W.Johnson Jr., Phys.Rev.B 37, 3745 (1988).
11. J.Cava. B.Batlogg, J.J.Krajewski, R.Farow, L.W.Rupp Jr., A.E.White, K.Short, W.F.Peck, and T.Kometani, Nature 332, 814 (1988).
12. Y.Tokura, H.Takagi, and S.Uchida, Nature 337, 345 (1989).
13. R.S.Liu, J.M.Liang, Y.T.Huang, W.N.Wang, S.F.Wu, H.S.Koo, P.T.Wu, and L.J.Chen, Physica C 162-164, 869 (1989).
14. J.Cava, B.Batlog, R.B.vanDover, D.W.Murphy, S.Sunshine, T.Siegrist, J.P.Remeika, E.A.Rietman, S.M.Zahurak, and G.Espinosa, Phys.Rev.Lett. 58, 1676 (1987).
15. M.A.Subramanian, J.C.Calabrese, C.C.Torardi, J.Gopalakrishnan, T.R.Askew, R.B.Flippen, K.J.Morrissey, U.Chowdhry, and A.M.Sleight, Nature 332, 420 (1988).
16. K.K.Pan, H.Mathias, C.M.Rey, W.G.Moulton, H.K.Ng, L.R.Testard,

and Y.L.Wang, Phys.Lett. A 125, 147 (1987).

17. J.G.Bednorz, A.K.Muller, and M.Tagasige, Science 230, 93 (1987).

18. Z.Z.Sheng, A.M.Hermann, D.C.Vier, S.Schultz, S.B.Oseroff, D.J.George, and R.M.Hazen, Phys.Rev.B 38, 7074 (1988).

19. Z.Z.Sheng, L.Sheng, X.Fei, and A.M.Hermann, Phys.Rew.B 39, 2918 (1989).

20. M.A.Subramanian, C.C.Torardi, J.Gopalakrishnan, P.L.Gai, J.C.Calabrese, T.R.Askew, R.B.Flippen, and A.M.Sleight, Science 242, 249 (1988).

21. R.J.Cava, B.Batlogg, J.J.Krajewski, L.W.Rupp, L.F.Schneemeyer, T.Siegrist, R.B.vanDover, P.Marsh, W.F.Peck,Jr, P.K.Gallagher, S.H.Glarum, J.H.Marshall, R.C.Farrow, J.V.Waszczak, R.Hull, and P.Trevor, Nature 336, 211 (1988).

22. M.A.Subramanian, J.Gopalakrishnan, C.C.Torardi, P.L.Gai, E.D.Boyes, T.R.Askew, R.B.Flippen, W.E.Farneth, and A.M.Sleight, Physica C 157, 124 (1989).

23. Z.Z.Sheng, C.Dong, X.Fei, Y.H.Liu, L.Sheng, and A.M.Hermann, Appl.Phys.Commun. 9, 27 (1989).

24. H.Ihara, M.Hirabayashi, N.Terada, M.Jo, K.Hayashi, M.Tokumoto, K.Murada, Y.Kimura, R.Sugise, T.Shimomura, and S.Ohashi., Proc. Intern. Superconductivity Sympo. on Superconductivity (August 28-31, 1988), Springer-Verlag, Tokyo, 1989.

25. P.S.Lui, P.T.Wu, J.M.Liang, and L.J.Chen, Phys.Rev.B 39, 2792 (1989).

26. Handbook of Chemistry and Physics, 62nd edition (1981-1982), edited by R.C. Weast, CRC Press, Inc., Boca Raton, Florida.

27. J.T.Market, C.L.Seaman, H.Zhou, and M.B.Mape, Solid State Commun. 66, 387 (1987).

28. J.T.Market, E.A.Early, T.Bjornholm, S.Ghamaty, B.W.Lee, J.J.Neumeier, R.D.Price, C.L.Seaman, and M.B.Maple., Physica C 158, 178 (1989).

29. J.M.Tranguada et al., Nature 337, 720 (1989).

30. J.T.Market, and M.B.Maple, Solid State Commun.(to be published).

31. M.K.Wu, J.R.Ashburn, C.A.Higgins, B.H.Loo, D.H.Burns, A.Ibrahim, T.D.Rolin, F.Z.Chien, and C.Y.Huang, Phys.Rev.B 37, 9765 (1988).

32. A.W.Sleight, J.L.Gillson, and P.E.Bierstedt, Solid State Commun. 17, 27 (1975).

33. S.S.P.Parkin, V.Y.Lee, A.I.Nazzal, R.Savoy, R.Beyers, and S.J.La Placa, Phys.Rev.Lett. 61, 750 (1988).

34. S.Matsuda, S.Takeuchi, A.Soeta, T.Suzuki, K.Aihara, and T.Kamo (to be published).

35. P.Haldar, S.Sridhar, A.Roig-Janicki, W.Kennedy, D.H.Wu, C.Zahopoulos, and B.C.Giessen, J.Superconductivity 1 211 (1988).

36. Z.Z.Sheng, C.Dong, Y.H.Lui, X.Fei, L.Sheng, J.H.Wang, and A.M.Hermann, Solid State Comm. 71, 739 (1989).

37. C.C.Torardi, M.A.Subramanian, J.C.Calabrese, J.Gopalakrishnan, E,M.McCarron, K.J.Morrissey, T.R.Askew, R.B.Flippen, U.Chowdhry, and A.M.Sleight, Phys.Rev.B 38, 225 (1988).

38. L.Gao, Z.J.Huang, R.L.Meng, P.H.Hor, J.Bechtold, Y.Y.Sun, C.W.Chu, Z.Z.Sheng, and A.M.Hermann, Nature 332, 623 (1988).

39. C.C.Torardi, M.A.Subramanian, J.C.Calabrese, J.Gopalakrishnan, K.J.Morrissey, T.R.Askew, R.B.Flippen, U.Chowdhry, and A.M.Sleight, Science 240, 631 (1988).

40. S.S.P.Parkin, V.Y.Lee, E.M.Engler, A.I.Nazzal, T.C.Huang, G.Gorman, R.Savoy, and R.Beyers, Phys.Rev.Lett. 60, 2539 (1988).

41. C.Michel, M.Hervieu, M.M.Borel, A.Grandin, F.Deslandes, J.Provost, and B.Raveau, Z.Phys. B 68, 421 (1987).

42. M.R.Hazen, C.T.Prewitt, R.J.Angel, N.L.Ross, L.W.Finger, C.G.Hadidiacos, D.R.Veblen, P.J.Heaneay, P.H.Hor, R.L.Meng, Y.Y.Sun, Y.Q.Wang, Y.Y.Xue, Z.J.Huang, L.Gao, J.Bechtold, and C.W.Chu, Phys.Rev.Lett. 60, 1174 (1988).

43. P.V.P.S.S.Sastry, I.K.Gopalakrishnan, J.V.Yakhmi, and R.M.Iyer, Physica C 157, 491 (1989).

44. V.I.Ozhogin (private communication).

45. See, for example, Z.Z.Sheng, L.A.Burchfield, and J.M.Meason (to be publishe in Mord.Phys.Lett.B).

MASS-SPECTROSOPIC STUDIES
ON THE BIASED LASER ABLATION OF $Ba_2Y_1Cu_3O_{7-\delta}$

H. Izumi, K. Ohata, T. Morishita, and S. Tanaka
Superconductivity Research Laboratory, ISTEC
10-13, Shinonome 1-chome, Koto-ku, Tokyo 135, Japan

ABSTRACT

Mass-spectroscopic studies on biased laser ablation of $Ba_2Y_1Cu_3O_{7-\delta}$ have been carried out by using time-of-flight mass-spectroscopy. The energies of the ions were hardly influenced by the electric field created by substrate biasing. This is probably because the ions in the plume are shielded from the external electric field by electrons. Energy differences between the atomic, monoxide and cluster ions have been observed. While the former has an energy of about 200 eV or more, the latter two have only a few eV. This implies that atomic ions are accelerated during the initial laser ablation process and monoxide and cluster ions are produced through plasma reactions in the plume. The multi-valenced ion species (Cu^{2+} and Cu^{3+}) were observed under positively biased conditions. They should be one of the key factors in preparing high quality superconductive thin films at lower deposition temperatures due to their high internal energies.

INTRODUCTION

Pulsed laser deposition is one of the most promising methods for *in-situ* preparation of superconducting oxide thin films[1-13]. Several efforts to reduce the substrate temperature have been carried out from the point of view of electronics applications. Witanachchi et al. have successfully reported a lower temperature deposition technique using a DC plasma induced in a coil set between the target and the substrate[14]. Koren et al. have reported that the decomposition of N_2O by a second excimer laser pulse, time delayed by a suitable period from the initial laser pulse, is effective in allowing the use of a lower deposition temperature[15]. Substrate bias voltages were also found to be significantly effective in the preparation of high quality superconducting $Ba_2Y_1Cu_3O_{7-\delta}$ films at the reduced deposition temperatures by the present authors[16].

Pulsed laser deposition is accompanied with a flash plasma on the target surface, which is usually called a "*plume*", caused by laser irradiation. Since the relation between the film formation and the effects of the enhancement techniques as mentioned above on laser ablation dynamics is still not clear, it is important to clarify the characteristics and nature of the excited states of the species in the plume in order to prepare high-quality superconducting thin films at lower deposition temperatures.   There are many efforts in laser plume diagnostics using time-resolved optical observation [17-23] and mass-spectroscopy [24-26] in order to clarify the nature of the laser plume.   While Koren et al. did not refer to the proof of the creation of the atomic oxygen [15], Witanachchi et al. also reported the presence of continuous plasma during the deposition [14].   We have revealed by a time-resolved optical observation the presence of the optical emission accompanying the laser plume [16].   We believe that this emission should indicate the occurrence of some activation reaction of ablated species in it, being effective in film formation.   There are, however, very few reports concerning the relationship between the creation of activated species and the film formation mechanism at lower deposition temperatures.   Therefore, the investigation of the activated states of the species in the laser plume with substrate biasing is of great interest, being relevant to the deposition mechanism.

In this paper, we have carried out mass-spectroscopic studies on biased laser ablation and investigated the energy distribution of ions in the laser plume to clarify the substrate biasing effect of the ionic species.

EXPERIMENTTAL

ArF (193 nm) excimer laser was used for ablation with a duration of 16 ns. The laser beam was introduced into the vacuum chamber through a fused silica window and focused on the target.   The energy density at the target surface was about $1 J/cm^2$.

Since the laser plume involves very complicated mass and energy distribution of the species and it is well known that there are clusters of the mass number up to several thousands in it [24], we have employed a time-of-flight mass-spectrometry for the mass analysis of the

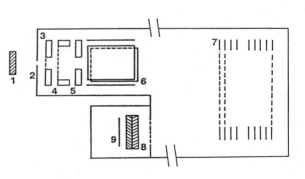

Figure 1   A schematic diagram of the time-of-flight mass spectrometer. 1: target, 2: sampling orifice, 3, 4, 5: retarding circuit, 6: bending electrode, 7: reflectron circuit, 8: multi channel plate, 9: collector.

plume.   This method is also able to obtain directly the ion energies.   As a time-of-flight mass-spectrometer (TOF-MS), we made use of a reflectron type TOF-MS (MSLT-2000: ULVAC JAPAN LTD.), not a simple time-of-flight arrangement, in which the ions introduced from the sampling orifice are turned to the opposite way by the reflection potential and go to the multi channel plate (MCP) set beside the sampling orifice. A schematic diagram is shown in Fig.1. This arrangement enables the attainment of a long drift distance of the ions even using a short analyzing tube and enhances the sensitivity in mass and energy separation. The effective drift distance in our tube is 2.067 m. The resulting mass spectra were recorded with a digital storage oscilloscope (DS-8631: Iwatsu Electronics Co.) synchronized with the laser pulse.

Mass-spectroscopic study has been carried out by replacing the substrate holding mechanism of our deposition chamber[16] with the TOF-MS. The sampling orifice was set at the position of the substrate, so we were able to analyze the ions impinging onto the substrate in our setup and determine directly the effects of ions on the film formation process. The identification of ions was carried out by the comparison with the results from the calculation for the each species versus the potential of each target, retarding and reflection. Most of signals has been successfully identified.

The energy distribution for the ions was also determined by varying the retarding potential applied to the grid (retarding circuit in Fig.1) set just behind the sampling orifice. The retarding also has an effect on reducing the noise from the plasma of the laser plume and sharpening the mass spectra. For the better spectra sampling, we usually fixed the target potential at +1 kV. Bias voltages were applied by balancing the potential difference between the target and the grid set in front of the sampling orifice.

RESULTS AND DISCUSSION

Mass analysis of the laser plume is very difficult due to its wide energy and mass distributions of the included species, and the most serious problem is that the laser plume is a kind of plasma in which very complicated reactions take place. For example, the result in Fig.2 seems to be quite strange for the ion velocity against the electric field. This figure shows the mass spectra of the plume of $Ba_2Y_1Cu_3O_{7-\delta}$ for the various target potentials. Even though the target potential is negative, the detected ion current is as large as those for the grounded and positive potentials. It seems that the motion of the ablated ions is hardly influenced by the external electric field.

By the way, when we measured the transient current between the target and ground during laser ablation, it was observed that very large number of electrons were emitted from the target[27]. This emission was suppressed in the case of negative grid potentials against the target and greatly enhanced by positive potentials. There are as many electrons in the plume emitted from the target by laser ablation as there are ions and these electrons are expected to have the ability to shield the inside of the plume from the external field due to their large mobility. The mass and energy analysis for the ablated ions, therefore, should be done by extracting ions from the laser plume to avoid the plasma effect on the mass spectra.

This problem were solved by equipping the retarding grid just behind the sampling orifice. The ions passing through the orifice feel the retarding potential and the low energy ones are repulsed. Only the ions getting over the retarding potential reach the grounded electrode set behind the retarding one and begin again to drift to the detector. In general, TOF-MS does the mass separation to let the ions drift a certain distance between the orifice and the detector then accelerating them up to an adequate energy for the analyzing tube geometry. Though the sampling orifice is helpful in reducing the obstacle noise from the plume by itself, there still remains an effect from the plasma in the mass spectra as long as a conventional arrangement is used. The retarding circuit was equipped in our TOF-MS just behind the orifice and this device enables us to decrease effectively the obstacle noise coming from the plume, and it serves as a means of sharpening the mass spectra. Figure 3 shows a typical spectra obtained from the plume of $Ba_2Y_1Cu_3O_{7-\delta}$ at the target, retarding and reflection

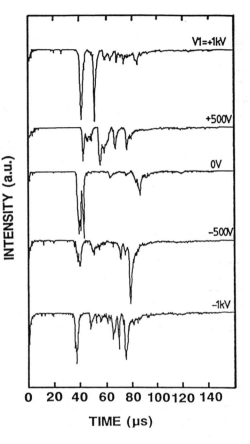

Figure 2   The mass spectra of the plume of $Ba_2Y_1Cu_3O_{7-\delta}$ target without retarding for the target potentials of +1, +0.5, 0, -0.5 and -1 kV, respectively. The reflection potential was +1.3 kV and the other potentials were grounded.

potentials of +1, +0.8 and +1.3 kV, respectively. It can be seen that the noise from the plasma is greatly reduced and each peak is sharpened compared with Fig.2.

We are also able to analyze the ion energies easily by varying the retarding potential. Fig.4 shows the mass spectra of the plume of $Ba_2Y_1Cu_3O_{7-\delta}$ for the various retarding potentials. The target and reflectron potential were fixed at +1.0 kV and +1.3 kV, respectively. The signals from

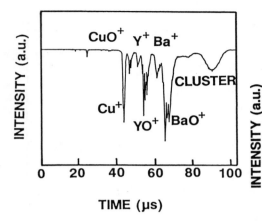

Figure 3   A typical mass spectrum of the plume of the $Ba_2Y_1Cu_3O_{7-\delta}$ target. The target, retarding and reflection potentials were +1, +0.8 and +1.3 kV, respectively. The pressure in the process chamber was 1.3 Pa and that in the analyzing tube was 2×10^{-4} Pa. The noise from the plasma of the plume are greatly reduced and each peak is sharpened.

the monoxides of each metal and the cluster ions can be seen in Fig.4 (a) as well as those from the atoms. The clusters observed in this figure are estimated to correspond to mass numbers of several hundreds and heavier ones were not detected. The detailed composition of the clusters is not clear, but there may be some reason for these mass numbers to be dominant. While the signals from monoxide ions have larger intensities than those from atomic ones for Y and Ba, Cu^+ is dominant over CuO^+. This may be due to the characteristics of copper in the $Ba_2Y_1Cu_3O_{7-\delta}$ target and it implies that some additive oxidation process may be useful in preparing high quality films.

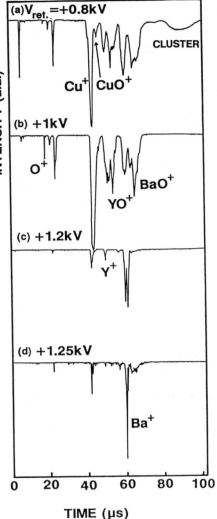

Figure 4   The mass spectra of the plume of the $Ba_2Y_1Cu_3O_{7-\delta}$ target for the retarding potentials of +0.8, +1 and +1.2 kV, respectively. The target and reflection potentials were +1 and +1.3 kV, respectively. The pressure in the process chamber was 1.3 Pa and that in the analyzing tube was 2×10^{-4} Pa.

Increasing the retarding potential to +1 kV, the signals from cluster ions fall off rapidly (Fig.4 (b)). This indicates that the cluster ions have come into the sampling orifice, which corresponds to the substrate in the case of film deposition, with very low energies. Monoxide ions are also diminished at the higher retarding potential of +1.2 kV. The detail examination has revealed that monoxide ions have energies of the order of a few tens of eV. On the other hand, atomic ions have energies more than 200 eV so that the signals from them still remain, even at the retarding potential of 1.2 kV, as large as those in Fig.4 (a) and (b). At the retarding potential of +1.25 kV, only very weak signals were detected except for Ba^+, so the atomic ions' energies are mainly distributed between 200 to 250 eV but Ba^+ has an energy tail in the energy distribution higher than those of the other metals. The obtained value for the atomic ion energies here are slightly high compared with the previously reported ones which are about 100 eV or less[17,18,25]. On the other hand, the resulting energy values of our experiments are confirmed by the observation of the change in

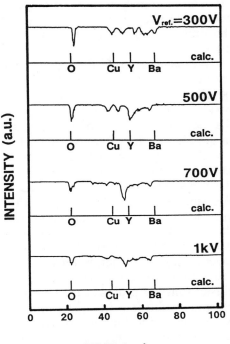

Figure 5   The mass spectra of the plume of the $Ba_2Y_1Cu_3O_{7-\delta}$ target for the reflection potentials of +0.3, +0.5, +0.7 and +1 kV, respectively. The pressure in the process chamber was 1.3 Pa and that in the analyzing tube was 2×10^{-4} Pa. The detection times for O^+, Cu^+, Y^+ and Ba^+ calculated with the assumption of an initial energy of 200 eV are also shown for each spectrum.

mass spectra against the reflection potentials. The detected times of the ions becoming faster as the reflection potential is increased because the penetration depth of ions into the reflection potential is shortened.   Figure 5 shows the mass spectra of the plume of the $Ba_2Y_1Cu_3O_{7-\delta}$ target for various reflection potentials, and the calculated values for each atomic ion with the assumption that the ion energy is 200 eV are also seen in this figure. All potentials except the reflection were grounded, so the spectra are not so clear. There is, nevertheless, a good agreement between the experimentally observed peaks and the calculated values of the detected times for atomic ions. Further prudent examination must be made to discuss the differences in the reported energy values, but we believe that time-of-flight mass-spectroscopy is one of the most suitable ways to obtain the information about the ion energies directly.

We have reported the value of the order of several eV for the emissive species from a time-resolved optical observation[16] and they were thought to be electrically neutral. Saenger has also reported an energy difference between ions and neutrals[18]. In addition to this, our experiments indicate that neutrals, atomic, monoxide and cluster ions each have different energies in the plume. It is implied by such a wide energy distribution that the ablation mechanism generating these species is very complicated. We suppose that atomic ions with very high energy are generated at the initial stage of the ablation process and there is some mechanism to accelerate ions. On the other hand, monoxide and cluster ions are probably created by plasma reactions in the plume. Since monoxide ions are expected to be directly created by the collision between the high energy atomic ions and oxygen molecules, they should have higher energies than clusters. It is also possible that some of the monoxide ions are created from the decomposition of large clusters. However, we believe that the latter case is not dominant because the various fragments should be observed at the same time and they are not. Clusters are thought to be emitted from the target into the plume with very low energies and they are probably charged in plasma afterward.

We applied TOF-MS for analysis of the ions in a biased laser ablated plume. Figure 6 shows the results of the plume of $Ba_2Y_1Cu_3O_{7-\delta}$ for the bias voltages

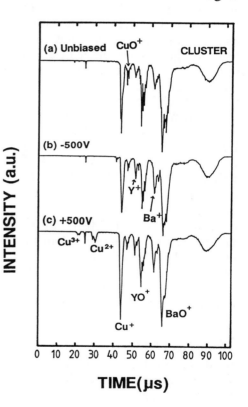

Figure 6   The mass spectra of the plume of $Ba_2Y_1Cu_3O_{7-\delta}$ biasing with 0, -500 and +500 V, respectively. The target, retarding and reflection potentials were +1, +0.8 and +1.3 kV, respectively. The pressure in the process chamber was 1.3 Pa and that in the analyzing tube was 2×10^{-4} Pa.

of 0, -500 and +500 V, respectively. The retarding and reflectron potentials were fixed at +0.8 and +1.3 kV, respectively. While the detected times for each ion seem not to be influenced by the bias voltages, it should be noted that the multi-valenced ions, Cu^{2+} and Cu^{3+}, are obviously observed in the case of the positive bias voltage.

Since the ion energies are expected not to be affected by the external electric field due to plasma shielding from the results in Fig.2, it is natural that the ion energies are hardly influenced in this figure. On the other hand, it is very interesting that positive bias voltages have the effect of generating multi-valenced ions. It has been also reported by the present authors that the electrons emitted from the target to the laser plume are greatly increased when positive bias voltages are applied[27]. When positive bias voltages are applied, the electrons generated by ablation probably are torn off rapidly from the target surface. It can be qualitatively explained that the multi-valenced ions tend to be generated because excess positive charges are left there. It is also expected that the multi-valenced ions would have a higher internal energy than the other ions (for example, ionization potentials for Cu^+ and Cu^{2+} are 7.724 and 20.29 eV, respectively), so they should be helpful for the film formation and be one of the key factors in preparing high quality superconducting thin films at lower deposition temperatures in the biased laser ablation method.

CONCLUSIONS

Mass-spectroscopic studies of biased laser ablation plume have been carried out by using a time-of-flight mass analysis method. Equipping the retarding circuits in our TOF-MS, it has been possible to reduce the noise from plasma and to investigate successfully the energies of ions. Atomic, monoxide and cluster ions were observed in laser plume. While the monoxide ions are more plentiful in the plume than the atomic ones for the cases of Y and Ba, Cu^+ is dominant over CuO^+. Therefore, copper is transported from the target to the substrate under the less oxidized condition compared with the other metals.

Energy differences between the atomic, monoxide and cluster ions have been observed from the energy analysis. While the former have energies of about 200 eV or more, the latter have energies of a few eV. This indicates that the atomic ions are accelerated in the initial laser ablation process and the monoxide and cluster ions are probably produced by plasma reactions in the plume.

Ions were hardly influenced by the external electric field formed by substrate biasing. This is because the ions in the laser plume are shielded by the electrons from the external electric field. On the other hand, the multi-valenced ions (Cu^{2+} and Cu^{3+}) were observed under positively biased conditions. It can be qualitatively explained that the multi-valenced ions tend to be generated because excess positive charges are left there. They are also expected to have a higher internal energy than the other ions, so they should be one of the key factors in preparing high quality superconducting thin films at lower deposition temperatures in the biased laser ablation method.

ACKNOWLEDGEMENT

A part of this work was supported by New Energy and Industrial Technology Development Organization under the management of the R&D Basic Technology for Future Industries. We also thank Mr. Kubo from ULVAC JAPAN LTD. for the assistance in mass-spectroscopy analysis and fruitful discussions.

REFERENCES

1) A. Inam, M. S. Hedge, X. D. Wu, T. Venkatesan, P. England, P.F. Miceli, E. W. Chase, C. C. Chang, J. M. Tarascon, and J. B. Watchman, Appl. Phys. Lett., 53, 908(1988).
2) G. Koren, A. Gupta, E. A. Giess, A. Segmuller, and R. B. Laibowitz, Appl. Phys. Lett., 54, 1054(1989).
3) J. Narayan, N. Buinno, R. Singh, O.W. Holland, and O. Auciello, Appl. Phys. Lett., 51, 1845(1987).
4) W. A. Weimer, Appl. Phys. Lett., 52, 2171(1988).
5) S. Deshmuck, E. W. Rothe, G. P. Reck, T. Kushida, and Z. G. Wu, Appl. Phys. Lett., 53, 2698(1988).
6) H. S. Kwok, P. Mattocks, L. Shi, X. W. Wang, S. Witanachchi, Q. Y. Ying, J. P. Zheng, and D. T. Shaw, Appl. Phys. Lett., 52, 1825(1988).
7) R. E. Muenchausen, K. M. Hubbard, S. Foltyn, R. C. Estler, N. S. Nogar, and C. Jenkins, Appl. Phys. Lett., 56, 578(1990).
8) E. Fogarassy, C. Fuchs, P. Siffert, J. Perriere, X. Z. Wang, and F. Rochet, Solid State Comm., 67, 975(1988).
9) D. K. Fork, J. B. Boyce, F. A. Ponce, R. I. Johnson, G. B. Anderson, G. A. N. Connell, C. B. Eom, and T. H. Geballe, Appl. Phys. Lett., 53, 337(1988).
10) L. Lynds, B. R. Weinberger, G. G. Peterson, and H. A. Krasinski, Appl. Phys. Lett., 52, 320(1988).
11) A. M. Desantolo, M. L. Mandich, S. Sunshine, B. A. Davidson, R. M. Fleming, P. Marsh, and T. Y. Kometani, Appl. Phys. Lett., 52, 1995(1988).

12) H. U. Habermeier, and G. Mertens, Physica, C **153-155**, 1429(1988).

13) O. Eryu, K. Murakami, K. Takita, K. Masuda, H. Uwe, H. Kudo, and T. Sakudo, Jpn. J. Appl. Phys., **27**, L628(1988).

14) S. Witanachchi, H. S. Kwok, X. W. Wang, and D. T. Shaw, Appl. Phys. Lett., **53**, 234(1988).

15) G. Koren, A. Gupta, and R. J. Basemann, Appl. Phys. Lett., **54**, 1920(1989).

16) H. Izumi, K. Ohata, T. Hase, K. Suzuki, T. Morishita, and S. Tanaka, J. Appl. Phys., (1990), in printing.

17) P. E. Dyer, R. D. Greenough, A. Issa, and P. H. Key, Appl. Phys. Lett., **53**, 534(1988).

18) K. L. Saenger, J. Appl. Phys., **66**, 4435(1989).

19) O. Eryu, K. Murakami, K. Masuda, A. Kasuya, and Y. Nishina, Appl. Phys. Lett., **54**, 2716(1989).

20) X. D. Wu, B. Dutta, M. S. Hedge, A. Inam, T. Venkatesan, E. W. Chase, C. C. Chang, and H. Howard, Appl. Phys. Lett., **54**, 179(1989).

21) O. Auciello, S Athavale, O. E. Hopkins, M. Sito, A. F. Schreiner, and N. Biunno, Appl. Phys. Lett., **53**, 72(1988).

22) C. Girault, D. Damiani, J. Aubreton, and A. Catherinot, Appl. Phys. Lett., **55**, 182(1989).

23) J. P. Zheng, Q. Y. Ying, S. Witanachchi, Z. Q. Huang, D. T. Shaw, and H. S. Kwok, Appl. Phys. Lett., **54**, 954(1989).

24) C. H. Becker and J. B. Pallix, J. Appl. Phys., **64**, 5152(1988).

25) T. Venkatesan, X. D. Wu, A. Inam, Y. Jeon, M. Croft, E. W. Chase, C. C. Chang, J. B. Watchman, R.W. Odom, F. R. di Brozolo, and C. A. Magee, Appl. Phys. Lett., **53**, 1431(1988).

26) L. Lynds, B. R. Weinberger, D. M. Potrepka, G. G. Peterson, and M. P. Lindsay, Physica, C **159**, 61(1989).

27) H. Izumi, K. Ohata, T. Morishita, and S. Tanaka, to be published in IEEE Trans. on Mag, **27**, March 1991.

EFFECT OF MAGNETIC FIELD ON THE RESISTIVE TRANSITION AND CRITICAL CURRENT DENSITY OF Y-Ba-Cu-O/Ag ROLLED TAPES

G.Swaminathan, S.Rajendra Kumar, N.Ramadas, K.Venugopal,
K.A.Durga Prasad, M.V.T.Dhananjeyan and R.Somasundaram
Corporate Research & Development Division,
Bharat Heavy Electricals Ltd., Hyderabad-500593, INDIA

ABSTRACT

Critical current density is an important parameter for large scale application of superconductors. Oxide superconductors however carry only small critical currents though intra-grain currents are estimated to be high. Rolling of oxide superconductors is one of the promising processes to achieve higher current densities as it improves the density, grain contact and texture of the material. Silver addition to the basic composition is known to improve the electrical properties of the oxide superconductors. Rolled tapes of Y-Ba-Cu-O with 15% and 30% Ag additions were investigated for the effect of magnetic field and its orientation on resistive transition and critical current density. The measurements were carried out at 77 K and for fields up to 90 Oe in directions perpendicular and parallel to the sample. The observed shift in resistive transition for field orientation perpendicular and parallel to the current flow gives an understanding of the percolation path of currents. The nature of shift and Jc(B) curves is discussed in terms of the effectiveness of coupling across the grains. These measurements could lead to optimization of additive concentration for achieving maximum critical current density for a given process.

INTRODUCTION

There have been considerable efforts to improve critical current density in High Tc Superconductors. With the basic granular nature of these superconducters the critical current density is limited by the intergrain contact effectiveness. The very short coherence length makes electrical and magnetic properties exceedingly sensitive to crostructural inhomogenieties and intergrain junction dimensions [1-4]. Silver addition has been observed to improve the intergrain contacts, hence electrical and mechanical properties in Y-Ba-Cu-O systems [5-9]. The nature of resistive transition is an indicator of effectiveness of intergrain coupling which can be inferred based on the n values of logV - logI plots [4]. The resistive transition is an extrinsic property in these materials and depends on the fabrication process, additive concentration etc. The resistive transition variation for field orientation has a bearing on the percolation paths, intergrain contact effectiveness and intergrain junction dimensions.

In the present study we report our measured data of resistive transition in low magnetic fields, up to 90 Oe with field orientation perpendicular and parallel to the rolled tape samples of 123 with 30% Ag. and 15% Ag. additions. The predominance of percolation paths and the critical current density variation as a function of magnetic field Jc (B) is analysed to assess the dimensions of intergrain contact introduced due to the additive concentration.

EXPERIMENTAL PROCEDURE

$YBa_2Cu_3O_x$ powder was prepared by the usual solid state reaction technique. The powder was filled in copper tube of ID 6mm without and with 15% and 30% Ag additives and rolled into tapes. From the final tape 650 microns thick, 200mm long and 10mm in width, several samples of 30-40mm long were prepared by leaching out copper in HNO_3 acid and sintering them at $930^{\circ}C$ followed by low temperature oxygenation. The sintered samples were lightly emeried. Silver paste was applied and fired at $450^{\circ}C$ for 30 min. followed by slow cooling in O_2 atmosphere for making electrical contacts. A small glass cryostat was placed inside a long thin solenoid with a large uniform zone for measuring the resistive transitions and Jc determination at low magnetic fields. The measurements were carried out using dc four probe technique at zero field and for field levels of 2 to 90 Oe. The samples were mounted on fixture to orient the samples in either parallel or perpendicular direction with respect to the magnetic field. Twisted and shielded cables were used for measurements and the contacts were made by direct soldering of measurement leads on to the silver paste by using 2% Ag lead tin solder.

RESULTS

The critical currents at 77 K for rolled tapes without Ag additions were less than 10 A/cm^2 and dropped to negligible values even in the presence of 2 Oe field. In comparison critical current densities were in the range 100-130 A/cm^2 for 30% Ag 123 (30 Ag) tapes and 150-180 A/cm^2 for 15% Ag 123 (15 Ag) tapes.

Figs.1a and 1b are the resistive transition curves of 30 Ag sample for perpendicular and parallel field orientations respectively. Figs.2a & 2b are the corresponding curves for 15 Ag sample. The Jc (B) curves for both 30 Ag and 15 Ag samples are given in Fig.3. The n values from zero field logV – logI plots for 30 Ag and 15 Ag tapes are calculated to be 11 and 4.5 respectively. 1 microvolts/cm voltage criterion was used for Jc determination.

The Jc values for the samples at 88 Oe field and 77 K are as shown in Table 1:

TABLE 1

Sample	Area reduction	Jc (ZF) A/cm^2	Jc (88 Oe)* A/cm^2	n
Y-Ba-Cu-O	2	10	< 0.1	4
30 Ag	2	128	8	11
15 Ag	2	148	20	4.5

* Field perpendicular to the sample

ANALYSIS AND DISCUSSIONS

From Table 1 it can be seen that for the given processing conditions, the Jc values at 88 Oe field for all the samples are reduced to nearly 10% of the zero field values. The nature of weak link behaviour is more evident from the resistive transition and Jc (B) curves for fields in the range of 0 to 10 Oe.

The take off current density from resistive transition and rate of shift of resistive transition as illustrated by Jc (B) curves in 30 Ag sample (Figs.1a, 1b and 3) shows remarkable difference for parallel and perpendicular field orientations. Up to 8 Oe, the Jc value reduced to 40% of zero field value and the take off current reduced in similar proportion for perpendicular orientation. In comparison 10% reduction was observed for parallel orientation. This indicates that percolation paths are predominantly in the rolling direction. The large Jc (B) variation can be attributed to the inability of the grain boundaries to transmit super currents in the presence of magnetic field indicating larger inter grain junction dimension. One of the contributing factor being silver additive. This can be verified from Figs.2a, 2b and 3 for sample with reduced additive concentration, 15% Ag. The Jc (B) variation is much smaller, 18% for 8 Oe perpendicular field, indicating a smaller junction dimension. This conclusion could be held reasonable as the processing conditions are the same for both samples and Jc (zero field) has also shown an increase for 15 Ag sample. From the micrograph (Fig.4) it was observed that the extent of grain alignment was poor with pockets of silver regions. The resistive transition curve for 15 Ag sample also indicate poor inter grain contact. From this analysis it is observed that in spite of improved grain contact, excess silver addition leads to larger junction dimension leading to large variations in Jc (B). This could effectively be used to optimize additive concentration for a given processing and heat treatment conditions. In our study the reduction ratio employed was very low. Higher reduction ratios would lead to higher Jc due to improved grain contacts, grain alignment and reduced junction dimensions because of the ductility of silver. This should be appreciated in the light of higher Jc (B) values obtained for conventional superconductor, Nb-Ti, for reduction ratios in excess of 10,000 [10]. Presently 10% silver sheathed tapes are processed with different reduction ratios and for a reduction ratio of 55 Jc in the range of 300-550 A/cm^2 is being obtained.

CONCLUSIONS

From the study of resistive transition for low fields and for different field orientations in Y-Ba-Cu-O/Ag rolled tapes. We find that for a given processing condition one can estimate the predominance of percolation path to estimate the extent of alignment. Also the influence of additive concentration upon the junction dimension is estimated from Jc (B) curves. These results can be used for optimizing the additive concentration as well as modifying the process to improve the extent of grain alignment to get higher critical current densities in these materials.

ACKNOWLEDGMENTS

The authors wish to thank the BHEL management for giving permission to publish this work. The authors also wish to thank Department of Science and Technology for funding this work.

REFERENCES

1   D.M.Kroeger, J. Minerals, Metals & Mater. Soc. 41, 14 (1989)

2   H.Obara, H.Yamasaki, Y.Kimura and S.Kosaka, Appl. Phys. Lett. 55, 2342 (1989)

3   C.M.Gilmore, Mater. Sci. and Engg. B1, 283 (1988)

4   J.E.Evetts, B.A.Glowacki, P.L.Sampson and M.G.Blamire, IEEE Trans. Mag. 25, 2041 (1989)

5   B.Dwir, M.Affronte and D.Pavuna, Appl. Phys. Lett. 55, 399 (1989)

6   D.Pavuna, H.Berger, J.L.Tholence, M.Affronte. R.Sanjines, A.Dubas, Ph.Bugnon and F.Vasey, Physica C 153–155, 1339 (1988)

7   J.P.Singh, D.Shi and D.W.Capone II, Appl. Phys. Lett. 53, 1113 (1988)

8   P.Strobel, C.Paulsen and J.L.Tholence, Solid State Commun. 65, 585 (1988)

9   J.W.Ekin, T.M.Larson, N.F.Bergren, A.J.Nelson, A.B.Swartzlander, Kazmerski, A.J.Panson and B.A.Blankenship Appl. Phys. Lett. 52, 1819 (1988)

10   D.A.Collin, T.A.de Winter, W.K.McDonald and W.C.Turner, IEEE Trans. Mag. MAG–13, 848 (1977)

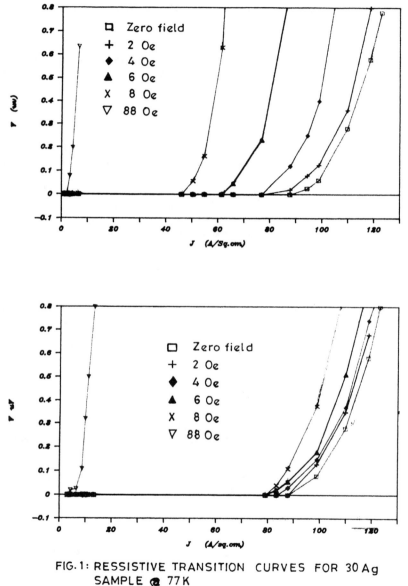

FIG.1: RESSISTIVE TRANSITION CURVES FOR 30 Ag
    SAMPLE @ 77K
  a: FIELD PERPENDICULAR TO THE ROLLING DIRECTION
  b: FIELD PARRALEL TO THE ROLLING DIRECTION

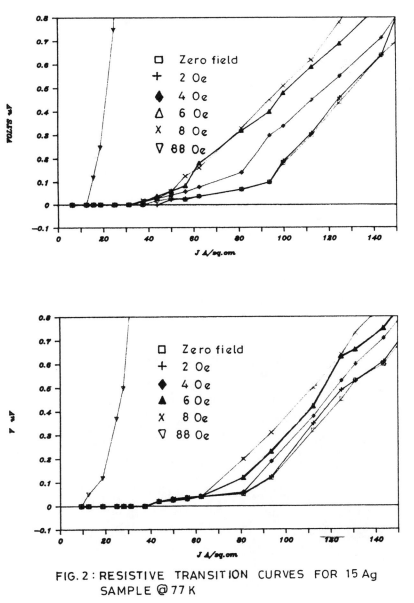

FIG. 2 : RESISTIVE TRANSITION CURVES FOR 15 Ag
SAMPLE @ 77 K

a: FIELD PERPENDICULAR TO THE ROLLING DIRECTION
b: FIELD PARRALEL TO THE ROLLING DIRECTION

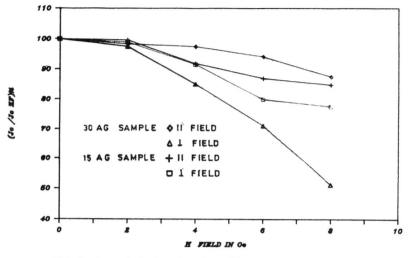

FIG.3: Jc(B) CURVES FOR 30 Ag SAMPLE AND 15 Ag
SAMPLE FOR DIFFERENT FIELD ORIENTATION
TO ROLLING DIRECTION.

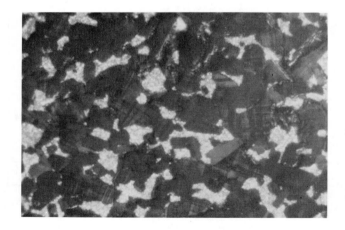

FIG.4: MICROGRAPH OF 30 Ag SAMPLE

AC SUSCEPTIBILITY OF HIGH-Tc SUPERCONDUCTORS
IN A MODEL OF A GRANULAR JOSEPHSON MEDIUM

V.A.Khlus, A.V.Dyomin

Institute for Single Crystals, Kharkov, USSR 310141

ABSTRACT

Low-field ac magnetic properties of high-Tc superconductors
are studied using a model of a granular system with weak links
between grains. The existence of closed loops consisting of grains
connected by small area Josephson contacts is assumed. The periodicity condition of the superconducting order parameter phase leads
to the family of the energy branches corresponding to different
phase 'winding' numbers. When the loop self-inductance L is taken
into account, the hysteresis occurs for the applied magnetic flux
close to half-integer values of flux quantum Φ_0. The width of the
hysteresis range is proportional to $LI_0/N\Phi_0$, where N is the number
of weak links in the loop, I_c is the contact critical current. In
this model the ac susceptibility is found, its imaginary part is
due to the energy dissipation in the branch-switching processes.
The dependence of the susceptibility on the ac field amplitude, the
second harmonic power of the magnetization vs. dc magnetic field,
temperature variation of the dissipative part of the susceptibility
are calculated. Statistical properties of the medium are taken into
account by an average in loop areas.

INTRODUCTION

Many properties of the new high-Tc oxide superconductors may
be understood assuming the existence of weak Josephson links both
between grains in ceramic samples and at twin boundaries inside the
grains or crystallites[1,2]. Strong nonlinearities observed in low-
field magnetic characteristics of these materials may be explained
in terms of Josephson properties of a granular medium formed by
superconducting grains embedded in a nonsuperconducting host. In
this paper we present some theoretical results on ac magnetic
susceptibility of such composite systems.

Following the approach suggested by Ebner and Stroud[3] we focus
our attention on closed loops formed by superconducting grains
weakly connected by Josephson junctions. The relevance of this loop
model to real disordered composites with random values of intergrain weak coupling energies was confirmed[3] by Monte Carlo
numerical simulations. Closed loops with a great number of weak
contacts may play an important role in determining magnetic
properties of disordered granular systems near the percolation
threshold, where a network of one-dimentional clusters of strongly
coupled grains is formed[4]. For these reasons it seems instructive
to consider a simple model of independed multicontact closed loops
in an applied magnetic field. We shall find the magnetic

susceptibility dependence on the ac field amplitude and the dc magnetic field bias, and the temperature dependence of the susceptibility assuming the Ambegaokar-Baratoff[5] relation between the critical current I_c and the temperature.

The periodicity condition imposed on the superconducting order parameter phase leads to the appearance of multivalued energy function of the loop vs. the magnetic flux. Different branches of that function correspond to different phase 'winding' numbers. We suggest that hysteresis effects may be important in this situation when the final loop self-inductance is taken into account. The hysteresis leads to energy dissipation when jumps between different energy states occur. These effects may be responsible for the dissipative part of the ac susceptibility.

THE MODEL

Loops with large number of Josephson contacts ($N \gg 1$) will be considered below, and the loop self-inductance L enters the results through the parameter

$$\lambda = 2\pi L I_c / N\Phi_0, \tag{1}$$

where I_c is the critical current, and Φ_0 is the flux quantum. The assumption of equal critical currents made here is, of cause, an oversimplification of the situation. Effects of disorder, which are important in real systems, are simulated by averaging over the loop areas. The value of L in (1) is proportional to the loop perimeter and scales as the number of grains in the loop, N, assuming the grain sizes to be nearly equal. This means that λ may be taken constant at the averaging procedure.

For typical granular high-Tc superconductors[6] the grain size is much larger than the magnetic field penetration depth in the grains, and we may also assume that the intergrain contacts are small compared to the Josephson penetration length. In this case only the gauge-invariant phase difference between neighboring grains

$$\varphi_{i,j} = \theta_i - \theta_j - \frac{2\pi}{\Phi_0} \int_{x_i}^{x} A dx \tag{2}$$

are the relevant variable, where $\theta_{i,j}$ are the phases of the order parameter at the grain centers, and the integral of the magnetic vector potential **A** is taken along the line connectingthe grain centers.

The requirement of single-valuedness of the order parameter results in the equation

$$\sum_{j=1}^{N} \varphi_j + \frac{2\pi\Phi}{\Phi_0} = 2\pi n , \tag{3}$$

where φ_j is the phase difference at the jth contact, and the total magnetic flux Φ is related to the external flux Φ_e and the loop current I by

$$\Phi = \Phi_e + LI \tag{4}$$

The Hamiltonian of the system is the sum of the weak coupling energy and the energy of the magnetic field

$$\mathcal{H} = E_0 \sum_{j=1}^{N}(1-\cos\varphi_j) + \frac{1}{2L}(\Phi-\Phi_e)^2, \tag{5}$$

where $E_0 = \Phi_0 I_c/2\pi$ is the Josephson coupling energy.

The external magnetic flux is $\Phi_e=SH(t)$, where $H(t)=H_0+H_1\sin\omega t$. The diamagnetic shielding in disconnected grains may be taken into account introducing the effective magnetic permeability of the medium (see e.g.[ ]) depending on the volume fraction of the superconducting material and on the ratio of the temperature dependent field penetration depth to the grain size.

THE THERMODYNAMIC EQUILIBRIUM SOLUTION

The thermodynamic properties of our model are obtained from the partition function which involves integration over all phase variables φ_j and the total magnetic flux Φ,

$$\mathcal{Z} = C\int\frac{d\Phi}{\Phi_0}\int\prod_j\frac{d\varphi_j}{2\pi}\exp\left(-\frac{\mathcal{H}}{T}\right)\sum_n \delta\left(\sum\varphi_j+\frac{2\pi\Phi}{\Phi_0}-2\pi n\right) \tag{6}$$

where the periodically continued δ-function is introduced to satisfy the eq.(3). The exact value of the dimensionless normalization constant C is unimportant for our purposes. Using the Fourier representation of the δ-function we obtain the Gaussian integral over Φ. The phase integrations can be performed easily in the limit $E_0 \gg T$, and the final result for the field-dependent part of the loop free energy is

$$F = -T\ln\left[1 + 2\sum_{m=1}^{\infty}\cos\left(m\frac{2\pi\Phi_e}{\Phi_0}\right)\exp\left\{-m^2\frac{NT}{2E_0}(1+\lambda)\right\}\right] \tag{7}$$

The magnetic moment density may be expressed as

$$M = n_L\frac{4\pi T}{\Phi_0}\left\langle S\frac{\sum_{m=1}^{\infty}m\sin\left(m\frac{2\pi\Phi_e}{\Phi_0}\right)\exp\left\{-m^2\frac{NT}{2E_0}(1+\lambda)\right\}}{1 + 2\sum_{m=1}^{\infty}\cos\left(m\frac{2\pi\Phi_e}{\Phi_0}\right)\exp\left\{-m^2\frac{NT}{2E_0}(1+\lambda)\right\}}\right\rangle_S \tag{8}$$

where brackets mean an average over the loop area S, and n_L is the number of loops in the unit volume. The magnetization is periodic function in Φ_e with the period Φ_0. For the applied external field of the frequency ω higher harmonics appear and $M(t)$ may be written in the form

$$M(t) = \sum_{k=1}^{\infty}\left(M_k'\sin(k\omega t) + M_k''\cos(k\omega t)\right) \tag{9}$$

In the following we shall calculate the magnetic susceptibility components,

$$\chi' = M_1'/H_1, \qquad \chi'' = M_1''/H_1,$$

where H_1 is the ac field amplitude.

One can use eq.(8) for ac magnetization calculations only in the limit of zero frequency, because the energy of the system has many local minima in the phase space separated by energy barriers. The thermally activated fluctuations can establish the full thermodynamic equilibrium only after a long enough time, but the characteristic time values may be very large. This means that even at low field frequencies one has to consider metastable states of the system and transitions between such states.

NONLINEAR MAGNETIC SUSCEPTIBILITY

In the case of identical contacts minimum values of the energyof the system are achieved at $\varphi_i = \varphi$, where φ is related to themagnetic flux by

$$N\varphi = 2\pi \left(n - \Phi/\Phi_0 \right) \tag{10}$$

Under this restriction the loop energy becomes the function of the variable Φ only, and the final minimization in Φ should be made to find the stable states of the system. The energy is the multivalued function of Φ, and its branches may be indexed by the 'winding' number n

$$E_n(\phi) = NE_0 \left[1 - \cos\frac{2\pi}{N}(\phi - n) \right] + \frac{\Phi_0^2}{2L}(\phi - \phi_e)^2 , \tag{11}$$

where the dimensionless variables are introduced, $\phi = \Phi/\Phi_0$, $\phi_e = \Phi_e/\Phi_0$. The functions $E_n(\phi)$ are shown in fig.1a.

When the inductance L is neglected , the magnetic energy in (11) disappears ($\phi = \phi_e$) and the total loop energy vs. ϕ_e is determined by the lowest energy value available among $E_n(\phi)$ at $\phi = \phi_e$.

Different energy branches cross at $\phi_e = (n+1/2)$ and transitions between them are possible. For a non-zero L a degenerate energy minimum, say at $\phi_e = 1/2$, is splitting into two minima corresponding to the different internal flux values,

$$\phi_{\pm} = \frac{1}{2}\left(1 \pm \frac{\lambda}{1+\lambda}\right), \tag{12}$$

where it is assumed that $N \gg 1$. The energy barrier exists between the neighboring minima preventing branch-crossing transitions at $\Phi_e = (n+1/2)\Phi_0$ as e.g. in the Ebner-Stroud's theory[3]. We assume that the transition does take place when the branch-crossing point coincide with the energy minimum on one of the branches involved, and the energy barrier height becomes zero (fig.1b,c). After the transition having occurred the contact phase differences and the total flux relax to the new stable minimum values. The characteristic relaxation time is of the order L/R, R is the intergrain normal resistance. The relaxation process leads to energy dissipation in the loop. When the external magnetic field is increased from zero and there is no trapped flux in the loop, the first transition appears at the critical value of ϕ_e which is $\phi_{c1} = (1+\lambda)/2$ (fig.1b). For field decreasing the reverse transition occurs at $\phi_{c2} = (1-\lambda)/2$ (fig.1c). The hysteresis appearing in the dependence of the minimum energy of the system and the loop current

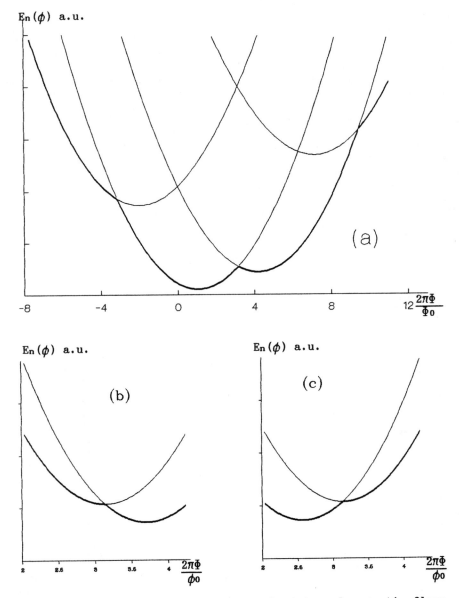

Fig. 1. The energy of the loop vs. the internal magnetic flux:
(a) - for an arbitrary external flux;
(b),(c) - for critical values ϕ_{e1}, ϕ_{e2}.

on Φ_e has the width $\lambda\Phi_0$.

Using the picture described above we have calculated the real and imaginary components of the nonlinear magnetic susceptibility at the applied ac field frequency ω . We assume that the condition $\omega L \ll R$ holds and the dissipative current in the loop can be neglected excluding only the short time intervals in which the transitions take place. The results are shown in fig.2a,b (note the difference in the vertical scale on these pictures). The temperature dependence is included through the $I_c(T)$ [5] entering the parameter λ .The field amplitude dependence of the dissipative part of the ac susceptibility for zero dc bias, similar to that represented in fig.2b, has been recently observed experimentally [8]. In considering the tempeature behavior of the ac energy losses resulting from the present model (fig.3), one should remember that the intragrain Meissner diamagnetism leads to the T-dependent effective magnetic permeability of the granular medium [7]. The effective applied magnetic flux through the loop embedded into the granular array may have, therefore, the additional temperature dependence. This may lead to decreasing of the dissipative susceptibility component at low temperatures even for large ac field amplitudes.

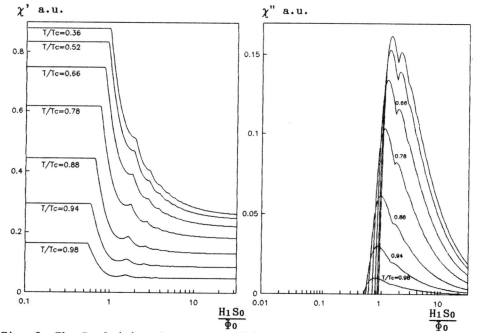

Fig. 2. The Real (a) and imaginary (b) components of the ac susceptibility for different temperatures as the functions of
the ac field amplitude H1. The low-temperature value of the
parameter λ is $\lambda(T=0)=1$.

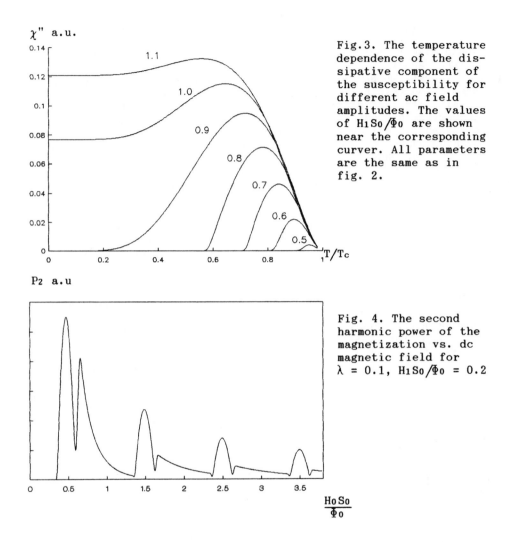

Fig.3. The temperature dependence of the dissipative component of the susceptibility for different ac field amplitudes. The values of $H_1 S_0/\Phi_0$ are shown near the corresponding curver. All parameters are the same as in fig. 2.

Fig. 4. The second harmonic power of the magnetization vs. dc magnetic field for $\lambda = 0.1$, $H_1 S_0/\Phi_0 = 0.2$

The essential consequence of the nonlinearity of our model is the higher-harmonic structure of the time-dependent magnetic response (see,e.g.[8] and references therein). When the dc magnetic field is non-zero the even-order harmonics appear. The characteristic example of the dependence of the second harmonic power on the dc magnetic flux is shown in fig.4.

DISCUSSION

The simple model presented in this paper is hoped to be

relevant to the description of ac magnetic properties of high-Tc granular materials in the low-field and low-frequency region. We would like to point out some problems which may be of interest for further analysis.

First, we have assumed above that all energy dissipation is due to the relaxation to the local energy minimum state, and the possibility of dissipation in the branch-switching events themselves has not been considered. The related problem has been analyzed [9] previously. It was assumed that the change of the quantum number n by unity is accompanied by the abrupt increase of the contact phase differences by $2\pi/N$. The corresponding voltage impulse appears and the normal dissipative current flows in the loop. The resulting expression for the dissipated energy depends explicitly on the normalresistance R. We believe that the change of n occurs through a phase-slip process in one of the contacts, for example, a contact with very large normal resistance and a critical current much less than the critical current Ic of all other junctions. For N>>1 the loop current is always much less than Ic but in the "weakest" contact of the loop the phase slip is possible, and the energy dissipation in this process may be small.

Second, we mention the recent work [10], in which the microwave radiation absorption experiments in single-crystal samples were interpreted in a model close to ours. The presence of twin boundary Josephson contacts was assumed, but it was argued that transitions between different states appear when the loop critical current Ic is reached [11].

In conclusion, we note that the dynamics of transitions may be strongly affected by thermal fluctuations, especially near the grain critical temperature. The relation between the ac field period and characteristic fluctuation times is important in determining the probabilities of brunch-switching processes.

REFERENCES

1. K. A. Muller, M. Takashige, and J. G. Bednorz, Phys. Rev. Lett. 58, 1143 (1987).
2. G. Deutscher, Physica C 153-155, 15 (1988).
3. C. Ebner and D. Stroud, Phys. Rev. B31, 165 (1985).
4. S. John and T. C. Lubensky, Phys. Rev. B34, 4815 (1986).
5. V. Ambegaokar and A. Baratoff, Phys. Rev. Lett. 10, 486 (1963); E 11, 104 (1963).
6. R. Marcon, R. Fastampa, M. Giura, and C. Matacotta, Phys. Rev. B39, 2796 (1989).
7. J. R. Clem, Physica C 153-155, 50 (1988).
8. C. D. Jeffries, Q. H. Lam, Y. Kim, C. M. Kim, A. Zettl, and M. P. Klein, Phys. Rev. B39, 11526 (1989).
9. T. Xia and D. Stroud, Phys. Rev. B39, 4792 (1989).
10. H. Vichery, F. Ruller-Albenque, F. Beuneu, and P. Lejay, Physica C 162-164, 1583 (1990).
11. J. A. Blackburn, J. Appl. Phys. 56, 1474 (1984).

SINGULARITIES OF PHONON SPECTRA OF SUPERCONDUCTORS OBSERVED WITH POINT CONTACTS: A THEORY FOR DIRTY S-N CONTACTS

V. A. Khlus and A. V. Dyomin
Institute for Single Crystals, 310141
Lenin Avenue, 60, Kharkov, USSR

ABSTRACT

Nonlinearities in the conductance of point contacts between a superconductor and a normal metal are considered for applied voltages larger than the energy gap Δ. Voltage dependence of the conductance and the second derivative of the current-voltage characteristic, resulting from the local electron-phonon interaction in the contact region, are obtained when electrons move diffusively through the contact, their mean free path being much shorter than the contact dimension d. The conductance nonlinear structure corresponding to peaks of the Eliashberg function is calculated for simple model examples. These results are valid when the inequality $\lambda d^2\Delta/nD<1$ holds, where λ is the electron-phonon coupling constant and D is the electron diffusion coefficient. The second derivative, d^2I/dV^2, is obtained for voltages near the values corresponding to Van Hove singularities of the phonon density of states.

INTRODUCTION

The discovery of new superconducting materials, especially the high T_c oxide superconductors, has stimulated further interest in tunneling and point-contact measurements to study the density of states and the energy gap in the electron spectrum. Point-contacts are also used to obtain the energy dependence of electron interactions with other excitations, especially with phonons. For normal metals, the point-contact spectroscopy has been reviewed elsewhere [1,2]. To develop a theoretical basis for point-contact experiments with superconductors, it seems of interest to calculate the current-voltage characteristic I(V) for a constriction junction between a superconductor and a normal metal (ScN-type contact) in the voltage region corresponding to characteristic phonon frequencies. In our previous works [3,4] this problem has been treated for pure contacts, when the electron elastic mean free path l, and the superconductor coherence length ξ are longer than the contact dimension d. The influence of the finite reflection coefficient of the S-N boundary on phonon structures of I(V) characteristics has been also considered.

In the present paper, we extend our analysis to the case of diffusive electron motion through the contact, taking into account the possibility of structural disorder on the superconductor surface, as well as high impurity concentration.

Some experimental evidence also exists that suggests that in many respects the properties of the new oxide superconductors, with the exception their high T_c's, are well sketched by the BCS dirty limit model. Therefore, we shall consider the limit in which the following condition holds

$$l_i \ll d \tag{1}$$

In normal dirty contacts, the amplitudes of point-contact spectra, reflecting the effects of inelastic electron-phonon scattering, are reduced by the factor l_i/d as compared with the case of ballistic electron motion [5].

THE MODEL AND THE BASIC RESULTS

In what follows, we shall investigate the influence of the local electron-phonon interaction (EPI) in the contact on the I(V) characteristics. To do this, we start with the theory developed for low T_c superconducting microbridges of ScN and ScS type [6], which explains such experimental results as the excess current at high voltages and the maximum of conductivity (the gap singularity) at the voltage corresponding to the superconducting energy gap .
The simplified model of the contact is represented in Figure 1.

The quantity of interest is the one-electron quasiclassical matrix Green's function [7] in Gor'kov-Keldysh representation [8]

$$G_p(\varepsilon,R) = \begin{pmatrix} \widehat{G}^R & \widehat{G} \\ 0 & \widehat{G}^A \end{pmatrix} \tag{2}$$

It depends on the electron energy, the direction of the momentum vector p on the Fermi surface, and the electron position R. This function obeys the equation [7,8]

$$V_F\frac{\partial G}{\partial R} + \left[\frac{1}{2\tau}\langle G \rangle + i\Sigma_{ph} - i\varepsilon\tau_3, G\right] = 0 \tag{3}$$

where Σ_{ph} is the electron-phonon self-energy operator, the angular brackets mean the p-direction average, and $\tau = 1_i/V_F$ is the momentum-relaxation time. The matrix τ_3 in the same representation as (2) is defined by

$$\tau_3 = \begin{bmatrix} \hat{\tau}_3 & 0 \\ 0 & \hat{\tau}_3 \end{bmatrix} \tag{4}$$

where τ_3 is the Pauli matrix. The first and the second terms in the commutator in eq. (3) describe the impurity scattering and EPI, respectively. We write G_p in the form

$$G_p(\varepsilon,x) = G_o(\varepsilon,x) + \cos\vartheta_p \, G_a(\varepsilon,x), \tag{5}$$

where G_o and G_a are the isotropic and anisotropic parts of the Green's function. The greatly simplified one-dimensional model of the contact is assumed, $0<x<d$, and ϑ_p is the angle between the x-axis and the momentum vector. The voltage V is applied to the normal electrode and the chemical potential of the superconductor is taken to be zero. The electric field distribution is time independent and for this reason it does not enter eq. (3). The voltage drop V through the contact appears only due to boundary conditions on the bridge ends.

In the dirty limit, $\Delta\tau \ll 1$, eq. (3) can be reduced to the following pair of equations

$$V_F\frac{\partial G_o}{\partial x} + \frac{1}{2\tau}[G_o,G_a] = 0,$$

$$\frac{1}{3}V_F\frac{\partial G_a}{\partial x} - i\varepsilon\,[\tau_3, G_o] = -i\,[\Sigma_o,G_o]. \tag{6}$$

In our model, the following boundary conditions on the symmetric part G_o should be imposed:

$$G_o(\varepsilon,0) = G_N(\varepsilon), \qquad G_o(\varepsilon,d) = g(\varepsilon) \tag{7}$$

The quasiclassical Green's function G_N in the normal region depends on the voltage V, which shifts the chemical potential $g(\varepsilon)$ which is the Green's function of the uniform superconductor in thermodynamic equilibrium, its definition can be found in [7,8]. The explicit form of these functions is the same as in [3,4].

The right-hand side of the second equation contains the isotropic part of the EPI operator. Taking into account that the phonon nonlinearities of $I(V)$ are small, we treat the EPI term as a perturbation and find the corresponding contribution to the nondiagonal component of the matrix (2). The additional current term due to EPI is of the form

$$I(V) = - \frac{d}{8eRD} \int d\varepsilon \; Sp \; \tau_3 \langle g_a(\varepsilon) v_{Fx} \cos \vartheta_p \rangle \tag{8}$$

Here R is the normal state resistance, and $D = v_F l_i / 3$ is the electron diffusion coefficient.

The EPI current contribution, eq. (8), appears due to the isotropic part of the electron-phonon self-energy operator, this part also defines the charge imbalance relaxation rate and the electric field penetration length in dirty superconductors [9]. In the point contact, the electric field strength and current density decrease rapidly far from the constriction and the electron-phonon processes. This contributes to the electric potential relaxation and gives only a small correction to the high-voltage conductivity. However, this term contains the important dependence on V, which is sensitive to the EPI spectral function behavior. In finding the contact conductance we must consider, along with the electron-phonon relaxation, the virtual processes described by the real part of the phonon Green's function appearing in the self-energy Σ_{ph}. Finally, we can represent the EPI-induced current, eq. (8), as the sum of two terms

$$I_{ph}(V) = I_1(V) + I_2(V) \tag{9}$$

where the first term results from the virtual processes, and the seond is due to inelastic relaxation of the electric field. The expression for the conductance may be written in a rather simple form for voltages much greater than the gap.

For $\sigma(V) = dI/dV$ we obtain the sum of the two terms:

$$\sigma_1\,(V) = \frac{d^2\,\Delta}{RD} \int_0^\infty d\omega\,G(\omega) \int \frac{d\varepsilon_1}{4T}\,\mathrm{ch}\left(\frac{\varepsilon_1 - V}{2T}\right) \times$$

$$\int_0^\Delta d\varepsilon\, f_1\,(\varepsilon/\Delta) \left[\frac{1}{(\varepsilon_1 - \omega)^2 - \varepsilon^2} - \frac{1}{(\varepsilon_1 + \omega)^2 - \varepsilon^2}\right]$$

(10)

$$\sigma_2\,(V) = \frac{d^2}{RD} \int_0^\infty d\omega\,G(\omega) \int \frac{d\varepsilon}{4T}\,\mathrm{ch}^{-2}\left(\frac{\varepsilon - V + \omega}{2T}\right) \times$$

$$\vartheta\,(|\varepsilon| - \Delta)\,\mathrm{th}\left(\frac{\varepsilon}{2T}\right)\,f_2\left(\frac{\varepsilon}{\Delta}\right)$$

(11)

Here the spectral EPI function is defined,

$$G(\omega) = \frac{1}{2}\,N(0) \sum_\lambda \frac{\left\langle |g^\lambda_{pp_1}|^2\,w^\lambda_{p-p_1}\,\delta(w - w^\lambda_{p-p_1})\cos^2\vartheta_p\right\rangle}{\left\langle\cos^2\vartheta_p\right\rangle}$$

(12)

which is very much like the Eliashberg function of $a^2(\omega) - F(\omega)$. In eq. (12) $N(0)$ is the electron density of states at the Fermi surface, $g^\lambda_{pp_1}$ is the EPI matrix element, ω^λ_q is the phonon frequency in the polarization state λ, with the wave vector q. It should be noted that the local electron and phonon characteristics in the contact enter the above equations. In this respect the situation differs from the case of high quality tunnel junctions, where the energy dependence of the gap function and the tunneling density of states on the superconducting side result in the nonlinear features of the conductance (see for a review, e.g. [10]). The functions $f_{1,2}(\varepsilon/\Delta)$ in eqs. (10), (11) are given by the following equations:

$$f_1\left(\frac{\varepsilon}{\Delta}\right) = 2\int_0^1 dz\,(1-z)\,\sin\left(\frac{\pi z}{2}\right)\left[\left(\frac{\Delta}{\varepsilon}\right)\mathrm{sh}\,(zB_\varepsilon) - z\,\mathrm{ch}\,(zB_\varepsilon)\right],$$

$$f_2\left(\frac{\varepsilon}{\Delta}\right) = \pi \int^1 dz\,(1-z)\,z\left[\left(\frac{\varepsilon}{\Delta}\right)\mathrm{sh}\,(zB_\varepsilon) - z\,\mathrm{ch}\,(zB_\varepsilon)\right],$$

(13)

where
$$B_\varepsilon = \begin{cases} \mathrm{arcth}\left(\frac{\varepsilon}{\Delta}\right) & \text{for}\quad |\varepsilon| < \Delta \\ \mathrm{arccth}\left(\frac{\varepsilon}{\Delta}\right) & \text{for}\quad |\varepsilon| > \Delta \end{cases}$$

These functions are shown in Figure 2.

It can be seen that the contribution to $\sigma(V)$ comes from the EPI processes in which one of the electron energies involved is close to the voltage V, while the second is near the energy gap edge. The former is obvious because the energy distribution of injected electrons (from the N-side) is shifted by V above the superconductors chemical potential level, and the latter means that the gap edge region, which produces the gap singularity of $\sigma(V)$, [6], also gives the nonlinearities connected with the EPI.

In Figure 3a,b the EPI-induced conductance term is shown for the function $G(\omega)$ of the Lorentzian form, with the mean frequency ω_o and the half-width γ. In Figures 4 and 5, the conductance and second derivative of the current-voltage characteristic are represented, including the tails of the gap singularity in the large V region for different model choices of the function $G(\omega)$. The conductance minima correspond to the peaks of $G(\omega)$. These minima are shifted to lower voltages relative to the peak's frequencies. It can be seen that the nonlinear contributions to $\sigma(V)$ due to EPI are small. This is the result of the narrow ε – integration region in eqs (10), (11). The dimensionless parameter, appearing before the integrals, is

$$\lambda d^2 \Delta / nD, \tag{14}$$

where λ is the dimensionless EPI coupling constant. In all our calculations this value is taken to be 0.5.

We also consider the possible effects of Van Hove singularities associated with the critical points of phonon spectrum on point-contact characteristics. For tunnel superconducting junctions, this problem was considered in (11).

In a three-dimensional lattice we can approximate the function $G(\omega)$ near the critical frequency ω_o as

$$
\begin{aligned}
G(\omega) &= G(\omega_o) + \vartheta(\omega_o - \omega)\ A\sqrt{\omega_o - \omega} && \text{(a)} \\
G(\omega) &= G(\omega_o) + \vartheta(\omega - \omega_o)\ A\sqrt{\omega - \omega_o} && \text{(b)}
\end{aligned}
\tag{15}
$$

With the superposition of the square-root singularity in $dG/d\omega$ and singularities in integrals (10), (11) produces the contribution to d^2I/dV^2 near ω_o of the following form:

a: $d^2I/dV^2 = -Ad^2\sqrt{\Delta F_A}\,(v)/2RD$ (16)

where $v = (V-\omega_o)/\Delta$ and

$$F_A(v) = \frac{\pi}{2}\int_0^1 dx\, f_1(x)\,/x\left[\frac{\vartheta(v-x)}{\sqrt{v-x}} - \frac{\vartheta(v+x)}{\sqrt{v+x}}\right] +$$

$$+\int_V^\infty \frac{dx}{\sqrt{x-v}}\,\vartheta\,(|x|-1)\,\text{sgn}(x)\,f_2(x)$$ (17)

b: $d^2I/dV^2 = -Bd^2\sqrt{\Delta F_A}\,(v)/2RD$ (18)

$$F_B(v) = \frac{\pi}{2}\int_0^1 dx\, f_1(x)\,/x\left[\frac{\vartheta(x-v)}{\sqrt{x-v}} - \frac{\vartheta(-v-x)}{\sqrt{-v-x}}\right] +$$

$$+\int_V^\infty \frac{dx}{\sqrt{x+v}}\,\vartheta\,(|x|-1)\,\text{sgn}(x)\,f_2(x)$$ (17)

The singularities of the second derivative of I(V) associated with the Van Hove critical points are shown in Figure 6a,b.

CONCLUSION

The present results agree reasonably with the experimental data obtained on dirty contacts Ag-T_c [12]. The superconductivity of technetium shows strong coupling EPI origin. The relations betweeen $\sigma(V)$ and $G(\omega)$ obtained here fit well the neutron-diffraction data for the phonon density of states. As to point-contact measurements on high T_c superconductors, we can mention the work of K. E. Gray et. al. [13] on Au-YBaCuO contacts where a conductance structure very similar to that shown in Figure 3,4 has been obtained at energies above 30-40 meV. It is tempting to relate this structure to the phonon mode at 310-330 cm⁻, observed in Raman spectra, which seems to be characteristic of the superconducting state [14].

Our treatment is also valid in the case of a non-phonon mechanism of superconducting pairing responsible for the gap value in the S-bank of the contact. Electron-phonon interaction can be important, in explaining the normal-state properties, such as normal resistance, of the new superconductors. Point-contact measurements could give some information regarding the strong interaction of the phonon modes with charge carriers.

REFERENCES

1. A.M. Duif, A. G. M. Jansen, P. Wyder, J. Phys.: Condens. Matter 1, 3157 (1989).
2. I.K. Yanson, Sov. J. Low Temp. Phys. 9, 343 (1983), Fiz. Nizk. Temp. 9, 676 (1983).
3. V.A. Khlus, Sov. J. Low Temp. Phys., Fiz. Nizk. Temp. 9, 985 (1983).
4. V. A. Khlus, Sov. Phys. JETP, Zh. Eksp. Teor. Fiz. 94, 341 (1988).
5. I. O. Kulik, I. K. Yanson, Sov. J. Low Temp. Phys. 4, 596 (1978); Fiz. Nizk. Temp. 4, 1267 (1978).
6. S. N. Artemenko, A. F. Volkov, A. V. Zaitsev, Solid State Commun. 30, 771 (1979).
7. G. Eilenberger, Z. Phys. 214, 195 (1968).
8. A. I. Larkin, Yu.N. Ovchinnikov, Sov. Phys. JETP, 41, 960 (1975); Zh. Eksp. Teor. Fiz. 68, 1915 (1975).
9. S. N. Artemenko, A. F. Volkov, A. V. Zaitsev, J. Low Temp. Phys. 30, 487 (1978).
10. E. L. Wolf, Principles of Electron Tunneling Spectroscopy (Oxford Univ. Press, New York, Clarendon Press, Oxford, 1985).
11. D. J. Scalapino, P. W. Anderson, Phys. Rev.132, 921 (1964).
12. A. A. Zakharov, M. B. Zeitlin, Sov. Phys. JETP Letters, Pis'ma Zh. Eksp. Teor. Phys. 41, 11 (1985).
13. K. E. Gray, E. R. Moog, M. B. Hawley, in Int. Meeting on High-T_c Superconductivity, Mauterndorf (1988), pp. 158-161.
14. R. Feile, U. Schmitt, P. Leiderer, Physica C. 153-155, 293 (1988).

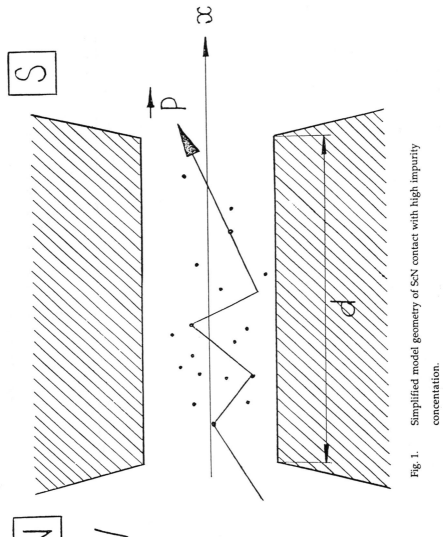

Fig. 1.   Simplified model geometry of ScN contact with high impurity concentation.

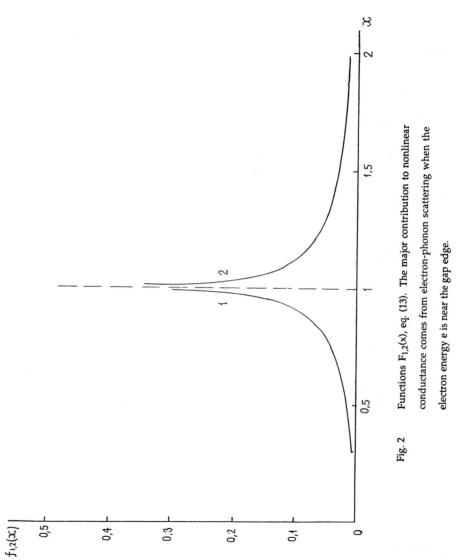

Fig. 2   Functions $F_{1,2}(x)$, eq. (13). The major contribution to nonlinear conductance comes from electron-phonon scattering when the electron energy e is near the gap edge.

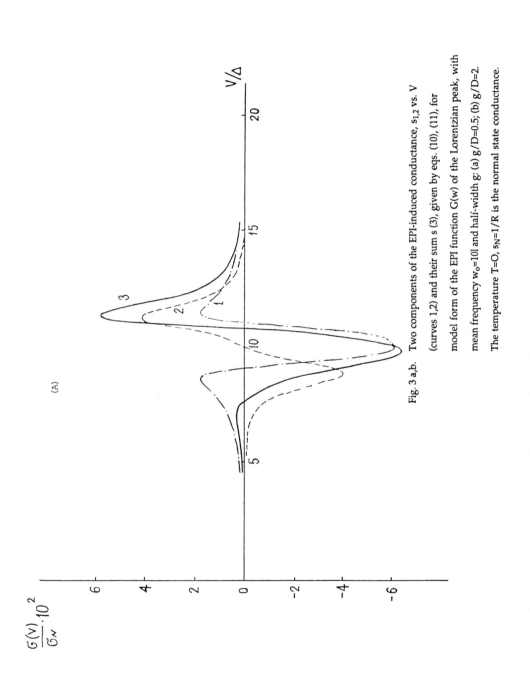

Fig. 3 a,b.   Two components of the EPI-induced conductance, $s_{1,2}$ vs. V (curves 1,2) and their sum s (3), given by eqs. (10), (11), for model form of the EPI function G(w) of the Lorentzian peak, with mean frequency $w_o=10l$ and half-width g: (a) g/D=0.5; (b) g/D=2. The temperature T=O, $s_N=1/R$ is the normal state conductance.

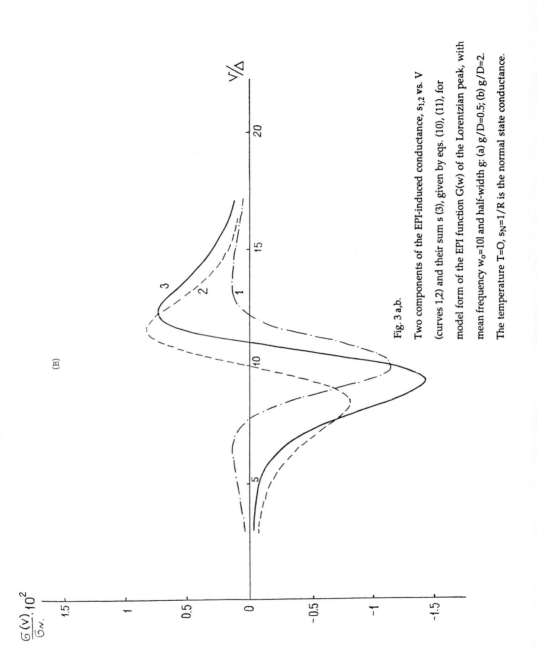

(B)

Fig. 3 a,b.

Two components of the EPI-induced conductance, $s_{1,2}$ vs. V (curves 1,2) and their sum s (3), given by eqs. (10), (11), for model form of the EPI function G(w) of the Lorentzian peak, with mean frequency $w_o = 10l$ and half-width g: (a) g/D=0.5; (b) g/D=2. The temperature T=O, $s_N = 1/R$ is the normal state conductance.

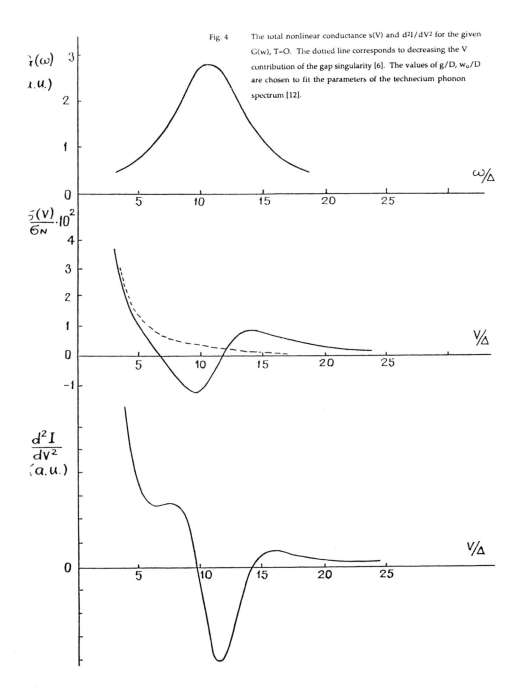

Fig. 4   The total nonlinear conductance s(V) and d²I/dV² for the given G(w), T=O. The dotted line corresponds to decreasing the V contribution of the gap singularity [6]. The values of g/D, w₀/D are chosen to fit the parameters of the technecium phonon spectrum [12].

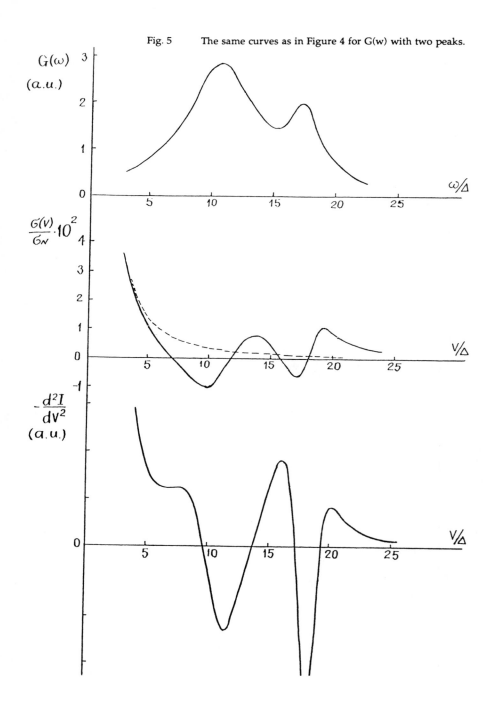

Fig. 5   The same curves as in Figure 4 for G(w) with two peaks.

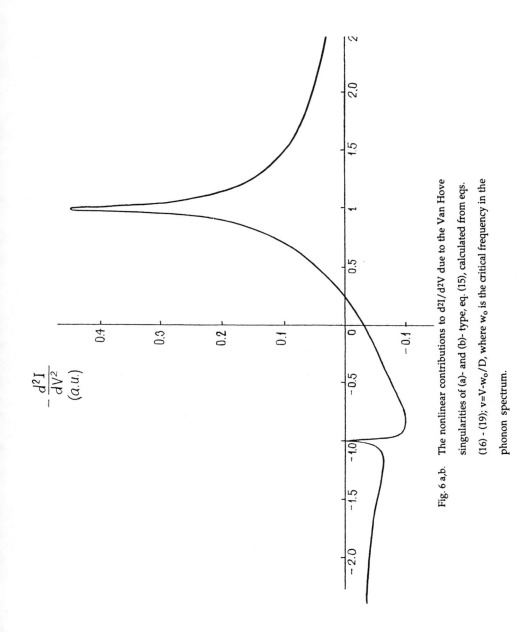

Fig. 6 a,b.   The nonlinear contributions to d^2I/d^2V due to the Van Hove singularities of (a)- and (b)- type, eq. (15), calculated from eqs. (16) - (19); $v=V-w_0/D$, where w_0 is the critical frequency in the phonon spectrum.

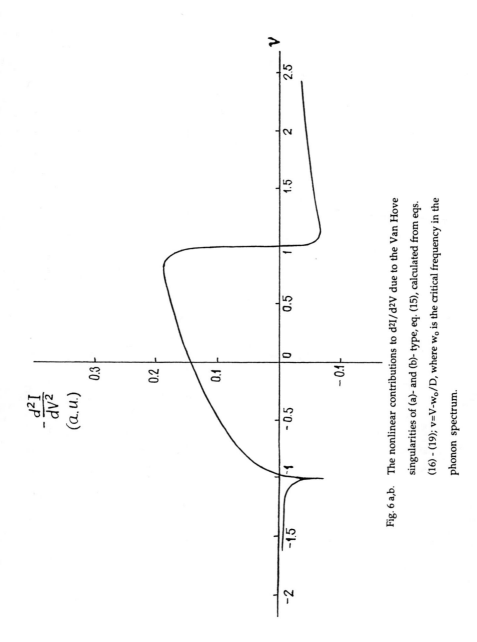

Fig. 6 a,b.   The nonlinear contributions to d^2I/d^2V due to the Van Hove singularities of (a)- and (b)- type, eq. (15), calculated from eqs. (16) - (19); $v=V-w_o/D$, where w_o is the critical frequency in the phonon spectrum.

POLARIZED OXYGEN K EDGE XANES STUDIES IN $Tl_2Ba_2Ca_2Cu_3O_{10}$ AND $Bi_2Sr_2CaCu_2O_8$

M. Faiz[+*], G. Jennings[+], J. C. Campuzano[+*] and E. E. Alp[+]

+ Argonne National Laboratory, Argonne, Illinois 60439

* University of Illinois at Chicago, Chicago, Illinois 60680

ABSTRACT

We have measured oxygen K edge absorption spectra in $Tl_2Ba_2Ca_2Cu_3O_{10}$ and $Bi_2Sr_2CaCu_2O_8$ layered copper oxide superconductors. We have tried to isolate the contribution of Tl-O and Bi-O bonds to the spectra by measuring O K edge absorption spectra in Tl_2O_3 and Bi_2O_3. Since Tl and Bi are determined to be in +3 valence state in both of the superconducting compounds, we attribute the peaks below 530 eV in the absorption spectra to holes in the Cu-O bonds. We compare the angular dependence of holes in oxygen by using polarized x-rays.

INTRODUCTION

Superconductivity in the cuprate superconductors is related to the oxygen 2p hole concentration in the hybridized Cu 3d-O 2p valence band.[1] Therefore a detailed study of the hole concentration in these high-Tc superconductors is essential to understand the underlying physics. X-ray absorption spectroscopy is one of the most direct methods for measuring hole concentration in which core electrons are excited to empty states.[2,3] With synchrotron radiation sources, X-ray absorption spectroscopy has become a powerful technique to investigate the photoexcitation of core electrons into empty states.

In this paper we report polarized x-ray absorption spectra at the O K edge in $Tl_2Ba_2Ca_2Cu_3O_{10}$ and $Bi_2Sr_2CaCu_2O_8$ single crystals. Tl and Bi are determined to be close to +3 valence state in the layered cuprate superconductors.[4,5] We have included our Bi L_3 edge absorption measurements in $Bi_2Sr_2CaCu_2O_8$ and the reference compounds, Bi_2O_3 and $KBiO_3$, to reinforce the statement. Since Tl and Bi are in the +3 valence state in Tl_2O_3 and Bi_2O_3 respectively, we assume that the oxygen environment in the Tl-O and Bi-O layers in the superconductors is similar to that in the respective oxides. Therefore, we have also measured O K edge absorption spectra in Tl_2O_3 and Bi_2O_3 in order to isolate the contribution of Tl-O and Bi-O bonds to the O K edge absorption spectra of the superconductors.

EXPERIMENT

The experiments were carried out with polarized x-rays, at the Synchrotron Radiation Center (Stoughton, Wisconsin), monochromatized by an extended range Grasshopper monochromator. Single crystals of $Tl_2Ba_2Ca_2Cu_3O_{10}$ and $Bi_2Sr_2CaCu_2O_8$ were oriented so that the electric field vector of the x-rays is in the ab-plane of the crystal at normal incidence. The crystals were cooled to 20K before cleaving them in a vacuum of about 10^{-10} T in order to minimize any possible oxygen loss from the crystal surfaces.[6] Oxygen K edge absorption spectra were taken in the total electron yield mode[3] for different values of angle (θ) between the electric field vector of the x-rays and the c-axis of the crystals. O K edge absorption spectra were taken in Tl_2O_3 and Bi_2O_3 at room temperature. The energy resolution of these measurements was 0.5 eV. Bi L_3 edge measurements were carried out in the transmission mode at the X-18 beamline at NSLS (Brookhaven National Laboratories, New York). The energy resolution of the Si (220) monochromator was estimated to be about 3 eV at 13419 eV.

RESULTS AND DISCUSSION

Fig. 1 shows the Bi L_3 edge of $Bi_2Sr_2CaCu_2O_8$, Bi_2O_3 and $KBiO_3$. The position of the edge is assumed to be at the mid point of the main step in the edge. The zero energy has been chosen as the Bi_2O_3 edge. The Bi L_3 edge in $Bi_2Sr_2CaCu_2O_8$ is very close to that in Bi_2O_3, while there is about 1 eV shift between Bi_2O_3 and $KBiO_3$. Furthermore, a pronounced pre-edge peak is present only in $KBiO_3$. Since Bi is in an outer shell configuration of $6s^2 6p^3$, it can be in the +3 valence state when only the 6p states are empty, as well as in the +5 state when both the 6s and 6p states are empty. When the 6s states are empty, 2p to 6s transition is also allowed as well as 2p to 6d, according to the selection rules of photoabsorption. The 2p to 6d transition corresponds to the L_3 edge, while the 2p to 6s corresponds to the pre-edge peak.[7,8] Therefore, the pre-edge peak is a clear indication of higher valency. Note that Bi is in the +5 valence state in $KBiO_3$ and +3 in Bi_2O_3. With this theoretical background and the close resemblance of Bi L_3 edge in $Bi_2Sr_2CaCu_2O_8$ and Bi_2O_3, we are convinced that the formal valency of Bi in $Bi_2Sr_2CaCu_2O_8$ is close to +3, and the oxygen environment in the Bi-O layers in $Bi_2Sr_2CaCu_2O_8$ is similar to that in Bi_2O_3. A similar conclusion has been reached for Tl-O layers in $Tl_2Ba_2Ca_2Cu_3O_{10}$.[4]

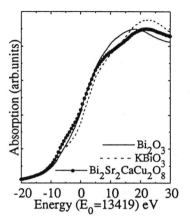

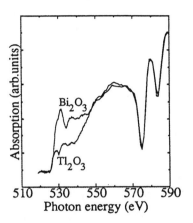

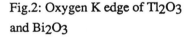

Fig.1: Bismuth L_3 edge of Bi_2O_3, $KBiO_3$ and $Bi_2Sr_2CaCu_2O_8$

Fig.2: Oxygen K edge of Tl_2O_3 and Bi_2O_3

The O K edge of Tl_2O_3 and Bi_2O_3 at room temperature is shown in Fig. 2. Fig. 3(a & b) show the O K edge of $Tl_2Ba_2Ca_2Cu_3O_{10}$ and $Bi_2Sr_2CaCu_2O_8$ single crystals at 20K as a function of the angle θ. A straight line, fitted to the region below 525 eV, has been subtracted as background.The spectra also contain chromium L_2 and L_3 edges resulting from the coating on the monochromator grating (shown only in Fig.2). We have exploited them to calibrate the energy scale and to normalize the spectra. Since the Cr L edges are caused by the monochromator, they are independent of the sample orientation. Therefore we have matched the Cr L edge portion of each spectrum in order to normalize them. The spectra shown in the figures have not been smoothed in any way. Note that the onset of O K edge is at about 530.0±0.5 eV for all the samples. Since we are observing the transitions starting from the O 1s state, the peaks preceding the O K edge are a measure of O 2p holes according to the selection rules of photoabsorption, and their intensity is proportional to the O 2p hole concentration. It is clear from the figures that $Tl_2Ba_2Ca_2Cu_3O_{10}$ shows a pre-edge peak at 529.0 eV while $Bi_2Sr_2CaCu_2O_8$ shows two (labeled A and B) at 525.0 and

527.5 eV. Tl_2O_3 also shows a pre-edge peak, the intensity of which is comparable to that of $Tl_2Ba_2Ca_2Cu_3O_{10}$ at $\theta=70°$, at about 529.0 eV. However it is not clear whether this feature is present in Bi_2O_3.

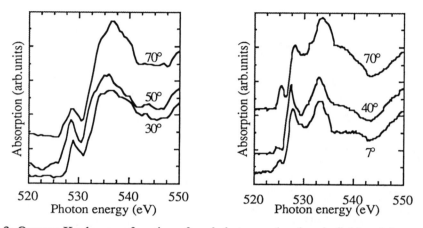

Fig.3: Oxygen K edge as a function of angle between the electric field and the c-axis of (a) $Tl_2Ba_2Ca_2Cu_3O_{10}$ and (b) $Bi_2Sr_2CaCu_2O_8$

The angular dependance of the pre-edge peak in $Tl_2Ba_2Ca_2Cu_3O_{10}$ can be seen in Fig. 3(a), which shows a maximum intensity at $\theta=50°$. We interpret this result as follows: in $Tl_2Ba_2Ca_2Cu_3O_{10}$ not only the O(2) sites in the Cu-O layers but also the apex O(3) sites contribute to O 2p holes (Fig. 4(a)). Since the Cu(2)-O(2) and Cu(2)-O(3) bond lengths are 1.9 Å and 2.5 Å respectively,[9] the O 2p hole concentration can add up to a maximum at about $\theta = \tan^{-1}(2.5/1.9) \approx 50°$. Here we have assumed that the O 2p holes are of σ symmetry. On the other hand, in the case of $Bi_2Sr_2CaCu_2O_8$, the intensity of peak A shows a minimum at $\theta=40°$ (Fig. 3(b)). Peak B is not as sensitive to the polarization of the x-rays as peak A. We interpret this result as follows: while the holes in the ab-plane come from the Cu-O layers as in $Tl_2Ba_2Ca_2Cu_3O_{10}$, the holes along the c-axis come from the Bi-apex O(3) bonds (Fig. 6(b)).

Fig.4: Crystal structure of (a) $Tl_2Ba_2Ca_2Cu_3O_{10}$ and (b) $Bi_2Sr_2CaCu_2O_8$

The Cu-O(3) bonds in $Bi_2Sr_2CaCu_2O_8$ do not seem to contribute significantly to the O 2p holes. This is not so surprising because the sum of ionic radii of Ba^{2+} and O^{2-} is nearly equal to the Ba-O(3) bond length in $Tl_2Ba_2Ca_2Cu_3O_{10}$. Therefore the Ba-O plane can actively contribute to transport properties and hence to the O 2p holes. However, this is not quite true in Sr-O planes. Another remarkable difference between the two samples is that Bi-O(3) bonds contribute to O 2p holes significantly, whereas Tl-O(3) bonds do not. We think that the reason might be the following: although the valency of Tl in $Tl_2Ba_2Ca_2Cu_3O_{10}$ is determined to be exactly +3,[4] that of Bi in $Bi_2Sr_2CaCu_2O_8$ is slightly less than +3.[5] This slight deviation makes a significant difference between $Bi_2Sr_2CaCu_2O_8$ and Bi_2O_3, as we have now seen. Also, since the ionic radii of Bi^{3+} and Ca^{2+} are nearly the same, it is estimated that about 20% of Bi^{3+} occupy Ca^{2+} sites.[10] Therefore the Bi-O layers appear to play the role of electron sinks for the Cu-O layers.[5] The pre-edge peak observed in Tl_2O_3 seems to agree with the metallic behavior of Tl-O layer predicted by band structure calculations.[11-14]

It is also informative to note that , in an earlier paper,[3] we have reported a similar study in $YBa_2Cu_3O_{6.9}$ single crystals in which we have identified, without ambiguity, the reason for the polarization dependence of the pre-edge peaks as the buckled Cu-O layers. In $Tl_2Ba_2Ca_2Cu_3O_{10}$ and $Bi_2Sr_2CaCu_2O_8$, however, the Cu-O layers are almost flat.

CONCLUSION

We have reported a systematic O K edge absorption study in $Tl_2Ba_2Ca_2Cu_3O_{10}$ and $Bi_2Sr_2CaCu_2O_8$ single crystals. In contrast to what one might expect, they show a remarkable difference as far as the O 2p hole contribution is concerned, although they belong to the same group of high-T_c superconductors. While the Cu(2)-O(3) bonds contribute significantly to the O 2p holes in $Tl_2Ba_2Ca_2Cu_3O_{10}$, the Bi-O(3) bonds do so in $Bi_2Sr_2CaCu_2O_8$ along the c-axis in addition to the Cu-O layer contribution in both samples. Neither of them shows a simple $\sin^2\theta$ variation as one might expect if only the Cu-O layers contribute to the O 2p holes.[15] More quantitative work is in progress, and will be published elsewhere.

ACKNOWLEDGEMENTS

We would like to acknowledge M. Ramanathan and S. Mini for their help in obtaining Bi L3 edge data. We thank F. J. Himpsel and G. V. Chandrashekhar for $Bi_2Sr_2CaCu_2O_8$ single crystals and D. M. Ginsberg and E. D. Bukowski for $Tl_2Ba_2Ca_2Cu_3O_{10}$ single crystals. We also thank K. Gofron for his help in drawing the crystal structures. This work was supported by the US DOE-BES under contract # W-31-109-ENG-38 and NSF grant # DMR-8914120.

REFERENCES

1. V. J. Emery, Phys. Rev. Lett. 58, 2794 (1987)

2. P. Kuiper, G. Kruizinga, J. Ghijsen, M. Grioni, P. J. W. Weijs, F. M. F. de Groot, G. A. Sawatzky, H. Verweij, L. F. Feiner, and H. Petersen, Phys. Rev. B 38, 6483 (1988)

3. E. E. Alp, J. C. Campuzano, G. Jennings, J. Guo, D. E. Ellis, L. Beaulaigue, S. Mini, M. Faiz, Y. Zhou, B. W. Veal, and J.Z. Liu, Phys. Rev. B 40, 9385 (1989)

4. F. Studer, R. Retoux, C. Martin, C. Michel, B. Raveau, E. Dartyge, A. Fontaine, and G. Tourillon, Int. J. Mod. Phys. B 3, 1085 (1989)

5. R. Retoux, F. Studer, C. Michel, B. Raveau, A. Fontaine, and E. Dartyge, Phys. Rev. B 41, 193 (1990)

6. R. S. List, A. J. Arko, Z. Fisk, S. W.Cheong, S. D. Conradson, J. D. Thompson, C. B. Pierce, D. E. Peterson, R. J. Bartlett, N. D. Shinn, J. E. Schirber, B. W Veal, A. P. Paulikas, and J. C. Campuzano, Phys. Rev. B 38, 11966 (1988)

7. K. J. Rao and J. Wong, J. Chem. Phys. 81, 4832 (1984)

8. S. Salem-Sugui, Jr., E. E. Alp, S. M. Mini, M. Ramanathan, J. C. Campuzano, G. Jennings, M. Faiz, S. Pei, B. Dabrowski, Y. Zheng, D. R. Richards, and D. G. Hinks, submitted to Phys. Rev. B

9. Klaus Yvon and Michel Francois, Z. Phys. B 76, 413 (1989)

10. David P. Matheis and Robert L. Snyder, J. Powder Diffraction, Nov.1989

11. D. R. Hamann and L. F. Mattheiss, Phys. Rev. B 38, 5138 (1988)

12. J. Yu, S. Massidda and A. J. Freeman, Physica C 152, 278 (1988)

13. R. V. Kasowski, W. Y. Hsu, and F. Herman, Phys. Rev. B 38, 6470 (1988)

14. P. Marksteiner, J. Yu, S. Massidda, A. J. Freeman, J. Redinger, and P. Weinberger, Phys. Rev. B 39, 2894 (1989)

15. M. Faiz et al., to be published.

SYNCHRONIZED EXCITATIONS WITHIN HIGH T_c OXIDE SUPERCONDUCTORS

R. E. CLAPP

The MITRE Corporation, Bedford, Massachusetts 01730

Abstract

Ion channeling measurements on 90-K and 60-K YBCO provide evidence of an abrupt synchronization of certain lattice vibrations as T is lowered through T_c. This evidence supports a theoretical model in which kinetic energy is stored in pi-electron holes circulating in Möbius orbits; for $T < T_c$ these orbital excitations become locked together with the aid of synchronous lattice vibrations. A pattern of electric potential peaks and valleys driven by the synchronized orbiting holes carries mobile charges as singlet pairs, whose net flow in the presence of a vector potential $\mathbf{A}$ constitutes the observed supercurrent, $\mathbf{J} = -(e^2 n/mc)\mathbf{A}$.

1. Introduction: ion channeling

Ion channeling experiments on the 1:2:3 oxide superconductors have been carried out by Sharma et al [1,2]. Because of the atomic alignments in a single crystal of $YBa_2Cu_3O_x$, an incident beam of 6.5 MeV helium ions, directed parallel to a crystal axis, will be channeled into the interior. This results in a deep reduction in Rutherford backscattering (RBS) and also in K_α x-ray yield. Angular scans, as in figure 1, give dip profiles that can be translated into measures of atomic vibrational amplitudes. For c-axis channeling, the x-ray method permits separate identification of the Y–Ba columns and the Cu–O columns.

As the temperature is lowered through T_c, there is an abrupt reduction in an average vibrational amplitude, u_1, for the Cu–O columns but not for the Y–Ba columns [1,2]. Here u_1 is a quantity extracted from a channeling scan; it is clear from the way u_1 is obtained that it really represents the transverse displacement of individual atoms in a column, relative to the momentary mean location of the other atoms in that column.

The time required for a helium ion to traverse its channel is so short that the lattice vibrations are momentarily frozen in place. Thus an abrupt reduction in u_1 could arise either (a) from an abrupt reduction in thermal vibrations or (b) from an abrupt synchronization of vibrations whose amplitudes are undiminished. If (a) were the explanation, then we would expect a large jump in Debye properties including the specific heat, but that is not observed. We are left with (b), an abrupt synchronization in the vibrational movements in the Cu–O columns, but no such abruptness in the Y–Ba columns.

Figure 2 establishes that the abruptness is truly correlated with passage through T_c. For an oxygen stoichiometry (x = 6.6) giving T_c = 54 K, the abrupt change in u_1 comes as T crosses 54 K. For full oxygen content (x ~ 7) giving T_c = 92 K, the abrupt change is at T ~ 92 K. For nonsuperconducting YBCO (x = 6), no discontinuity is observed [2].

2. Synchronized Möbius orbits

The crystal structure of YBCO is shown in figure 3, from David et al [3]. With the 90–K stoichiometry (x ~ 7) the chain connectivity permits us to draw orbits of the form in figure 4. A migrating pi-electron hole, accompanying a traveling pi bond between an oxygen P state and copper D state, can traverse this orbit. There will be a topological phase reversal, illustrated in figure 5. For phase continuity around the orbit, we need an additional 180° phase shift, which will be provided by travel around the orbit if the orbital perimeter equals one-half de Broglie wavelength. The resulting Möbius orbit has the orbital angular momentum quantum number L = 1/2 and is metastable, since radiative decay needs $\Delta L = 1$ and is not accessible from L = 1/2. Möbius orbits have been discussed by Heilbronner [4], Zimmerman [5], and Clapp [6].

For the 36-bond orbit in figure 4, the perimeter of 70.9 Å gives a de Broglie wavelength of λ_{deB} = 141.8 Å. From the momentum, $p = h/\lambda_{deB}$, and the free-electron mass, $m = 9.11 \times 10^{-28}$ gram, we can solve for the orbital period, τ = 138.3 femtoseconds (fs). This is just twice the 69.1 fs period of the 483 cm^{-1} Raman line in figure 6, from Hemley and Mao [7]. The 483 cm^{-1} line corresponds to the vibration of the Cu(1)–O(4) bonds in figure 4. Thus an orbiting hole in figure 4 will find the long steps from O(4) to Cu(2) momentarily shortened, if the orbit is synchronized to the 483 cm^{-1} lattice vibration and appropriately timed.

A circulating hole in an elongated orbit will undergo a momentum reversal at each end of the orbit, and there will be a momentum recoil taken up by the lattice. If two adjacent orbital excitations are timed so that the momentum recoils are equal, opposite, and simultaneous, then the balanced recoils will minimize the propagation of vibrational energy away from the pair of loops. The phonon energy thus confined and saved can provide positive attraction and energy for pairing, but only if T is below the temperature, T_c, at which a thermal quantum is just sufficient to disturb the synchronization.

That is, we define T_c through

$$k_B T_c = p^2/2m, \tag{1}$$

where $p^2/2m$ is the kinetic energy of a circulating hole in its Möbius orbit. For the 36-bond orbit in figure 4, synchronized to the 483 cm^{-1} vibration in figure 6, we solve equation (1) and obtain

$$T_c = 86.8 \text{ K}, \tag{2}$$

which is close to the observed T_c for the 90–K YBCO.

The momentum recoils within the lattice, timed to match the 483 cm^{-1} vibration, will tend to modify the longitudinal vibrational motion within a Cu(1)–O(4)–Cu(2) bond system by introducing a transverse component. This transverse motion will be driven by the circulating pi-electron hole passing through that bond system. For $T > T_c$ the orbiting holes are not synchronized with each other, and the transverse motion in these bond systems will accordingly be generally uncorrelated. However, as T is lowered through T_c the orbiting holes will be drawn into synchronism. As a result, the lattice vibrations at 483 cm^{-1}, particularly those vibrations incorporating transverse motion driven by recoiling orbiting holes, will become correlated.

In this way we can understand the abrupt correlation shown by the sequence of solid circles in figure 2 as the temperature is lowered through 92 K. But what about the sequence of open circles in figure 2?

When the oxygen content in $YBa_2Cu_3O_x$ is reduced to $x = 6.6$, the chains lose their ability to control the superconductivity. The control then shifts to the planes, where we are no longer dealing with the Raman line at 483 cm^{-1} but with the line at 338 cm^{-1}, shown in the lower curve in figure 6. This vibration is associated with the bonds that link $Cu(2)$ to $O(2)$ and $O(3)$, and has a period of about 99 fs. There is an approximately square Möbius orbit in the planes, 44 bonds in perimeter, whose orbital period of 200 fs allows for a synchronization between the momentum recoils at the orbital corners and the lattice vibration at 338 cm^{-1}. The orbital perimeter of 85.5 Å gives a de Broglie wavelength of 171.0 Å and a transition temperature of $T_c = 59.7$ K, close to the observed $T_c = 54$ K for the oxygen-depleted crystal whose u_1 values are plotted as open circles in figure 2.

That is, we can understand the discontinuity in the sequence of open circles as an abrupt correlation in the 338 cm^{-1} vibrations, resulting from the abrupt synchronization of a distribution of 44-bond Möbius excitations in the planes as T is lowered through the temperature (54 K) at which these excitations become locked to each other.

Let us now return to a discussion of the 90–K YBCO with its elongated orbits.

3. Traveling potential patterns

When viewed from the c-direction, the orbit of figure 4 takes the form of a simple rectangle. In figure 7 we have drawn an array of twelve such rectangles containing positive charges, carrying out synchronized Möbius orbital excitations. For convenience in description, we further simplify the bridging structures into straight crossing arms, and the simplified timing then lets us use figure 8 to show the same charges at a moment one-eighth period later. For this synchronization to be maintained, the temperature T is held below T_c.

These are orbital excitations in a layer of chains in a YBCO crystal (see figure 3). In an adjacent CuO plane, there will be an electrostatic potential pattern generated by the twelve positive charges, together with the compensating distributions of negative charge through which the positive charges move. The positive charges are shown as localized points, but they will actually be somewhat delocalized within their orbits, so that the potential patterns will have a distributed character.

These distributed potential patterns have been drawn in figures 9 and 10, to correspond to the moments depicted in figures 7 and 8. The solid contours represent potential peaks, while the dashed contours represent potential valleys.

Let us examine the top row in figures 9 and 10. When we study this row, we see that the potential peaks and valleys are traveling from left to right. The central peak in this row follows the motion of the positive charge at the top of figures 7 and 8. Furthermore, it is evident that the central peak will be passed to the positive charge in the upper right orbit, and then to the charge in the next orbit to the right (not shown), and so forth. The top row in figures 9 and 10 thus constitutes a continuously traveling potential pattern that can serve as a channel for carrying mobile electrical charges within the CuO planar layer.

4. Singlet charge pairs

Along the channel formed by the top row in figures 9 and 10, there will be a roughly sinusoidal traveling potential pattern. If the mobile charges in the CuO planar layer are holes, then they will be carried most effectively if they form a sinusoidal charge distribution whose maxima travel along with the potential valleys, as shown in figure 11. A beat pattern of this form can be constructed from the antisymmetrical combination of the following pair of single-particle wave functions:

$$\psi_{rh}(z_j t_j) = \chi_j(+)\exp[+ik_+z_j-i\omega_+t_j] + \chi_j(-)\exp[-ik_-z_j-i\omega_-t_j], \tag{3a}$$

$$\psi_{lh}(z_j t_j) = \chi_j(-)\exp[+ik_+z_j-i\omega_+t_j] + \chi_j(+)\exp[-ik_-z_j-i\omega_-t_j]. \tag{3b}$$

The subscript j indicates that a wave function applies to the jth particle. The subscripts rh and lh refer to right-handed and left-handed helicities. From the antisymmetrical combination of two particles with opposite helicities,

$$\psi = \psi_{rh}(z_1,t_1)\psi_{lh}(z_2,t_2) - \psi_{rh}(z_2,t_2)\psi_{lh}(z_1,t_1)$$

$$= [\chi_1(+)\chi_2(-)-\chi_2(+)\chi_1(-)]f_{sym} + [\chi_1(+)\chi_2(+)-\chi_1(-)\chi_2(-)]f_{anti}, \tag{4}$$

we select the singlet portion. We remove the triplet portion by setting $z_1 = z_2 = z$ and $t_1 = t_2 = t$, which defines z and t and results in the vanishing of f_{anti}.

We then are left with a traveling charge distribution proportional to $|f_{sym}|^2$, which matches the desired sinusoidal curve shown in figure 11. Furthermore, this singlet pair is a boson (resembling a Cooper pair), so that many pairs can condense and contribute charges to the same sinusoidal distribution.

These charges are then carried forward with the traveling potential pattern in the top row in figures 9 and 10. Similar charge motion from left to right will occur with rows four, five, and eight, while charge motion from right to left will take place in the channels defined by rows two, three, six, and seven. In each case the moving charge pairs are in their lowest energy state. That is, it requires an energy input to detach them from the moving potential pattern and halt their motion.

5. London equation

With the potential patterns in figures 9 and 10, there will be a balance between singlet charge pairs carried to the right and singlet charge pairs carried to the left. To get a net current flow in one direction or the other, we need an imbalance, which will result if we introduce a vector potential $\mathbf{A}$. The momentum of a circulating pi-electron hole in its Möbius orbit will then change to

$$m\mathbf{v} = m\mathbf{v}_0 - (e/c)\mathbf{A}, \tag{5}$$

where $m\mathbf{v}_0$ is the momentum in the absence of the vector potential.

For this discussion, let us consider the case in which $\mathbf{A}$ has only the component A_y, where the y-axis is directed from left to right in figure 7, and let us examine the effect of A_y on the positive charges in the lowest row of orbits. We will assume that A_y is negative, so that the velocity of each hole is increased from v_0 to $(v_0+\delta)$ when the hole is moving to the right, and is reduced to $(v_0-\delta)$ when the hole moves to the left, where $\delta = -(e/mc)A_y$. The perimeter of each Möbius orbit is $\lambda_{deB}/2$, so that the original orbital period, $\tau_0 = \lambda_{deB}/2v_0$, is modified to

$$\tau = (\lambda_{deB}/4)/(v_0+\delta) + (\lambda_{deB}/4)/(v_0-\delta) = \tau_0/(1-\delta^2/v_0^2). \tag{6}$$

We will choose as an initial time, $t_1 = 0$, a moment when the lower left hole has the position shown in figure 7. When the time has advanced to $t_2 = (\lambda_{deB}/8)/(v_0+\delta)$, this circulating hole will reach the right-hand end of its orbit. Local synchronization ensures that the same moment places the hole in the next orbit at the left-hand end of its path, starting its travel to the right. It will reach the right-hand end of its orbit at the time $t_3 = (3\lambda_{deB}/8)/(v_0+\delta)$. The third hole in the lowest row in figure 7 will then be at the left-hand end of its orbit; it will reach the center of the lower arm of its orbit at $t_4 = (\lambda_{deB}/2)/(v_0+\delta)$.

Meanwhile the hole in the first orbit in this lowest row (at the lower left of figure 7) has been tracing out its path, but it will not complete its full orbit until the time $t_5 = \tau = \tau_0/(1-\delta^2/v_0^2)$. The extra time that it will need is

$$\Delta t = t_5 - t_4 = (\lambda_{deB}/2)/(v_0-\delta^2/v_0) - (\lambda_{deB}/2)/(v_0+\delta)$$
$$= (\lambda_{deB}/2)\delta/(v_0^2-\delta^2). \tag{7}$$

Relative to the time τ for one modified orbit, given in equation (6), the increment Δt represents a phase shift of $\Delta\phi = 2\pi\delta/v_0$. The distance between the first and third orbits in this lowest row is $\Delta y = \lambda_{deB}/2$, so that the gradient of phase is $\Delta\phi/\Delta y = 4\pi\delta/v_0\lambda_{deB}$.

The physical locations of the three holes in the lowest row, at the moment $t_4 = (\lambda_{deB}/2)/(v_0+\delta)$, are shown in figure 12 for the particular case of $\Delta t/\tau \sim 1/8$. Note that the first and third holes are spread apart, relative to the positions shown in figure 7. Their force of repulsion on positive charge pairs will accordingly be reduced. At the same time, the distributed negative charge in the lowest channel is unchanged.

The result is that the lobe of the sinusoidal charge distribution bounded by those two holes (see figure 11) will carry an additional quantity of mobile charge to the right. A similar numerical analysis shows that the adjacent channel, just above, will carry a reduced quantity of mobile charge to the left. If n is the volume density of mobile charges, then we find that the net electrical current is given by $J_y = -(e^2 n/mc)A_y$, in accordance with the London equation, $\mathbf{J} = -(e^2 n/mc)\mathbf{A}$.

6. Discussion

The abrupt synchronization of certain lattice vibrations, observed by Sharma et al [1,2] when the temperature of an oxide superconductor is lowered through T_c, points to the involvement of lattice phonons in the mechanism of superconductivity in these high T_c materials. However, the involvement is unlikely to have the BCS form, since the BCS theory [8] does not have a role for an abrupt synchronization.

A theory with a central role for synchronization has been given a brief description here. The theory explains the results of the ion channeling experiments, and includes quantitative calculations of T_c values for the 1:2:3 materials used in these experiments. The theory provides a derivation of the London equation, based on an explicit picture of the physical processes taking place within the crystal layers. Treatments in more detail will address the magnetic and other properties of the oxide superconductors [9,10].

Acknowledgments

The writer is indebted to R. P. Sharma for many important discussions. This research has been supported in part by the internal program of MITRE Sponsored Research, and in part by a grant to The MITRE Corporation from The Boston Foundation.

References

[1] R.P. Sharma, L.E. Rehn, P.M. Baldo and J.Z. Liu, Phys. Rev. B 40 (1989) 11396.

[2] R.P. Sharma, L.E. Rehn, P.M. Baldo and J.Z. Liu, Materials Research Society
    Symposium proceedings series, Volume 169 (1990).

[3] W.I.F. David, W.T.A. Harrison, J.M.F. Gunn, O. Moze, A.K. Soper, P. Day,
    J.D. Jorgensen, D.G. Hinks, M.A. Beno, L. Soderholm, D.W. Capone II, I.K. Schuller,
    C.U. Segre, K. Zhang and J.D. Grace, Nature 327 (1987) 310.

[4] E. Heilbronner, Tetrahedron Letters 29 (1964) 1923.

[5] H.E. Zimmerman, J. Am. Chem. Soc. 88 (1966) 1564; J. Am. Chem. Soc. 88 (1966)
    1566; Accounts of Chemical Research 4 (1971) 272; Accounts of Chemical Research 5
    (1972) 393; Quantum Mechanics for Organic Chemists (Academic Press, New York,
    1975); in: Pericyclic Reactions, Vol. I, A.P. Marchand and R.E. Lehr, eds. (Academic
    Press, New York, 1977), pp. 53-107; Tetrahedron 38 (1982) 753.

[6] R.E. Clapp, Theoret. Chim. Acta 61 (1982) 105; Biophys. J. 53 (1988) 535a; in: High-
    Temperature Superconductivity — The First Two Years, R.M. Metzger, ed. (Gordon and
    Breach, New York, 1989), pp. 365-372; Biophys. J. 55 (1989) 570a; Biophys. J. 57 (1990)
    243a.

[7] R.J. Hemley and H.K. Mao, Phys. Rev. Lett. 58 (1987) 2340.

[8] J. Bardeen, L.N. Cooper and J.R. Schrieffer, Phys. Rev. 106 (1957) 162.

[9] R.E. Clapp, The coupling of lattice vibrations, orbiting holes, and singlet charge pairs in
    oxide superconductors, MITRE Report M89-76 (MITRE Corporation, Bedford,
    Massachusetts, 1989).

[10] R.E. Clapp, Möbius Theory for Oxide Superconductors, in preparation.

Figure captions

Figure 1. Channeling scans (c-axis) of Cu K_α x-ray yield from YBCO taken at 80 K, 100 K, and again at 80 K. The vertical displacement of the curves is for convenience in display, and the repeated curve for 80 K establishes that the ion bombardment has not altered the crystal significantly. From Sharma et al [2]. Used by permission.

Figure 2. Average vibrational amplitude, u_1, extracted from Cu K_α x-ray scans with two YBCO samples having different oxygen stoichiometry and therefore different transition temperatures. Open circles: T_c = 54 K. Solid circles: T_c = 92 K. From Sharma et al [2]. Used by permission.

Figure 3. Crystal structure of YBCO. From David et al [3]. Reprinted by permission from Nature, Vol. 327, No. 6120, pp. 310-312. Copyright (c) 1987 Macmillan Journals Limited.

Figure 4. An orbital loop formed from CuO chain segments linked by bridges utilizing the adjacent CuO planes.

Figure 5. The phase reversal in a Möbius orbit. (a) Part of a chain in figure 4. (b) Insertion of a positive charge at the central oxygen, with a pi bond coupling the O^+ to the lower Cu. (c) The O^+ coupled instead to the upper Cu. (d) The phasing in (b). (e) The phasing in (c). When the pi-hole structure in (b) and (d) migrates downward and around the loop, it returns to the original O atom where the p electron that is part of the structure rejoins the oxygen P state but with its quantum mechanical phasing reversed.

Figure 6. Raman spectra of single crystals of YBCO. From Hemley and Mao [7]. Used by permission.

Figure 7. An array of twelve coplanar Möbius excitations, coupled and synchronized.

Figure 8. The array of figure 7, seen at a moment $\tau/8$ later, where τ is the orbital period.

Figure 9. Potential peaks (solid) and valleys (dashed) generated by the array of charges in figure 7.

Figure 10. Potential peaks (solid) and valleys (dashed) generated by the array of charges in figure 8.

Figure 11. The relationship between a traveling sequence of potential peaks and valleys and a sinusoidal charge distribution carried forward by the traveling potential pattern.

Figure 12. The three orbiting holes from the lowest row of figure 7, with two of the locations shifted by the presence of the vector potential A_y.

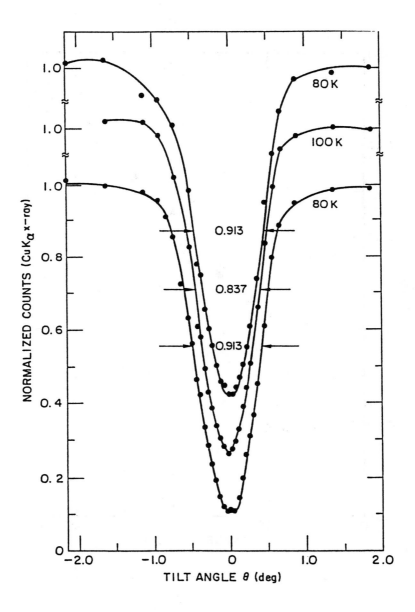

Figure 1

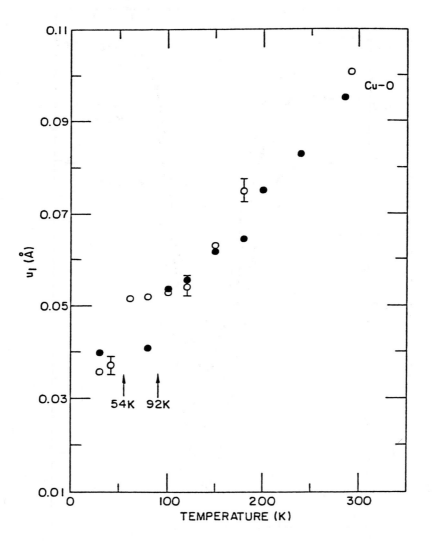

Figure 2

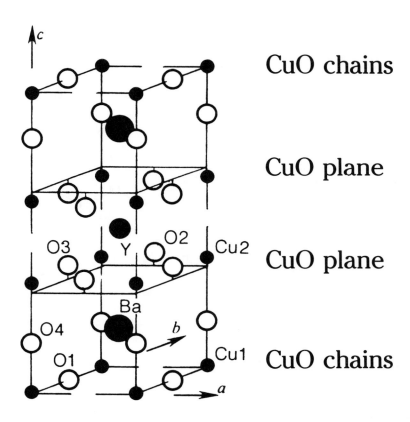

Figure   3

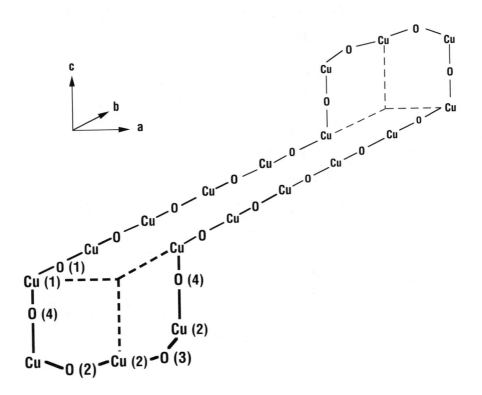

Figure 4

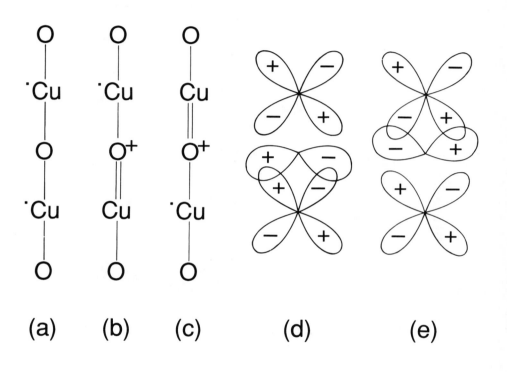

(a) (b) (c) (d) (e)

Figure 5

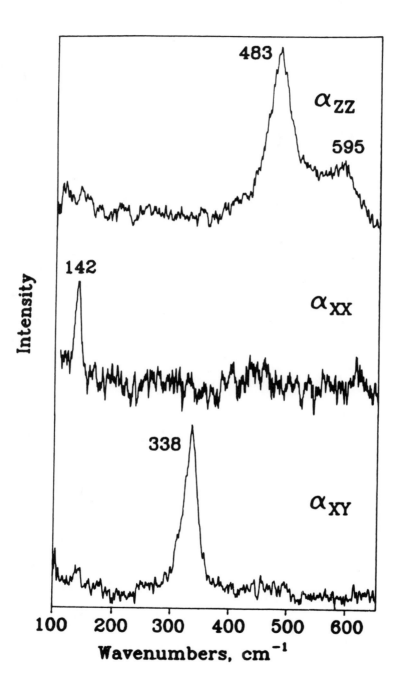

Figure 6

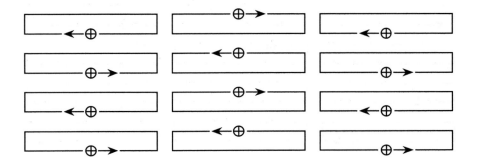

Figure   7

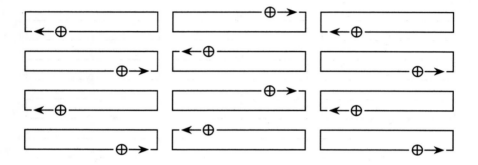

Figure   8

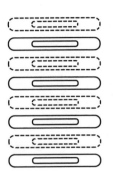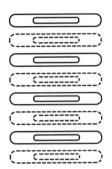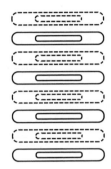

Figure   9

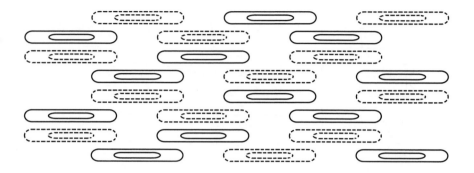

Figure 10

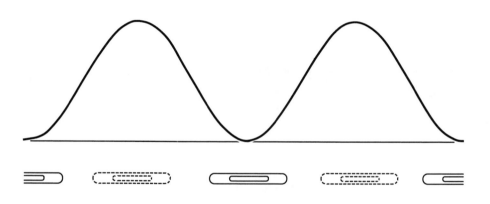

Figure   11

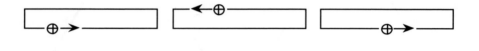

Figure   12

XANES AND EXAFS STUDY OF Au-SUBSTITUTED YBa$_2$Cu$_3$O$_{7-\delta}$

Mark W. Ruckman
Brookhaven National Laboratory, Upton, NY 11973-5000

Aloysius F. Hepp
NASA Lewis Research Center, Cleveland, OH 44135

ABSTRACT

The near-edge structure (XANES) of the Au L$_3$ and Cu K edges of YBa$_2$Au$_{0.3}$Cu$_{2.7}$O$_{7-\delta}$ has been studied. X-ray diffraction suggests that Au goes on the Cu(1) site and XANES shows that this has little effect on the oxidation state of the remaining copper. The gold L$_3$ edge develops a white line feature whose position lies between that of trivalent gold oxide (Au$_2$O$_3$) and monovalent potassium gold cyanide (KAu(CN)$_2$) and whose intensity relative to the edge step is smaller than in the two reference compounds. The L$_3$ EXAFS for Au in the superconductor resembles that of Au$_2$O$_3$. However, differences in the envelope of the Fourier filtered component for the first shell suggest that the local structure of the Au in the superconductor is not equivalent to Au$_2$O$_3$.

INTRODUCTION

Substitution of many metals in YBa$_2$Cu$_3$O$_{7-\delta}$ (abbreviated as 123 in the following)[1-5] usually depresses the transition temperature T$_c$ and has other negative effects on the superconducting properties. Trivalent metal ions like Fe, Co, or Al replace linear chain site copper, the Cu(1) site, and depress T$_c$ more slowly than Zn and Ni which replaces copper located on the CuO$_2^{2-}$ sheets, the Cu(2) site. XANES studies[6-8] indicate that transition metal substitutions on the Cu(1) site usually effect the oxidation state of the Cu(2) site and its associated oxygen, but substitution of zinc has little effect on the oxidation state.[9]

Noble metals like gold and silver are exceptional in that considerable amounts of Ag can be put into 123 before T$_c$ begins to decrease[10], Au/123 junctions have a small contact resistance[11,12] and show little evidence of solid state reaction[13]. This is of obvious practical benefit in the fabrication of any device requiring a junction with a normal metal. Streitz et al.[14] examined the microstructure of Au/123 composites and found that separate Au and 123 phases exist after heat treatment in oxygen. They also determined that a small amount of Au (x < 3 atomic %) went into the orthorhombic 123 phase.[14] Hepp et al. came to a similar conclusion for YBa$_2$(Au$_x$Cu$_{1-x}$)$_3$O$_{7-\delta}$.[15]

We report results of an examination of the Cu K and Au L$_3$ edges for YBa$_2$(Au$_x$Cu$_{1-x}$)$_3$O$_{7-\delta}$. Gold substitution of 8 mole % has little effect on the oxidation state of Cu in 123. The appearance of the L$_3$ edge suggests that the oxidation state of Au is lower in the superconductor than Au$_2$O$_3$ and that Au has fewer unoccupied d-states in 123 than in the trivalent oxide. The appearance of the extended x-ray absorption fine structure (EXAFS) of the Au L$_3$ edge supports earlier findings that Au is incorporated in the lattice at this concentration.

EXPERIMENTAL

The samples used in this investigation were synthesized and characterized as previously discussed.[15] For $YBa_2(Au_xCu_{1-x})_3O_{7-\delta}$, x-ray diffraction (XRD) and x-ray photoemission (XPS) data suggest that trivalent Au goes into the Cu(1) site. When this occurs, the a and b axes remain unchanged but the c axis expands from 11.69 to 11.75 Å. This is in accord with the well known structural chemistry of Au (i.e., Au(III) forms square planar complexes).[15]

No evidence was found for the formation of secondary phases in the XRD patterns for Au substitutions less than 10 mole %; formation of a second phase was readily detected in the XRD data when more than 10 mole % was put in 123. T_c was observed to be 89 K for an 8 mole % gold containing material and 91 K for the parent 123 material made without Au_2O_3. Very similar conclusions were drawn in a recent study by Cieplak et al.[16], but they were unable to produce single-phase material when substituting Au for Cu. We were only able to obtain single-phase material, as determined by XRD and microscopy when using BaO_2 in the synthesis of the material.[15]

X-ray absorption measurements were made in the transmission mode using powdered samples dispersed on adhesive tape. Au or Cu foil absorbers were placed after the samples to run in conjunction with the samples to maintain a calibrated energy scale. The work was done at the X-11A beamline at the Brookhaven National Synchrotron Light Source).[17] The monochromator resolution is estimated to be 1.0 eV at the Cu K-edge and 1.2 eV at the Au L_3 edge. Samples were crushed into powder, screened through 400 mesh and dispersed onto adhesive tape.

The near-edge and EXAFS data for the Cu K and Au L_3 edges was analyzed using standard procedures.[18] A linear background was removed from the edge before normalization. To extract the EXAFS, the atomic absorption background was approximated by a spline curve. The $\chi(k)$ data was converted from energy space to k-space assuming an E_0 for Cu of 8992 eV and 11,919 eV for Au. The Fourier transforms were computed using a Gaussian window function to obtain a quantity related to the radial distribution function around the absorbing atom.

RESULTS AND DISCUSSION

The Cu K edges for the 123 material (solid line) and the 8 mole percent Au sample (dashed line) are shown in Fig. 1. The Cu K edge is complex and several interpretations of it exist.[19-23] XANES results are now available for highly oriented powders or single crystal materials using polarized x-rays and these provide more reliable data on the Cu K edge.[24-27] The Cu K near-edge structure arises from dipole transitions from the Cu 1s core level to the low-lying copper valence or conduction band states with p or π symmetry and to transitions from the Cu 1s to continuum final states that are modified by multiple scattering (shape resonances). The transitions to bound final states are related to the electron density of states and are sensitive to changes in the chemical state of the Cu while the shape resonances are sensitive to structural modification. Heald et al.[24] examined the Cu K edge from 123 powders oriented such that the x-ray polarization

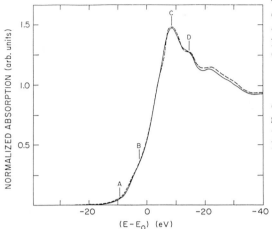

Fig. 1. Cu K edge XANES: $YBa_2Cu_3O_{7-\delta}$ (solid curve) and $YBa_2Au_{0.3}Cu_{2.7}O_{7-\delta}$ (dashed curve). The energy reference was maintained using a copper foil.

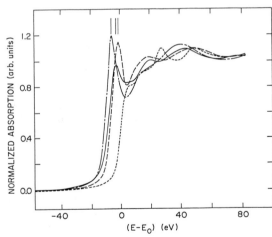

Fig. 2. The Au L_3 absorption edges for Au (dotted curve), a monovalent gold compound - $KAu(CN)_2$ (dashed curve), trivalent gold oxide - Au_2O_3 (dot-dashed curve) and $YBa_2Au_{0.3}Cu_{2.7}O_{7-\delta}$ (solid). A consistent energy reference was maintained by examining the L_3 edge of gold foil simultaneously.

vector ê was either parallel or perpendicular to the c axis. The position of a weak pre-edge feature marked "A" due to 1s to 3d quadrupole transitions is marked on figure 1 and it is directly related to the valence of the Cu. In oxygen deficient 123 material, this peak grows in proportion to the removal of holes from the Cu site and the formation of Cu.[28,29] Peak "B" is due to transitions from the 1s to $4p\pi$ band (Cu(1) and Cu(2) sites) accompanied by shake down transitions ($4p\pi^*$) and peak "C" contains contributions from the 1s to $4p\pi$ and 1s to $4p\sigma$ transitions from the Cu(2) and the Cu(1) sites. The feature marked "D" is identified as a shape resonance. Figure 1 shows the Cu K edges of $YBa_2Cu_3O_{7-\delta}$ and $YBa_2(Au_xCu_{1-x})_3O_{7-\delta}$ to be virtually identical. This suggests that Au substitution has little or no effect on the valence of copper on either site and that the Cu is still formally divalent.

The Au L_3 edges for $YBa_2(Au_xCu_{1-x})_3O_{7-\delta}$, Au foil, monovalent $KAu(CN)_2$ and trivalent Au_2O_3 are shown in figure 2. The spectra for the reference compounds and Au substituted 123 have been normalized to the edge step of the Au absorption edge. The near-edge structure of the L_2 and L_3 edges in 5d transition metal compounds is dominated by $2p_{1/2}$ to $5d_{3/2}$ and $2p_{3/2}$ to $5d_{1/2}$ transitions (white line feature).[30,31] The L_3 edge also has a contribution from the $2p_{3/2}$ to $5d_{3/2}$ but it is much weaker than the $2p_{1/2}$ to $5d_{3/2}$ and

$2p_{3/2}$ to $5d_{1/2}$ transitions. The intensity of this feature is thought to provide a good probe of the 5d-band occupation.[32] For Au (small dashed curve), the 5d band is filled and no white line feature is observed at the L_3 edge. The oxidation of Au to the mono- or trivalent state creates the white line feature. We find that the white line area is a little larger for Au_2O_3 (dot-dashed line) than $KAu(CN)_2$ (dashed line) and also find that the white line feature shifts to lower photon energy when the oxidation number increases from +1 to +3. It should be noted that the ratio of the white line areas for Au_2O_3 and $KAu(CN)_2$ is less than the ratio of d-electron removal suggested by the valence. This means that the actual d-band occupation changes less than expected from simple electron counting. The white line area for Au in 123 (solid line) is smaller than either of the reference compounds and the white line lies at lower energy than $KAu(CN)_2$.

From Au L_3 near-edge data presented for $YBa_2(Au_{0.1}Cu_{0.9})_3O_{7-\delta}$, it is clear that the electronic structure of Au in 123 differs from that of Au in trivalent Au_2O_3. Hepp et al.[15] made an assignment of the trivalent state for Au in the 123 material using XPS measurements of the Au 4f core level. We believe the L_3 near-edge data shows that oxidation state of Au in the 123 material is not equivalent to that of Au in the formally trivalent oxide. Au may be "formally" trivalent in this material but significant departures in the 5d band occupation seem to be taking place. Trivalent Fe also replaces Cu on the chain site. However, substitution of the same amount of Fe depresses T_c by 55 K[33] rather than the 2 K found for Au. Yang et al.[6] found that the Fe substitution modifies the O K edge and reduces the number of 2p holes on the oxygen. We speculate that the reduction in the apparent number of unoccupied Au 5d states for Au in 123 when compared to Au in Au_2O_3 and the small Au induced change in T_c implies that little or no change occurs in the number of the oxygen 2p holes. This is supported by our data for the Cu K edge which shows no change in the Cu-O bonding like that observed when the high T_c material becomes oxygen deficient and holes are removed from the copper and oxygen.[28] We also speculate that the placement of holes on the oxygen nearest neighbors reduces the hybridization of Au 5d states[32] and this is responsible for the smaller L_3 white line area for Au in 123 when compared to Au(III) oxide.

The $\chi(k)$ data extracted from the L_3 edge is shown in figure 3 for $YBa_2(Au_{0.1}Cu_{0.9})_3O_{7-\delta}$ (a), Au_2O_3 (b), and Au (c) . The data are k^2-weighted to enhance the amplitude of high-k oscillations. The Au L_3 EXAFS functions for Au substituted 123 and Au_2O_3 look similar and are very different from that obtained for Au. The reduction in the strength of the EXAFS oscillations as a function of k suggests that the Au in the superconductor is coordinated by light elements like oxygen rather than a heavy element like Au.[33]

The appearance of the Au L_3 EXAFS provides support for our interpretation of the Au L_3 edge discussed earlier. The reduced strength of the white line could be rationalized by assuming that the L_3 edge is the sum of an oxidized Au component and metallic Au. However, a significant fraction of the Au would have to be in the metallic phase to account for the XANES result. If the Au particles are so small as to preclude getting EXAFS from the Au, it is difficult to see how the gold could have the metallic electronic structure.[34] Hence,

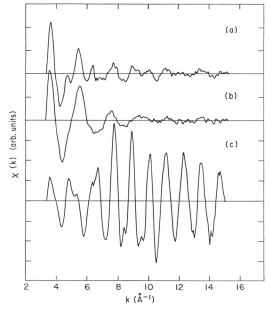

Fig. 3. Au L_3 $\chi(k)$ data for (a) $YBa_2Au_{0.3}Cu_{2.7}O_{7-\delta}$, (b) gold oxide, and (c) metallic gold. The chi data has been weighted by k^2 to enhance the oscillations at large k.

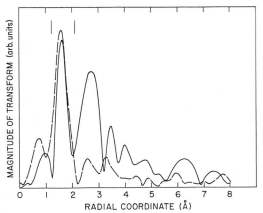

Fig. 4. The Fourier transforms of L_3 edge $k^2\chi(k)$ for Au_2O_3 (dashed curve) and $YBa_2Au_{0.3}Cu_{2.7}O_{7-\delta}$ (solid curve).

we conclude that the gold has not formed a separate metallic component as seen during the formation of 123/Au composites.

In figure 4, the Fourier transforms of the k^2-weighted L_3 $\chi(k)$ data for $YBa_2(Au_{0.1}Cu_{0.9})_3O_{7-\delta}$ (solid line) and Au_2O_3 (dashed line) are shown. The Fourier transformed EXAFS for both $YBa_2(Au_{0.1}Cu_{0.9})_3O_{7-\delta}$ and Au_2O_3 shows a peak located at a radial coordinate of 1.6 Å. This value is close to the first shell radial coordinate measured for copper by a number of groups[35-37] and ourselves for 123 (not shown); detailed analysis of the copper EXAFS[35] indicates a Cu-O spacing of 1.9 Å for the Cu(1)-O bond. Hepp et al.[15] found Au substitution causes little change in the a and b lattice constants and we think the Au-O spacing is nearly equal to that of Cu-O. Detailed analysis of the Au L_3 EXAFS data is hampered by the lack of a suitable Au-O standard.[38] To the best of our knowledge, a detailed structural study of Au_2O_3 has not been performed. We think that Au_2O_3 is highly disordered because the Fourier transformed EXAFS for Au_2O_3 shows only the single peak corresponding to the first shell. Figure 4 shows a second peak at 2.7 Å for the Au-substituted 123 material. X-ray[39] and neutron scattering[40] studies of the structure of $YBa_2Cu_3O_7$ indicate that Ba atoms are located 3.43 Å from the Cu(1) site and the next Cu(1) atom is located at 3.86 Å. Fourier transformed EXAFS

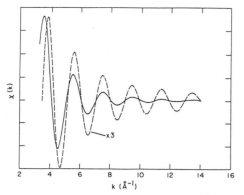

Fig. 5. The Fourier filtered components attributed to the first shells of Au_2O_3 (solid curve) and $YBa_2Au_{0.3}Cu_{2.7}O_{7-\delta}$ (dashed curve).

data for the Cu K edge shows a peak between 2 and 3 Å containing contributions from a higher order shell containing Ba.[35-37]

Figure 5 shows the Fourier filtered first shell contribution to EXAFS for the Au_2O_3 and $YBa_2Au_{0.3}Cu_{2.7}O_{7-\delta}$. The first shell contribution was obtained by back-transforming the region marked by vertical lines in figure 4. The $\chi(k)$ damps out rapidly for Au_2O_3; this can be attributed to the rapid decrease in backscattering amplitude of the oxygen nearest neighbors.[33] The static Debye-Waller term may also serve to diminish the EXAFS signal at higher k values. The $\chi(k)$ term does not decrease as rapidly for $YBa_2(Au_{0.1}Cu_{0.9})_3O_{7-\delta}$, this suggests that the Au-O first shell in 123 is less disordered or has a smaller Debye-Waller factor.

SUMMARY AND CONCLUSIONS

XANES and EXAFS features support earlier studies[15] which concluded that small amounts of gold can be incorporated in $YBa_2Cu_7O_{7-\delta}$ at the Cu(1) site. The gold shows significantly less 5d involvement in the Au-O bond in the superconductor than in Au_2O_3. This could be related to the doping of the oxygen with holes and this is similar to the change in the number of holes on the copper-oxygen component when the material is made superconducting by hole doping. Au has less effect on the superconducting properties than Fe presumably because the gold does not localize holes on itself. There are no obvious changes in the Cu K edge and we conclude that Au substitution has little or no effect on the chemical state of copper or oxygen. Our EXAFS data show that Au is coordinated by oxygen and is in a more ordered environment than Au_2O_3. The lack of a good Au-O standard hinders a more through analysis of the Au L_3 EXAFS. The Cu K edge EXAFS is virtually identical to that published for $YBa_2Cu_3O_{7-\delta}$.

ACKNOWLEDGEMENTS

The authors thank Professor M. Croft of Rutgers University for help in obtaining some of the EXAFS data and S. M. Heald, J. Jayanetti, and E. Barrera for help in performing and analyzing the EXAFS and XANES measurements. M. W. R. acknowledges the support of the U.S. Department of Energy, Division of Materials Sciences under Contract No. DE-AC02-CH00016. A. F. H. acknowledges support from the Space Electronics

Division of the NASA Lewis Research Center. The X-11 beamline and the National Synchrotron Light Source are supported by the U.S. Department of Energy, Contracts No. DE-AC05-80-ER10742 and DE-AC02-CH00016, respectively.

REFERENCES

1. Y. Maeno, T. Tomita, M. Kyogoku, S. Awaji, Y. Aoki, K. Hoshino, A. Minami, and T. Fujita, Nature 328, 512 (1987).

2. J. M. Tarascon, P. Barboux, P. F. Miceli, L. H. Greene, G. W. Hull, M. Eibschutz, and S. A. Sunshine, Phys. Rev B36, 8393 (1987); ibid B37, 7458 (1988).

3. G. Xiao, F. H. Streitz, A. Garvin, Y. W. Du, and C. L. Chien, Phys. Rev. B35, 8782 (1987).

4. Y. Tokura, J. B. Torrance, T. C. Huang, and A. Nazzal, Phys. Rev. B38, 7156 (1988).

5. M. Shafer, T. Penny, B. L. Olson, R. L. Greene, and R. H. Koch, Phys. Rev. B39, 2914 (1989).

6. C. Y. Yang, S. M. Heald, J. M. Tranquada, Y. Xu, X. L. Wang, A. R. Moodenbaugh, D. O. Welch, and M. Suenaga, Phys. Rev. B39, 6681 (1989).

7. H. Qian, E. A. Stern, Y. Ma, R. Ingalls, M. Sarikaya, B. Thiel, R. Kurosky, C. Han, L. Hutter, and I. Aksay, Phys. Rev. B39, 9192 (1989).

8. C. Y. Yang, A. R. Moodenbaugh, Y. L. Yang, Y. Xu, S. M. Heald, D. O. Welch, M. Suenaga, D. A. Fisher, and J. E. Penner-Hahn, (preprint).

9. M. L. den Boer, C. L. Chiang, H. Peterson, M. Schaible, K. Reilly, and S. Horn, Phys. Rev. B38, 6588 (1988).

10. C. A. Chang, Appl. Phys. Lett 52, 924 (1988).

11. R. Mizushima, M. Sagoi, T. Miura, and J. Yoshida, Appl. Phys. Lett. 52, 1101 (1988).

12. J. W. Ekin, T. M. Larson, N. F. Bergren, A. J. Nelson, A. B. Swartzlander, L. L. Kazmerski, A. J. Panson, and B. A. Blankenship, Appl. Phys. Lett. 52, 1919 (1988).

13. T. J. Wagnener, Y. Gao, I. M. Vitomirov, C. M. Aldo, J. J. Joyce, C. Capasso, J. H. Weaver, and D. W. Capone II, Phys. Rev. B38, 232 (1988).

14. F. H. Streitz, M. Z. Cieplak, G. Xiao, A. Garvin, A. Bakhshai, and C. L. Chien, Appl. Phys Lett. 52, 91 (1988).

15. A. F. Hepp, J. R. Gaier, J. J. Pouch, and P. D. Hambourger, J. Solid State Chem. 74, 433 (1988).

16. M. Z. Cieplak, G. Xiao, A. Bakhshai, D. Artymowicz, W. Bryden, C. L. Chien, J. K. Stalick, and J. J. Rhyne, Phys. Rev., in press.

17. D. E. Sayers, S. M. Heald, M. A. Pick, J. I. Budnick, E. A. Stern, and J. Wong, Nuc. Instrum. Meth. 208, 631 (1983).

18. X-ray Absorption: Principles, Applications, Techniques of EXAFS, SEXAFS and XANES, eds. D. C. Koningsberger and R Prins (Wiley, NY 1987).

19. G. Antonini, C. Calandra, F. Corni, F. C. Matacotta, and M. Sacchi, Europhys. Lett. 4, 851 (1987).

20. H. Oyanagi, H. Ihara, T. Matsubara, T. Matsushita, T. Tokumoto, M. Hirabayashi, N. Terada, K. Senzaki, Y. Kimura, and T. Yao, Jpn. J. Appl. Phys. 26, L638 (1987).

21. K. B. Garg, A. Bianconi, S. Della Longa, A. Clozza, M. De Santis and A. Marcelli, Phys. Rev. B38, 244 (1988).

22. F. W. Lytle, R. B. Gregor, and A. J. Panson, Phys. Rev. B37, 1550 (1988).

23. H. Tolentino, E. Dartyge, A. Fontaine, T. Gourieux, G. Krill, M. Maurer, M-F. Ravet, and G. Tourillon, Phys. Lett. A139, 474 (1989).

24. S. M. Heald, J. M. Tranquada, A. R. Moodenbaugh, and Y. Xu, Phys. Rev. B38, 761 (1988).

25. J. Whitmore, Y. Ma, E. A. Stern, F. C. Brown, R. L. Ingalls, J. P. Rice, B. G. Pazol, and D. M. Ginsberg, Physica B158, 440 (1989).

26. N. Kosugi, H. Kondoh, H. Tajima, and H. Kuroda, Physica B158, 450 (1989).

27. J. Guo, D. E. Ellis, G. L. Goodman, E. E. Alp, L. Soderholm, and G. K. Shenoy, Phys. Rev. B41, 83 (1990).

28. J. M. Tranquada, S. M. Heald, A. R. Moodenbaugh, and Y. Xu, Phys. Rev. B38, 8893 (1988).

29. H. Oyanagi, H. Ihara, T. Matsubara, M. Tokumoto, T. Matsushita, M. Hirabashi, K. Murata, N. Terada, T. Yao, H. Iwasaki, and Y. Kimura, Jpn. J. Appl. Phys. 26, L1561 (1987).

30. T. K. Sham, Solid State Commun. 64, 1103 (1987).

31. B. Qi, I. Perez, P. H. Ansari, F. Lu, and M. Croft, Phys. Rev. B36, 2972 (1987).

32. I. Perez, B. Qi, G. Liang, F. Lu, M. Croft, and D. Uieliczka, Phys. Rev. B38, 12233 (1988).

33. Y. Xu, M. Suenaga, J. Tafto, R. L. Sabatini, A. R. Moodenbaugh, and P. Zolliker, Phys. Rev. B39, 6667 (1989).

34. B. K. Teo and P. A. Lee, J. Amer. Chem. Soc. 101, 2815 (1979).

35. E. B. Crozier, N. Alberding, K. R. Bauchspiess, A. J. Seary, and S. Gygax, Phys. Rev. B37, 8288 (1987).

36. K. Zhang, G. B. Bunker, G. Zhang, Z. X. Zhao, L. Q. Chen, and Y. Z. Huang, Phys. Rev. B37, 3375 (1988).

37. J. B. Boyce, F. Bridges, T. Cleason, and M. Nygren, Phys. Rev. B39, 6555 (1989).

38. P. A. Thiessen and H. Schutza, Z. Anorg. Allg. Chem. 243, 32 (1939).

39. R. J. Cava, B. Batlogg, R. B. van Dover, D. W. Murphy, S. Sunshine, T. Siegrist, J. P. Remeika, E. A. Reitman, S. M. Zahurak, and G.P. Espinosa, Phys. Rev. Lett. 58, 1676 (1987); E. M. Engler, V. Y. Lee, A. Nazzai, R. B. Beyers, G. Lim, P. M. Grant, S. S. P. Parkin, M. L. Ramirez, J. E. Vasquez, and R. J. Savoy, J. Amer. Chem. Soc. 109, 2848 (1987).

40. M. A. Beno, L. Soderholm, D. W. Capone II, D. G. Hinks, J. D. Jorgensen, I. K. Schuller, C. U. Segre, K. Zhang, and J. D. Grace, Appl. Phys. Lett. 51, 57 (1987).

EXAFS STUDIES OF $Bi_2Sr_2CaCu_2O_8$ GLASS AND GLASS-CERAMIC

S. S. Bayya, R. Kudesia and R. L. Snyder
Institute for Ceramic Superconductivity
NYS College of Ceramics
Alfred, NY 14802

Abstract

EXAFS measurements were done on $Bi_2Sr_2CaCu_2O_8$ glass and glass-ceramic to study the local environment of copper and strontium. X-ray absorption measurements at the Cu K-edge indicate that copper atoms are surrounded by four oxygen atoms in the glass and six oxygens in the glass-ceramic with Cu-O bond distances of 1.83 Å and 1.91 Å, respectively. Measurements at the Sr K-edge indicate that the Sr-O distance in the glass-ceramic is 2.70 Å with a coordination of nine.

INTRODUCTION

Since the discovery of superconductivity in the Bi-Sr-Ca-Cu-O system[1] scientists have been trying to improve the current carrying capacity of this material by adopting different processing techniques. Common ceramic fabrication processes in this system, result in superconductors with poor strengths, very low current carrying capacity and are prone to atmospheric degradation. Observed low current capacities are due to the high porosity in samples, poor connectivity of the grains[2], and giant flux creep. The glass-ceramic process has the potential of overcoming a number of these problems. Other advantages of the glass-ceramic techniques have been discussed by Bhargava et al.[3] Bismuth oxide has the capability to form glass in the presence of other modifiers. These glasses can then be devitrified to crystallize superconducting phases. This study is focused on the change in local environment of Cu and Sr atoms from the glass to the glass-ceramic state.

The extended X-ray absorption fine structure (EXAFS) is a well known technique to probe the local structure around an absorbing atom. EXAFS is a final state interference effect involving scattering of the outgoing photoelectron from the neighboring atoms. It refers to the oscillatory variation of the X-ray absorption as a function of the photon energy beyond an absorption edge. EXAFS spectra generally ranges from 40 to 1000 eV above the absorption edge. Several EXAFS studies have been made to determine the local structure around cations in Y-Ba-Cu-O[4,5,6,7,8] and Bi-Sr-Ca-Cu-O[9] superconductors. In the present work, Cu and Sr K-edge absorption spectra of $Bi_2Sr_2CaCu_2O_8$ glass and glass ceramic were investigated.

EXPERIMENTAL PROCEDURE

Reagent grade $SrCO_3$, CuO (J.T. Baker Chemical Co., Philipsburg, NJ) Bi_2O_3 and CaO (Fisher Scientific Co., Fair Lawn, NJ) were used in this study. Ten gram batches were prepared with stoichiometry of Bi-Sr-Ca-Cu in 2-2-1-2 ratio. Batches were initially mixed with a mortar and pestle using acetone. The batch was melted in a platinum-gold crucible at 1100°C for 30 min in air. The melt was poured on a steel plate and quenched with another similar plate. The samples thus obtained were analysed by X-ray powder diffraction and were found to be amorphous. These glass samples were heat treated at 800°C for 72 hours to fully crystallize $Ba_2Sr_2CaCu_2O_8$.

EXAFS experiments were performed at Brookhaven National laboratory using National Synchrotron Light Source facility. Spectra were collected on SUNY beamline X3B1 with synchrotron beam at an energy of 2.6 GeV. A Si (111) crystal monochromator was used. EXAFS samples were prepared by brushing a fine powder of the sample onto a Scotch tape. Data was collected on four or five such layers stacked together, in transmission mode. The monochromator was calibrated using a copper metal foil before collecting the spectra. Copper K-edge spectra was collected on the glass, glass-ceramic and CuO (reference material). Nitrogen gas was used in the ion chambers. For Sr K-edge spectra, argon was used in the ion chambers. $SrCO_3$ was used as the reference material for strontium K-edge analysis. For all samples, five data sets were collected at room temperature and the signal was averaged.

EXAFS DATA ANALYSIS

The EXAFS spectra can be described by a single scattering theory[10]:

$$\chi(k) = \frac{1}{k} \sum_i \frac{N_i}{R_i^2} F_i(k) \exp(-2\sigma_i^2 k^2) \sin[2kR_i + \phi_i(k)]$$

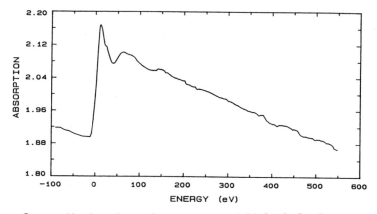

Figure 1: Copper K-edge absorption spectrum of $Bi_2Sr_2CaCu_2O_8$ glass-ceramic.

where N_i is the number of atoms in ith coordination shell at a distance R from the absorbing atom. F_i is the backscattering amplitude from neighboring atoms and ϕ is the phase shift experienced by the ejected photoelectron. σ_i is the mean relative displacement between the absorbing and scattering atoms (Debye-Waller type factor).

EXAFS data was analysed using a software package originally developed by E.C. Marques and currently in use at beamline X18 B at BNL. EXAFS oscillations were extracted from the absorption spectra (Figures 1 and 4) and normalized using a spline fit.[10] The interference function $\chi(E)$ in energy space was converted to $\chi(k)$ in k-space. $\chi(k)$ was multiplied by k^3 to compensate for the diminishing amplitude at high k. A Fourier transform of these oscillations were taken to obtain the r-space data. Figures 2 and 5 show the Fourier transformed data of Cu K-edge and Sr K-edge respectively. The distances in this transformed data are displaced from the actual distances by a phase shift. The peak corresponding to the cation-oxygen coordination shell was back-transformed into k-space. Non least squares single shell fitting was applied to the Fourier filtered data using theoretically calculated back scattering amplitude factor for oxygen.[11]

RESULTS AND DISCUSSION

Figure 2 shows the Fourier transform of Cu K-edge spectra in CuO, glass and glass-ceramic. The first major peak in these spectra correspond to the Cu-O distance. In order to determine the structural parameters of glass and glass-ceramic, CuO was used as the model compound. Using the known values of the

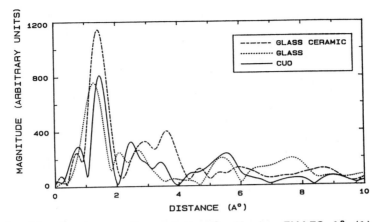

Figure 2: The Fourier-transform of the Cu K-edge EXAFS $k^3\chi(k)$ in CuO, $Bi_2Sr_2CaCu_2O_8$ glass and glass-ceramic.

Cu-O bond distance and coordination number in CuO[12], the phase function for Cu-O pair was determined. Theoretically calculated values of back scattering amplitudes were used. The phase function determined from CuO and the theoretical amplitude function were then transferred to the Fourier spectra in the glass and glass-ceramic for curve fitting in order to determine actual Cu-O distances and the coordination numbers. Figure 3 shows the curve fitting of $k^3\chi(k)$ in these samples. The Standard deviation of these fits range from 2 to 4 %.

The best fit least-squares refined bond-distances and coordination numbers are listed in table I.

Table I: Cu K-edge results for Cu-O bond distance and number of coordinating oxygens

Compound	Cu-O Length $R(\mathring{A})$	Coordination Number, N
CuO (model compound)	1.96	4.0
$Bi_2Sr_2Ca_1Cu_2O_8$ glass	1.83	3.9
$Bi_2Sr_2Ca_1Cu_2O_8$ glass-ceramic	1.91	6.0

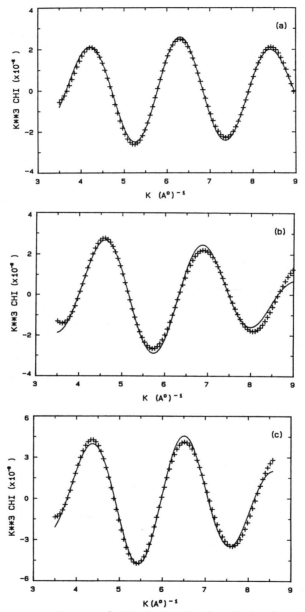

Figure 3: The Fourier-filtered $k^3\chi(k)$ (solid lines) and simulated curve (points) of Cu K-edge EXAFS oscillations in (a) CuO, (b) $Bi_2Sr_2CaCu_2O_8$ glass and (c) glass-ceramic.

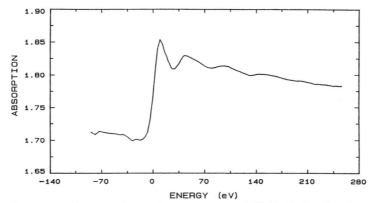

Figure 4: Strontium K-edge absorption spectrum of $Bi_2Sr_2CaCu_2O_8$ glass-ceramic.

Cu-O bond distance in the $Bi_2Sr_2CaCu_2O_8$ glass-ceramic was found to be 1.91 Å with 6 oxygens surrounding each copper atom. Considering the error margin in the coordination number determination, these values are in good agreement with those reported by Meada *et al.*[9] ($R_{Cu-O} = 1.91Å, N = 5.1$). However, a shorter Cu-O bond distance, 1.83 Å, was observed in the glass. This reduction of Cu-O bond distances in glass may be attributed to the absence of the crystal field energy which, if present, would pull the atoms apart from their minimum energy positions. Further, these glass samples were not annealed for structural relaxations. The stresses present in these glass samples may also contribute to the reduction in Cu-O bond distance.

Figure 5 shows the Fourier transform of Sr K-edge spectra in $SrCO_3$ and the glass-ceramic. The strontium K-edge spectra was analysed in a similar manner as the Copper K-edge spectra was. The phase function was determined from curve fitting the back transformed correlation function $k^3\chi(k)$ in $SrCO_3$ (the model compound), using a theoretically calculated back scattered amplitude function[11] and known bond distances & coordination numbers.[13] These phase functions and amplitude functions were transferred to the $k^3\chi(k)$ data in the glass-ceramic to determine the Sr-O bond distances and the coordination number using a curve fitting technique. Figure 6 shows the curve fitting of $k^3\chi(k)$ in these samples. The standard deviation of these fits range from 2 to 3 %. Average Sr-O bond distance in the $Bi_2Sr_2CaCu_2O_8$ glass-ceramic was found to be 2.70 Å with 9 oxygens surrounding each strontium atom (Table II). These values are comparable to the coordination number of 9 and average Sr-O bond distance of 2.67 Å in $Bi_2Sr_2CaCu_2O_8$ as reported by Tarascon *et al.* [14]

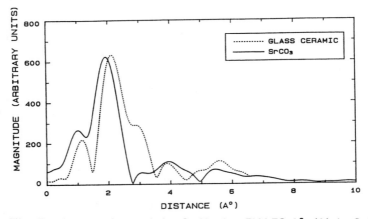

Figure 5: The Fourier-transform of the Sr K-edge EXAFS $k^3\chi(k)$ in $SrCO_3$ and $Bi_2Sr_2CaCu_2O_8$ glass-ceramic.

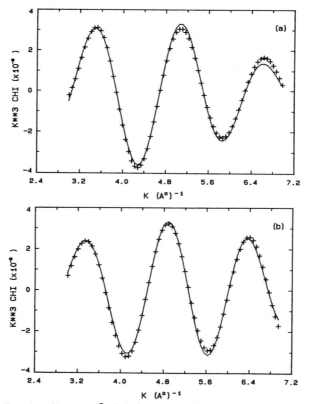

Figure 6: The Fourier-filtered $k^3\chi(k)$ (solid lines) and simulated curve (points) of Sr K-edge EXAFS oscillations in (a) $SrCO_3$ and (b) $Bi_2Sr_2CaCu_2O_8$ glass-ceramic.

Table II: Sr K-edge results for Sr-O bond distance and number of coordinating oxygens

Compound	Sr-O Length $R(\overset{\circ}{A})$	Coordination Number, N
$SrCO_3$ (model compound)	2.62	8.0
$Bi_2Sr_2Ca_1Cu_2O_8$ glass-ceramic	2.70	9.0

CONCLUSIONS

- In $Bi_2Sr_2CaCu_2O_8$ glass, copper atoms are coordinated with 4 oxygen atoms at an average distance of $1.83\overset{\circ}{A}$.

- In $Bi_2Sr_2CaCu_2O_8$ glass-ceramic, copper atoms are coordinated with 6 oxygen atoms at an average distance of $1.91\overset{\circ}{A}$.

- In $Bi_2Sr_2CaCu_2O_8$ glass-ceramic, strontium atoms are coordinated with 9 oxygen atoms at an average distance of $2.70\overset{\circ}{A}$.

ACKNOWLEDGMENTS

The authors would like to thank the New York State Science and Technology Foundation and their Center for Advanced Ceramic Technology along with the New York State College of Ceramics for the financial support towards the study. Support from the New York State Institute on Superconductivity and the Kodak corporation is gratefully acknowledged. Technical assistance by Drs. Alex Darovski and Yan Gao at National Synchrotron Light Source, Brookhaven National Laboratories, NY is gratefully acknowledged.

References

[1] H. Maeda, Y. Tanaka, M. Fukutomi and T. Asano, *Jpn. J. Appl. Phys.* **27**, L209 (1988).

[2] W. Herkert, H.W. Neumuller and M. Wilhelm, *Sol. State Comm.* **69**, 183-185 (1989).

[3] A. Bhargava, A.K. Varshneya and R.L. Snyder, *Proceedings of the Conference on Superconductivity and Its Applications*, H.S. Kwok and D.T. Shaw editors, Elsevier New York (1988), 124-129.

[4] J.B. Boyce, F. Bridges, T. Claeson and M. Nygren, *Phys. Rev. B* vol **39**, No **10**, 6555-6566 (1989).

[5] H. Oyanagi *et al. Jpn. J. Appl. Phys.* **26**, L638 (1987).

[6] J.B. Boyce, F. Bridges, T. Claeson, R.S. Howland and T.H. Geballe, *Phys. Rev. B* **36**, 5251 (1987).

[7] H. Oyanagi *et al. Jpn. J. Appl. Phys.* **26**, L1233 (1987).

[8] E.D. Crozier, N. Alberding, K.R. Bauchspiess, A. J. Seary and S. Gygax, *Phys. Rev. B* **36**, 8288 (1987).

[9] H. Meada *et al.*, *Jpn. Jou. Appl. Phy.* vol. **27**, No **5**, L807-L810 (1988).

[10] B.K. Teo, *EXAFS: Basic Principles and Data Analysis*, Springer-Verlag, Heidelberg (1986).

[11] B.K. Teo, P.A. Lee, A.L. Simons, P. Eisenberger and B.M. Kineaid, *J. Am. Chem. Soc.* **99**, 3854-3857 (1977).

[12] S. Asbrink and L.J. Norrby, *Acta. Cryst.* **26**, 8-15 (1970).

[13] V.W. Pannhorst and J. Löhn, *Z. Kristallogr.* **131**, S. 455-459 (1970).

[14] J.M. Tarascon *et al.*, *Physical Review B* Vol. **37**, No **16**, 9382-9389 (1988).

Thin Films and Synthesis

SYSTEMATIC TRENDS IN THE SUPERCONDUCTING PROPERTIES OF ION-BEAM THINNED HIGH-T_c FILMS

A. F. Hebard, M. P. Siegal, Julia M. Phillips, R. H. Eick
AT&T Bell Laboratories, Murray Hill, NJ 07974

A. Gupta
IBM T. J. Watson Research Center, Yorktown Heights, NY 10598

ABSTRACT

Glancing angle ion beam milling has been used to successively thin c-axis oriented $Ba_2YCu_3O_7$ films fabricated using an *in situ* pulsed laser deposition process. The systematics of the thinning process are characterized by scanning electron microscopy, ion channeling by Rutherford backscattering spectrometry, and measurements of the resistance transitions. Comparisons are made with the results of previously published work in which *ex situ* annealed films, fabricated in a three-source evaporator and having a significantly different surface morphology, were subjected to a similar thinning treatment.

INTRODUCTION

The usefulness of ion beams in the post-deposition processing of high-T_c films has been demonstrated in a variety of ways. These include etching with focused ion beams to obtain micron-size features,[1,2] ion damage of selected areas to obtain altered, usually reduced, superconducting properties,[3,4] ion beam thinning to obtain weak links,[5] and ion beam thinning and polishing to obtain ultrathin superconducting films.[6-8] In these latter thinning and polishing applications c-axis oriented films are mounted on a rotating substrate and exposed to a glancing-angle broadbeam ion flux. This previous work[6-8] on milling of $Ba_2YCu_3O_7$ (BYCO) films, which were *ex situ* annealed to give a c-axis orientation, showed uniform thinning and planarization, enhanced surface smoothness, and negligible ion-induced surface damage. Ultrathin superconducting films having an effective thickness on the order of a few lattice constants and a normal-state sheet resistance with an upper bound set by the combination of fundamental constants $h/4e^2 = 6450\Omega$ were obtained.[8]

In this paper, we present new results on the ion-mill induced changes in the electrical and structural properties of pulsed laser deposited (PLD) *in situ* annealed BYCO films and make comparisons to previously published results on similar films fabricated in a three-source evaporator and *ex situ* annealed. The primary advantages of PLD are the reproducibility of film stoichiometry and the ability to deposit films in relatively high O_2

partial pressures. However, these advantages can be compromised by the undesired presence of ablated particulates of unknown composition which, by becoming embedded in the film, can adversely affect the surface smoothness and cause the film to have a dusty appearance. Accordingly, PLD films have a distinctly different surface morphology and internal structure than do the *ex situ* annealed co-evaporated films previously studied.[6-8] The comparison presented below should be useful in assessing the applicability of these ion beam milling and polishing techniques to a variety of BYCO films with different characteristic morphologies.

EXPERIMENTAL

The Xe^+ ions used for milling were extracted from a Kaufman ion source[9] and had typical energies of 1 keV and current densities of 150 $\mu A/cm^2$. The ions are directed at a $10°$ glancing angle onto a substrate platform rotating at ~ 2 Hz. The ambient pressure of 1.5×10^{-4} Torr is sufficient to assure stable operation of the ion source. Two pieces of the same film were milled simultaneously: one piece for electrical characterization and the other for structural characterization by scanning electron microscopy (SEM) and Rutherford backscattering spectrometry (RBS). For the electrical characterization, the four indium leads, attached at the edges of the sample in a Van der Pauw configuration, remained in place for the entire sequence of milling and measurement operations.

The basic experimental system used for pulsed laser deposition has been described in detail previously.[10] The third-harmonic pulses of a Nd-YAG laser (wavelength of 355 nm and 8 nsec pulse duration) were used for ablation. The laser beam was incident on the Ba-Y-Cu-O ceramic target at $45°$ and was scanned over it in a zig-zag pattern to increase the spatial thickness uniformity of the films. The laser beam was focused on the target to produce a fluence of about 2 J/cm^2. The films were deposited on optically polished (100) $SrTiO_3$ substrates at 720-730$°$C in 200 mTorr of oxygen. At the end of the deposition process 700 Torr of oxygen was added to the chamber and the films were cooled down slowly with a dwell time of 30 min at 425$°$C.

The results presented below are on a single *c-axis* oriented laser-deposited 1400 Å-thick film which was chosen from a set of four separate deposition runs to have the sharpest superconducting transition and best crystallinity as determined by RBS channeling.

ELECTRICAL PROPERTIES

If a constant thickness-independent resistivity ρ is assumed, then the sheet conductance, $G = d/\rho$, becomes a direct measure of the effective film thickness d, a thickness which we call the electrical thickness[7] because it is directly calculated from a conductance measurement and its observed linear dependence on total milling time t_m. The relation is

$$d = \frac{G}{G_o} d_o \qquad (1)$$

where G_o and d_o are the initial $(t_{\mathrm{m}} = 0)$ sheet conductance and thickness of the unmilled film.

Shown in Fig. 1 is the dependence of the room-temperature sheet conductance on total milling time t_{m}. The linear dependence, seen also in earlier work on evaporated films,[6-8] indicates the uniform removal of constant resistivity material. The endpoint time of 96 min denotes the thickness at which the film no longer conducts. We emphasize that the electrical thickness[7] is zero at this point but that there might well be an additional residual layer of insulating material at the substrate-film interface.

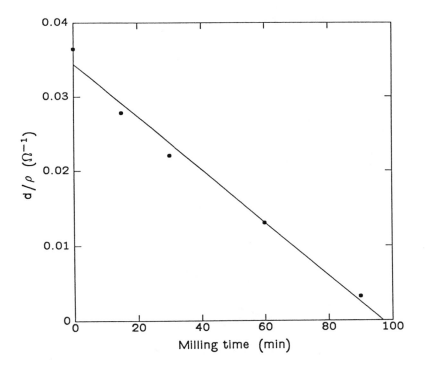

Fig. 1. Dependence of room-temperature sheet conductance on milling time for a film with initial thickness of 1400 Å.

The average milling rate of $5.6\times10^{-4}\Omega^{-1}\mathrm{min}^{-1}$ ($15\,\text{Å}/\mathrm{min}$) for this laser-deposited film compares favorably with the rates for *ex-situ* annealed evaporated films of $2.7\times10^{-4}\Omega^{-1}\mathrm{min}^{-1}$ ($14\,\text{Å}/\mathrm{min}$) with $d_0 = 2400\,\text{Å}$ [6] and $2.2\times10^{-4}\Omega^{-1}\mathrm{min}^{-1}$ ($13\,\text{Å}/\mathrm{min}$) with $d_0 = 4000\,\text{Å}$. [7-8] The thinnest film milled ($d_0 = 1000\,\text{Å}$) was an exception to this agreement, showing a positive curvature in the dependence of G on t_m and an average milling rate of $31\,\text{Å}/\mathrm{min}$ for the same ion-beam voltage and current densities. [7]

The resistance transitions of Fig. 2 show how the 90 K transition temperature, the sharpness of the transition, and the linearity of the temperature dependence of the normal-state resistance R_n are preserved during milling. The legend indicates the milling times associated with each curve. The electrical thickness for curve (d) is calculated to be $500\,\text{Å}$, indicating that 64% of the original film has been removed. The transition temperature begins to noticeably decrease and the transition region broadens with further milling. For example, the T_c of the $t_m = 90$ point in Fig. 1 was 79 K. The decrease in T_c and its dependence on R_n is in

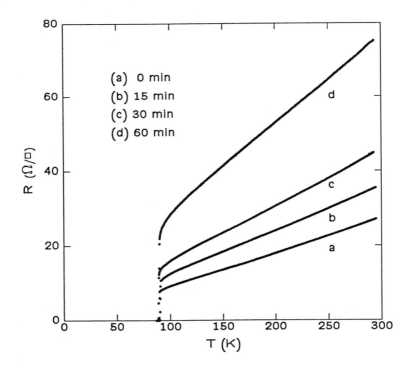

Fig. 2. Temperature dependence of sheet resistance for the milling times indicated in the legend. These data correspond to four of the five points shown in Fig. 1.

agreement with trends noted in earlier work[8] where it was observed that $T_c \rightarrow 0$ when $R_n(100\,\mathrm{K})$ approaches a value $h/4e^2 = 6450\,\Omega$. Similar observations have since been made on ultrathin c-axis $DyBa_2Cu_3O_7$ films[11] and $YBa_2Cu_3O_{7-x}/PrBa_2Cu_3O_{7-x}$ superlattices.[12]

TOPOGRAPHY, COMPOSITION AND CRYSTALLINITY

Shown in Fig. 3 are SEM micrographs at two different magnifications with milling times identified in the caption. The large boulder-like particulates having a lateral extent approaching $1\,\mu$m are typical of laser-deposited films and are usually associated with the ablation of non-stoichiometric material. The sequence of the SEM micrographs reveals a decrease in the density of these precipitates as milling proceeds. Concomitant with the reduction in particulate density is an increase in the density of cavity-like depressions or pinholes. At the highest magnification (lower panel of Fig. 3) the areal density of these pinholes seen in frame D is comparable to the areal density of particulates seen in frames A-C. These observations are consistent with the notion that most of the particulates are embedded in the film, leaving cavity-like remnants of their vanishing presence when the film becomes thin enough. Since the particulates are large compared to film thickness, none are uncovered by ion milling. We conclude that the presence of such embedded particulates in laser-ablated films can severely compromise the continuity of the CuO_2 planes and thus may be a severe obstacle in reproducibly obtaining small-scale structures.

The high magnification micrograph of frame A in the lower panel of Fig. 3 also reveals the presence of numerous rod-like protrusions oriented at right angles to each other in a "basket-weave" configuration. In previous work[13] this "basket-weave" morphology has been identified as an a-axis component growing on or near the surface of predominantly c-axis material. Fifteen minutes of milling (frame B) is sufficient to remove this component. Ion milling at glancing angles is thus seen to remove the initially predominant a-axis material on both evaporated and laser-ablated films.

Rutherford backscattering spectrometry (RBS) and channeling analysis using $1.8\,\mathrm{MeV}$ $^4He^+$ ions has provided complimentary information about microstructure. In addition, film composition and thickness are determined to within $\pm 10\%$ by comparing random yield spectra to simulations with the RUMP computer program.[14] For example, the as-received film was measured to have a Y-rich composition $Y_{1.3}Ba_{2.0}Cu_{3.0}O_{7-\delta}$ and a RBS thickness of $1400\,\mathrm{\AA}$. Assuming the applicability of the equilibrium ternary phase diagram, the material will grow as Y_2BaCuO_5 and CuO.[15] After removal of the first $400\,\mathrm{\AA}$, the measured composition, $Y_{1.1}Ba_{1.8}Cu_{3.0}O_{7-\delta}$, has converged more closely to the ideal 1:2:3 ratio. This noticeable reduction with milling of the Y component implies that an Y-rich phase near the surface (most probably the particulates shown in Fig. 3) has been preferentially removed by the ion-milling process.

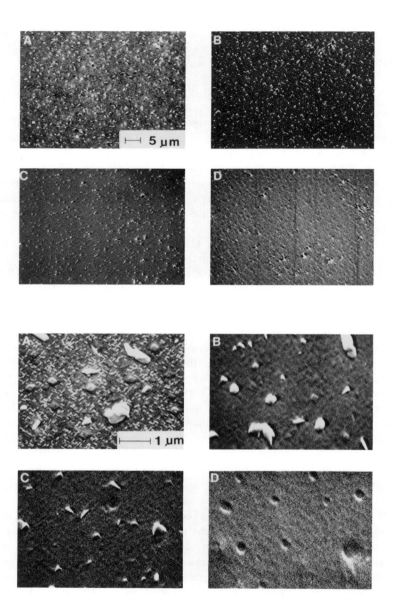

Fig. 3. SEM micrographs at low ($5\,\mu$m bar, upper panel) and high ($1\,\mu$m bar, lower panel) magnifications. Frames A-D correspond to milling times of t_m = 0, 30, 60, and 90 min in the upper panel and milling times of t_m = 0, 15, 30, and 90 min in the lower panel respectively.

Figure 4 shows the thickness dependence of X_{min}, a quantity calculated as the ratio of the $<001>$ channeled yield to the random yield taken just below the Ba surface peak region of the spectrum. Low values of X_{min} ($\sim 2\text{-}3\%$) imply good crystalline order. The thicknesses on the abscissa are calculated in two ways: electrically as from Eq. 1 (•), and from RBS ($\times$). We note that the RBS thicknesses are consistently higher than the electrical thicknesses, implying a significant difference between the actual and theoretical density of BYCO and/or the existence of an appreciable dead layer of non-conducting film. Most interesting, however, is the constancy of $X_{min} = 11\%$ during the removal of the top third of the film. Thus ion milling does not introduce sub-surface crystalline disorder which would be expected to show up as an immediate increase in X_{min} beginning at $t_m = 0$. Furthermore, the size of the Ba surface peak also remains constant as the top part of the film is removed. We attribute the increase in X_{min}, which does develop after approximately one third of the film has been removed, to pre-existing strain and mismatch effects which have their origin at the film-substrate interface.

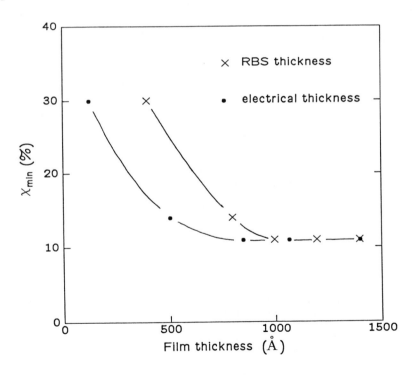

Fig. 4. Plot of X_{min} as a function of effective thickness determined in two ways for the same film: by RBS ($\times$) and electrically (•). The initial thickness for the electrical calculation, $d_o = 1400\,\text{Å}$, was determined from RBS.

CONCLUSIONS

In this paper we have presented results on ion mill thinning of pulsed laser deposited c-axis oriented BYCO thin films which have been *in-situ* annealed. The results have been compared to previously published work on ion mill thinning of evaporated films which have been *ex-situ* annealed. In both cases the electrical thickness decreases approximately linearly with milling time and there is a correspondingly similar dependence of T_c on film thickness. In like manner, for both types of film, there is prompt removal of an a-axis component on the surface of the films. The large particulates associated with the PLD films appear to contain a Y-rich phase which decreases as the film is thinned and the density of particulates decreases. Crystalline quality, as measured by the thickness dependence of χ_{min}, can actually improve if there is a significant amount of misoriented surface material;[6] although, for the results reported here, it remains constant until disorder associated with the substrate-film interface becomes apparent. At some level of sensitivity, glancing angle milling is expected to introduce surface disorder and an off-stoichiometric surface composition. The physical depth to which such deleterious effects occur and how this depth depends on incident ion energy is an interesting subject for further study.

REFERENCES

1. H. Tsuge, S. Matsui, N. Matsukura, Y. Kojima, and Y. Wada, Jpn. J. Appl. Phys. 2, Lett. **27**, 2237(1988).

2. S. Matsui, Y. Ochiai, Y. Kojima, H. Tsuge, N. Takado, K. Asakawa, H. Matsutera, J. Fujita, T. Yoshitake, and Y. Kubo, J. Vac. Sci. Technol. B, Microelectron. Process. Phenom. (USA) **6**, 900(1988).

3. G. J. Clark, A. D. Marwick, R. H. Koch, and R. B. Laibowitz, Appl. Phys. Lett. **51**, 139(1987).

4. A. E. White, K. T. Short, R. C. Dynes, A. F. J. Levi, M. Anzlowar, K. W. Baldwin, P. A. Polakas, T. A. Fulton, and L. N. Dunkleberger, Appl. Phys. Lett. **53**, 1010(1988).

5. G. C. Hilton, E. B. Harris, and D. J. Van Harlingen, Appl. Phys. Lett., vol. 53(12), pp.1107-1109, Sept. 1988.

6. A. F. Hebard, R. M. Fleming, K. T. Short, A. E. White, C. E. Rice, A. F. J. Levi, and R. H. Eick, Appl. Phys. Lett. **55**, 1915(1989).

7. A. F. Hebard, T. Siegrist, E. Coleman, and R. H. Eick, SPIE Symposium on *Processing of Films for High T_c Superconducting Electronics*, Oct 1989, Santa Clara, CA (to be published SPIE Proceedings, Vol. 1187).

8. A. F. Hebard, R. H. Eick, T. Siegrist and E. Coleman, to be published in the Proc. of Fall MRS Symposium on High-T_c Superconductors, Boston MA, 1989.

9. Ion Tech, Inc., Fort Collins, Colorado

10. G. Koren, A. Gupta, E. a. Giess, A. Segmuller, and R. B. Laibowitz, Appl. Phys. Lett. **54**, 1054(1989).

11. T. Wang, K. M. Beauchamp, D. D. Berkley, J-X. Liu, B. R. Johnson, and A. M. Goldman, to be published in Phys. Rev. B.

12. Q. Li, X. X. Xi, X. D. Wu, A. Inam, S. Vadlamannati, W. L. McLean, T. Venkatesan, R. Ramesh, D. M. Hwang, J. A. Martinez, and L. Nazar, Phys. Rev. Lett. **64**, 3086(1990).

13. G. J. Fisanick, P. M. Mankiewich, W. Skocpol, R. E. Howard, A. Dayem, R. M. Fleming, A. E. White, S. H. Liou, and R. Moore, Proc. of Fall MRS Symposium on High-T_c Superconductors, Boston MA, Nov. 30 - Dec. 4, 1987, 703(1987).

14. L. R. Doolittle, Nuc. Instrum. Meth. Phys. Res. B**9**, 344(1985).

15. R. S. Roth, C. J. Rawn, F. Beech, J. D. Whitler, and J. O. Anderson, in *Ceramic Superconductors II*, ed. M. F. Yan (The American Ceramics Society, Inc., Westerville, Ohio, 1988), 303.

IN SITU RHEED AND XPS STUDIES ON CERAMIC LAYER EPITAXY IN UHV SYSTEM

Hideomi Koinuma, Mamoru Yoshimoto, Hirotoshi Nagata,[*]
Takuya Hashimoto,[**] Tadashi Tsukahara, Satoshi Gonda,
Shunji Watanabe, Maki Kawai, and Takashi Hanada
Research Laboratory of Engineering Materials, Tokyo
Institute of Technology, Nagatsuta-cho 4259, Midori-ku,
Yokohama, Kanagawa 227, Japan

ABSTRACT

Ceramic films relating to high-Tc superconductors
were grown in a UHV deposition-analysis system. Against
our intuition that large oxygen deficiency is inevitable
in the growth of oxide films in UHV, required oxidation
state was achieved either by using such a highly oxida-
tive gas as NO_2 or by choosing thermodynamically stable
oxides. Two-dimensional growth of ceramics were verified
in MBE growth of Bi-Sr-Cu-O as well as in laser MBE
growth of various ceramic thin films. The growth mode
and the oxidation state of the films were investigated
by *in situ* RHEED and XPS analyses. Discussion is extended to the possibility of our method for constructing new
ceramics layers which may exhibit higher Tc.

1. INTRODUCTION

Artificial construction of well defined ceramic
layers is important not only to fabricate tunnel junction
devices of high-Tc superconductors with very short coherence lengths but also to investigate the possibility of
new high-Tc materials. There have been reported many
attempts for layer-by-layer deposition of high-Tc superconducting films by such methods as multitarget sputtering,[1,2] laser deposition,[3-6] and conventional molecular beam epitaxy (MBE)[7]. However, *in situ* analysis on
the growth manner and interface structure of the layers
have been scantly reported.

For the verification of two-dimensional layer-by-
layer growth, observation of RHEED intensity oscillation
during the deposition is essential as it is in the case
of semiconductor growth by conventional MBE.[8] Another
key-technology for oxide film epitaxy is the control of

---

[*]on leave from Central Research Laboratory, Sumitomo
Cement Co. Ltd., Toyotomi-cho 585, Funabashi, Chiba 274,
Japan
[**]on leave from Department of Industrial Chemistry,
Faculty of Engineering, University of Tokyo, Hongo 7-3-1,
Bunkyo-ku, Tokyo 113, Japan

oxidation state in the as-grown film. In general, ultra-high vacuum (UHV) atmosphere ($<10^{-8}$Torr) and high sub-strate temperature (300-800°C) applied in the convention-al MBE system are thermodynamically unfavorable for incorporating sufficient oxygen (hole) in the film. This contradictory condition, oxidation in UHV, could be diverted by the use of high oxygen-affinity elements[9] and/or highly reactive oxidants.[7,10]

In the present study, we have investigated the conditions and methods for fabricating layered oxide structures by employing the two types of MBE apparatus. One is a MBE using Knudsen cells to evaporate metal sources with an oxidant gas beam. The other is a laser MBE of our original design, in which an oxide molecular beam is generated directly by ablating a sintered oxide pellet by the irradiation with pulsed laser beam. The thermodynamic consideration is also made with respect to the effective oxidization of cuprate films.

2. UHV DEPOSITION AND ANALYSIS SYSTEM

Figure 1 illustrates our deposition system (assembl-ed by ULVAC Co. Ltd.) connected with X-ray photoelectron (XPS: JEOL model JPS-80) and Auger electron spectroscopy (AES) analyzers *via* a vacuum line ($<10^{-8}$Torr). Two depo-sition chambers for the synthesis of ceramic thin films are both equipped with reflection high energy electron diffraction (RHEED) and quadrapole mass spectrometers (QMS).

One of the deposition chambers is a reactive MBE apparatus equipped with four Knudsen cells for evapora-tion of Bi, Sr, Ca and Cu metals. Surface of the $SrTiO_3$ substrate was cleaned by the *in situ* method developed by Kawai *et al.*[11] After the surface cleaning, depositions

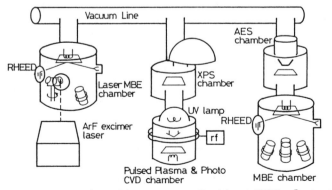

Fig.1. Schematic diagram of the UHV depostion system used in this study. Two MBE chambers were used to synthesize ceramic films.

of metallic multilayers in UHV (10^{-10}Torr) at room temperature and NO_2 flushing at 1×10^{-7}Torr onto the deposited metallic film at 350°C were iterated.   The atomically controlled multilayered films were designed to construct $Bi_2Sr_{1+x}Cu_xO_y$ (x=1,2) structures.

The other deposition chamber is a laser MBE[12] apparatus in which as many as sixteen   sintered oxide targets can be ablated with an ArF excimer laser in such a way as programmed by a computer to deposit films successively on a heated substrate.

3. THERMODYNAMIC CONSIDERATION ON OXIDATION IN UHV

Figure 2 shows   equilibrium lines for gas phase reactions between Cu and various oxidants.   Calculations were done   by assuming that ΔH and ΔS values were kept constant at $\Delta H_f{}^O$ and $\Delta S_f{}^O$ values in this temperature range.   When N_2O, O_3, NO_2 and NO were used as oxidants (Eq. 2, 3, 4 and 5 in Fig.2), it was further assumed that all the reactants were in the gas phase.   Fig.2 indicates that O· and NO_2 have an oxidation ability as strong as O_3.   Berkley et al.[13] and Schlom et al.[7] reported the growth of fully oxidized high-Tc superconducting

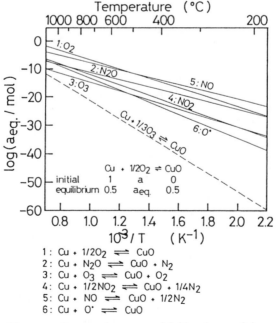

Fig.2. Thermodynamic equilibrium lines of gas phase reactions between Cu and various oxidants into CuO. The broken line in the Figure is for the reaction: $Cu+1/3O_3 \rightarrow CuO$.

films using pure O_3 in UHV-MBE systems. Effect of O⁻
species upon superconducting film formation has been
frequently tested in sputtering.[14] We employed NO_2 to
deposit fully oxidized cuprate films.

As soon as NO_2 was introduced into the UHV deposi-
tion chamber at a pressure of $1 \times 10^{-8} - 10^{-7}$ Torr under irradiation of laser beam, a strong QMS peak of #30(NO) and
weak peaks of #14(N), #16(O) and #32(O_2) were detected in
addition to #46(NO_2). The peak intensity of these decomposed species was synchronizing with the laser pulses.
Hence, NO_2 could be decomposed effectively into *in
situ* oxygen species by the laser beam during the film
deposition. Almost fully oxidized cuprate films were
deposited with NO_2 flushing at a pressure of $1-2 \times 10^{-7}$
Torr, as described later.

4. EXPERIMENTAL

In the laser MBE, CeO_2, Nd_2O_3 and $(Sr_xCa_{1-x})CuO_2$
films were deposited on substrates at 600-800°C by ablating corresponding sintered targets with an ArF excimer
laser (about $1J/cm^2$, 2Hz). The deposition pressure was
kept at mid-10^{-9} Torr during the deposition of CeO_2 and
Nd_2O_3 films on clean Si(111) surfaces. When (Sr_xCa_{1-x})-
CuO_2 films were deposited on $SrTiO_3$(001) substrates,
NO_2 was flushed onto the substrate surface and chamber
pressure was kept at $1-2 \times 10^{-7}$ Torr during the deposition
and cooling. Distance between the substrate and the NO_2
feeder was 20mm.

Deposited films were analyzed *in situ* by RHEED, XPS
and AES. *Ex situ* analysis of films was also performed
with an X-ray diffractometer (XRD).

5. RESULTS AND DISCUSSION

5-1. Low Temperature Synthesis of $Bi_2Sr_2Cu_1O_x$ Epitaxial Films by Reactive MBE using NO_2

On a $SrTiO_3$(100) substrate, one monolayer of metallic Sr was supplied at first followed by additional monolayers of Cu and Sr (Sr/Cu/Sr) at room temperature in
UHV. Then, NO_2 was introduced into the chamber through
the variable leak valve to a pressure of 1×10^{-7} Torr, and
the substrate was heated gradually to 350°C and kept for
10-20min. This treatment made the RHEED pattern of the
film streaky, as shown in Fig. 3. On the other hands,
use of O_2 as the oxidant did not make the RHEED streaky
but spotty, as also shown in Fig.3. It is clear that the
surface crystallinity in the case of O_2 oxidant is much
less oriented compared with the crystallinity in the case
of NO_2. NO_2 enhances not only the oxidation of the
elements but also the crystallization of the cuprate

films.   After the formation of the first Sr-Cu-Sr-O layer, a second layer was formed at a room temperature in
the order of double monolayers of Bi, monolayers of Sr,
Cu, and Sr (Bi/Bi/Sr/Cu/Sr).   By the subsequent oxidation with NO_2 at $350°C$, the RHEED streak pattern due to
the epitaxially grown phase appeared.   The procedure of
this second layer formation was repeated   four times to
give two units of the $Bi_2Sr_2Cu_1O_x$ layered-crystal.
      The film was blackish and electrically conductive.
Figure 4 shows XRD pattern of this ultra-thin film, indicating successful formation of the $Bi_2Sr_2Cu_1O_x$ phase with
its c-axis perpendicular to the substrate.   The thickness
of the film was calculated from a width of the (008)
diffraction peak using the Scherrer's equation[15] to be
5.7nm, which agrees well with the designed thickness of
5.7nm (four Bi-Bi-Sr-Cu-Sr-O and the additional Sr-Cu-
Sr-O layers).   Divalent state of Cu in the deposited
film was confirmed by XPS from its typical satellite
structure of Cu 2p for Cu^{2+}.[10]

5-2. Epitaxial Growth of CeO_2 Films on Si by UHV Laser MBE

      CeO_2 film was confirmed to be grown epitaxially by
laser MBE on clean surface of Si(111) in the temperature
range from 600 to $700°C$ at a pressure of mid-10^{-9}Torr
without any oxidant gas supply. During the film growth,
RHEED intensity oscillation corresponding to the unit
distance of CeO_2(111) was observed, as shown in Fig.5,

RHEED PATTERNS

θ = 4 degree

[011] azimuth

oxidation by NO_2

OXIDATION

T_{SUB}= 350°C

SR   P = 10^{-5} PA

CU

SR

SrTiO₃(100)

oxidation by O_2

Fig.3.   RHEED patterns after the oxidation of
Bi/Sr/Cu with molecular O_2 and NO_2. Beam azimuth
and incidence angle are [011] and $4°$ with respect
to $SrTiO_3$(100) substrate, respectively.

indicating two-dimensional growth of the CeO_2(111) film.[16] Nd_2O_3 (an oxygen-deficient fluorite type) was also expected to grow epitaxially on Si(111), since the lattice mismatch is small (2.2%). Nd_2O_3 film was also verified to grow epitaxially on Si(111). Then, we attempted to form a CeO_2(111)/Nd_2O_3(111) superlattice on the Si(111). The superlattice was designed to be ten CeO_2(1.5nm thick)/Nd_2O_3(1.5nm thick) bilayers which contain about (5+5) unit cells in each bilayer along <111> direction.

Figure 6 shows an intensity variation of the central RHEED streak observed during the deposition of the superlattice. Although the intensity oscillation due to the two-dimensional growth of each unit cell layer was not clear, the oscillation depending on the shuttering was seen clearly. The thickness of the superlattice was measured to be 3.0nm by the small angle XRD peak, being in good agreement with the designed thickness of 3.0nm. These results indicate that the laser MBE is a promising method for the formation of ceramic superlattices.

5-3. Construction of New Layered Cuprate Structures

Since high-Tc superconductors have layered perovskite structures, we can expect the possibility of constructing the same structures by piling up subunit layers along the c-axis. The subunit layers could be divided into three types; a $(CuO_2)^{2-}$ layer working as a conductive layer, a coupling layer always sandwiched by two

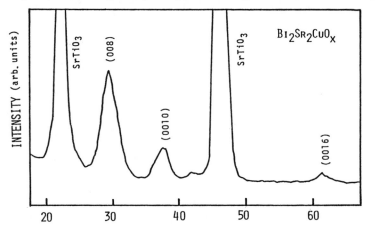

Fig.4. XRD pattern of an ultra-thin film of Bi_2Sr-Cu_1O_x formed on $SrTiO_3$(100) substrate. Peaks are assigned to (00*l*) of $Bi_2Sr_2Cu_1O_x$ phase.

$(CuO_2)^{2-}$ layers, and a blocking layer composed of more than two atomic layers of oxides and separating two $(CuO_2)^{2-}$ layers. The lattice of $Bi_2Sr_2CaCu_2O_8$, for example, is constructed from a $(SrO-Bi_2O_2-SrO)^{2+}$ blocking layer, a $(CuO_2)^{2-}$ layer, a Ca^{2+} coupling layer, and

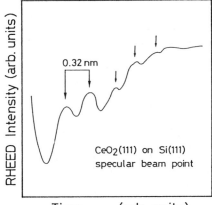

CeO2(111) on Si(111)
specular beam point

0.32 nm

RHEED Intensity (arb. units)

Time   (arb. units)

Fig.5. Typical RHEED intensity oscillation observed during $CeO_2(111)$ growth on Si(111) at 650°C. The oscillation periodicity calculated from the deposition rate agreed well with the interplane distance of $CeO_2(111)$ of 0.312nm.

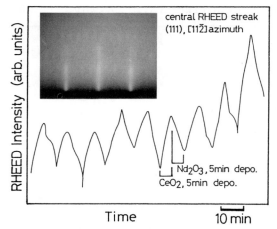

central RHEED streak
(111), [11$\bar{2}$]azimuth

RHEED Intensity (arb. units)

Nd2O3,5min depo.
CeO2,5min depo.

Time   10 min

Fig.6. Intensity oscillation depending on the shuttering observed in the formation of CeO_2-(111)/Nd_2O_3(111) superlattice on the Si(111) substrate. The streaky RHEED pattern with Kikuchi lines was seen throughout the superlattice formation. The oscillation was monitored at the central RHEED streak.

Table I Possible combination of blocking and coupling layers to form layered perovskite structures.

		Blocking layer	
		$(La_2O_2)^{2+}$	$(Nd_2O_2)^{2+}$
	none	LSCO (1)	NCCO (1)
Coupling layer	Ca^{2+}	LSCCO (2)	---- (2)
	RE^{3+}	(5/2)	(5/2)
	$2Ca^{2+}$	---- (3)	---- (3)
	$2RE^{3+}$	---- (4)	---- (4)
	$3Ca^{2+}$	---- (4)	---- (4)
	$3RE^{3+}$	(11/2)	(11/2)
	$4Ca^{2+}$	---- (5)	---- (5)
	$4RE^{3+}$	---- (7)	---- (7)

Number of $(CuO_2)^{2-}$ layer to get charge neutrality is shown in the parentheses. The combination to give integer number of $(CuO_2)^{2-}$ layers could be filled by new materials.

LSCO, NCCO and LSCCO are abbreviations for $(La,Sr)_2CuO_4$, $(Nd,Ce)_2CuO_4$ and $(La,Sr)_2$-$CaCu_2O_8$, respectively.

another $(CuO_2)^{2-}$ layer, respectively. This three elementary sublayers model is valid not only for constructing the already discovered high-Tc superconductors but also for providing us an idea to design new superconductors. As shown in Table I, there are a lot of combinations of blocking and coupling layers to be explored for the possibility of new layered structures. One of these new materials, $(La, Sr)_2CaCu_2O_6$, was discovered recently by Cava *et al.*[17]

To construct such new structures, atomically regulated layer-by-layer technique should be developed. The deposition unit layers may be different from the constitutional sublayers described above, since the sublayers are frequently charged and would not be stably grown by themselves. A charged sublayer can be combined with the adjacent atomic layer(s) to form neutral deposition unit.

One of the deposition units thus assembled is Nd_2CuO_4 ($a=b$=0.394nm), which is composed of a $(Nd_2O_2)^{2+}$ blocking layer and a $(CuO_2)^{2-}$ layer, and another example is $SrCuO_2$ ($a=b$=0.393nm). In these cuprates, lattice constants can be adjusted by partial substitution at the rare earth (RE) and alkaline earth (AE) sites.

(AE)CuO_2 films were deposited on $SrTiO_3$(001) substrates by laser MBE with NO_2 flushing. Figure 7 shows the intensity oscillation of the central RHEED streak observed during the epitaxial growth of $(Sr_{0.5}Ca_{0.5})CuO_2$ film at 700°C. The oscillation cycle (0.33nm) was very close to the c-axis length (0.332nm) of the film measured by XRD. The valence state of Cu in the as-grown film was evaluated to be 2+ by *in situ* XPS shown in Fig. 8. The full width at half maximum of rocking curve of (002) XRD peak was 0.43°, indicating a very good crystallinity. Alternating epitaxial growth of (AE)CuO_2 and $(RE)_2CuO_4$ layers is now in progress to construct new ceramic superlattices.

6. CONCLUSION

Two-dimensional epitaxial growth of ceramic films in UHV deposition system was confirmed by *in situ* RHEED and XPS analyses. To oxidize cuprate films in UHV, pure NO_2 flushing onto the substrate was as effective as pure O_3. These results demonstrate that new layered structures relating to high-Tc superconductors could be constructed by our laser MBE method.

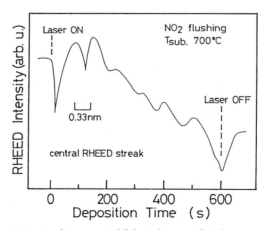

Fig.7. Intensity oscillation of the central RHEED streak observed during the deposition of the $(Sr_{0.5}Ca_{0.5})CuO_2$ film. The film grew epitaxially with its c-axis normal to the $SrTiO_3$(001) substrate.

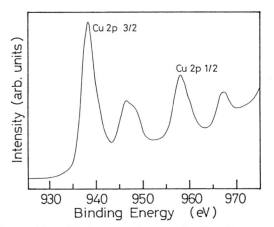

Fig.8. In situ XPS spectra of Cu *2p* measured on the 30nm thick $(Sr_{0.5}Ca_{0.5})CuO_2$ film. The profile with clear satellite peaks is assigned to that of CuO, indicating divalent state of Cu in the as-grown film.

ACKNOWLEDGMENT

The research was supported in part by a Grant-in-Aid for Scientific Research from Ministry of Education, Science and Culture of Japan.

REFERENCES

1. K.Nakamura, J.Sato, M.Kaise, and K.Ogawa, Jpn.J.Appl.Phys., **28**, L437 (1989).
2. H.Adachi, S.Kohiki, K.Setsune, T.Mitsuyu, and K.Wasa, Jpn.J.Appl.Phys., **27**, L1883 (1988).
3. M.Kanai, T.Kawai, S.Kawai, and H.Tabata, Appl.Phys.Lett., **54**, 1802 (1989).
4. M.Kanai, T.Kawai, and S.Kawai, Appl.Phys.Lett., **57**, 198 (1990).
5. J.J.Kingston, F.C.Wellstood, P.Lerch, A.H.Miklich, and J.Clarke, Appl.Phys.Lett., **56**, 189 (1990).
6. C.T.Rogers, A.Inam, M.S.Hegde, B.Dutta, X.D.Wu, and T.Venkatesan, Appl.Phys.Lett., **55**, 2032 (1989).
7. D.G.Schlom, A.F.Marshall, J.T.Sizemore, Z.J.Chen, J.N.Eckstein, I.Bozovic, K.E. von Dessonneck, J.S.Harris, Jr., and J.C.Bravman, to be published in J.Crystal Growth.
8. P.J.Dobson, B.A.Joyce, J.H.Neave, and J.Zhang, NATO ASI Ser. B, **191**, 185 (1988).
9. M.Yoshimoto, H.Nagata, T.Tsukahara, and H.Koinuma, Jpn.J.Appl.Phys., **29**, L1199(1990).
10. S.Watanabe, M.Kawai, and T.Hanada, Jpn.J.Appl.Phys., **29**, L1111 (1990).
11. S.Watanabe, M.Kawai, and T.Hanada, unpublished.
12. J.T.Cheung and H.Sankur, CRC Critical Reviews in Solid State and Materials Sciences, 15, Issue 1, (CRC Press Inc., 1988), p63.
13. D.D.Berkley, B.R.Johnson, N.Anand, K.M.Beauchamp, L.E.Conroy, A.M.Goldman, J.Maps, K.Mauersberger, M.L.Mecartney, J.Morton, M.Tuominen, and Y-J.Zhang, Appl.Phys.Lett., **53**, 1973 (1988).
14. I.Yazawa, N.Terada, K.Matsutani, R.Sugise, M.Jo, and H.Ihara, Appl.Phys.Lett., **29**, L566 (1990).
15. B.D.Cullity, "Elements of X-ray Diffraction," (Addison-Wesley, Massachusetts, 1977).
16. H.Koinuma, H.Nagata, T.Tsukahara, S.Gonda, and M.Yoshimoto, Ext.Abst. of the 22nd (1990 Int'l) Conf. on Solid State Devices and Materials, (The Japan Society of Applied Physics), p.933.
17. R.J.Cava, B.Batlogg, R.B.van Dover, J.J.Krajewski, J.V.Waszczak, R.M.Fleming, W.F.Perk, Jr.,L.W.Rupp, Jr.,P.Marsh, A.C.W.P.James, and L.F.Scneemeyer, Nature, **345**, 602 (1990).

ORIENTATION-DEPENDENT CRITICAL CURRENTS IN $Y_1Ba_2Cu_3O_{7-x}$ EPITAXIAL THIN FILMS: EVIDENCE FOR INTRINSIC FLUX PINNING?*

D. K. Christen, C. E. Klabunde, R. Feenstra, D. H. Lowndes, D. P. Norton, J. D. Budai, H. R. Kerchner, J. R. Thompson, and S. Zhu
Solid State Division, Oak Ridge National Laboratory, Oak Ridge, TN 37831

A. D. Marwick
IBM Thomas J. Watson Research Center, Yorktown Heights, NY 10598-0218

ABSTRACT

For YBCO epitaxial thin films the basal plane transport critical current density J_c, flowing perpendicular to an applied magnetic field H, depends sensitively on the orientation of the crystal with respect to H. In particular, J_c is sharply peaked and greatly enhanced when H is precisely parallel to the copper-oxygen planes. Experiments on a series of epitaxial monolithic and superconductor-insulator multilayer thin films provide clear evidence that the enhancement is a bulk, rather than surface or thin sample, phenomenon. Measurements of the orientation dependence are presented and compared with a model of "intrinsic flux pinning" by the layered crystal structure.

INTRODUCTION

There have been several reports of large anisotropies in the basal plane critical current densities of epitaxial, high-temperature superconducting thin films,[1-3] with large enhancements in J_c when the applied magnetic field is precisely parallel to the copper-oxygen planes of the layered lattice structure. For these films, with the crystalline c-axis oriented perpendicular to the substrate surface, this special geometry also corresponds to the field being oriented nearly parallel to the film surface. The phenomenon has been difficult to confirm in single crystals, largely due to the experimental difficulties in performing transport J_c measurements on these materials. (J_c determinations by magnetic hysteresis are complicated by the combined influence of induced supercurrents flowing both in the ab basal plane and along the c-axis; the latter must be deconvoluted by comparison of measurements performed on crystals of different geometrical aspect ratios).[4] As a result, one may question whether the effect is associated with the crystal anisotropy, or is rather a "thin film" phenomenon, such as surface pinning, or enhancements arising from quasi-one-dimensional flux line (FL) arrays for film thicknesses comparable to the flux lattice spacing.[5] Theoretical arguments favoring a bulk phenomenon predict "intrinsic flux pinning" to arise from the layered crystal structure, whereby the modulated superconducting order parameter provides a minimum energy when the vortex cores are located in the weak- or non-superconducting regions between the (double) Cu-O layers.[6,7] This configuration poses an energy barrier to FL motion past the Cu-O planes.

*Research sponsored by the Division of Materials Sciences, U.S. Department of Energy under contract DE-AC05-84OR21400 with Martin-Marietta Energy Systems, Inc.

We have measured the field, temperature, and orientation dependence of the transport J_c for a series of high quality epitaxial films of various thicknesses, and deposited on several different types of single crystal oxide substrates. Two of the films were multilayer superlattices of $Y_1Ba_2Cu_3O_{7-x}$ and $PrBa_2Cu_3O_{7-x}$.[8] In the following we present experimental results that strongly support the bulk nature of the pinning mechanism, and provide a test of models for the intrinsic flux pinning mechanism.

EXPERIMENT

The single-layer films were deposited by coevaporation (post-annealed) and by laser ablation (*in situ*) onto (001) substrate surfaces of single crystal $SrTiO_3$, $KTaO_3$, $LaAlO_3$, or Y-stabilized ZrO_2. Thicknesses ranged from 50 nm to 380 nm, and the films showed uniformly ideal resistive properties with zero-resistance at $T_{co} \simeq 90$ K. The multilayer films were deposited by pulsed laser ablation onto $SrTiO_3$, and are asymmetric superlattices, with alternating epitaxial layers of M unit cells YBCO by N unit cells of PrBCO. The present films each were composed of 5 layer periods, and may be designated 16 x 16 and 32 x 16 unit cells YBCO/PrBCO; they had T_{co} values of 88.5 K and 89.5 K, respectively.[8] All films were fully epitaxial, having the orthorhombic crystalline c-axis normal to the substrate surface and in-plane alignment of the film and substrate [110] axes.[9]

The films were patterned either by photolithography and wet etching or deposition through a mask to 3 mm long bridges of 50, 100, or 500 μm width. J_c was defined at an electric field criterion of 1 μV/cm, and a sample rotator provided *in situ* orientation in magnetic fields to 8 T.

RESULTS and DISCUSSION

First we discuss the dependence of J_c on magnetic fields aligned parallel to the high symmetry directions, $H \| c$ and $H \perp c$ (i.e. in the basal ab plane). Except at temperatures within a few percent of T_c, the observed $J_c(H \| ab)$ is large, relatively insensitive to the field, and remarkably reproducible from sample to sample.[10] These characteristics are illustrated in Fig. 1, which shows the normalized critical current density $J_c(H)/J_c(0)$ at 77 K for eight different epitaxial c-oriented films. In zero applied magnetic field, all samples had J_c in the range of a few MA/cm^2, with some of the variation attributable to uncertainties in thickness of films on the highly-twinned $LaAlO_3$ substrates.

For $H \| ab$ and in fields to 8 T, the data can be represented as a universal curve. In Fig. 1 the symbols

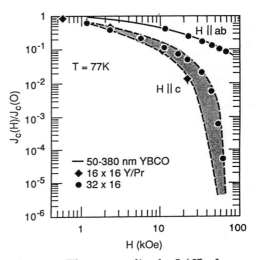

Fig. 1. The normalized $J_c(H)$ for a series of single layer epitaxial films with thicknesses in the range indicated, and for two multilayer superlattices.

are data for the 32x16 superlattice film, while the solid curve depicts J_c of *all* the six single-layer films, to within a spread that is about the size of the symbols. This uniformity is indicative of a strong flux pinning mechanism that is common to all seven samples, and is nearly independent of the film thickness over a range from 37 nm to 380 nm (implying a bulk effect). The relatively weak field dependence, with $J_c > 10^5$ A/cm^2 at 8 *T*, implies that the flux pinning energy barrier U_o is large compared to thermal energies $k_B T$. (In fact, these J_c values are comparable to results obtained for the best Nb$_3$Sn operating at 4.2 *K*). At low temperatures, where thermally activated processes are unimportant, $J_c(0)$ typically is several tens of MA/cm^2 and decays by only a few percent in fields to 8 T

In contrast, with $H\|c$ the observed J_c decreased abruptly for fields greater than 2 - 4 *T*, and displayed orders-of-magnitude variation from sample to sample, as indicated in Fig. 1 by the shaded region. The rapid decay of J_c has been associated with flux lattice melting[11], thermally-activated flux motion,[12] or a vortex glass transition.[13] The large, sample-dependent variations in J_c for $H\|c$ are not directly correlated with the sample thickness, and are probably related to variations in the flux pinning defect structure which results during the film growth process. To what extent J_c can be enhanced by optimized defect tailoring is a matter of fundamental and practical importance.

In Fig. 2 we present further evidence that the J_c enhancements are associated with the layered lattice structure, rather than with the film morphology. Single layer films were deposited onto SrTiO$_3$ substrates having surfaces cut at small angles with respect to the (001) lattice planes. XRD showed the films were epitaxial with the *c*-axis parallel to the substrate [001] direction, in a plane orthogonal to the current flow. With a sample biased at a current density $J > J_c$, the normalized dissipative flux flow voltage V is plotted as a function of the angle θ between the direction of H and the film surface ($\theta=90°$ corresponds to $H\|$ *surface*). Clearly, the minimum dissipation (maximum J_c) occurs when H is parallel to the Cu-O planes, tilted away from the film surface by the measured miscut angle $\Delta\theta$. The apparent asymmetry in the angular dependence for finite miscuts may indicate some residual effects of surface pinning.

Intrinsic flux pinning mechanisms have been proposed on theoretical grounds by treating the superconductive state either as an array of microscopic superconductor-superconductor (SS'S) or superconductor-insulator (SIS) layers.[6,7] In both cases, the predicted field dependence is weak (neglecting thermally activated processes) and estimates of the scale of J_c are consistent with the

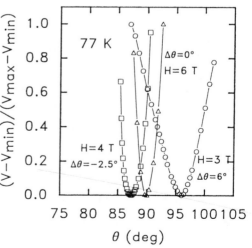

Fig. 2. The flux flow voltage at fixed bias current for field orientations near to the film surface ($\theta=90°$). $\Delta\theta$ is the measured surface miscut angle.

observations ($J_c(77K) \sim 1\text{MA/cm}^2$ and $J_c(0K) \lesssim 6 \times 10^7 \text{ A/cm}^2$).

For applied field direction away from the high-symmetry axes, intrinsic flux pinning may be expected to strongly influence the angular dependence of J_c. Tachiki and Takahashi [14] (TT) have modeled this angular dependence based upon a picture whereby FL segments jog at right angles, alternately pinned independently either by the weak superconducting interlayers for a FL segment ∥ *ab*, or by presumed extended defects along the *c*-axis for a FL segment ∥ *c*. The result can be stated simply according to,

$$\frac{J_c(\theta)}{J_c(90°)} = smaller \ of \ \begin{bmatrix} 1, \\ \dfrac{J_c(0)/J_c(90°)}{(cos\theta)^{1/2}} \end{bmatrix}, \tag{1}$$

where, $\theta=0$ now corresponds to $H \parallel c$.

Figures 3*a* and 3*b* illustrate this observed angular-dependent ratio, measured for an epitaxial film at 79.9 K, and for the 32x16 superlattice film at 4.2 K. The family of data curves correspond to different applied field intensities, and the solid curves are plotted from Eq. 1, using representative values of $J_c(0°)/J_c(90°)$. Overall agreement with the TT intrinsic pinning model is only qualitative, but appears to be best at low temperatures and in high magnetic fields. It might be argued that Eq. 1 is strictly a zero-temperature description, since the high-temperature effects of flux creep for field directions away from the basal planes would distort the idealized kinks in the FL's as they jog from the *ab*- to the *c*-directions. Recently, Feinberg and Villard [7] have carefully accounted for the vortex line energy in an SIS layer model. For $H_{c1} << H << H_{c2}$, their development leads to the prediction of vortex trapping between the Cu-O layers for applied field directions within a

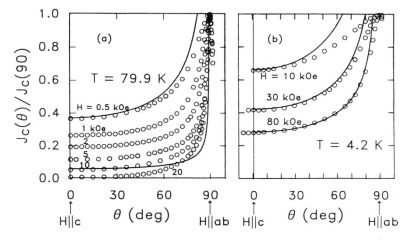

Fig. 3. The dependence of J_c on the sample orientation in the applied field H. The solid curves are derived from the theoretical model of Tachiki and Takahashi (Ref. 13).

few degrees of $\theta=90°$. Accordingly, J_C should assume the intrinsic value within this small angular range, rather than the much wider saturation range of the TT model that is illustrated in Fig 3. We will consider these points more thoroughly in a future publication.

It is important to establish whether the observed enhancements can arise simply from the large anisotropies in the superconductive properties. That is, an alternative explanation for the sharp angular dependence might invoke a conventional flux pinning mechanism, but include effects of the large anisotropies in superconducting parameters; the latter are known to be adequately described by a three-dimensional effective mass model.[15] Assuming isotropic pinning defects, one would expect a generalization of the scaling behavior for J_C that is typified by,

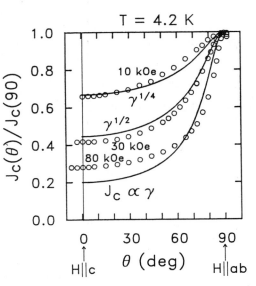

Fig. 4. The orientation dependence of J_C compared to that predicted by a simple flux pinning scaling model, and the effective mass model of the anisotropy. γ is described in the text.

$$J_C \propto B_{c2}^n(\theta)\ B^p\ f(b), \quad (2)$$

where $b=B/B_{c2}$, and for simple pinning models $f(b)=(1-b)^q \simeq 1$ at low temperatures for the fields attainable.[16] Note that the exponents n, p, and q are not necessarily integers. The angular dependent upper critical field may be written,[15]

$$B_{c2}(\theta)=B_{c2}(0)\ \gamma(\theta),$$

where,

$$\gamma(\theta)=[cos^2\theta + (\tfrac{m_{ab}}{m_c})\ sin^2\theta]^{-1/2}.$$

(3)

For YBCO, the effective mass ratio $m_c/m_{ab} \simeq 25$.[15] Using Eqs. 2 and 3 we obtain for the angular dependence of the scaled critical current density,

$$\frac{J_C(\theta)}{J_C(90°)} = \left[\frac{\gamma(\theta)}{\gamma(90°)}\right]^n. \quad (4)$$

Fig. 4 compares the observed $J_C(\theta)$ at 4.2 K with that predicted from Eq. 4, for the three indicated fields and exponents n. That the data cannot be described by a single value of n is a violation of Eq. 4, and indicates that J_C does not obey simple scaling behavior. This observation reflects a fundamental fact already observed — that the field dependence of J_C is quite different for the two high symmetry directions, and therefore does not follow the field-scaled form of Eq. 2. This failure does not rule out the viability of a conventional flux pinning mechanism, however, since Eq. 2 does not account for the possibility, for example, of highly anisotropic pinning defect structures. Of course, the layered crystal structure may be regarded as the latter in the extreme limit of intrinsic pinning.

As a final note, we mention briefly some additional experimental findings that may have relevance in confirming the existence of intrinsic pinning. It is well established that energetic particle irradiations produce dense, flux pinning defects that dramatically increase the basal plane critical current density of high quality single crystals, as determined magnetically for the geometry where $H \parallel c$.[17,18] It is also true that for optimal doses of protons or neutrons, for example, these J_c values of single crystals approach the values observed in unirradiated, as-formed epitaxial thin films (e.g. $J_c \gtrsim 10^5$ A/cm^2 at 77 K

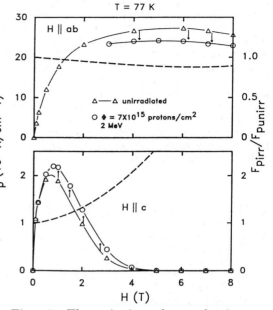

Fig. 5. The pinning force density $F_p = J_c B$ of an epitaxial film on LaAlO$_3$, before and after irradiation with 2 MeV protons. The dashed curves show the relative change in F_p on the right hand axis.

and $H = 1$ *T*). Fig. 5 shows the effect on J_c of irradiating a single layer epitaxial YBCO film with 7×10^{15} H$^+$/cm^2 of energy 2 MeV, where this dose corresponds to the energy deposited for optimum enhancement of flux pinning in single crystals.[17] In Fig. 5 we have plotted the field-dependent pinning force density $F_p = J_c B$ at 77 *K*, before and after irradiation, for H parallel to the high symmetry directions. While the overall effect is not dramatic, it is clear that the incremental defect flux pinning has <u>increased</u> for $H \parallel c$ and <u>decreased</u> for $H \parallel ab$. Similar effects have been observed in epitaxial films irradiated with Xe ions.[19] From the point of view of intrinsic pinning, a possible interpretation of this result is that the same irradiation-induced material inhomogeneities that marginally increase the pinning for FL $\parallel c$, only serve to locally disrupt the lattice periodicity, thereby reducing the total length of FL that can be effectively, intrinsically pinned. A possible counter example to this argument comes from recent neutron irradiation experiments that showed enhancements for both orientations.[20]

CONCLUSIONS

The critical current density J_c for a series of YBCO epitaxial single-layer and superlattice thin films shows strong enhancements when the applied magnetic field is parallel to the copper-oxygen layers of the lattice structure, suggesting a possible intrinsic flux pinning mechanism. The film thickness dependence and the angular dependence observed for films deposited on miscut substrate surfaces strongly suggest that the enhancements arise from the lattice structure, and are not simply a thin film or surface influence. The detailed angular dependence of $J_c(\theta)$ on the field direction with respect to the lattice planes is in qualitative agreement with the model of intrinsic flux pinning by the layered crystal structure. The critical current fails to scale in a way that is consistent with a *simple* effective mass anisotropy influence on J_c, but could reflect a conventional, but more complicated anisotropic flux pinning mechanism. Defects induced by proton irradiation decrease the flux pinning only when the field is parallel to the Cu-O planes, and may indicate a reduction in intrinsic pinning due to disruptions in the lattice periodicity.

REFERENCES

1. D.K. Christen, *et al.*, Physica C **162-164**, 653 (1989).
2. B. Roas, L. Schultz, G. Saemann-Ischenko, Phys.Rev. Lett. **64**, 479 (1990).
3. Y. Kuwasawa, *et al.*, Physica C **169**, 39 (1990).
4. E. M. Gyorgy, *et al.*, Appl. Phys. Lett. **55**, 283 (1989).
   F. M. Sauerzopf, *et al.*, Cryogenics **30**, 650 (1990).
   D. C. Cronemeyer, *et al.*, to appear in the Proceedings of the International Conference on Transport Properties of Superconductors, ed. R. Micolsky, (World Scientific, Singapore, 1990).
5. I. Hlásnik, *et al.*, Cryogenics **25**, 558 (1989).
6. M. Tachiki and S. Takahashi, Physica C **162-164**, 241 (1989).
7. D. Feinberg and C. Villard, Phys. Rev. Lett. **65**, 919 (1990).
8. D. H. Lowndes, *et al.*, Phys. Rev. Lett. **65**, 1160 (1990).
9. J.D. Budai, *et al.*, Phys. Rev. B **39**, 12355 (1989).
10. D.K. Christen, *et al.,* in: "High Temperature Superconductors: Fundamental Properties and Novel Materials Processing," eds. D. Christen, J. Narayan, and L. Schneemeyer (Materials Research Society, Pittsburgh, Vol 169, 1990) p. 883.
11. D. R. Nelson and S. Seung, Phys. Rev. B **39**, 9153 (1989).
12. Y. Yeshurun and A. P. Malozemoff, Phys Rev. Lett. **60**, 2202 (1988).
13. M. P. A. Fisher, Phys. Rev. Lett. **62**, 1415 (1989).
14. M. Tachiki and S. Takahashi, Sol. State Comm. **72**, 1083 (1989).
15. U. Welp, *et al.*, Phys. Rev. B **40**, 5263 (1989).
16. D.C. Larbalestier, *et al.*, in Proceedings of the International Symposium on Flux Pinning and Electromagnetic Properties of Superconductors, eds. K. Yamafuji and F. Irie (Matsukuma, Fukuoka, 1985) pp. 58-67.
17. L. Civale, *et al.*, Phys. Rev. Lett. **65**, 1164 (1990).
18. R. B. van Dover, *et al.*, Nature **342**, 55 (1989).
19. B. Roas, *et al.*, preprint.
20. W. Schindler, *et al.*, Physica C **169**, 117 (1990).

CRITICAL CURRENTS AND rf PROPERTIES OF LASER-DEPOSITED HIGH-T$_c$ FILMS

L. Schultz

Siemens AG, Research Laboratories, Erlangen, F.R.G.

ABSTRACT

High T$_c$ superconductor films were prepared by laser deposition using a SIEMENS XP 2020 excimer laser. Epitaxial YBaCuO films of high quality are obtained for a uniform laser beam energy density of about 4 J/cm^2, a substrate temperature above 730 $^\circ$C and an oxygen pressure of about 0.4 mbar. These films show high DC critical current densities as 5×10^6 A/cm^2 at 77 K or 6×10^7 A/cm^2 at 4.2 K and zero field and excellent microwave properties. The performance of first passive microwave devices made from these films is promising. Also epitaxial Bi$_2$Sr$_2$CaCu$_2$O$_x$ films were prepared on SrTiO$_3$ substrates. The inductively measured T$_{c,onset}$ reached 83 K. Critical current densities up to 10^7 A/cm^2 at 4.2 K and 10^6 A/cm^2 at 50 K were obtained. In magnetic fields, these films show a dramatic dependence on the field direction, with j$_c$ being almost field-independent (up to 7 T) at low temperatures, when the field lies within the film plane.

1. INTRODUCTION

High quality thin YBa$_2$Cu$_3$O$_{7-x}$ films can be produced by several techniques, such as electron beam co-deposition[1-3], sputtering[4], molecular beam epitaxy[5] or laser deposition[6,7]. Their superconducting properties still differ widely, especially if technologically relevant properties are regarded. To investigate the physical properties and to evaluate the technological feasibility of these films it is necessary to prepare single-crystalline and single-phase films of good homogeneity. As a consequence they will show a sharp inductively measured transition to superconductivity at about 90 K, high critical current densities and excellent rf properties.

In this paper, details of the preparation of YBa$_2$Cu$_3$O$_{7-x}$ and Bi$_2$Sr$_2$CaCu$_2$O$_x$ films by laser deposition, their current carrying capability and their rf properties are reported.

2. SAMPLE PREPARATION

For the film preparation by laser deposition we used a Siemens XP 2020 XeCl excimer laser with 308 nm wavelength, 60 ns pulses, a repetition rate of 5 Hz and a pulse energy of 2 J. For initial experiments we also used a Nd:YAG laser (λ = 1064 nm, two-fold frequency doubling possible, 1.2 J/pulse, 10 Hz). The laser beam is focused onto a rotating, sintered YBa$_2$Cu$_3$O$_{7-x}$ or Bi$_2$Sr$_2$CaCu$_2$O$_x$ pellet in a vacuum chamber (for details see Ref. 7). The material evaporated perpendicularly from the target surface is deposited onto the substrate at a distance of about 30 mm. The substrates can be heated up to 850°C, controlled by a pyrometer. The heater is an insulated resistance wire (Philips Thermocoax 1NCAC15) wound into a spiral with a diameter of 30 mm. Oxygen can be added during the deposition and cooling processes. The chamber is evacuated by a turbomolecular pump, the base pressure is less than 10^{-6} mbar.

As substrates we used <100> orientated $SrTiO_3$, MgO and $LaAlO_3$ single crystals. They were Syton polished and heat treated in-situ before the deposition. The deposition itself took about 5 minutes, corresponding to 1500 laser pulses. With a typical deposition rate of 2 Å/pulse the film thickness ranges from 0.2 μm to 0.3 μm.

2.1. YBaCuO FILMS

The basic requirements to obtain ideal films are: (1) an exact composition with a 1:2:3 atomic ratio of the metallic elements, (2) an epitaxial film growth and (3) an optimized oxygen content. Laser deposition allows the adjustment of the preparation conditions in a way that these three requirements can be met sequentially and independently of each other.

The first basic requirement to obtain high-quality $YBa_2Cu_3O_{7-x}$ films is to assure the exact composition of the metallic elements, with the 1:2:3 ratio. The film composition was investigated as a function of several deposition parameters, as substrate temperature, oxygen pressure, target-substrate distance and laser beam energy density. The laser beam energy density on the target turned out to be the most important parameter. Whereas at low energy densities the yttrium content of the films is understoichiometric, the highest energy density (7.5 J/cm^2) produces a copper understoichiometry in the film and a copper-rich target spot. At 4.5 J/cm^2 the compositions are reproducibly close to the ideal stoichiometry. As a consequence for laser deposition, it is essential to have a laser beam with a homogeneous spatial distribution and a highly reproducible pulse energy. Therefore, we use an optical set-up, where a rectangular mask is imaged on the target[7]. Also the excellent pulse-to-pulse stability of the Siemens XP 2020 excimer laser (less than 1 % deviation) is very helpful.

Besides the composition, the quality of the films (i.e. the superconducting properties like T_c, critical currents and rf surface resistance) is determined by the structure and the oxygen content of the grown films. Non-destructive techniques to evaluate the film quality are X-ray diffraction and inductive T_c measurements. X-ray diffraction gives information on the epitaxial growth of the film on the substrates (at least on texturing), on the c-axis parameter which is closely related to the oxygen content and on the degree of order for which the width (full width at half maximum: FWHM) and the intensity of the diffraction peaks are the measures. For the 123 material the (005) diffraction peak gives the best information. Fig. 1 shows the characteristic parameters of the (005) peak for films prepared at different substrate temperatures and Fig. 2 shows the corresponding T_c values. The films were prepared under an oxygen atmosphere of 0.4 mbar and subsequently cooled in 1 bar oxygen. Figs. 1 and 2 demonstrate that high T_c's are achieved when the FWHM of the (005) peak is below 0.3° and the peak intensity is suffi-

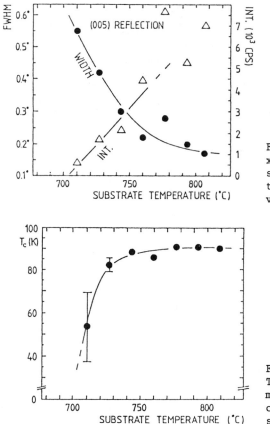

Fig. 1. Full width half maximum (FWHM) and peak intensity of the (005) diffraction peak of YBaCuO films vs. substrate temperature.

Fig. 2. Critical temperature Tc (inductively measured midpoint of the transition) of YBaCuO films vs. substrate temperature.

ciently high. A minimum substrate temperature of about 730 ^{O}C is required to provide the deposited atoms with sufficient mobility for layer-by-layer growth instead of island growth.

The oxygen pressure during the film preparation and the way how the films are cooled afterwards strongly influence the oxygen content of the films, which is inversely proportional to the c-axis parameter (derived from the (005) peak position), and, therefore, also the critical temperature (Figs. 3 and 4). In the experiment, two samples were prepared at the same time. After the film deposition at a substrate temperature of 750 ^{O}C, one of the films was in-situ removed from the substrate holder, thus rapidly cooling under the oxygen pressure used for the film preparation ("quenched sample"). The second film was slowly cooled in 1 bar oxygen. The dependence of the c-axis parameter of both the quenched and the slowly cooled samples on the oxygen pressure during the preparation is quite similar but it is always about 0.2 Å larger for the quenched samples. This reflects the filling of the oxygen chains and the tetragonal to orthorhombic phase transition during the slow cooling. On the other hand, the large c-axis parameter of films prepared at low oxygen pressures (below 0.1 mbar) indicates the presence of vacancies on oxygen sites of the tetragonal phase after the film preparation - probably in the CuO layers. They can not be occupied by annealing at 1 bar oxygen at temperatures below about 800 $^{\circ}$C. The superconducting transition width ΔT_c of the films prepared at low oxygen pressures and slowly cooled in 1 bar oxygen is still very small but the T_c is considerably depressed (50 K at 10^{-2} mbar oxygen pressure; see Fig. 3). Therefore, the oxygen pressure during the preparation determines the occupancy of the oxygen sites of the tetragonal phase, whereas the cooling process determines the occupancy of the oxygen chains of the orthorhombic phase. Both effects seem to be independent from each other. In order to prepare high T_c films (T_c above 90 K) in-situ, the oxygen pressure must be above 0.3 mbar during the preparation. This corresponds with thermodynamic measurements by Bormann and Nölting[8] predicting stability of the tetragonal phase only at these oxygen pressures at 750 $^{\circ}$C.

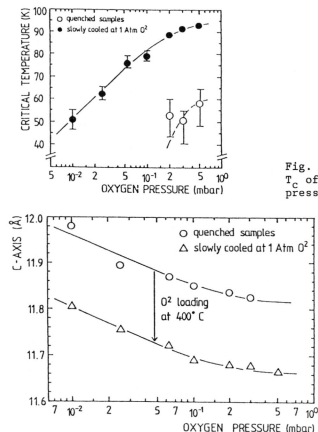

Fig. 3. Critical temperature T_c of YBaCuO films vs. O_2 pressure during preparation.

Fig. 4. c-axis parameter of the 123 structure in YBaCuO films vs. O_2 pressure during preparation.

The results given in Figs. 3 and 4 also show that samples prepared at 750 °C at relatively high oxygen pressures are tetragonal.
To obtain a high T_c, the films must be allowed to perform the
tetragonal-to-orthorhombic transition which takes place during a
"slow" cooling (slow compared to the kinetics of the tetragonal-to-
orthorhombic transition) under a higher oxygen pressure.

2.2. YBaCuO-LaAlO$_3$ HETEROSTRUCTURES

For many electronic applications, multilayer structures of
superconducting and insulating layers are needed. For this aim the
preparation chamber was equipped with a target exchanger providing
three different rotating targets which could be successively moved
into the laser beam. In a first attempt we prepared a three-layer
structure of YBaCuO/LaAlO$_3$/YBaCuO on a SrTiO$_3$ substrate[9]. Each layer
is about 2000 Å thick. Fig. 5 shows the corresponding X-ray diffraction pattern exhibiting only the (001) peaks of YBaCuO, SrTiO$_3$ and
LaAlO$_3$ demonstrating the heteroepitaxial growth of these layers. The
inductive T_c measurement gives a sharp transition at 90 K, but it
cannot distinguish between the bottom and the top superconducting
layer. Because of shorts between the two superconducting layers
through the LaAlO$_3$, which are probably caused by the spheres deposited on the first YBaCuO layer during the preparation, the T_c of the
top layer could not be determined by resistive measurements. These
preliminary results show that epitaxial LaAlO$_3$ layers can be easily
laser-deposited on YBaCuO films and will be useful for insulating
layers.

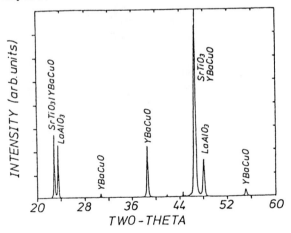

Fig. 5. X-ray
diffraction pattern of
an heteroepitaxial
YBaCuO/LaAlO$_3$/YBaCuO
multilayer on SrTiO$_3$
substrate.

2.3. Bi$_2$Sr$_2$CaCu$_2$O$_x$ FILMS

Thin Bi$_2$Sr$_2$CaCu$_2$O$_x$ films were prepared onto <100> SrTiO$_3$ and
LaAlO$_3$ single crystalline substrates from a rotating stoichiometric
sinter target at an energy density of about 3 J/cm^2 in 0.4 mbar O$_2$.
Details of the preparation have been published elsewhere (Ref. 10).

Without any annealing step, the inductively measured T_c did not exceed 76 K and the narrowest full transition width ΔT_c was around 8 K. Both T_c and T_c could be improved by an in-situ post-anneal step in 0.6 mbar or 1.0 mbar O_2 without changing the temperature applied during the deposition. In case of $SrTiO_3$ substrates, a strong substrate-film reaction seems to limit the quality of the films, which was shown by annealing at different temperatures for different periods of time. As turned out, the best value of $\Delta T_c \approx 4$ K has been obtained by a 25 min anneal, whereas for achieving the highest T_c of about 83 K a 1 hour anneal was necessary.

As can be seen from Fig. 6 in case of $LaAlO_3$ substrates, sharp phase transitions with $\Delta T_c \approx 4$ K have been obtained with $T_{c,midpoint} \approx 78$ K by applying the same 1 hour post-anneal step at a temperature about 20 K below the melting point of the 80 K phase in 0.4 mbar O_2, which corresponds to about 760°C. For both substrates a FWHM of the (00$\underline{10}$) reflection rocking curve of 0.4° is reproducibly measured showing the very strong c-axis texture of the films. The Laue photograph shown in Fig. 7 of one of the films on $SrTiO_3$ with the incident beam almost parallel to the c-axis of the $SrTiO_3$ substrate demonstrates the strong in-plane texture by showing spots instead of rings.

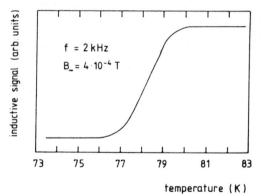

Fig. 6. Inductively measured phase transition for a film prepared on $LaAlO_3$.

Fig. 7. Laue photograph for a film prepared on $SrTiO_3$.

3. CRITICAL CURRENTS

After investigating the structural features and achieving high critical temperatures the next property to be optimized is the current carrying capability of the films, both in zero magnetic field and its field and field direction dependence.

3.1. YBaCuO FILMS

For the critical current measurements, the films were patterned using standard photolithography. Strip lines of 10 μm width and 180

μm length were etched by diluted phosphoric acid. These small strip lines allowed the measurement of critical current densities up to the 5 x 10^7 A/cm^2 range by four-probe technique.

Fig. 8 shows the magnetic field dependence of the critical current density j_c at 77 K for two magnetic field directions for an YBaCuO film deposited on a MgO substrate. The zero field j_c is 3.5 x 10^6 A/cm^2. This is a little bit smaller than the 5 x 10^6 A/cm^2 we obtained for films on SrTiO$_3$.

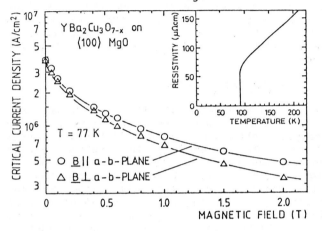

Fig. 8. Magnetic field dependence of j_c at two field directions and 77 K for an YBaCuO film deposited on a MgO substrate.

In order to get an idea about the mechanisms determining the flux pinning in these films we determined j_c in detail as a function of temperature T, magnetic field B and magnetic field direction denoted by the angle Θ between the field direction and the c-axis (perpendicular to the film plane)[11]. The $j_c(\Theta)$ behavior is given in Fig. 9 for 4.2 K and 7 T. Extremely high j_c values (4.5 x 10^7 A/cm^2 at 7 T) are observed when the magnetic field (i.e. the flux lines) is aligned parallel to the CuO planes ($\underline{B} \perp \underline{c}$). $j_c(\Theta)$ peaks can be

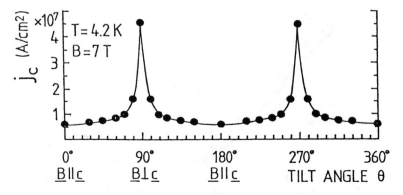

Fig. 9. Angular dependence of j_c for an YBaCuO film at 4.2 K and 7 T.

seen at 90° and 270°. The Lorentz force acts parallel to the c-axis. In this case, two mechanisms might be responsible for the strong pinning. One is intrinsic pinning between the CuO planes assuming a modulation of the order parameter along the c-axis. The maximum pinning is only achieved when the flux lines are exactly aligned along the film plane. If the field direction is slightly tilted from this condition, the flux lines are only pinned at a few interaction points. This results in rather sharp $j_c(\Theta)$ peaks. A second contribution, especially at high magnetic fields, when the flux line lattice becomes more rigid, is the anisotropy of the shear modulus C_{66} of the flux line lattice.[12] For the $\underline{B} \parallel \underline{c}$ case extended two-dimensional defects in-line with the c-axis, like twins or stacking faults act as pinning centers.[11]

3.2. $Bi_2Sr_2CaCu_2O_x$ FILMS

Transport critical current densities of films, which had been prepared without a post-anneal step, were measured across a 1.8 mm long and 44 μm wide stripline obtained by a photolithographic process using diluted HCl or H_2SO_4 for etching. The applied voltage criterion of 0.5 μV corresponds to an electric field of 3 μV/cm. This patterning technique, however, caused serious degradation of T_c, which was sometimes as large as 10 K. For this reason, the annealed films were patterned by scratching using a HP 7550A plotter equipped with a stainless steel needle. By this method a 4 mm long and 140 μm wide stripline could be obtained. Despite the degradation in T_c, the wet chemically etched samples showed values of the critical current density up to 1×10^7 A/cm^2 at 4.2 K. Fig. 10 shows the temperature dependence of the normalized critical current density for differently annealed samples. The samples prepared without in-

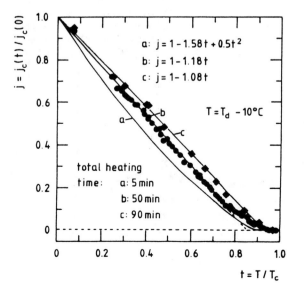

Fig. 10. Temperature dependence of j_c for differently annealed BSCCO samples.

situ annealing have been shown to be well described by a parabolic fit of the form

$$j_c(T/T_c) = j_c(0)x(1 - 1.58xT/T_c + 0.5x(T/T_c)^2) \qquad (Ref.10),$$

whereas the annealed samples showed a linear temperature dependence of j_c of the form

$$j_c(T/T_c) = j_c(0)x(1 - \alpha xT/T_c)$$

with α decreasing with increasing annealing time. This flatter temperature dependence is ascribed to an improved pinning due to a better ordering of the atoms.

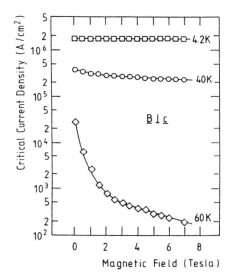

Fig. 11. Magnetic field dependence of j_c of a film prepared on SrTiO$_3$ without in-situ annealing for the $\underline{B} \perp \underline{c}$ direction.

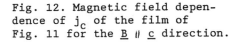

Fig. 12. Magnetic field dependence of j_c of the film of Fig. 11 for the $\underline{B} \parallel \underline{c}$ direction.

The magnetic field dependence of j_c was measured for the $\underline{B} \perp \underline{c}$ and the $\underline{B} \parallel \underline{c}$ directions up to 7 T. Fig. 11 and 12 show a logarithmic plot of j_c as a function of the magnetic field for these two directions. For the $\underline{B} \perp \underline{c}$ direction, j_c is almost independent of the external magnetic field up to 7 Tesla for temperatures below about $T_c/2$. For the $\underline{B} \parallel \underline{c}$ direction, an exponential decay of j_c with increasing field is observed at elevated temperatures. As can be seen from Fig. 13 the normalized critical current density $j_c(T,B)/j_c(0,0)$ in external magnetic fields $\underline{B} \perp \underline{c}$ is improved by the in-situ anneal step. However, for the direction $\underline{B} \parallel \underline{c}$ the magnetic field dependence remains catastrophic, revealing again the fact,

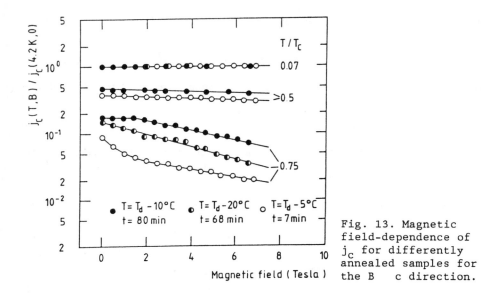

Fig. 13. Magnetic field-dependence of j_c for differently annealed samples for the B   c direction.

that the improved pinning is due to a better in-plane ordering of the atoms.

5. rf PROPERTIES AND FIRST DEVICES

With the exception of dc SQUIDs, all possible applications of high T_c thin films will be at high frequencies, as applications in microelectronics or high frequency devices. Therefore, a good T_c and even excellent j_c values of thin films are not sufficient for applications, unless the films exhibit good rf properties. Since $SrTiO_3$ has extremely high dielectric losses, we prepared single-crystalline epitaxial YBaCuO films on the better suited $LaAlO_3$ substrates. The rf properties of the films are easiest characterized by their surface resistance, as plotted in Fig. 14 as a function of frequency for the 0.1 to 100 GHz range. The obtained values at 77 K for our films are 8 ± 2 mΩ at 87 GHz (Klein et al.[13]) or 0,1 mΩ at 10 GHz (Hammond et al.[14]), which is one or even two orders of magnitude below the value of Cu cooled to 77 K. Results for the films prepared by other groups are shown for comparison in Fig. 14. The values for differently prepared films can easily differ by half an order of magnitude or even more.

The low values of R_s(77 K) at microwave frequencies are promissing for applying these films for rf devices. Valenzuela et al.[15] prepared a coplanar waveguide $\lambda/2$ resonator of $YBa_2Cu_3O_{7-x}$ on a $LaAlO_3$ substrate for 6.5 GHz. The obtained quality factor of this resonator was 3850 ± 180 at 77 K. This is about 40 times better than the value measured for an identical but thicker copper resonator. The quality factor did not depend on the dissipated power in the resonator up to 1.5 mW, indicating that the critical current was not

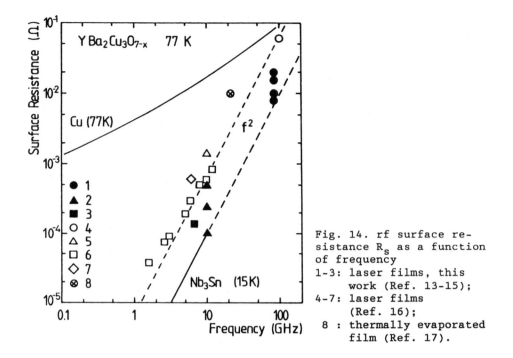

Fig. 14. rf surface re-
sistance R_s as a function
of frequency
1-3: laser films, this
work (Ref. 13-15);
4-7: laser films
(Ref. 16);
8 : thermally evaporated
film (Ref. 17).

yet reached. From these results they also calculated a surface resistance of 0.14 mΩ for this film at 6.5 GHz and 77 K[15] (see Fig. 14).

6. CONCLUSIONS

Laser deposition is a useful tool to prepare YBaCuO and BiSrCaCuO films with excellent properties on various substrates. Criteria for the quality of these films are a high transition temperature T_c, a high dc critical current density j_c (B), a low rf surface resistance R_s and an independence of R_s of the dissipated power up to high values. Whereas high T_c's and, in many cases, also high j_c's are obtained by different preparation techniques, the low R_s values and especially their power independence are only rarely achieved, but necessary for applications of these films.

7. ACKNOWLEDGEMENTS

The author gratefully acknowledges technical assistance by M. Kühnl and the excellent cooperation and stimulating discussions with B. Roas, P. Schmitt, H.E. Hoenig and G. Saemann-Ischenko. This work has been supported by the Bundesministerium für Forschung und Technologie.

7. REFERENCES

1. P. Chaudhari, R.H. Koch, R.B. Laibowitz, T.R. McGuire, and R.J. Gambino, Phys. Rev. Lett. 58, 2684 (1987).
2. M. Naito, R.H. Hammond, B. Oh, M.R. Hahn, J.W.P. Hsu, P. Rosenthal, A.F. Marshall, M.R. Beasley, T.H. Geballe, and A. Kapitulnik, J. Mat. Res. 2, 713 (1987).
3. T. Terashima, K. Iijima, K. Yamamoto, Y. Bando, and H. Mazaki, Jap. J. Appl. Phys. 27, L91 (1988).
4. H. Itozaki, S. Tanaka, K. Higaki, K. Harada, N. Fujimori, and S. Yazu, Proc. MRS Int. Meeting on Adv. Mat., Tokyo, May 30 - June 3, 1988.
5. C. Webb, S.L. Weng, J.N. Eckstein, N. Missert, K. Char, D.G. Schlom, E.S. Hellmann, M.R. Beasley, A. Kapitulnik, and J.S. Harris, Jr., Appl. Phys. Lett. 51, 1191 (1987).
6. D. Dijkkamp, T. Venkatesan, X.D. Wu, S.A. Shaheen, N. Jisrawi, Y.H. Min-Lee, W.L. McLean, and M. Croft, Appl. Phys. Lett. 51, 619 (1987).
7. B. Roas, L. Schultz, and G. Endres, Appl. Phys. Lett. 53, 1557 (1988); B. Roas, L. Schultz, and G. Endres, J. Less-Common Metals 151, 413 (1989).
8. R. Bormann and J. Nölting, Appl. Phys. Lett. 54, 2148 (1989).
9. L. Schultz, B. Roas, P. Schmitt, and G. Endres, in: "Processing of Films for High T_c Superconducting Electronics", SPIE Proc. Vol. 1187 (1989) p. 204.
10. P. Schmitt, L. Schultz, and G. Saemann-Ischenko, Physica C, 168, 475 (1990).
11. B. Roas, L. Schultz, and G. Saemann-Ischenko, Phys. Rev. Lett. 64, 479 (1990).
12. V.G. Kogan and L.J. Campbell, Phys. Rev. Lett. 62, 1552 (1989).
13. N. Klein, G. Müller, H. Piel, H. Chaloupka, B. Roas, L. Schultz, U. Klein, and M. Peiniger, Appl. Phys. Lett. 54, 757 (1989).
14. R. Hammond et al., to be published (1990).
15. A.A. Valenzuela, G. Daalmans, and B. Roas, Electronic Letters 25, 1434 (1989).
16. A. Inam, X.D. Wu, L. Nazar, M.S. Hedge, C.T. Rogers, T. Venkatesan, R.W. Simon, K. Daly, H. Padamsee, J. Kirchgessner, D. Moffat, D. Rubin, Q.S. Shu, D. Kalokitis, A. Fathy, V. Pendrick, R. Brown, B. Brycki, E. Belohoubek, L. Drabeck, G. Gruner, R. Hammond, F. Gable, B.M. Lairson, and J.C. Bravman, Appl. Phys. Lett., in press, 1990.
17. D.W. Cooke, E.R. Gray, R.J. Houlton, B. Rusnak, E.A. Meyer, J.G. Beery, D.R. Brown, F.H. Garzon, I.D. Raistick, A.D. Rollet, and R. Bolmaro, Appl. Phys. Lett. 55, 914 (1989).

PREPARATION OF HIGH-T_c SUPERCONDUCTING OXIDES THROUGH A DECOMPOSITION OF CITRIC ACID SALT

T. Asaka, Y. Okazawa and K. Tachikawa
Faculty of Engineering, Tokai University, Kanagawa 259-12, JAPAN

ABSTRACT

Y-base and Bi-base high-T_c oxide superconductors have been successfully synthesized by the citric acid salt decomposition process. T_c's of 92K and 109K have been obtained after a relatively short sintering in the Y-system and Bi-system, respectively. The fine dispersion of particles Ag in the Y-base oxide appreciably decreases the sintering temperature and enhances the transport J_c. The intermediate press increases both T_c and J_c in Bi-base oxide. A J_c (77K) of 1500A/cm^2 has been obtained in the sintered Bi-base specimen.

INTRODUCTION

The chemical processes via liquid phase reaction, e.g. the co-precipitation process, the sol-gel process and the citric acid salt process, can keep the homogeneity and the chemical stoichiometry of high-T_c superconducting oxide specimens much easier than the conventional powder metallurgical process. These chemical processes have a potential for preparing fine powders, thin films and fibers.

Among them, the co-precipitation process was widely used for preparing oxide superconductors in early days, consisting of co-precipitation reaction between metal ions and oxalates or carbonates [1,2]. Precipitates of corresponding oxalate or carbonate with particle size of submicrons are formed after the reaction.

The sol-gel process using alkoxides or acetates as raw materials, is also well-known for preparing superconducting oxides, consisting of an inorganic polymerization of alkoxides or acetates by hydrolysis. This process has been adapted to the preparation of fine powders, films and fibers of Y-Ba-Cu-O system [3,4] and Bi-Sr-Ca-Cu-O system [5,6].

However, the co-precipitation process needs a strict pH control, and a great deal of rinsing water. Meanwhile, the sol-gel process needs not only expensive metal alkoxides as raw materials but also long aging treatment for the inorganic polymerization.

In comparison with above chemical processes, the citric acid salt process does not require strict pH control and aging treatment, resulting an easy preparation of oxide superconductors. In addition, this process allows addition of different macroelements. Thus, the citric acid salt process has been applied for the synthesis of La-Sr-Cu-O system [1], Y-Ba-Cu-O system [7], Bi-Sr-Ca-Cu-O system [8,9,10] and Tl-Ba-Ca-Cu-O system [11].

In this paper, we will describe the preparation procedures,

structures and superconducting properties of high-T_c superconducting oxides synthesized by the citric acid salts decomposition process.

EXPERIMENTAL PROCEDURE

(a) Preparation of Y-base precursors

Figure 1 shows the sample preparation process. The precursor
was prepared in the following way. 99.9% pure Y_2O_3, $BaCO_3$ and CuO
powders were used as the starting materials. 99.99% pure Ag
powder (grain size < 26μm) was used for the preparation of Ag
added $YBa_2Cu_3O_y$. Additional ratios of Ag were 3, 10 and 30wt% to
the objective composition. These powders were mixed in the
molecular ratio of $YBa_2Cu_{3+x}O_y$ (X=0, 1.0), dissolved with
nitric acid, and diluted by deionized water. Citric acid
monohydrate and ethylene glycol were added to the mixed solution.
Subsequently, the solution was heated and stirred continuously on
a hotplate stirrer around 90°C for 2–3h. Then condensation of
solution and polymerized reaction between citric acid and metal
cations were simultaneously proceeded to form a rigid gel. This
gel exhibits a thermoplasticity, providing a possibly of drawing
it into a fiber like form as shown in Fig. 2.

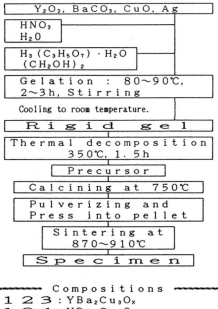

Fig. 1. Process of sample treatment.

Fig. 2. Drawing fiber from gel.

The gel was decomposed at 350°C for 1.5h in an electric furnace. Then the decomposed gel was ground for 1h to obtain an ash like precursor.

Subsequently, the resulting powder was calcined at 750°C for 8h and ground for 1h. The calcining and grinding were repeated twice. Then the powder was pressed into rectangular bars, 4mm in width, 2mm in thickness and 22mm in length. The bars were sintered at 860–940°C for 2–20h in an electric furnace. All heat treatments were performed in air.

(b) Preparation of Bi-base precursor
99.9% pure Bi_2O_3, $SrCO_3$, $CaCO_3$, CuO and Pb_3O_4 powders were used as the starting materials. These powders were mixed in the molecular ratio of $Bi_{1.8}Pb_{0.2}Sr_2Ca_2Cu_3O_z$. Preparation detail was the same as that of Y-base precursor described above. Obtained gel was also exhibits a thermoplasticity. The gel was decomposed and ground to obtain an ash like precursor.

Subsequently, the resulting powder was calcined at 780°C for 8h and ground for 1h. The calcining and grinding were repeated twice. Then the powder was pressed into rectangular bars, 4mm in width, 2mm in thickness and 22mm in length. The bars were sintered at 860–865°C for 10–200h in an electric furnace. Again all heat treatments were performed in air.

(c) Specimen characterization
X-ray diffraction analyses were made on the sintered specimens. Crystal morphology and element distribution in sintered specimens were studied by scanning electron microscope (SEM) and electron probe micro analyzer (EPMA). Offset critical temperature T_c and critical current I_c of the specimens were measured resistively by a DC four-probe method. Four lead wires were connected onto specimens using an ultrasonic soldering technique. Critical current density J_c was calculated by dividing I_c by cross-sectional area of the specimen. The magnetic hysteresis of the specimens was also measured by means of a vibrating sample magnetometer (VSM).

RESULTS AND DISCUSSION

(a) Results in Y-system
First, all specimens were sintered at a temperature between 860°C and 940°C every 10°C for 8h. In 123 specimens, a partial melt was observed at sintering temperatures above 920°C. Thus, sintering temperature of 123 specimen was fixed at 910°C. Figure 3 shows T_c vs sintering temperature for specimens with different Ag content. High T_c's were observed in specimens with 10 and 30wt% Ag content at 880°C. The optimum sintering temperature is decreased by ca. 50°C by the Ag addition. Figure 4(a) shows J_c vs sintering temperature for Ag doped specimens. The highest J_c was observed in all specimens at 880°C. The decreases of the sintering temperature in Ag doped specimens may be attributed to the decrease in the melting point of specimens.

Figure 4(b) shows J_c vs sintering time for Ag doped and undoped specimens. A high J_c is obtained after a relatively short sintering time of 4h in the Ag doped specimens. J_c was appreciably improved by the Ag doping. From these results, the highest T_c and J_c of 92K and 200A/cm^2, respectively, were obtained in the 30wt% Ag doped specimen. The best sintering condition for Ag doped specimens is 880°C for 4h.

Figures 5(a)–5(d) shows the SEM and EPMA analysis images for 10wt% Ag doped 123 specimen with different sintering time. From Figs. 5(a) and 5(c), the grain size of the 20h sintered specimen is larger than that of 4h sintered specimen. Figures 5(b) and 5(d) shows the distributions of Ag in the EPMA mapping.

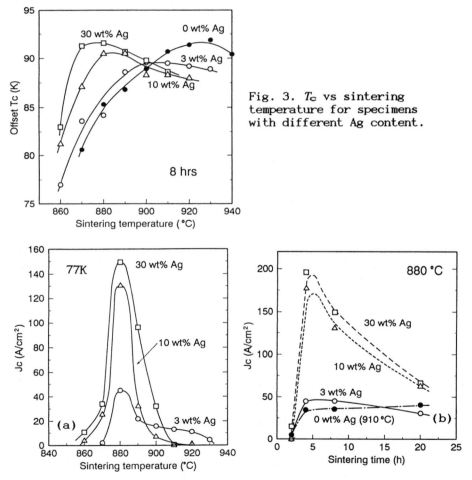

Fig. 3. T_c vs sintering temperature for specimens with different Ag content.

Fig. 4. (a) J_c vs sintering temperature for specimens with different Ag content sintered for 8h, (b) J_c vs sintering time for specimens with different Ag content.

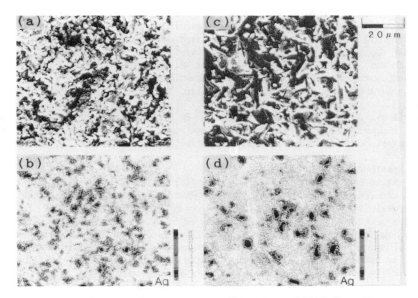

Fig. 5. 10wt% Ag–123 specimen sintered at 880°C for 4h:
(a) SEM, (b) EPMA (Ag), sintered 20h: (c) SEM, (d) EPMA (Ag).

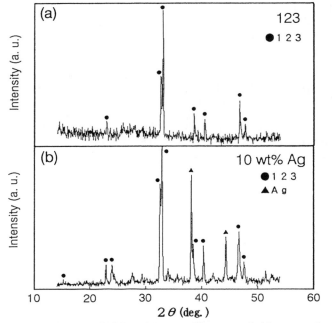

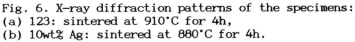

Fig. 6. X–ray diffraction patterns of the specimens:
(a) 123: sintered at 910°C for 4h,
(b) 10wt% Ag: sintered at 880°C for 4h.

These images indicates that the number of Ag precipitates is decreased and the size of Ag precipitates is increased with increasing sintering time. The fine dispersion of Ag particles seems to enhance the transport J_c. The EPMA analysis also indicates that other constituent elements than Ag distribute rather homogeneously in sintered specimens.

Figures 6(a) and 6(b) shows the X-ray diffraction patterns for 123 specimen sinterd at 910°C for 8h and 10wt% Ag doped specimen. The peaks indicated by ● and ▲ are of 123-phase and Ag, respectively. In the 124 specimen, dispersion of CuO was observed besides the 123 phase in the X-ray diffraction pattern.

(b) Results in Bi-system

From our preliminary experimental data, the best T_c of 102K was obtained in the specimen heat treated at 860°C for 20h. Intermediate pressing is well-known to improve J_c in Bi-base oxides [11]. Then, the effects of intermediate pressing and grinding between heat-treatment cycles on T_c were studied.

Figure 7 shows the T_c dependence on the total sintering time for specimens with different intermediate pressing pressures. T_c's of specimens increase with increasing sintering time, i.e. a number of intermediate pressing, and pressing pressure. The appreciable increase in T_c by the intermediate press in Bi-base system was reported elsewhere [10]. The intermediate press may enhance the formation of the high-T_c phase, though the exact mechanism is not yet clear.

Figure 8(a) shows the resistance-vs-temperature curve for the pressed specimen sintered at 860°C for 60h, of which T_c is 109K. Figures 8(b) and 8(c) show the temperature dependence of the real (X') and imaginary (X'') part of the ac magnetic susceptibility, respectively. In Figure 8(b), the signal, i.e. Meissner signal, increases rapidly below 110K, and a weak step is observed around 80K, indicating the presence of a small amount of low-T_c phase. The imaginary-part signal changes sharply only at 110K, as shown in Fig. 8(c). This sharp signal change implies that the high-T_c phase is formed predominantly in this specimen.

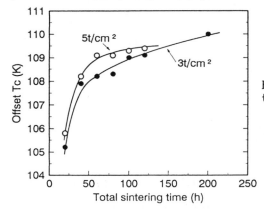

Fig. 7. T_c vs total sintering time at 860°C.

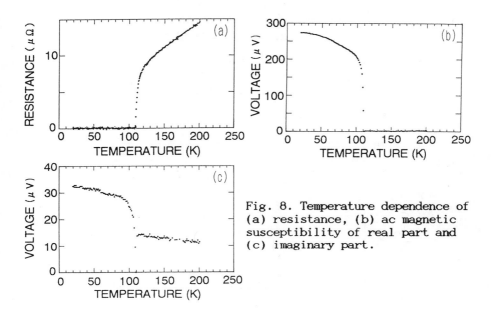

Fig. 8. Temperature dependence of (a) resistance, (b) ac magnetic susceptibility of real part and (c) imaginary part.

Figure 9 shows the J_c measured at 77K under zero magnetic field for pressed specimen sintered at 860°C. The J_c increases with increasing sintering time and intermediate pressing pressure. The best J_c of 1540A/cm^2 has been observed. Figures 10(a) and 10(b) shows the current transition at 77K, and the magnetic hysteresis curve of the specimen measured by VSM at 4.2K, respectively. The J_c value estimated from this magnetization curve using Bean's equation is about 2.9×10^4A/cm^2.

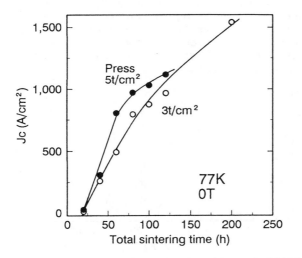

Fig. 9. J_c vs total sintering time at 860°C.

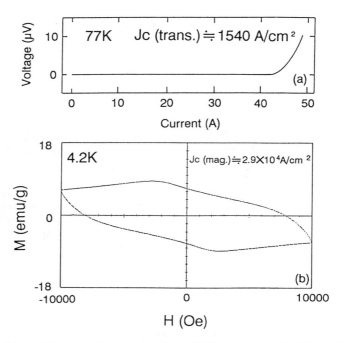

Fig. 10. (a) J_c measured at 77K in zero field, and (b) magnetic hysteresis curve at 4.2K.

Figure 11 shows the X-ray diffraction pattern of specimen sintered at 860°C for 60h. The peaks indicated by H(hkl) and L(hkl) are of high-T_c phase Bi$_2$Sr$_2$Ca$_2$Cu$_3$O$_x$ and low-T_c phase Bi$_2$Sr$_2$CaCu$_2$O$_y$, respectively. Strong H(00l) diffraction peaks are observed. Therefore, it is again evident that the high-T_c phase is predominant in this sample. From these results, the citric acid salt decomposition process reduces the required sintering time and improves T_c and J_c after intermediate pressing.

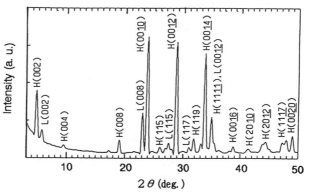

Fig. 11. X-ray diffraction pattern of pressed specimen.

CONCLUSIONS

1. T_c's of 92K and 109K have been obtained after a relatively short sintering in the Y–base and Bi–base oxides prepared by the citric acid salt decomposition process.
2.   The dispersion of 10–30wt% Ag in the Y–base oxide appreci-ably decreases the sintering temperature and enhances the transport J_c.   The finer dispersion of Ag precipitates may produce further increase in J_c.
3.   The intermediate press increases both T_c and J_c of Bi–base oxide.   A J_c (77K) of 1540A/cm^2 has been obtained in the sintered specimen.
4.   The citric acid salt decomposition process is promising for preparing high–T_c oxides after the optimization of processing conditions.

ACKNOWLEDGMENTS

The authors would like to thank Mr. Yasuo Miyamoto for his help in SEM observation and EPMA analysis, and Prof. Toru Hirayama for his encouragement.

REFERENCES

1)   J. G. Bednorz and K. A. Müller, Z. Phys. B – Condensed Matter 64, 189 (1986).
2)   T. Nakamori, H. Abe, Y. Takahashi, T. Kanamori and S. Shibata, Jpn. J. Appl. Phys. 27, L649 (1988).
3)   S. Shibata, T. Kitagawa, H. Okazaki, T. Kimura and T. Murakami, Jpn. J. Appl. Phys. 27, L53 (1988).
4)   H. Murakami, S. Yaegashi, J. Nishino, Y. Shiohara and S. Tanaka, Jpn. J. Appl. Phys. 29, L445 (1990).
5)   T. Kobayashi, K. Nomura, F. Uchikawa, T. Masumi and Y. Uehara, Jpn. J. Appl. Phys. 27, L1880 (1988).
6)   H. Zhuang, H. Kozuka and S. Sakka, Jpn. J. Appl. Phys. 28, L1805 (1989).
7)   C. T. Chu and B. Dunn, J. Am. Ceram. Soc. 70, C–375 (1987).
8)   R. S. Liu, W. N. Wang, C. T. Chang and P. T. Wu, Jpn. J. Appl. Phys. 28, L2155 (1989).
9)   A. Aoki, Jpn. J. Appl. Phys. 29, L270 (1990).
10) T. Asaka, Y. Okazawa, T. Hirayama and K. Tachikawa, Jpn. J. Appl. Phys. 29, L280 (1990).
11) T. Asano, Y. Tanaka, M. Fukutomi, K. Jikihara, J. Machida and H. Maeda, Jpn. J. Appl. Phys. 28, L380 (1989).

THALLIUM-BASED SUPERCONDUCTORS – A REVIEW

Allen M. Hermann, Hong-min Duan, and W. Kiehl
Department of Physics, Campus Box 390, University of Colorado, Boulder, CO 80309-0390

Lin Shu-yuan, Lu Li, and Zhang Dian-lin
Institute of Physics, Chinese Academy of Sciences, Beijing 100080, China

ABSTRACT

Stable bulk superconductors in the Tl-Ca-Ba-Cu-O system with zero resistance up to 125K are discussed. Structural, magnetic, and electronic transport properties, including thermo-electric power data, are presented on a variety of phases in this system. The anisotropic resistivity of $Tl_2Ba_2CaCu_2O_8$ single crystals is discussed. The in-plane and c-axis resistivities are $\sim 7 \times 10^{-4}$ Ω-cm and 2×10^{-1} Ω-cm, respectively, at room temperature. The c-axis resistivity decreases with decreasing temperature in the temperature range from room temperature down to the superconduct-ing transition temperature, and the second derivative of ρ_c is negative, unlike that of $YBa_2Cu_3O_7$ and $Bi_2Sr_{2.2}Ca_{0.8}Cu_2O_8$ single crystals. The anisotropic thermopower of $Tl_2Ba_2CaCu_2O_8$ single crystals is also described. The room temperature value of S_{ab} is about $14\mu V/K$. S_{ab} increases lin-early with decreasing temperature at a rate of $0.06 - 0.065\mu V/K^2$ down to about 160K, below which S_{ab} flattens off and drops to zero at T_c. S_c at room temperature is around $30\mu V/K$ and depends only weakly on temperature. Conventional scattering mechanisms cannot explain these normal state properties self-consistently.

I. INTRODUCTION

Discoveries of 30K La-Ba-Cu-O superconductors[1] and 90K Y-Ba-Cu-O superconductors[2] have stimulated a worldwide race for new and even higher temperature superconductors. Break-throughs were made by the discoveries of the 90K Tl-Ba-Cu-O system,[3,4] 110K Bi-Sr-Ca-Cu-O system[5,6] and 120K Tl-Ba-Ca-Cu-O system.[7-9] High temperature superconductivity was also observed in the Tl-Sr-Ca-Cu-O system,[10-12] and in the M-Tl-Sr-Ca-Cu-O with M = Pb[13,14] and rare earths.[15] In this paper, we present structural and electronic characterization of several phases of bulk polycrystalline and single crystal Tl-Ba-Ca-Cu-O superconductors.

II. STRUCTURE

The Tl-Ba-Ca-Cu-O system can form a number of superconducting phases. Two phases, $Tl_2Ba_2Ca_2Cu_3O_{10+x}$ (2223) and $Tl_2Ba_2Ca_1Cu_2O_{8+x}$ (2212), were first identified.[16] The 2223 super-conductor has a 3.85 x 3.85 x 36.25A tetragonal unit cell. The 2212 superconductor has a 3.85 x 3.85 x 29.55A tetragonal unit cell.[16,17] The 2223 phase is related to 2212 by addition of extra calcium and copper layers. In addition, the superconducting phase in the Ca-free Tl-Ba-Cu-O system is $Tl_2Ba_2CuO_{6+x}$ (2201).[16,18] Fig. 1 shows schematically the arrangements of metal atom planes in these three Tl-based superconducting phases. The 2201 phase has a zero-resistance temperature of about 80K, whereas the 2212 and 2223 phases have zero-resistance temperatures 108K and 125K respectively.[16-21]

A series of superconducting compounds with a single Tl-O layer, which we denote by

TlBa$_2$Ca$_{n-1}$Cu$_n$O$_{2n+2.5}$, was subsequently reported.[22, 23] The Tl-Ba-Ca-Cu-O superconducting system should be represented using a general formula of Tl$_m$Ba$_2$Ca$_{n-1}$Cu$_n$O$_{1.5m+2n+1}$ with m = 1 and 2, and n = 1, 2, 3, 4, and 5. The T_c of the single Tl-O layer compounds also increases with the number of Cu-Ca layers (up to n = 3) and is slightly lower than that of the corresponding double Tl-O layer compounds.

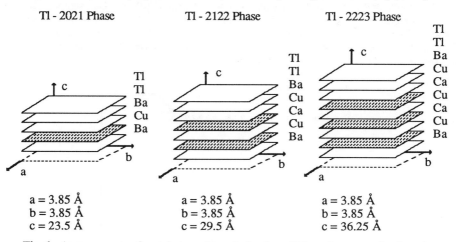

Tl - 2021 Phase	Tl - 2122 Phase	Tl - 2223 Phase
a = 3.85 Å	a = 3.85 Å	a = 3.85 Å
b = 3.85 Å	b = 3.85 Å	b = 3.85 Å
c = 23.5 Å	c = 29.5 Å	c = 36.25 Å

Fig. 1   Arrangements of metal atom planes in the three Tl-based superconducting phases.

III. BULK POLYCRYSTALLINE MATERIALS

Fig. 2 shows the resistance-temperature variation for a nominal Tl$_{2.2}$Ba$_2$Ca$_2$Cu$_3$O$_{10.3+x}$ sample.[9] This sample has an onset temperature near 140K, midpoint of 127K, and zero resistance temperature at 122K.

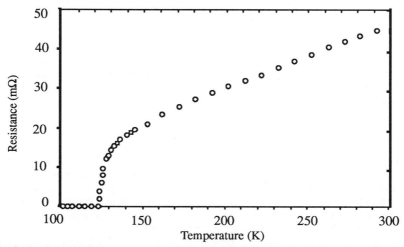

Fig. 2   Resistance-temperature variation for a nominal Tl$_{2.2}$Ba$_2$Ca$_2$Cu$_3$O$_{10.3+x}$ sample.

A. MAGNETIZATION

Fig. 3 shows[24] dc magnetization (field cooled and zero field cooled) as a function of temperature for an applied field of 1 mT for a nominal $Tl_2Ca_4BaCu_3O_{11+x}$ sample. The insert of Fig. 3 shows data traces for the same sample, and also for a well-prepared $EuBa_2Cu_3O_{7-x}$ sample, where the vertical axis represents the dX"/dH signal of an EPR spectrometer. As is seen from the insert, the onset temperature for the sample A ($Tl_2Ca_4BaCu_3O_{11+x}$) is 118.3K, 23.9K higher than that of the $EuBa_2Cu_3O_{7-x}$ (sample B), whose onset temperature is 94.4K. This onset temperature is consistent with those measured by resistance-temperature variations.

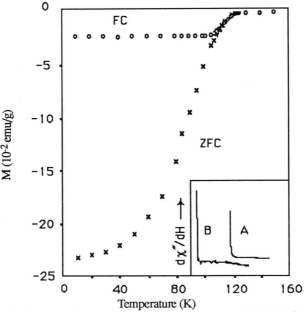

Fig. 3   dc magnetization as a function of temperature for an applied field of 1mT for a nominal $Tl_2Ca_4BaCu_3O_{11+x}$ sample. The insert of Fig. 3 shows data traces for the same sample, and also for a well-prepared $EuBa_2Cu_3O_{7-x}$ sample, where the vertical axis represents the dX"/dH signal of an EPR spectrometer.

B. THERMOELECTRIC POWER

The thermoelectric power as a function of temperature for a nominal $Tl_2Ca_2Ba_2Cu_3O_{10+x}$ sample[25] is shown in Fig. 4. The normal-state thermoelectric power is positive, indicating dominant hole conduction. At least three separate ranges of temperature-dependent behavior are apparent. Below the transition (the midpoint of the transition was determined to lie at 118K), the thermoelectric power is an increasing function of temperature. Finally, from 175K to room temperature, the thermoelectric power decreases linearly with increasing temperature. The temperature dependence of the Tl-Ca-Ba-Cu-O superconductor is qualitatively similar to that of Y-Ba-Cu-O sample,[26] and is not characteristic of a typical normal metal whose thermoelectric power increases with increasing temperature.

Fig. 5 shows the resistivity-temperature and thermoelectric power-temperature variations for a ceramic sample of the 2201 phase. While this phase is the lowest T_c phase for the double Tl-O compounds, its unusual thermoelectric power-temperature variation with a cross-over from negative to positive thermopower as temperature is lowered (above T_c) warrants substantial study of the 2201 phase.

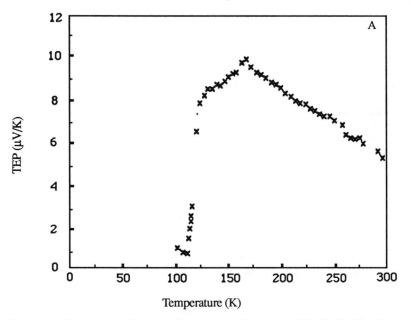

Fig. 4 Thermoelectric power as a function of temperature for a nominal $Tl_2Ca_2Ba_2Cu_3O_{10+x}$ sample.

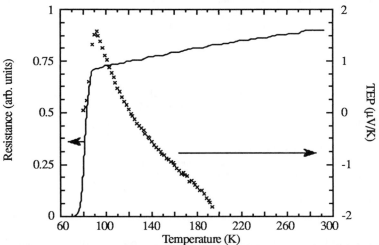

Fig. 5 Resistivity-temperature and thermoelectric power-temperature variations for a ceramic single phase sample of Tl 2201.

IV. SINGLE CRYSTALS

In this section we report recent results of anisotropic resistivity and thermoelectric power on single crystals of TBCCO.

A. CRYSTAL GROWTH

Two different processes were used to grow the single crystals. In the first process (A), we used a $Ba_2Ca_{22}Cu_3O_x$ precursor. Appropriate amounts of BaO, CaO and CuO (each 99.99% pure) were mixed and ground to form a mixture with a nominal stoichiometry of $Ba_2Ca_2Cu_3O_x$. The mixture was heated in air for 24 hours at 920°C, with an intermediate grinding carried out in order to facilitate a complete reaction. The reacted mixture was then ground again to form the precursor powder. Proper amounts of precursor and Tl_2O_3 powders were mixed and ground to form a powder with a nominal stoichiometry of $Tl_2Ba_2Ca_2Cu_3O_x$. This powder was then pressed in pellet form under a pressure of 460 kg/cm^2 (6500 lb/in^2). Pellets of mass 25 grams were placed in an Al_2O_3 crucible which was then loosely covered by an Al_2O_3 plate to protect the Tl_2O_3 from direct evaporation. The crucible was then put into a quartz tube which was located in a horizontal tube furnace. Because the mixture melts at ~938°C, a 15-minute heating period at 940°C to 945°C melted most of the mixture. The melt was then cooled to 900°C at a rate of 0.2°C/min. When it was cooled, crystals were formed. The nucleation was followed by a 1°C/min cooling to 700°C, 2° C/min to 400°C and five or more hours of furnace cooling to room temperature. This procedure was designed to relieve some inner stress and optimize the oxygen distribution and crystal structure. The entire process of heating the pellet occured in flowing oxygen. In the second process (B), the precursor preparation step was skipped. Appropriate amounts of Tl_2O_3, BaO, CaO, and CuO were ground to form a powder with a nominal composition of $Tl_2Ba_2Ca_2Cu_3O_x$. The remaining procedures were the same as in process A.

Like other Cu-O based high T_c superconducting single crystals, the crystals were rectangular plates with typical dimensions of 2mm x 2mm x 0.2mm. X-ray dot maps and XPS results show the crystals were chemically uniform with stoichiometries of $Tl_2Ba_2CaCu_2O_8$ (2212) and $Tl_2Ba_2Ca_2Cu_3O_{10}$ (2223). The X-ray four-circle diffraction verified the 2212 single crystals had high quality, and no minor misoriented phase could be found. However, in the 2223 phase grains, sometimes two or three subgrains were observed with a few degrees of misorientation. X-ray diffraction confirmed the c-axis was perpendicular to the large rectangular dimensions. We present results of measurements of 2212 crystals in this report.

B. RESISTIVITY

The standard Montgomery method[27] was employed to measure the anisotropic resistivities. Contacts, with resistance less than 2Ω, were made by pure indium or other low melting point indium alloys such as Ag-In, Sn-In or Pb-Sn-In. The contacts were located on the corners of the ab-plane. Since the alloy pads were less than 50μm in diameter (less than one-tenth of the size of the smallest sample in these experiments), the deviation from the ideal condition was negligible.[28] Platinum fibers, 20μm in diameter, were used to electrically connect the metallic pads to copper leads. A lock-in amplifier measured the voltage on two contacts when an 500μA RMS current at 87Hz passed through the other two contacts (see top left of Fig. 6). Samples were ohmic up to 10mA. A thermometer with 0.2K accuracy was used to measure the temperature; the temperature ramp rate was kept to 1 K/min. The resistance vs. temperature curves were repeatable with thermal cycling to within one percent.

The insert of Fig. 6 shows the typical temperature dependence of the Montgomery resistances. The Rc (Rp) represents the ratio of the voltage cross the contacts A, B (D, B) and the excitation current through C, D (A, C). Unlike the results of YBCO[29, 30] and BSCCO[31] single crystals, the ratio of Rc/Rp of the 2212 single crystal changes are smooth in the temperature range studied, especially near transition point. The resistance components Rp and Rc may be transformed into the resistivity components, ρ_c and ρ_p, by using an equivalent rectangle that maps the internal electric field. The resistivity components are shown in Fig. 6. In more than 10 samples studied, the room temperature in-plane resistivities ranged from 6.5 to 8.5 x 10^{-4} Ω-cm, while the c-axis room temperature resistivity had a relative higher spread from 0.1 to 0.4 Ω-cm. Similar behavior was observed in the thermoelectricpower (TEP) measurements discussed next. A wider spread (Δ S/S ~ 25%, T = 300K) was observed along c-axis, while in-plane TEP data from different crystals was in good agreement (Δ S/S < 3%, T = 300K).[32]

We note that the actual temperature dependence of ρ_c is not the same as that of Rc, the Montgomery resistance. This feature is a reflection of the anisotropy in the resistivity tensor. Similar phenomena have also been observed in the YBCO system in a limited temperature range.

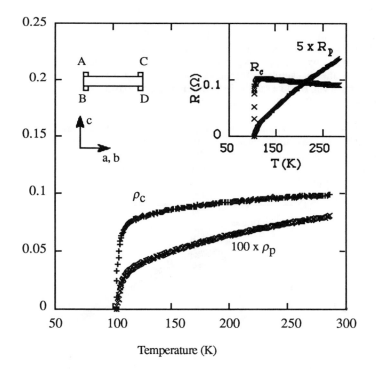

Fig. 6 Temperature dependence of resistivities parallel to (ρ_p) and perpendicular to (ρ_c) the Cu-O planes. The ρ_p values are multiplied by 100. The inset shows the measured resistances by using the top left configuration.

C. THERMOELECTRIC POWER

The method for thermopower measurements was described earlier.[33] At each constant reference temperature a voltage difference was recorded in response to a sweeping temperature differential monitored by a constantan-copper thermocouple, which resulted in a straight line (except at the superconducting transition) without thermal hysteresis. The slope, normalized to the sensitivity of the constantan-copper thermocouple, gives the thermopower of the sample after correcting for the contribution of the leads. The thermopower was measured for several 2212 crystals from room temperature down to 80K.

Fig. 7 shows the thermopower in ab plane, S_{ab}. The room temperature value of S_{ab} is about 14μV/K, quite reproducible for the three samples measured. As the temperature is lowered, S_{ab} increases linearly down to 160K with slightly different slopes for three different samples. The thermopower peaks around 140K and drops precipitously at T_c. Above 160K, S_{ab} can be fairly well fitted by the linear expression

$$S_{ab} = A + BT \qquad (1)$$

where the intercept A ranges between 31 and 34μV/K and the slopes B range from -0.06 to -0.05μV/K². This behavior is similar to that observed in BiSrCuO single crystals.[34] In BiSrCaCuO it was found that B was sample-independent (-0.05μV/K2), but A depended strongly on the oxygen deficiency.[35] Somewhat similar behavior was found in YBa$_2$Cu$_3$O$_7$ single crystals for which, below a certain doping level, B had a sample-independent value of -0.015μV/K².[33] It seems very likely that the linear increase of S_{ab} with decreasing temperature is a common feature of these oxide superconductors.

The thermopower along c axis, S_c, is plotted in Fig. 8. We see that S_c depends only weakly on temperature. At lower temperature S_c increases to form a small hump before dropping to zero at T_c. This behavior is somewhat similar to that for BiSrCaCuO,[34] but is quite different from that observed in YBa$_2$Cu$_3$O$_7$[33, 36] for which S_c increases markedly with temperature. In contrast to S_{ab}, S_c shows some sample-dependence.

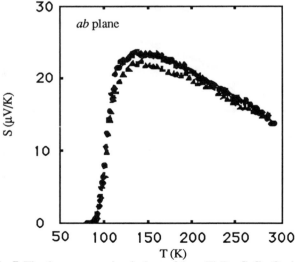

Fig. 7 The thermopower in ab plane of three Tl$_2$Ba$_2$CaCu$_2$O$_8$ single crystals.

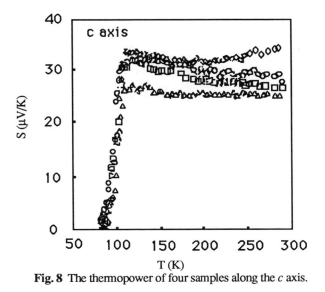

Fig. 8 The thermopower of four samples along the *c* axis.

V. DISCUSSION

It's not yet clear at this point whether the coupling between hole pairs in the superconducting state of Cu-O based high T_c superconducters is due to a Bose-Einstein condensation or to a BCS- like condensation (with Cooper pair coupling not likely due to phonon mediation).[37] In view of this significant uncertainty, we confine our remarks in this discussion to the properties of the TBCCO normal state.

For the following analysis, we assume that the TBCCO system is metallic. Clayhold et al.[38] have found a relatively temperature-insensitive Hall coefficient for bulk ceramic samples of the 2223 phase (the Hall coefficient measured a factor of about two from 120K to 280K).

We concentrate on the normal state transport properties of the Cu-O planes of the TBCCO system. In particular, for 2212 crystals, Fig. 6 shows the in-plane resistivity to be a linearly increasing function of temperature. While much has been made of the fact that the $T^{3/2}$ dependence (expected for acoustical phonon scattering in three dimensions) is not seen; here we consider the carrier mobility calculated in two dimensions to be appropriate for transport in the ab plane.

Using the relaxation time formalism[39] and a two dimensional density of states, Fivav and Mooser[40] have calculated a mobility temperature dependence of T^{-1} for scattering by (low energy) acoustical phonons. The same two dimensional calculation gives a steeper functional dependence for optical phonon scattering. Hence the resistivity data shown in Fig. 6 can be interpreted in terms of conventional scattering by acoustical phonons (for a two dimensional system).

The difficulty with this interpretation is the temperature dependence of the thermoelectric power S shown in Fig. 7. This data can be fit well to Eq. 1. The negative coefficient of the temperature, however, defies interpretation within conventional theory. Conventional theory gives a linear dependence for thermopower on temperature with a positive coefficient for metals, and the positive nature of this coefficient is independent of dimensionality. Calculations of diffusive motion in disordered systems also show a similar linear temperature dependence with a positive coefficient, along with additonal terms in T_n, where n is a non-integral exponent.

A thermopower that decreases with increasing temperature can be explained by electron-correlation effects for electrons in non half-filled bands. The temperature behavior seen in Fig. 7 can be fitted to an Extended Hubbard Model.[41, 42] However, this model does not explain the magnetic field *independence* of the TEP observed in the normal state in La-and Bi-based high Tc superconducters.[43] When a magnetic field is applied to an electronic system with a free-spin degree of freedom, a spin entropy term is added to the TEP that has an exponential dependence on magnetic fields. While the magnetic field dependence of the TEP in the Tl-system compounds has not yet been measured, it would at least be surprising if the magnetic field dependence is substantially different from that (i.e. no field dependence) seen in the Bi-system. It therefore appears that conventional scattering even in two dimensions cannot explain the transport properties of TBCCO, and a good theoretical understanding of the normal state properties of TBCCO (and other Cu-O superconducters) is presently lacking.

VI. REFERENCES

1. J.G. Bednorz and K.A. Müller, Z. Phys. B 64, 189 (1986)
2. M.K. Wu, J.R. Ashburn, C.T. Torng, P.H. Hor, R.L. Meng, L. Gao, Z.J. Huang, Y.Z. Wang, and C.W. Chu, Phys. Rev. Lett. 58, 908 (1987)
3. Z.Z. Sheng and A.M. Hermann, Nature 332, 55 (1988)
4. Z.Z. Sheng, A.M. Hermann, A. El Ali, C. Almason, J. Estrada, T. Datta, and R.J. Matson, Phys. Rev. Lett. 60, 937 (1988)
5. H. Maeda, Y. Tanaka, M. Fukutomi, and T. Asano, Jpn. J. Appl. Phys. Lett. 27, L207 (1988)
6. C.W. Chu, J. Bechtold, L. Gao, P.H. Hor, Z.J. Huang, R.L. Meng, Y.Y. Sun, Y.Q. Wang, and Y.Y. Xue, Phys. Rev. Lett. 60, 941 (1988)
7. Z.Z. Sheng and A.M. Hermann, Nature 332, 138 (1988)
8. Z.Z. Sheng, W. Kiehl, J. Bennett, A. El Ali, D. Marsh, G.D. Mooney, F. Arammash, J. Smith, D. Viar, and A.M. Hermann, Appl. Phys. Lett. 52, 1738 (1988)
9. A.M. Hermann, Z.Z. Sheng, D.C. Vier, S. Schultz, and S.B. Oseroff, Phys. Rev. B 37, 9742 (1988)
10. Z.Z. Sheng, A.M. Hermann, D.C. Vier, S. Schultz, S.B. Oseroff, D.J. George, and R.M. Hazen, Phys. Rev. B 38, 7074 (1988)
11. W.L. Lechter, M.S. Osofsky, R.J. Soulen, Jr., V.M. LeTourneau, E.F. Skelton, S.B. Qadri, W.T. Elam, H.A. Hein, L. Humphreys, C. Skowronek, A.K. Singh, J.V. Gilfrich, L.R. Toth, and S.A. Wolf (submitted)
12. S. Matsuda, S. Takeuchi, A. Soeta, T. Suzuki, K. Aihara, and T. Kamo (submitted)
13. M.A. Subramanian, C.C. Torardi, J. Gopalakrishnan, P.L. Gai, J.C. Calabrese, T.R. Askew, R.B. Flippen, and A.M. Sleight (submitted)
14. Z.Z. Sheng and A.M. Hermann (unpublished)
15. Z.Z. Sheng, L. Sheng, X. Fei, and A.M. Hermann, Phys. Rev. B 39, 2918 (1989)
16. R.M. Hazen, L.W. Finger, R.J. Angel, C.T. Prewitt, N.L. Ross, C.G. Hadidiacos, P.J. Heaney, D.R. Veblen, Z. Z. Sheng, A. El Ali, and A.M. Hermann, Phys. Rev. Lett. 60, 1657 (1988)
17. L. Gao, Z.J. Huang, R.L. Meng, P.H. Hor, J. Bechtold, Y.Y. Sun, C.W. Chu, Z.Z. Sheng, and A.M. Hermann, Nature 332, 623 (1988)
18. C.C. Torardi, M.A. Subramanian, J.C. Calabrese, J. Gopalakrishnan, E.M. McCarron, K.J. Morrissey, T.R. Askew, R.B. Flippen, U. Chowdhry, and A.M. Sleight, Phys. Rev. B, 225 (1988)

19. M.A. Subramanian, J.C. Calabrese, C.C. Torardi, J. Gopalakrishnan, T.R. Askew, R.B. Flippen, K.J. Morrissey, U. Chowdhry, A.M. Sleight, Nature 332, 420, (1988)
20. C.C. Torardi, M.A. Subramanian, J.C. Calabrese, J. Gopalakrishnan, K.J. Morrissey, T.R. Askew, R.B. Flippen, U. Chowdhry, and A.M. Sleight, Science 240, 631 (1988)
21. S.S.P. Parkin, V.Y. Lee, E.M. Engler, A.I. Nazzal, T.C. Huang, G. Gorman, R. Savoy, and R. Beyers, Phys. Rev. Lett. 60, 2539 (1988)
22. Y. Luo, Y.L. Zhang, J.K. Liang, and K.K. Fung (submitted)
23. S.S.P. Parkin, V.Y. Lee, A.I. Nazzal, R. Savoy, R. Beyers, and S.J. La Placa, Phys. Rev. Lett. 61, 750 (1988)
24. A.M. Hermann, Z.Z. Sheng, D.C. Vier, S. Schultz, and S.B. Oseroff, Phys. Rev. B 37, 9742 (1988)
25. N. Mitra, J. Trefny, B. Yarar, G. Pine, Z.Z. Sheng, and A.M. Hermann, Phys. Rev. B 38, 7064 (1988)
26. N. Mitra, J. Trefny, M. Young, and B. Yarar, Phys. Rev. B 36, 5581 (1987)
27. H.C. Montgomery, J. Appl. Phys. 42, 2971 (1971).
28. B.F. Logan, S.O. Rice, and R.F. Wick, J. Appl. Phys. 42, 2975 (1971).
29. S.W. Tozer, A.W. Kleinsasser, T. Penney, D. Kaiser, and F. Holtzberg, Phys. Rev. Lett. 59, 1768 (1987).
30. S. Hagen, T.W. Jing, Z.Z. Wang, J. Horvath, and N.P. Ong, Phys. Rev. B37, 7928 (1988).
31. S. Martin, T. Fiory, R.M. Fleming, L.F. Schneemeyer, and J.V. Waszczak, Phys. Rev. Lett. 60, 2194 (1988); J.H. Wang, G. Chen, X. Chur, Y. Yan, D. Zheng, Z. Mai, Q. Yang, and Z. Zhao, Supercond. Sci and Technol. 1, 27 (1988).
32. Lin Shu-yuan, Lu Li, Zhang Dian-lin, H.M. Duan and A.M. Hermann, Europhys. Lett. (in press).
33. Lu Li, Ma Bei-hai, LIn Shu-yuan, Duan Hong-min, and Zhang Dian-lin, Europhys. Lett. 7, 555 (1988).
34. M.F. Crommie, G. Briceno, and A. Zettl, Physica C 162-4, 1397 (1989).
35. The authors of reference did not give notation to the phase of these crystals. From the dependence on annealing, the samples should be 2223 phase.
36. Z.Z. Wang, and N.P. Ong, Phys. Rev. B38, 7160 (1988).
37. W.A. Little, Science 242, 1390 (1988).
38. J. Clayhold, N.P. Ong, P.H. Hor, and C.W. Chu, Phys. Rev. B38, 7016 (1988).
39. See, e.g. A.C. Beer, *Galvanomagnetic Effects in Semiconductors* (Academic Press, New York, 1963), Ch. 25.
40. R. Fivav and E. Mooser, Phys. Rev. 163, 743 (1967).
41. G. Beni, Phys. Rev. B10, 2186 (1974).
42. B. Fisher and A. Ron, Sol. St. Comm. 40, 737 (1981).
43. R.C. Yu, M.H. Naughton, X. Yan, P.M. Chaikin, F. Holtzberg, R.L. Greens, J. Stuart, and P. Davies, Phys. Rev. B37, 7963 (1988).

UNIAXIALLY AND BIAXIALLY ALIGNED HIGH-T_c SUPERCONDUCTOR OXIDES: PREPARATION AND CHARACTERIZATION

R.S. Markiewicz, F. Chen, B. Zhang, J. Zhang, J. Sigalovsky, S. Wang and B.C. Giessen

Northeastern University, Boston, MA 02115

ABSTRACT

Magnetically aligned polycrystalline samples of high-T_c superconductors have been shown to display improved critical currents, but are still limited by the presence of intergranular weak links. We discuss a new technique, involving biaxial granular alignment, which can in principle overcome the weak link problem. We discuss recent experimental progress in demonstrating biaxial alignment.

INTRODUCTION

The new high-T_c superconducting oxides show a remarkable sensitivity of the critical current j_c to the form in which a sample is prepared. For $YBa_2Cu_3O_{7-\delta}$ (YBCO), values of j_c at 77K range from greater than $10^6 A/cm^2$ for epitaxial films to just $100 - 250 A/cm^2$ for bulk polycrystalline samples. Whereas the former values are suitable for applications, the latter are much too low, and much effort has gone into improving these values. Since the low values of j_c are known to be associated with weak transmission of current across grain boundaries, efforts have been made to eliminate these grain boundaries. For instance, much higher j_c's are found in single crystals, although these would not be of commercial interest. At present, the most promising technique of j_c enhancement is melt-textured growth[1-4], which works predominantly by allowing the growth of very large grains. For the highest j_c's, it is important to introduce extra pinning. Thus, as the quality of single crystals has improved, the values of critical current have dropped (less pinning); and Murakami's melt-textured growth samples probably owe their high-j_c values to microscopic inclusions of an insulating phase[2].

In the layered high-T_c materials based on Bi or Tl, there is an additional, intrinsic limitation to j_c, not related to intergranular connections. This effect may be associated with flux lattice melting or flux creep. However, these problems are not present at low temperatures, and at 4.2K, polycrystalline $Bi_2Sr_2CaCu_2O_x$ (Bi-2212) can be prepared with very high j_c values, which persist out to high magnetic fields[5].

In this paper, we discuss an alternate route for overcoming the problem of weak intergranular links. This route is based on the idea of aligning the axes of individual grains of superconductor before they are pressed into a pellet. Initial realizations of this idea have used a magnetic field to orient a single axis of the grains[6-10]. This technique has produced moderate improvements, to $\sim 1000 - 2000 A/cm^2$, but the strong magnetic field dependence of j_c indicates that weak links are still a major problem. We propose a modification of this procedure, which would allow the orientation of two independent crystalline axes, and hence effectively of all three axes. This will

374

in principle eliminate the weak link problem, and may allow the introduction of much higher pin densities.

ORIGIN OF WEAK LINK BEHAVIOR

Fundamentally, the weak link arises because the superconducting coherence length is so short that any perturbation at the grain boundary can strongly suppress superconductivity. Dimos, et al.[11] have shown that there is an intrinsic weak link coupling between two grains misaligned in the a,b-plane. This a-b plane misalignment can in principle account for most of the difference between the polycrystalline j_c's and those of epitaxial films[12,13]. Shin, et al.[14] have shown that the weak links persist in samples with clean grain boundaries, and hence are not an impurity effect.

The field dependence of j_c in polycrystalline samples provides further evidence that misorientation is the main source of weak links. It is found that above $\sim$100G, j_c stops falling off rapidly with field. This suggests that there are percolating paths through the sample which do not contain any weak links. An obvious interpretation is that two grains have a weak link junction only if they are misoriented by more than some critical angle. The high-field current is carried by a backbone of well oriented grains[10,15].

Weak links also affect normal state transport, causing giant deviations from Matthiessen's rule[16,17], which could be explained if most of the grain boundaries were insulating, but with small conducting 'pinholes.' Similar pinholes may arise at the interface between a superconductor and a normal metal[18]. Such pinholes could be formed if there is an insulating phase which forms on the dislocations associated with large-angle grain boundaries. A similar intrinsic insulating phase has recently been found at free surfaces in these materials[19]. Kogan[20] has suggested an alternate mechanism for misalignment weak links to arise. In either case, it is clear that improving the orientation of the grains should eliminate the weak link problem and improve j_c, but both c-axis *and a,b-plane orienting* are essential.

MAGNETIC ORIENTATION

Farrell, et al.[6], first demonstrated that grains of the high-T_c superconductors could be oriented by an intense magnetic field, with the direction of highest magnetic susceptibility aligning along the field. In these experiments, the grains were dispersed in an epoxy, thereby minimizing intergranular forces, so a high degree of alignment could be achieved and maintained as the epoxy hardened in the field. Subsequent experiments[7-10] prepared dense pellets by suspending the grains in a fluid which evaporated in the field, leaving a pellet which could be pressed and/or annealed, for subsequent transport or optical[21] studies.

Whereas most high-T_c materials orient with their c-axis parallel ('parallel aligner') to the field, in some rare earth (RE) substituted compounds, this direction actually lies within the (a,b)-plane ('normal aligner'), due to crystal-field effects[8]. The RE is substituted for Y in YBCO or for Ca in the Tl or Bi superconductors. An essential finding for the development of biaxial alignment is that the moments of normal aligners are associated with *specific* directions within the a-b plane[22]. It is found that Er, Eu, Tm, Yb, and Gd tend to align along the [020] axis in YBCO or along a [110] axis in the Tl, Bi systems. In the Bi compounds, care must

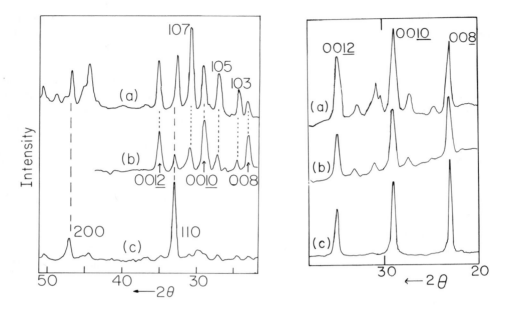

Fig. 1 (left) X-ray diffractograms (Cu-K$_\alpha$ radiation) of Bi-2212: 50% RE compounds, for field-oriented (b,c) samples. (a) Unoriented; RE = Er. (b) Parallel aligner; RE = Ho. (c) Normal aligner; RE = Er. [from Ref. 22].

Fig. 2 (right) X-ray diffractograms (2θ-scans) for a variety of shape-oriented samples of Bi-2212. (a) Orientation due to gravity, as in Fig. 3; (b) due to rolling; (c) due to pressing [from Ref. 26].

be taken when substituting RE's, since they can suppress T_c. However, we have been able to achieve a high degree of orientation in the a-b plane for samples containing 30% RE, for which T_c remains close to its undoped value.

Figure 1 shows typical results[22] of field orientation on RE-doped Bi-2212 compounds, comparing x-ray diffractograms of an unoriented specimen (Fig. 1a) with both a parallel (1b) and a normal (1c) aligner, both of which were aligned in epoxy. Since, in the absence of the RE, Bi-2212 and YBCO are both parallel aligners, doping with a 'normal aligning' RE leads to a crossover of the orientation axis as the RE concentration is increased[23]. This is most clearly seen in the case of Gd, which has the smallest magnetic moment. It is found[23] that a high concentration is needed to make YBCO:Gd a normal aligner, whereas Bi-2212, even with 100% substitution of Gd for Ca, remains a parallel aligner.

SHAPE ORIENTATION

In principle, the magnetic field is not the only way to generate a force capable of aligning grains of high-T_c material. One can couple a force to any anisotropic property of the grains. We have utilized the *shape anisotropy* of these materials – their tendency to form in the shape of plates, with much smaller thickness parallel to the c-axis. To optimize the degree of platyness in the starting materials, it is necessary to heat them close to or above the melting temperature, T_m. This is particularly important for the YBCO:RE materials, which are less two-dimensional, and care must be taken to note how T_m varies with RE[24]. Indeed, the tendency to shape-align is so strong in Bi-2212 that even the 'unoriented' specimen in Fig. 1a shows considerable [001] texturing, which causes the (00l) reflections to be stronger than in the calculated spectrum.

Shape orientation can be achieved by utilizing any of a number of mechanical forces or torques, including gravity and pressing. Fig. 2 shows the x-ray reflection patterns of Bi-2212 samples aligned by several such forces. These may be compared with the 'unoriented' sample of Fig. 1a. In all cases, the diffractograms show strong texturing, with predominantly the [001] direction perpendicular to the sample surface. The degree of alignment may be further improved by subsequent annealing[9,13]. The sample in Fig. 2a was aligned by gravity, as illustrated in Fig. 3 (but with B=0). The grains were suspended in fluid, and then allowed to settle slowly under the influence of gravity. It was found that the degree of alignment could be improved by vibrating the suspension. In Fig. 2b, the grains were dispersed in an epoxy binder, and alignment was achieved by rolling the epoxy into a thin flate before the epoxy had set. The c-axis orients perpendicular to the plane of the epoxy. Fig. 2c illustrates the alignment which can be obtained by pressure: the granular sample was pressed into a pellet. This technique orients only a thin surface layer, and polishing off a few microns of the pellet reveals the bulk to be completely unoriented.

BIAXIAL ORIENTATION

Magnetic and shape orientation can be combined into a single process[25,26], to produce grains aligned along two mutually perpendicular axes. Note, however, that shape orientation always leads to alignment of the grains' c-axes. Hence, the samples must be doped with a suitable RE, so that magnetic alignment will produce orientation within the a-b plane. Figure 3 illustrates a technique for preparing dense pellets of biaxially oriented materials. A magnetic field is applied

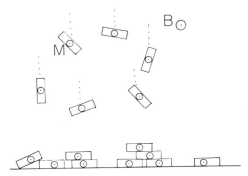

Fig. 3 Principle of biaxial alignment. Magnetic field B is used to orient grain magnetization M, assumed to point along a [110] axis; shape anisotropy (in this example, via gravity) to align the [001] axis, assumed perpendicular to planes of grains.

horizontally, orienting a particular direction in the a,b-plane; then, as the fluid evaporates and the grains settle out, their shape anisotropy causes the plates to lie flat, orienting the c-axis along the perpendicular direction.

For demonstration purposes, we have initially aligned the grains in an epoxy binder. Figure 4 shows the biaxial alignment of a single Bi-2212:30% Yb sample. The grains were suspended in epoxy, and magnetically oriented in a 5T horizontal magnetic field. Subsequently, the sample was removed from the field, and shape alignment was achieved by rolling the epoxy into a thin plate (as in Fig. 2b). The sample was then cut up and stacked, allowing x-ray powder diffractometer scans to be taken along the [001], [110], and [020] axes, as shown in Fig. 4. There is a high degree of orientation along both the [001] and [110] axes, along which the forces were applied. However, there is also significant orientation along the [020] axis, *even though no forces were applied along this axis*. The orientation is completely a consequence of biaxial orientation along the two other axes, forcing this axis into alignment. To the degree that the [020] axis is aligned, the technique of biaxial alignment has succeeded in mechanically reconstructing a single crystal (*pace* twinning) from a polycrystalline array.

These samples should be useful in a variety of experiments in which single crystals are either unavailable or are too small or too thin. Making dense pellets should allow us to demonstrate the degree to which the weak link behavior is suppressed. The technique of biaxial alignment should be amenable to bulk production techniques.

The field alignment was carried out in the horizontal field facilities of the Francis Bitter National Magnet Lab. We would like to thank L. Rubin and B. Brandt for their hospitality there. This research has been supported by the DOE under subcontract from Intermagnetics General Corporation.

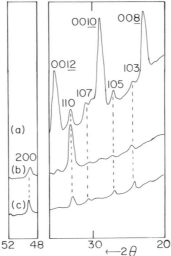

Fig. 4 X-ray diffractograms (2θ-scans) along three axes of biaxially-oriented sample of Bi-2212: 30% Yb. (a) [001], (b) [110], (c) [020] [from Ref. 26].

REFERENCES

1. S. Jin, T.H. Tiefel, R.C. Sherwood, M.E. Davis, R.B. van Dover, G.W. Kammlott, R.A. Fastnacht, and H.D. Keith, Appl. Phys. Lett. **52**, 2074 (1988).

2. M. Morita, M. Murakami, K. Miyamoto, K. Sawano, and S. Matsuda, Physica **C162-4**, 1217 (1989).

3. K. Salama, V. Selvamanickam, L. Gao, and K. Sun, Appl. Phys. Lett. **54**, 2352 (1989).

4. K. Chen, S.W. Hsu, T.L. Chen, S.D. Lan, W.H. Lee, and P.T. Wu, Appl. Phys. Lett. **56**, 2675 (1990).

5. K. Heine, J. Tenbrink, and M. Thöner, Appl. Phys. Lett. 55:2441 (1989); J. Tenbrink, K. Heine, and H. Krauth, Cryogenics **30**, 422 (1990).

6. D.E. Farrell, B.S. Chandrasekhar, M.R. De Guire, M.M. Fang, V.G. Kogan, J.R. Clem, and D.K. Finnemore, Phys. Rev. **B36**, 4025 (1987).

7. R.H. Arendt, A.R. Gaddipati, M.F. Grabauskas, E.L. Hall, H.R. Hart, Jr., K.W. Lay, J.D. Livingston, F.E. Luborsky, and L.L. Schilling, in "High-Temperature Superconductivity", ed. by M.B. Brodsky, R.C. Dynes, K. Kitazawa, and H.L. Tuller, Amsterdam, North-Holland (1988), p.203.

8. J.D. Livingston, H.R. Hart, Jr., and W.P. Wolf, J. Appl. Phys. **64**, 5806 (1988).

9. K. Chen, B. Maheswaran, Y.P. Liu, B.C. Giessen, C. Chan, and R.S. Markiewicz, Appl. Phys. Lett. **55**, 289 (1989).

10. J.E. Tkaczyk and K.W. Lay, J. Mater. Res. **5**, 1368 (1990).

11. D. Dimos, P. Chaudhari, J. Mannhart, and F.K. Le Goues, Phys. Rev. Lett. **61**, 219 (1988).

12. J. Rhyner and G. Blatter, Phys. Rev. **B40**, 829 (1989).

13. R.S. Markiewicz, K. Chen, B. Maheswaran, Y.P. Liu, B.C. Giessen, and C. Chan, presented at NYSIS Conf. on Superconductivity, Sept. 19-21, 1989, Buffalo, N.Y., to be published.

14. D.H. Shin, J. Silcox, S.E. Russek, D.K. Lathrop, B. Moeckly, and R.A. Buhrman, Appl. Phys. Lett. **57**, 508 (1990).

15. R.L. Peterson and J.W. Ekin, Phys. Rev. **B37**, 9848 (1988); J.W. Ekin, unpublished.

16. R.S. Markiewicz, Sol. St. Commun., **67**, 1175 (1988).

17. J. Halbritter, Int. J. Mod. Phys. **B3**, 719 (1989).

18. Y.P. Liu, K. Warner, C. Chan, K. Chen, R. Markiewicz, and R.L. Moore, J. Appl. Phys., **66**, 5514 (1989) and unpublished.

19. R. Claessen, G. Mante, A. Huss, R. Manzke, M. Skibowski, Th. Wolf, and J. Fink, unpublished.

20. V.G. Kogan, Phys. Rev. Lett. **62**, 3001 (1989).

21. F. Lu, C.H. Perry, K. Chen, and R.S. Markiewicz, J. Opt. Soc. Amer. **B6**, 396 (1989).

22. F. Chen, R.S. Markiewicz, and B.C. Giessen, I: in "Proc. 3d Ann. Conf. on Superconductivity and Applics.", Buffalo, N.Y. 1989, , (NY, Plenum, 1990); II: in "High Temperature Superconductors", 1989 MRS Conf. Proc. Vol. (1990), to be published.

23. F. Chen, J. Chen, R.S. Markiewicz, and B.C. Giessen, this conference.

24. S.Q. Wang, R.S. Markiewicz, and B.C. Giessen, this conference.

25. B.C. Giessen, R.S. Markiewicz, and F. Chen, patent applied for.

26. F. Chen, B. Zhang, R.S. Markiewicz, and B.C. Giessen, unpublished.

NEW Pb-BASED LAYERED CUPRATE SUPERCONDUCTORS

T. Maeda, H. Yamauchi, K. Sakuyama, S. Koriyama
and S. Tanaka
Superconductivity Research Laboratory
International Superconductivity Technology Center
10-13, Shinonome 1-chome, Koto-ku, Tokyo 135, JAPAN

ABSTRACT

Two kinds of superconducting Pb-based layered copper oxides, $(Pb,Cu)Sr_2(Y,Ca)Cu_2O_7$ (the "1212" phase) and $(Pb,Cu)(Ln,Ce)_2(Sr,Ln)_2Cu_2O_9$ (Ln: rare-earth element; the "1222" phase), are synthesized and characterized. Although most of previously-known Pb-based cuprate superconductors form in a mildly reducing atmosphere, these compounds form in an oxidizing atmosphere. Single-phase ceramics of the "1212" compounds are obtained in the compositional range of $(Pb_{(1+x)/2}Cu_{(1-x)/2})Sr_2(Y_{1-x}Ca_x)Cu_2O_z$ ($x = 0 \sim 0.35$ and $z = \sim 7$). The Pb-based "1222" compounds are the first "1222" superconductor as well as the first superconducting lead cuprate containing "fluorite-type" block layers. The maximum values of the observed superconducting onset temperatures are 65 K and 28 K for the "1212" phase and for the "1222" phase, respectively.

INTRODUCTION

Since the first superconducting lead cuprate, $Pb_2Sr_2(Y,Ca)Cu_3O_8$ of the "2213" phase, was discovered by Cava et al.,[1] there have been synthesized a variety of new lead cuprates having crystallographic structures similar to those of Bi- and Tl-based superconducting cuprates,[2-12] including lead cuprates reported to be non-super-conducting. Crystallographically, these lead cuprates characteristically contain a block layer consisting of both Pb and O together with Cu. Such layers correspond to those consisting of both Bi and O for the Bi-based cuprate superconductors or to those consisting of both Tl and O for the Tl-based cuprate superconductors.

Since most of those lead cuprates form in a mildly reducing atmosphere typically for the "2213" phase,[1] Pb ions are considered to be in the divalent state and Cu ions constituting the block layers together with Pb ions are considered to be in the monovalent state.   We recently reported[13-15] that both superconducting Pb-based layered copper oxides, $(Pb,Cu)Sr_2(Y,Ca)Cu_2O_z$ (the "1212" phase) and $(Pb,Cu)(Eu,Ce)_2(Sr,Eu)_2Cu_2O_z$ (the "1222" phase), formed in an oxidizing atmosphere in contrast to the "2213" phase.   In this paper, synthesis and characterization of the two kinds of recently discovered superconducting lead cuprates are reported.

EXPERIMENTAL PROCEDURES

Samples were prepared by a solid state reaction using high pure (> 99.9 %) powders of PbO, $SrCO_3$, Y_2O_3, $CaCO_3$, CeO_2 and CuO for the "1212" phase and PbO, $SrCO_3$, Ln_2O_3 (Ln = Sm, Eu or Gd), CeO_2 and CuO for the "1222" phase. These source powders were mixed into nominal compositions of $(Pb_{(1+x)/2}Cu_{(1-x)/2})Sr_2(Y_{1-x}Ca_x)Cu_2O_z$ (x = 0 ~ 0.5) or $(Pb_{0.5}Cu_{0.5})(Ln_{0.75}Ce_{0.25})_2(Sr_{0.875}Ln_{0.125})_2Cu_2O_z$.   The mixed powders were calcined in air at 850° C for 10 h and then pressed into parallelepiped bars of 3 x 3 x 20 mm^3. The bars were fired in air or O_2 gas flow at 1000 ~ 1120° C for 1 h being followed by rapid cooling to 900° C and further slow cooling to room temperature at a rate of 60 deg/h.   Some of the "1212" samples were annealed at 800° C in air for 1 h and then quenched into liquid nitrogen while some of the "1222" samples fired in air were heat-treated in O_2 atmosphere.   The temperature profile during the heat-treatment were shown elsewhere.[15]

The samples were analyzed by means of x-ray diffractometry (XRD) using CuK_α radiation reflected by a graphite monochromater.   Both electrical resistivity and dc-magnetic susceptibility were measured with respect to temperature employing respectively a conventional four probe method and a SQUID magnetometer.

RESULTS AND DISCUSSIONS

$(Pb,Cu)Sr_2(Y,Ca)Cu_2O_z$ (the "1212" phase)

Figure 1 shows a schematic representation of a crystalographic structure of the Pb-based "1212" phase, which has a tetragonal unit cell with the lattice constants, a = ~ 3.8 Å and c = ~ 11.8 Å.[16]   This structure is similar to but slightly different from that of $Pb_2Sr_2(Y,Ca)Cu_3O_z$[1] and also that of the well-known "123" superconductor, $YBa_2Cu_3O_z$. Powder XRD patterns for as-sintered samples with x = 0 ~ 0.4 are shown in Fig. 2. It was shown that, for x up to 0.35, single-phase samples were obtaind. The XRD pattern for a quenched sample was practically the same as those of corresponding as-sintered samples except for slight shifts of diffraction lines due to change in lattice constants induced by quenching.

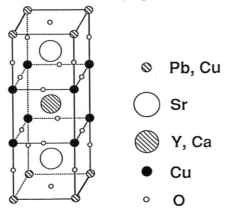

⊘ Pb, Cu

◯ Sr

⊘ Y, Ca

● Cu

○ O

Fig. 1. Schematic representation of the crystallographic structure of the Pb-based "1212" phase.[16]

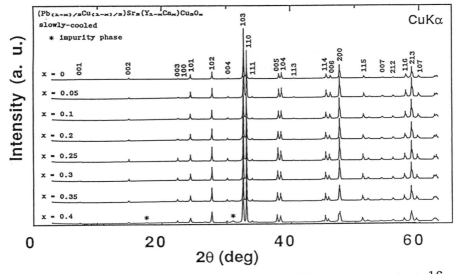

Fig. 2. XRD patterns for samples with x = 0 ~ 0.4.[16]

Figures 3(a) and 3(b) show the temperature dependence of electrical resistivity of as-sintered and quenched samples with x =0 ~ 0.4, respectively. For as-sintered samples, electrical resistivity tended to decrease as the Ca-content, x, increased, but none exhibited superconductivity. It was observed that the magnitude of electrical resistivity was lowered by quenching and that, for quenched samples, electrical resistivity also tended to decrease as x increased.

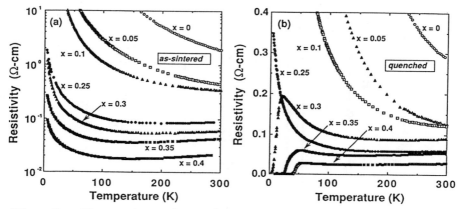

Fig. 3. Temprature dependence of electrical resistivity for (a) as-sintered and (b) quenched samples.[16]

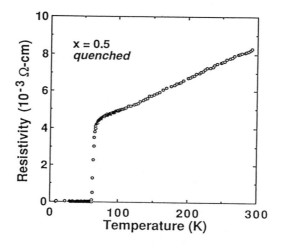

Fig. 4. Temperature dependence of electrical resistivity for quenched sample with x = 0.5.

Finally, superconductivity appeared only for quenched samples with x > 0.2.   The superconducting onset temperatures (T_C^{on}) for the quenched samples with x = 0.3, 0.35, 0.4 were 25 K, 45 K and 52 K, respectively.

Although sample with x = 0.5 contained a secondary phase, the highest T_C^{on} of 65 K was obtained for this sample when quenched, as shown in Fig. 4.   Figure 5 shows the temperature dependence of dc-magnetic susceptibility, which indicates that bulk superconductivity was observed for all of the superconducting samples.   The dependence of the lattice constants, a and c/3, on the Ca-content,

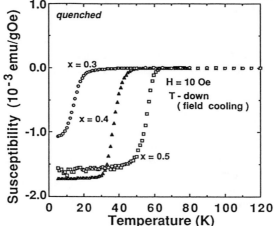

Fig. 5.   Temperature dependence of dc-magnetic susceptibility for samples with x = 0.3, 0.4 and 0.5.

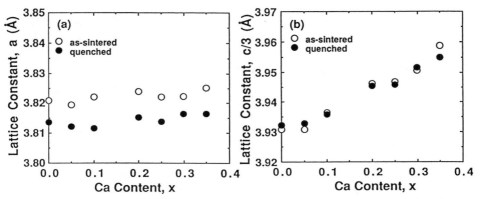

Fig. 6.   Dependence of the lattice constants, a and c/3, on the Ca-content, x.[16]

x, for both the as-sintered and the quenched samples are
shown in Fig. 6.   It was shown that the a-axis was nearly
independent of x for both as-sintered and quenched
samples, and that the a-axes were shrunken after quenching
from ~ 3.818 A to ~ 3.815 A.   In contrast, the c-axis
elongated as x increased and remained rather unchanged
after quenching.   Using a coulometric titration technique,[17,18] the oxygen content, z, was determined for as-
sintered (non-superconducting) samples and for quenched
(super-conducting) samples with x = 0.3 to be respectively
7.10(3) and 6.99(3).   The latter value was practically
equal to the stoichiometric value of 7.   It was strongly
suggested that the excess amount of oxygen contained in
the as-sintered samples were unpreferable for the appearance of superconductivity.   Similar observations were
reported for the case of the $La_2CaCu_2O_6$-related "326"
compounds.[19-21]

$(Pb,Cu)(Ln,Ce)_2(Sr,Ln)_2Cu_2O_z$ (the "1222" phase)

Compounds having the
"1222" structure were found
for the first time for the
Tl-Nd-Ce-Sr-Cu-O system,[22]
and later for the Pb-Pr-Sr-
Cu-O system.[11] However the
compounds were reported to be
non-superconducting.   We
successfully synthesized superconducting Pb-based "1222"
phase of the Pb-Eu-Ce-Sr-Cu-O
system,[15]   which was also
the first superconducting
lead cuprate containing
"fluorite-type" block layers
in its crystallographic
structure.   The structure of
$(Pb,Cu)(Eu,Ce)_2(Sr,Eu)_2Cu_2O_z$
is schematically represented
in Fig. 7 as drawn by Adachi
et al.[11]

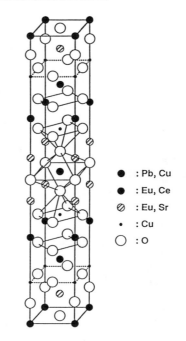

● : Pb, Cu

◗ : Eu, Ce

⊘ : Eu, Sr

· : Cu

○ : O

Fig. 7.   Schematic representation of crystallographic
          structure of the Pb-based "1222" phase.[15]

This structure characteristically contains (Pb,Cu)O-monolayers having "rocksalt-type" configuration instead of BiO- or TlO-bilayers observed respectively in Bi- or Tl-based "2222" phases which were recently reported.[23]

Figure 8 shows the XRD patterns for the Eu-, Sm- and Gd-containing samples with nominal compositions of $(Pb_{0.5}Cu_{0.5})(Ln_{0.75}Ce_{0.25})_2(Sr_{0.875}Ln_{0.125})_2Cu_2O_z$. It was indicated that all of the samples were of nearly single-phase, but in some samples, a small amount of secondary phase such as "1212" or "214" compounds was contained.

For the Pb-based "1222" phase containing Eu as Ln, we previously reported[15] that, although samples prepared by firing in air did not exhibit superconductivity, they became superconductive after being annealed in O_2 atmosphere. In the present study, superconducting samples containing Eu, Sm and Gd as Ln were prepared by firing in O_2 gas flow being followed by slow cooling. Figure 9 shows the temperature dependence of electrical resistivity for these samples together with that for a previously reported sample with Ln = Eu. Superconductivity transition at 25 ~ 28 K was observed for all the samples. However, the samples prepared in this study had rather

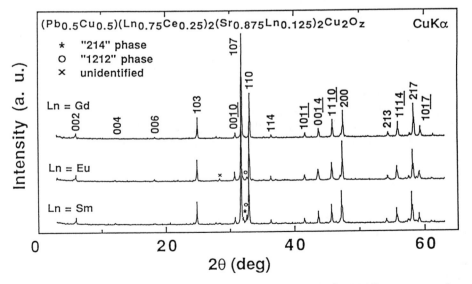

Fig. 8.   XRD patterns for the Pb-based "1222" compounds containing Eu, Sm and Gd.

high normal-state electrical resistivity.   Therefore, although the stoichiometric value of oxygen content, z, was
9, the Pb-based "1222" compounds tended to be in oxygen
deficient states in contrast to the case of the "1212"
compounds which contained excess amounts of oxygen.

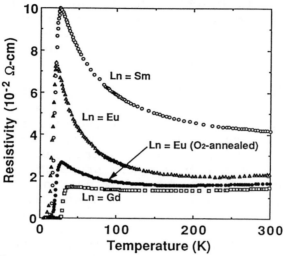

Fig. 9. Temperature dependence of electrical resistivity
for the Pb-based "1222" phase containing Eu, Sm
and Gd.

Summary

Two kinds of new superconducting lead cuprates,
$(Pb_{(1+x)/2}Cu_{(1-x)/2})Sr_2(Y_{1-x}Ca_x)Cu_2O_z$ (z = ~ 7, the
"1212" phase) and $(Pb_{0.5}Cu_{0.5})(Ln,Ce)_2(Sr,Ln)_2Cu_2O_z$ (z =
~ 9, the "1222" phase), were synthesized and characterized.   For the "1212" phase, single-phase ceramics
were obtained for x = 0 ~ 0.35 and superconductivity appeared in the samples containing appropriate amounts of Ca
after removing excess oxygen. T_C increased as x increased and the highest T_C^{on} was 65 K for multi-phase
samples with x = 0.5.   For the "1222" phase, superconducting samples were successfully synthesized with Ln =
Eu, Sm or Gd.   These samples exhibited superconductivity
transitions with T_C's around 25 K.   It was suggested that
the Pb-based "1222" phase was in an oxygen deficient state
in contrast to the Pb-based "1212" phase.

Acknowledgments

The authors would like to thank A. Ichinose of SRL-ISTEC for his collaboration in sythesizing "1222" samples and his helpful discussions. They also thank K. Kurusu of Furukawa Electric Co. Ltd. and T. Kawano and M. Kosuge of SRL-ISTEC for their kind help in coulometric titration.

REFERENCES

1. R. J. Cava, B. Batlogg, J. J. Krajewski, L. W. Rupp, L. F. Schneemeyer, T. Siegrist, R. B. van Dover, P. Marsh, W. F. Peck, Jr., P. K. Gallagher, S. H. Glarum, J. H. Marshall, R. C. Farrow, J. V. Waszczak, R. Hull and P. Trevor, Nature **336**, 211 (1988).

2. M. A. Subramanian, J. Goparakrishnan, C. C. Torardi, P. L. Gai, E. D. Boyes, T. R. askew, R. B. Flippen, W. E. Farneth and A. W. Sleight, Physica **C159**, 124 (1989).

3. O. Inoue, S. Adachi, Y. Takahashi, H. Hirano and S. Kawashima, Jpn. J. Appl. Phys. **28**, L60 (1989).

4. T. Rouillon, R. Retoux, D. Groult, C. Michel, M. Hervieu, J. Provost and B. Raveau, J. Solid State Chem. **78**, 322 (1989).

5. J. Y. Lee, J. S. Swinnea and H. Steinfink, J. Mater. Res. **4**, 763 (1989).

6. H. W. Zandbergen, W. T. Fu, J. M. van Ruitenbeek, L. J. de Jongh G. van Tendeloo and S. Amelinckx, Physica **C159**, 81 (1989).

7. T. Rouillon, J. Provost, M. Hervieu, D. Groult, C. Michel and B. Raveau, Physica **C159**, 201 (1989).

8. A. Tokiwa, T. Oku, M. Nagoshi, M. Kikuchi, K. Hiraga and Y. Syono, Physica **C161**, 459 (1989).

9. T. Rouillon, D. Groult, M. Hervieu, C. Michel and B. Raveau, Physica **C167**, 107 (1990).

10. R. J. Cava, P. Bordet, J. J. Capponi, C. Chaillout, J. Chenavas, T. Fournier, E. A. Hewat, J. L. Hodeau, J. P. Levy, M. Marezio, B. Batlogg and L. W. Rupp Jr., Physica **C167**, 67 (1990).

11. S. Adachi, O. Inoue, S. Kawashima, H. Adachi, Y. Ichikawa, K. Setsune and K. Wasa, Physica **C168**, 1 (1990).

12. S. Adachi, K. Setsune and K. Wasa, Jpn. J. Appl. Phys. 29, L890 (1990).
13. T. Maeda, K. Sakuyama, S. Koriyama and S. Tanaka, Advancec in Superconductivity II (Proc. 2nd Int. Symp. Superconductivity, November 14 - 17, Tsukuba, ed. T. Ishiguro and K. Kajimura, Springer Verlag, Tokyo, 1990), pp. 91 - 94.
14. S. Koriyama, K. Sakuyama, T. Maeda, H. Yamauchi and S. Tanaka, Physica C166, 413 (1990).
15. T. Maeda, K. Sakuyama, S. Koriyama, A. Ichinose, H. Yamauchi and S. Tanaka, Physica C169, 133 (1990).
16. T. Maeda, K. Sakuyama, S. Koriyama, H. Yamauchi and S. Tanaka, submitted to Phys. Rev. B.
17. K. Kurusu, H. Takami and K. Shintomi, Analyst 114, 1341 (1989).
18. M. Kosuge and K. Kurusu, Jpn. J. Appl. Phys. 28, L810 (1989).
19. J. B. Torrance, Y. Tokura, A. I. Nazzal and S. S. P. Parkin, Phys. Rev. Lett. 60, 542 (1988).
20. R. J. Cava, B. Battlog, R. B. van Dover, J. J. Krajewski, J. V. Waszczak, R. M. Fleming, W. F. Peck Jr., L. W. Rupp Jr., P. Marsh, A. C. W. P. James and L. F. Schneemeyer, Nature 345, 602 (1990).
21. M. Hiratani, T. Sowa, Y. Takeda and K. Miyauchi, Solid State Commun. 72, 541 (1989).
22. C. Martin, D. Bourgault, M. Hervieu, C. Michel, J. Provost and B. Raveau, Mod. Phys. Lett. B3, 993 (1989).
23. Y. Tokura, T. Arima, H. Takagi, S. Uchida, T. Ishigaki, H. Asano, R. Beyers, A. I. Nazzal, P. Lacorre and J. B. Torrance, Nature 342, 890 (1989).

TRANSPORT MEASUREMENTS OF $YBa_2(Cu_{0.98}M_{0.02})_3O_y$ (M=Fe,Co,Ni) SUPERCONDUCTING THIN FILMS

L.W. Song, E. Narumi, F. Yang, C.Y. Lee, H.M. Shao, M. Yang,
D.T. Shaw, and Y.H. Kao
New York State Institute on Superconductivity and
State University of New York at Buffalo, Amherst, NY 14260

ABSTRACT

Transport properties of $YBa_2(Cu_{0.98}M_{0.02})_3O_y$ (M = Fe, Co, Ni) superconducting thin films are studied and compared with the results measured in doped bulk materials. Critical current density as a function of magnetic field shows that Ni-doping has a larger effect on intragranular current while Fe-doping seems to have more influence on the intergranular coupling. All three doped samples show the same field-dependence on their volume pinning force F_p except in low field region. F_p can be scaled to a universal function $F_p/F_{p,max} = b(1-b)^2$ indicating that similar pinning mechanism is involved in all these samples. Finally, Hall effect measurements show that the effective carrier concentration is decreased in these doped films.

INTRODUCTION

Chemical doping in $YBa_2Cu_3O_y$ (YBCO) superconductors with 3d transition metals has long been a subject of considerable interest. In general, the substitution of these metals (e.g. Fe,Co,Ni) for Cu leads to a depression of T_c.[1,2] The mechanism for this change in T_c is still not well understood, and the effects on J_c is relatively unexplored. It has been known for bulk YBCO that Fe and Co doping induce an orthorhombic-to-tetragonal phase transition at ~ 3 at. % while an orthorhombic structure is retained with Ni doping.[1] The studies of site occupancies of these dopant elements have provided important information on the effects of chemical doping. It is believed that Fe and Co preferentially substitute into the Cu(1) chain sites[1,3] and superconductivity is depressed weakly by these substitutions. In Ni-doped material, however, both Cu(1) and Cu(2) substitutions have been suggested,[3,4] although other experimental evidence seems to favor substitution on Cu(2) sites only.[5] Another constructive study is to find out how the carrier concentration is affected by different dopants. It has been shown that the hole concentration decreases with increasing content of Fe, Co, and Ni.[6,8] Several studies have suggested that the holes are primarily located on the oxygen sites.[9,10] The changes of the hole concentration by metallic element doping may strongly affect the oxygen hole states, which is believed to be mainly responsible for high-T_c superconductivity. All the above-mentioned studies, however, are based on bulk superconducting materials. To the best of our knowledge, no study has been carried out in doped-YBCO thin films, this appears to be a challenging direction for the development of high-T_c superconductors.

In this paper, preliminary results of transport properties of Fe-, Co-, and Ni-doped YBCO superconducting thin films made by laser ablation technique are reported. For comparison, results of 2 at. % doping of Fe, Co, and Ni are presented. Transport critical current densities were studied as a function of temperature and magnetic field up to 5 Tesla. Pinning forces were extracted and compared for different dopants. The results can be interpreted by preferred-site substitution of the dopants. Carrier concentrations in these doped films were examined by Hall effect measurements and found to be consistent with those reported in doped bulk materials.

SAMPLE PREPARATION

Bulk specimens with Fe, Co, and Ni substitution were used as targets for thin film evaporation. Y$_2$O$_3$, BaCO$_3$, CuO powders were thoroughly mixed together with Fe$_2$O$_3$ (Co$_3$O$_4$ or NiO) powder in selected chemical proportion. The powder was then calcined at 900 ^{o}C for 10 hrs. It was ground and pressed into pellets (~80,000 psi) and sintered at 950 ^{o}C for over 24 hrs. in flowing oxygen. This procedure was repeated twice. The density of these pellets (> 75%) is comparable to those obtained from commercial source which was found to be essencial for preparing high-quality films.

The chemical composition and homogeneity of the targets were checked by SEM, EDX and x-ray diffraction (XRD). In all these targets, 123 phase remained as the dominant phase and no second phase could be detected within the resolution of our instruments. The superconducting transition was examined by standard four probe measurement and used as a reference to compare with the films made later from these targets.

Laser ablation technique has been successful in making high-quality superconducting films. Details of this method have been described elswhere.[11,12] The resulting films were checked by XRD for structural information and by SEM and EDX for chemical composition and homogeneity. Typical thickness of the films is about 200 nm. For transport critical current measurements, a microbridge of 10-20 μm wide and 100 μm long was inscribed by laser patterning.

EXPERIMENTAL RESULTS AND DISCUSSION

Some physical properties of the Fe-, Co-, and Ni-doped YBCO films are summarized in Table I. T_c was determined from R(T) vs. T measurement using a 1 μV onset voltage criterien for R=0. J_c(H,T) was measured from the I-V curve at 1 μV voltage drop across the microbridge sample. The Hall coefficient R_H(T) was determined by averaging the apparent Hall voltage in two opposite magnetic field directions. Hall number, defined as V_o/R_He where V_o is the unit cell volume, is also given in the Table which serves as an indication of the effective carrier concentration in the doped material.

From Table I we notice the change in T_c values seems to follow T_c(Co)

Table I. Values of transport superconducting properties in 2 at. % Fe-, Co-, Ni-doped and undoped YBCO thin films. Hall numbers (carrier/unit cell) are given both at maximum peak (M.P.) position and at room temperature(R.T.).

	Fe(0.02)	Co(0.02)	Ni(0.02)	Undoped
$T_c(R=0)$ (K)	78±3	80±2	76±3	85-90
ΔT_c (K)	~6	~4	~3	2-3
$J_c(0,5K)$ (A/cm^2)	$\geq 5 \times 10^6$	~5×10^6	$\geq 10^6$	~10^7
$J_c(0,60K)$ (A/cm^2)	~10^6	~5×10^5	~10^5	~10^6
Hall Number (M.P.)	~0.85	~0.78	~0.88	~1.05
Hall Number (R.T.)	~1.7	~1.9	~2.4	~4.0
H_{c2}^* (T=60K) (Tesla)	8.0	12.0	10.5	
$F_{p,max}$ (10^8 dyn/cm^3)	3.7	3.8	0.5	

$\geq T_c(Fe) \geq T_c(Ni)$, similar to those reported in Fe-, Co-, and Ni-doped bulk materials.[1,3] The change in zero field critical current density $J_c(0,T)$, however, show a different ordering, i.e. $J_c(0)_{Fe} \geq J_c(0)_{Co} \geq J_c(0)_{Ni}$. The low temperature (5K) values of $J_c(0)$ do not decrease much as compared to the undoped films. The temperature dependence of $J_c(0,T)$ was measured up to T_c. For $t \equiv T/T_c \leq 0.85$, J_c varies nearly linearly with temperature and can be described by a flux creep model[13] $J_c \sim 1-\alpha t-\beta t^2$ for $t<<1$ (ref. 14). In a region near T_c, $0.85 \leq t \leq 0.95$, J_c follows a power law dependence $J_c \sim (1-t)^n$ with n=1.5 or 2.0 as shown in Fig.1. This is in agreement with the Ginzburg-Landau description of bulk-like behavior[14] for n=1.5, and S-N-S weak link coupling[15] for n=2.0. In the very-near-T_c region, $t > 0.95$, $J_c - t$ dependence is n~1, as shown in Fig. 1. This behavior could only be interpreted as due to inhomogeneity of T_c's in the film which could give rise to a large error in determining the n value in narrow temperature region.[16] T_c inhomogeneity can also been seen from broadened transition width ΔT_c as compared to undoped films.

The field dependence of $J_c(H,T)$ may provide more insight into the effect of doping. In general, the transport critical current in magnetic field is controlled

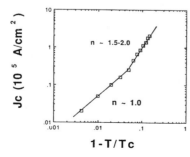

Fig.1   Typical $\ln J_c(0,T)$ vs.$\ln(1-T/T_c)$ of these doped films at near T_c ($t \geq 0.85$) region.

by two different mechanisms.[17]   An intergranular current which follows a weak-link behavior, therefore it can be described by a Josephson expression $J_c(H) = J_c(0)|\sin\pi x/\pi x|$ with x=H/H$_0$ where H$_0$ is the field corresponding to a flux quantum $\phi_0 = ch/2e$ in the weak-link area.[14] This field dependence leads

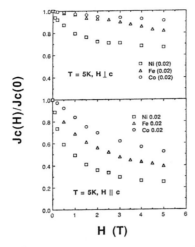

Fig.2   Normalized   critical   current density J_c(H)/J_c(0) vs. H at T=5K for Fe-, Co-, Ni-doped films.

to a 1/H behavior after averaging over different orientations and sizes of the weak link areas.[17,18] An intragranular effect, on the hand, is dominated by the superconducting properties inside the domains.   It has been found that a characteristic field H_c' can be defined to distinguish these two effects. H_c' can be extracted simply from a semilog plot of J_c vs. H at a crossover point where the slope changes.[17,19] In these doped YBCO films, H_c' is on the order of a few kOe at elavated temperatures (t $\geq$ 0.75).[20,21] Our measurements of J_c(H,T) in low temperatures seem to indicate that H_c' has shifted to a much lower field value, hence J_c(H,T) is mainly governed by the intragranular behavior of the sample. Fig. 2 shows the nomalized J_c(H,T) at T = 5K. It should be noted that J_c(H)/J_c(0) is depressed most in the Ni-doped films for both H ∥ c and H ⊥ c configurations, while Co-doped films seem to be less sensitive to the magnetic

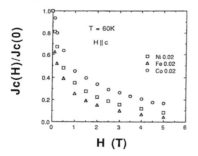

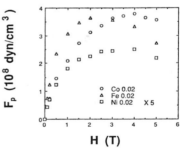

Fig.3 (a) J_c(H)/J_c(0) vs. H at T=60K   (b) F_p vs. H plot from the data (a).
for H∥c direction.

field.   This observation is consistent with the selective site occupancy of these impurities.   Fe and Co are believed to occupy the Cu(1) chain sites while Ni is believed to reside on the Cu(2) sites.   Therefore, Ni-doping leads to a large depression of superconductivity (both T_c and J_c).

At higher temperatures (e.g. 60K), the intergranular effect becomes more noticable.   The critical current density J_c(H,T) may contain more weak-link effects.   As shown in Fig. 3 (a), the normalized J_c for Fe-doped films decreases faster than the Ni-doped films in field at T=60K.   Two possibilities may be related to this reversed role of Fe-doped films at 60K.   One is that, besides the site preference, Fe impurities may somehow contribute to weak link coupling. It has been reported that Fe dopants can change the superconducting domain size.[22] Therefore it will alter the roles played by both the twin and grain boundaries in flux pinning. Another possible explanation, as has been reported,[22] is that the effective upper critical field H_{c2}^* of Fe-doped sample decreases with increaseing Fe concentration. The volume pinning force $F_p = |J_c \times H|$ is plotted against H in Fig. 3 (b) which indicates that there is a maxium at H < 5T for all three doped samples but the maximum occurs at a lower field value for Fe-doped than the Ni-doped films. It is suggestive to compare this field dependence with $F_p \propto b(1-b)^2$ where $b = H/H_{c2}^*$ (ref. 23). This F_p maximum in magnetic field can then be used to estimate H_{c2}^*. Values of H_{c2}^* and $F_{p,max}$ are included in Table I. It has already been suggested that a decrease of H_{c2}^* in Fe-doped material is caused by a strong reduction of the electron density of states at the Fermi surface.[22]

The pinning force F_p for the Fe-, Co-, and Ni-doped films can be scaled to follow the same field dependence. In Fig. 4 we plot the reduced pinning force $F_p/F_{p,max}$ vs. reduced magnetic induction b for three dopants. The fit is reasonably good at higher field (H ≥ 2T) which is known to be the region dominated by the intragranular effects as discussed earlier. In lower field region, intergranular current mixed in so that the fit deviates from such a scaling behavior. This is consistent with the results reported by Wordenweber *et al.*[22] in different concentration of Fe-doped YBCO bulk materials in which J_c's were determined from magnetization measurements. Detailed magnetization measurements of our doped films are necessary to support this argument. Our results suggest that, at least within a certain field and temperature region, the

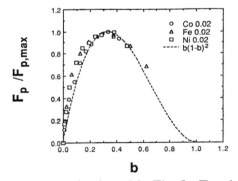

Fig.4 $F_p/F_{p,max}$ vs. $b = H/H_{c2}^*$ for all three doped films. Dashed line is a fit of $b(1-b)^2$.

pinning mechanism in 2 at.% Fe-, Co-, and Ni-doped YBCO thin films is about the same.

We have also measured Hall coefficients in these doped films in a field up to 1 Tesla. Typical Hall coefficient R_H as a function of temperature is plotted in Fig. 5. Two interesting features are observed: (i) R_H rises rapidly with increasing temperature at T ≥ T_c (R=0) and shows a sharp peak within a narrow ΔT_c (8~10K) region. This is different from those observed in undoped YBCO films[24] as well as in sigle crystals[25] and bulk superconductors[26] where R_H rises slowly and reaches a broad maximum at ~100K and then decreases slowly and follows an approximate 1/T dependence up to room temperature. (ii) Normal state R_H becomes less temperature-dependence after the sudden drop from transition region. This behavior is consistent with the interpretation that impurity scattering is largely increased in the normal state due

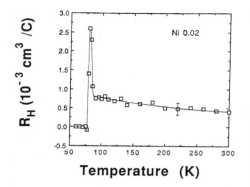

Fig.5   Hall coefficient R$_H$ as a function of temperature of a Ni-doped film. All three doped films have a very similar temperature dependence as shown in this figure.

to the presence of dopants. This was also observed in doped-YBCO bulk materials.[6-8]   The effective carrier concentration or Hall number are given in Table I for room temperature (R.T.) and that at the maximum peak (M.P.) position. By a comparison with undoped films, the effective carrier concentration is decreased near transition for all Fe-, Co-, and Ni-doped films. The overall feature of the Hall coefficient can be qualitatively explained in terms of a two band model.[27] The unusual behavior of R$_H$ in the narrow transition region, however, is not quantitatively understood at the present time.

CONCLUSION

We have made the first attempt to study the transport properties of Fe-, Co-, and Ni-doped YBCO thin films. Nearly epitaxially grown doped films offer a unique advantage to study the anisotropy of pinning mechanisms and microstructures over the doped bulk materials. From this study it is found that the microstructural defects introduced by dopants are more effective in thin films than in bulk materials, which is probabely due to the morphology of the thin films. Critical current density as a function of magnetic field shows that Ni-doping has a larger effect on intragranular current while Fe-doping seems to have more influrence on the intergranular coupling. The effective upper critical field H$_{c2}^*$ in Fe-doped films is lower than that of Co- and Ni-doped films. All three doped samples show the same field-dependnece on their volume pinning force F$_p$ except in low field region which was caused by a mixture of both intergranular and intragranular effects. F$_p$ can be scaled to a universal function F$_p$/F$_{p,max}$ = b(1-b)^2 indicating that similar pinning mechanism is involved in all these samples. Finally, Hall effect measurements show that the effective carrier concentration is decreased in these doped films.

We would like to thank J.P. Zheng for microbridge patterning on the films. This work was supported in part by DOE under grant number DE-FG02-87ER45283 and by AFOSR under grant number AFOSR-88-0095.

REFERENCES

1.   J.M. Tarascon et al., Phys. Rev. **B36**, 8393(1987); **B37**, 7458(1988), and

references therein.

2. Y. Maeno *et al.*, Nature **328**, 512(1987); Jpn. J. Appl. Phys. **26**, L1982(1987).
3. C.Y. Yang *et al.*, Phys. Rev. **B42**, 2231(1990), and references therein.
4. R.S. Howland *et al.*, Phys. Rev. **B39**, 9017(1989).
5. T. Kajtiani *et al.*, Jpn. J. Appl. Phys. **27**, L354(1988).
6. Y. Tokura *et al.*, Phys. Rev. **B38**, 7156(1988), and references therein.
7. M.W. Shafer *et al.*, Phys. Rev. **B39**, 2914(1989).
8. J. Clayhold, N.P. Ong, Z.Z. Wang, J.M. Tarascon, and P. Barboux, Phys. Rev. **B39**, 7324(1989).
9. N. Nueker *et al.*, Phys. Rev. **B37**, 5185(1987); **B39**, 6619(1989).
10. V.J. Emery, Phys. Rev. Lett. **58**, 2794(1987).
11. H.S. Kwok, P. Mattocks, L. Shi, X.W. Wang, S. Witanachchi, Q.Y. Ying, J.P. Zheng, and D.T. Shaw, Appl. Phys. Lett. **52**, 1825(1988).
12. E. Narumi, L.W. Song, F. Yang, S. Patel, Y.H. Kao, and D.T. Shaw, Appl. Phys. Lett. **56**, 2684(1990).
13. P.W. Anderson, Phys. Rev. Lett. **9**, 309(1962); P.W. Anderson and Y.B. Kim, Rev. Mod. Phys. **36**, 39(1964).
14. See, for example, M. Tinkham, *Introduction to Superconductivity* (McGraw Hill, New York, 1975).
15. P.G. de Gennes, *Superconductivity of Metals and Alloys* (Benjamin, New York, 1966).
16. J.W.C. de Vries, M.A.M. Gijs, G.M. Stollman, T.S. Baller, and G.N.A. van Veen, J. Appl. Phys. **64**, 426(1988).
17. L.W. Song and Y.H. Kao, Physica C, **169**, 107(1990).
18. R.L. Peterson and J.W. Ekin, Phys. Rev. **B37**, 9848(1988).
19. Y.H. Kao, Y.D. Yao, L.Y. Jang, F. Xu, A. Krol, L.W. Song, C.J. Sher, A. Darovsky, J.C. Phillips, J. Simmins, and R.L. Snyder, J. Appl. Phys. **67**, 353(1990).
20. L.W. Song, Y.H. Kao, Q.Y. Ying, J.P. Zheng, H.S. Kwok, Y.Z. Zhu, and D.T. Shaw, Mater. Res. Soc. Symp. **169**, (1990), in press.
21. L.W. Song, E. Narumi, F. Yang, D.T. Shaw, H.M. Shao, and Y.H. Kao, J. Appl. Phys. in press.
22. R. Wordenweber, G.V.S. Sastry, K. Heineman, and H.C. Freyhardt, J. Appl. Phys. **65**, 1648(1989).
23. J. van den Berg, C.J. van der Beek, P.H. Kes, J.A. Mydosh, A.A. Menovsky, and M.J.V. Menken, Physica C, **153-155**, 1465(1988).
24. Y. Iye, S. Nakamura, and T. Tamegai, Physica C, **159**, 616(1989).
25. See, for example, L. Forro, M. Raki, J.Y. Henry, and C. Ayache, Solid State Comm. **69**, 1097(1989).
26. See, for example, reference 6.
27. D.Y. Xing and C.S. Ding, preprint; D.Y. Xing, M. Liu, and C.S. Ding, Phys. Rev. **B37**, (1988).

ANGULAR DEPENDENCE OF THE TRANSPORT CRITICAL CURRENT DENSITY OF CVD YBaCuO FILMS

K. Watanabe, S. Awaji, N. Kobayashi,
H. Yamane, T. Hirai, and Y. Muto
Institute for Materials Research, Tohoku University
Katahira 2-1-1, Aoba-ku, Sendai 980, Japan.

H. Kawabe
TATSUTA Electric Wire and Cable Co., Osaka 578, Japan

H. Kurosawa
RIKEN Co., Kumagaya 360, Japan

ABSTRUCTS

$Y_1Ba_2Cu_3O_{7-\delta}$ films prepared by chemical vapor deposition indicate very strong anisotropic behavior of the upper critical field and the critical current density. It is found that the angular dependence of the critical current density predominantly comes from the anisotropy of the upper critical field.

INTRODUCTION

High-T_c Superconducting oxide $Y_1Ba_2Cu_3O_{7-\delta}$ films prepared by chemical vapor deposition (CVD) indicate excellent J_c properties in high magnetic field, and their J_c values at 77.3 K are much superior to those for conventional advanced superconductors at 4.2 K. Table 1 summarizes the representative J_c values obtained up to date for CVD processed YBaCuO films. All films were deposited onto single crystalline $SrTiO_3(100)$ substrates. We obtained the J_c value more than 6×10^4 A/cm2 at 77.3 K in high fields up to 27 T for our YBaCuO film[1], when the magnetic fields were applied perpendicular to the c-axis.

Recently, the J_c characters in high fields for Bi-oxide superconductors were also outstandingly improved. A serious problem of the flux lattice melting for the Bi-oxide system has been reported[2], so that the usage for the Bi-oxide system at temperature above 20-30 K has been

Table 1. Representative resistively measured J_c values in high field 77.3 K for $Y_1Ba_2Cu_3O_{7-\delta}$ processed by chemical vapor deposition.

$T_c(R=0)[K]$	$J_c^R[A/cm^2]$	
92	2.0×10^6 at B = 0 T = 77.3 K 6.5×10^4 at B = 27 T T = 77.3 K	H. Yamane et al. (Tohoku Univ.)
92	3.3×10^6 at B = 0 T = 77.3 K 3×10^4 at B = 30 T T = 77.3 K	S. Matsuno et al. (Mitsubishi)
94	8.5×10^5 at B = 0 T = 77.3 K 1×10^5 at B = 5.5 T T = 77.3 K	F. Schmaderer et al. (ABB)
92	2.0×10^5 at B = 0 T = 77.3 K 2.0×10^4 at B = 8 T T = 77.3 K	S. Aoki et al. (Fujikura)

considered to be difficult in spite of the higher T_c than the Y-oxide system. In fact, the J_c properties at 77.3 K in magnetic fields for the Bi-oxide system were very poor and the application of the very small magnetic field reduced abruptly the J_c value at zero field[3]. However, a BiPbSrCaCuO (2223) thin film prepared by sputtering achieved the high J_c value of 1 x 10^5 A/cm^2 at 1 T and 77.3 K, whose T_c was 105 K in a zero resistance criterion[4]. The J_c characters in magnetic fields at 77.3 K for the Y- and Bi-oxide films, as mentioned above, lead us to the existence of a strong pinning mechanism in high-T_c superconducting oxides. The introduction of the strong pinning enables us to solve the problems such as flux lattice melting. Table 2 lists the flux pinning candidates in high-T_c superconducting oxides reported so far. Intrinsic pinning[5] related to the dimensionality

Table 2.   Flux   pinnning   mechanism   in   oxide
          superconductors.

I . intrinsic pinning
II . extrinsic pinning
   1. twin boundary
   2. micro crack
   3. point defect
   4. lattice defect, dislocation
   5. precipitate
      (i)   CuO precipitate
      (ii) Cu-rich phase precipitate
      (iii) 211 phase precipitate

of the superconducting oxides and the very short coherence length, and extrinsic pinning such as twin planes, defects and precipitates have been considered as strong pinning mechanisms.   In high-T_c superconducting oxides, effective extrinsic pinning centers need to be finely dispersed because of the very short coherence length. It is necessary from physical and technical viewpoints to explore the riddle on the flux pinning mechanism.

Moreover, there exist many problems to be solved on the critical current density in high-T_c superconducting oxides.   Weak-links of grain boundaries[6], flux creep[7], anisotropy[8] and so on are pointed out as important problems.   In particular, the anisotropic behavior may become an important subject to be overcome for application.

This study reports on the flux pinning mechanism using high-J_c CVD-YBaCuO films.   Great interests are made for the angular dependences of both B_{c2} and J_c.   The caliculation of J_c using the flux pinning scaling law and the anisotropic Gizburg-Landau theory is intended.

EXPERIMENTAL

YBaCuO films have been made onto single crystalline $SrTiO_3(100)$ substrates by chemical vapor deposition.   The quality of the CVD films has been studied by X-ray diffraction (XRD), scanning electron microscope (SEM), electron probe micro-analyzer (EPMA) Auger electron spectroscope (AES), energy dispersive X-ray analyzer (EDX) and transmission electron microscope (TEM). Single crystalline growth with a preferred c-axis orientation perpendicular to substrates has been achieved. Many CuO precipitates with sizes of a few microns on the film surface which grew from the substrate surface were observed[10]. Moreover, many fine disk-shaped precipitates with typical dimensions of 250 Å in diameter and 10 Å thick among the ab plane were also observed[11].

The superconducting properties such as B_{c2} and J_c were measured resistively using patterned films with a narrow bridge (1.0 mm x 0.3 mm x 1.0 μm).

RESULTS AND DISCUSSION

The upper critical fields at 77.3 K determined by zero resistance for CVD-YBaCuO films were obtained to be 60 T for B $\perp$ c-axis and 12 T for B // c-axis as a typical example. Figure 1 shows the angular dependence of $-dB_{c2}$/dT. The upper part uses the definition of transition midpoint and the lower one the definition of zero resistance. The solid lines show the curves calculated from the Ginzburg-Landau effective mass model $dB_{c2}(\theta)/dT = dB_{c2}(0°)/dT(\varepsilon^2 \sin^2\theta + \cos^2\theta)^{-1/2}$. An interesting point is that the data points obtained by transition midpoints can be described well by the GL theory, but the data by zero resistance definition indicate a little bit scatter and a sharper anisotropy. We note that the zero resistance definition of B_{c2} is meaningful from the standpoint of J_c but the midpoint definition is more adequate to understand B_{c2}.

Figure 2 shows the angular dependence of J_c at 3.1 T and 77.3 K. We expect that the angular dependence of J_c should be related to the anisotropy of B_{c2}. We can estimate the $J_c(\theta)$ value assuming the flux pinning scaling law $F_p \propto [B_{c2}(\theta)]^m (B/B_{c2})^p (1-B/B_{c2})^q$. The solid line in the figure is derived from introducing the GL curve with $dB_{c2}(0°)/dT = -0.62$ T/K and $1/\varepsilon = 7.58$, which are obtained for the B_{c2} values defined by zero

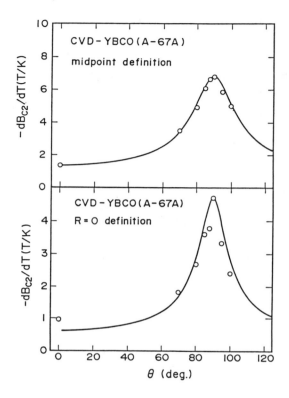

Fig. 1.   Slope of $B_{c2}(T)$ near T_c as a function of the angle θ between the magnetic field and the c-axis. The upper part indicates the data points by a midpoint criterion and the GL curve with $-dB_{c2}(0°)/dT = 1.3$ T/K and $1/\varepsilon = 5.1$. The lower part indicates the data points by a zero-resistance criterion and the GL curve with $-dB_{c2}(0°)/dT = 0.62$ T/K and $1/\varepsilon = 7.6$.

resistance. The dashed line defined by a midpoint of the resistive transition is also indicated for comparison. It is found that the attempt using the flux pinning scaling law combined with the anisotropic GL theory of $B_{c2}(\theta)$ is successful to explain the angular dependence of J_c. Therefore, the angular dependence of J_c in CVD-YBaCuO films mainly comes from the angular dependent $B_{c2}(\theta)$.

In conventional superconductors such as NbTi and Nb$_3$Sn,

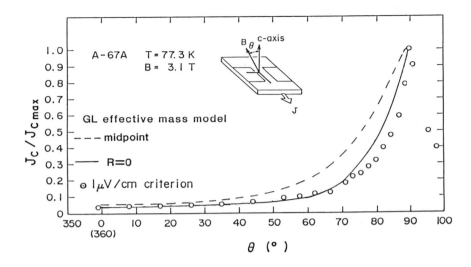

Fig. 2   Angular dependent J_c at 3.1 T and 77.3 K by a
1 μV/cm criterion. Both solid and dashed
curves are obtained by assuming the flux
pinning scaling law with scaling parameters of
m = 2.0, p = 0.5 and q = 2.0. The solid curve
uses the angular dependent B_{c2} derived from a
zero-resistance criterion, and the dashed one
from a midpoint criterion.

the value of B_{c2}^*, which is determined by the definition
of J_c=0, is almost close to that of B_{c2} obtained by zero
resistance. However, B_{c2}^* deviates from B_{c2} when the flux
pinning becomes weak. In particular, the value of B_{c2}^*
in high-T_c superconducting oxides seems to be smaller
than that of B_{c2}[12]. The flux creep phenomenon will
affect the J_c measurement at 77.3 K in the
superconducting oxides. This result suggests that we
should adopt the appropriate B_{c2}^* value in flux pinning
scaling law. Although the angular dependence of B_{c2}^* may
give disagreement with GL theory, the angular dependence
of J_c will be fitted more precisely by use of B_{c2}^* rather
than B_{c2}.

ACKNOWLEDGMENT

This work is supported by Grant-in-Aid for Scientific

Research on Priority Area from the Ministry of Education, Science and Culture, Japan.

REFERENCES

1. K. Watanabe, H. Yamane, H. Kurosawa, T. Hirai, N. Kobayashi, H. Iwasaki, K. Noto, and Y. Muto, Appl. Phys. Lett. 54, 575 (1989).
2. P. L. Gammel, L. F. Schneemeyer, J. V. Waszczak, and D. J. Bishop, Phys. Rev. Lett. 61, 1666 (1988).
3. K. Sato, T. Hikata, H. Mukai, T. Masuda, M. Ueyama, H. Hitotsuyanagi, T. Mitsui, and M. Kawashima, Advances in Superconductivity II, eds. T. Ishiguro and K. Kajimura (Springer-Verlag Tokyo, 1990) p335.
4. K. Hayashi, S. Okuda, S. Takano, O. Mizuguchi, S. Terai, and K. Hasegawa, 43rd Domestic Meeting on Cryogenics and Superconductivity (Yokohama, 1990) p82 (in Japanese).
5. M. Tachiki and S. Takahashi, Solid State Commun. 70, 291 (1989).
6. K. Watanabe, K. Noto, H. Morita, H. Fujimori, K. Mizuno, T. Aomine, B. Ni, T. Matsushita, K. Yamafuji, and Y. Muto, Cryogenics 29, 263 (1989).
7. Y. Yeshurun and A. P. Malozemoff, Phys. Rev. Lett. 60, 2202 (1988).
8. Y.Muto, K. Watanabe, N. Kobayahsi, H. Kawabe, H. Yamane, H. Kurosawa, and T. Hirai, Physica C 162-164, 105 (1989).
9. K. Watanabe, S. Awaji, N. Kobayashi, H. Yamane, T. Hirai, and Y. Muto, submitted to J. Appl. Phys.
10. H. Yamane, T. Hirai, H. Kurosawa, A. Suhara, K. Watanabe, N. Kobayashi, H. Iwasaki, E. Aoyagi, K. Hiraga, and Y. Muto, Advances in Superconductivity II, eds. T. Ishiguto and K. Kajimura (Springe-Verlag Tokyo, 1990) p767
11. K. Watanabe, T. Matsushita, N. Kobayashi, H. Kawabe, E. Aoyagi, K. Hiraga, H. Yamane, H. Kurosawa, T. Hirai, and Y. Muto, Appl Phys. Lett. 56, 1490 (1990).
12. K. Watanabe, N. Kobayashi, H. Yamane, T. Hirai, Y. Muto, H. Kawabe, and H. Kurosawa, Proc. International Cryogenic Materials Conference (ICMC' 90) (Garmisch-Partenkirchen, 1990).

LOW TEMPERATURE IN SITU GROWTH OF YBCO THIN FILMS ON SEMICONDUCTOR SUBSTRATE FROM SINGLE TARGET RF MAGNETRON SPUTTERING

Q.X.Jia and W. A. Anderson, State University of New York at Buffalo
Department of Electrical and Computer Engineering,
Institute on Superconductivity, Bonner Hall, Amherst, NY 14260

ABSTRACT

Superconducting YBCO thin films with a zero resistance temperature of around 80K and very sharp transition were obtained on Si substrates by RF magnetron sputtering from a stoichiometric oxide target. Metallic ruthenium oxide was used as a buffer layer to reduce the interdiffusion between YBCO and Si and to nucleate YBCO. The stoichiometry of the YBCO films was well controlled by arranging the substrates perpendicular with respect to the target. The introduction of oxygen into the sputtering chamber immediately after YBCO film deposition was very important to obtain such superconducting films as indicated by the in situ film resistance measurement.

INTRODUCTION

As commonly found before, the lack of success in depositing superconducting YBCO films on semiconductor substrates, such as Si, is mainly due to the interdiffusion between Si and YBCO. The different thermal expansion coefficient and lattice constant of these two materials make it much more difficult to grow YBCO directly on Si. Insulator materials, such as SiO_2 [1], MgO [2], ZrO_2 [3], and $SrTiO_3/MgAl_2O_4$ or $BaTiO_3/MgAl_2O_4$ [4] were accepted as the buffer layers to lessen the interdiffusion problem. We report recent success in depositing superconducting YBCO films on Si substrates totally in situ by RF magnetron sputtering, where a metallic oxide, RuO_2, sputtered by reactive DC magnetron, was used as a buffer layer to nucleate the superconducting film and to minimizing the reactions between Si and YBCO films.

EXPERIMENTAL DETAILS

Single crystal p-type Si wafers with a resistivity in the range of 1-4 Ω-cm and (100) orientation were used as the substrates. The substrate was subjected to ultrasonic cleaning using sequential chemical solutions of trichlorethylene, acetone, and methanol. The wafers were well rinsed in deionized (DI) water, and then etched by buffer HF. The Si substrates were immediately loaded into the sputtering system for buffer layer RuO_2 deposition after DI water rinsing and N_2 blowing dry of the wafers.

Metallic oxide RuO_2 was deposited using DC magnetron sputtering. Table 1 gives the parameters used for RuO_2 film deposition. The RuO_2 showed a metallic property of resistivity vs temperature in the range of 30-300K. Detailed relationships between electrical characteristics of RuO_2 and sputtering conditions can be found eleswhere [5].

Table 1.   DC and RF Magnetron Sputtering for Depositing Metallic Oxide RuO_2 and Superconducting YBCO Thin Films

Parameter	Ru Film	$YBa_2Cu_3O_{7-x}$ Film
Oxygen pressure	0.5mTorr	0.5mTorr
Total gas pressure $(Ar + O_2)$	10mTorr	30mTorr
Substrate temperature	500°C	500°C
DC voltage	450V	---
DC current	0.2A	---
RF Power	---	30W
Sputtering rate	60nm/min	3nm/min
Film thickness	200nm	600-900nm

Superconducting YBCO thin films were deposited by RF magnetron sputtering from a stoichiometric target. Highly controllable and reproducible results were obtained by arranging the substrate with respect to the target during sputtering [6-8]. Table 1 also gives the sputtering conditions used for depositing YBCO films. Oxygen was introduced into the sputtering chamber immediately after YBCO film deposition. The film was in situ annealed at 500°C in oxygen (2Torr) for 30min. The film resistance was in situ monitored as previously described [9].

The film quality was characterized by four probe resistance vs temperature (R-T) measurement to determine the zero resistance temperature of the YBCO films. The critical current was determined at a voltage drop across the film of $1\,\mu V/cm$. The morphology of the film was specified by scanning electron microscopy (SEM). The interfaces among three layers of YBCO/RuO$_2$/Si were characterized by Auger electron spectroscopy (AES) depth profiling.

RESULTS AND DISCUSSION

Four probe R-T measurement showed that the effective resistivity of the combination of YBCO/RuO$_2$/Si was relatively lower compared to that of YBCO on MgO/Si [2] and ZrO$_2$/Si [10]. Figure 1 shows a R-T curve of YBCO on RuO$_2$/Si, where the YBCO was in situ deposited without post-high temperature annealing. The typical value of the resistivity of YBCO on RuO$_2$/Si at room temperature was around $100\ \mu\Omega$-cm. A value of about 0.4 for the extrapolation of the slope of R/R_0 - T at T = 0K demonstrated the good metallic property of the film above the transition temperature, where the R and R_0 are the resistance at temperature T and 300K, respectively. In situ film resistance measurement during cooling down coincided well with the R-T curve measured below 300K [9]. The best film gave a zero resistance temperature of 80K and a critical current density greater than $5x10^3A/cm^2$ at 60K and zero field.

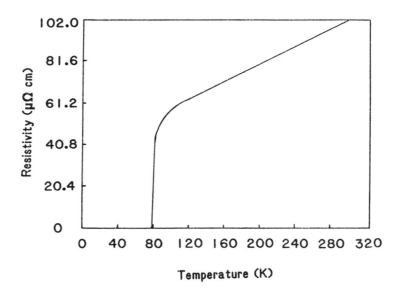

Fig. 1   R-T curve of a superconducting YBCO film deposited
on Si with RuO_2 as a buffer layer.

SEM surface survey showed the film to have smooth surface morphology [9]. No cracks could be observed through this investigation. A SEM micrograph also demonstrated the multiple stack nature of the film growth. The relatively low critical current density could be due to incomplete c-axis orientation of the films as revealed by x-ray diffraction analysis.

Experimental results have demonstrated RuO_2 to be a good barrier layer between Al and Si [11,12] used for very large scale integrated circuits. AES depth profiling on $YBCO/RuO_2/Si$ showed negligible interdiffusion between three layers. Figure 2 shows a AES depth profile of YBCO film on RuO_2/Si substrate, where the YBCO film was deposited at a substrate temperature of 500°C. The good thermal stability of RuO_2 at relatively high temperature also makes it very attractive to be used in superconductor technology.

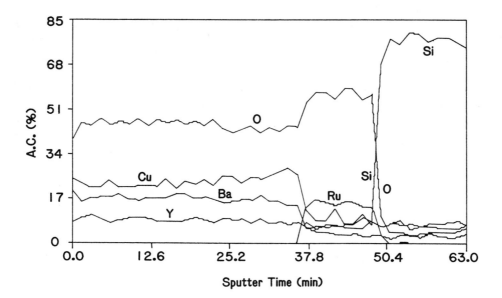

Fig. 2   AES depth profile of a superconducting YBCO film on a RuO2/Si substrate.

CONCLUSION

Superconducting YBCO thin films were deposited on Si at relatively low substrate temperature using RuO2 as a buffer layer. AES depth profiling showed negligible interdiffusion among three layers. RuO2 may find its wide applications in superconductor technology due to its high thermal stability and very high conductivity.

ACKNOWLEDGMENT

We gratefully acknowledge the assistance from P.Bush and R.Barone in material analysis. This work was supported by the New York Energy Research and Development Authority through the New York State Institute on Superconductivity.

REFERENCES

1. A. Mogro-Campero, B.D. Hunt, L.G. Turner, M.C. Burrell and W.E. Balz, Appl. Phys. Lett. 52, 584(1988).
2. S. Witanachchi, S. Patel, H.S. Kwok and D.T. Shaw, Appl. Phys. Lett. 54, 578(1989).
3. T. Venkatesan, E.W. Chase, X.D. Wu, A. Inam, C.C. Chang and F.K. Shokoohi, Appl. Phys. Lett. 53, 243(1988).
4. S. Miura, T. Yoshitake, S. Matsubara, Y. Miyasaka, N. Shohata and T. Satoh, Appl. Phys. Lett. 53, 1967(1988).
5. Q.X. Jia, Z.Q. Shi, K.L. Jiao, W.A. Anderson and F.M. Collins, Thin Solid Films, in press.
6. N. Terada, H. Ihara, M. Jo, M. Hirabayashi, Y. Kimura, K. Matsutani, K. Hirata, E. Ohno, R. Sugise and F. Kawashima, Jpn. J. Appl. Phys. 27, L639(1988).
7. R.L. Sanstrom, W.J. Gallagher, T.R. Dinger, R.H. Koch, R.B. Laibowitz, A.W. Kleinsasser, R.J. Gambino, B. Bumble and M.F. Chisholm, Appl. Phys. Lett. 53, 444(1988).
8. Q.X. Jia, D. Ren and W.A. Anderson, Int. J. Modern Phys. B3, 743(1989).
9. Q.X. Jia and W.A. Anderson, Appl. Phys. Lett. 57, 304(1990).
10. M. Migliuolo, A.K. Stamper, D.W. Greve and T.E. Schlesinger, Appl. Phys. Lett. 54, 859(1981).
11. E. Kolawa, F.C.T. So, E.T-S. Pan and M-A. Nicolet, Appl. Phys. Lett. 50, 854(1987).
12. L. Krusin-Elbaum, M. Wittmer and D.S. Yee, Appl. Phys. Lett. 50, 1879(1987).

USE OF 96Sn-4Ag ALLOY AS THE BINDER FOR
CARBON FIBER REINFORCED SUPERCONDUCTOR

C.T. Ho and D.D.L. Chung
Composite Materials Research Laboratory, Furnas Hall
State University of New York at Buffalo, Buffalo, NY 14260

U. Balachandran
Materials & Components Technology Division
Argonne National Laboratory, 9700 South Cass Avenue
Argonne, IL 60439

ABSTRACT

96Sn-4Ag was superior to Sn or 63Sn-37Pb as the binder for continuous carbon fiber reinforced superconductor. It allowed diffusion bonding to $YBa_2Cu_3O_{7-\delta}$ superconductor to take place at $85^\circ C$. By using 96Sn-4Ag as the binder and $YBa_2Cu_3O_{7-\delta}$ powder prepared by vacuum calcination, carbon fiber reinforced superconductor of J_c around 400 A/cm² at 77 K was obtained. The composite processing had negligible effect on the J_c of the superconductor.

INTRODUCTION

The high T_c superconductors are mechanically weak because (i) they are ceramics, (ii) they need to have large grains in order to enhance the critical current density J_c, and (iii) they have inherent microcracks resulting from the phase transformation stresses. The mechanical weakness is particularly severe in tension, as is the case for brittle materials in general. This weakness poses problems for practical applications.

Carbon fibers are well suited for the reinforcement of the high T_c superconductors because of their high strength (particularly in tension), nearly zero thermal expansion coefficient, high thermal conductivity, low electrical resistivity and chemical stability. It was recently reported that continuous carbon fiber reinforced $YBa_2Cu_3O_{7-\delta}$ with tin as the binder and 30 vol.% fibers had a tensile strength of 130 MPa at 300 K and 140 MPa at 77 K, and a tensile modulus of 15 GPa at 300 K and 27 GPa at 77 K [1]. However a J_c value of only 144 A/cm² at 40 K, as determined by transport measurements, was achieved because of the low J_c value (namely 147 A/cm² at 40 K) for the superconductor used. The objective of this work is to extend the carbon fiber reinforced superconductor technology to higher J_c values, i.e. around 400 A/cm² at 77 K. We found that the achievement of this objective required not only the use of a superconductor material of such a high J_c value, but also the use of 96Sn-4Ag alloy instead of pure Sn.

EXPERIMENTAL

The superconductor $YBa_2Cu_3O_{7-\delta}$ (123) powder was prepared in Argonne National Laboratory by vacuum calcination at $800^\circ C$ for 4 h, in flowing oxygen at a pressure of 2-4 mm Hg. Sintering of the powder

410

compact was carried out at 880°C for 12 h in an atmosphere containing 1% O_2 and 99% Ar. After that, annealing was performed at 450°C for 6 h in 100% O_2.

The tin and tin alloys used were in the form of foils. The tin used contained 99.8 wt.% Sn. The tin alloys used were 63Sn-37Pb and 96Sn-4Ag.

The carbon fibers used were continuous, PAN-based (Celion GY-70) and sized, with a tensile modulus of 517 GPa, a tensile strength of 1862 MPa, and a tensile ductility of 0.36%. (Desized fibers were found to give similar results as the sized fibers.)

The carbon fiber reinforced superconductor was prepared by using a two-step process. The first step involved the preparation of a metal-matrix unidirectional carbon fiber composite (abbreviated MMC) by laying up carbon fibers and metal foils in the form of alternate layers and consolidating by hot pressing (by using a hydraulic press) at 5°C above the melting temperature of the metal and at a pressure in the range from 8,000 to 10,000 psi (from 55 to 69 MPa) for 10-20 s. The minimum pressure for the first step is about 55 MPa. The second step involved hot pressing a layer of superconductor sandwiched by two layers of MMC at 170°C (or below) and 500 psi (4 MPa) for 20 min or more in order to achieve diffusion bonding. Note that 170°C is below the liquidus of tin (232°C), 63Sn-37Pb (183°C) and 96Sn-4Ag (227°C). The minimum pressure for the second step is about 500 psi (4 MPa).

The critical current density J_c was measured at 77 K. Its value was determined by taking the slope of the current-voltage plot at 0.1 μV and extrapolating this slope to zero volt. Table I lists the J_c values obtained for plain and reinforced superconductor (123). In order to investigate the effect of composite processing on J_c, comparison was made between plain and reinforced superconductor materials made from superconductor pieces sintered and annealed in the same run.

Sn-Pb was more detrimental to J_c than Sn was, even though Sn-Pb allowed diffusion bonding to the superconductor to take place at a lower temperature than Sn did. Sn-Ag was comparably detrimental to J_c as Sn-Pb when the diffusion bonding was performed for both at 100°C for 20 min. However, Sn-Ag allowed diffusion bonding to the superconductor to take place at lower temperatures than Sn-Pb or Sn. The lower the diffusion bonding temperature, the less detrimental was the composite processing to J_c. The least detrimental case (3% decrease in J_c) was obtained when Sn-Ag was used such that diffusion bonding to the superconductor took place at 85°C. For this case in particular, the effect of composite processing on J_c is actually negligible if the statistical significance is taken into consideration.

The beneficial effect of Sn-Ag is probably due to the fact that Ag is not detrimental to J_c and that Ag may even help the wetting of the alloy on the superconductor.

CONCLUSION

Carbon fiber reinforced superconductor of J_c around 400 A/cm² at 77 K was obtained by using $YBa_2Cu_3O_{7-\delta}$ powder prepared by vacuum calcination, 96Sn-4Ag alloy as the binder, and diffusion bonding at 85°C for 30 min. The composite processing had negligible effect on the J_c of the superconductor.

ACKNOWLEDGEMENT

This research was supported by an award from the New York State Institute on Superconductivity in conjunction with the New York State Energy Research and Development Authority.   Work at Argonne is supported by the U.S. Dept. of Energy, Conservation and Renewable Energy, as part of a program to develop electric power technology, under contract W-31-109-Eng-38.

REFERENCES

1. C.T. Ho and D.D.L. Chung, J. Mater. Res. 4(6), 1339 (1989).

Table I.   Effect of tin or tin alloy on J_c of $YBa_2Cu_3O_{7-\delta}$ at 77 K.

Material*	$J_c(A/cm^2)$	Decrease in J_c due to tin or tin alloy
123	425 ± 8.1	
123/Sn (120^oC^+, 20 min)	373 ± 5.4	~ 12%
123	403 ± 6.6	
123/Sn-Pb (100^oC^+, 20 min)	342 ± 4.1	~ 15%
123	415 ± 7.1	
123/Sn-Ag (100^oC, 20 min)	350 ± 5.2	~ 16%
123	402 ± 7.3	
123/Sn-Ag (90^oC, 25 min)	374 ± 6.3	~ 7%
123	402 ± 6.9	
123/Sn-Ag (85^oC^+, 30 min)	389 ± 4.6	~ 3%

*The diffusion bonding temperature and time are shown in parentheses.
+Minimum temperature for diffusion bonding.

EFFECT OF PARTICLE SIZE ON THE GRAIN ORIENTATION OF $YBa_2Cu_3O_{7-x}$ PREPARED BY MELT PROCESSING

Aristianto M.M. Barus and Jenifer A.T. Taylor
New York State College of Ceramics
Alfred University, Alfred, N.Y. 14802

ABSTRACT

The effect of particle size and shape on the orientation of bulk forms $YBa_2Cu_3O_{7-x}$ has been studied. Orientation was found in the green pressed sample made of relatively coarse powder(consisted of 50% -200 mesh and 50% -325 mesh), while there was no such orientation found in the sample prepared by shaker mill which consisted of fine particles ($1.10\mu m$). However, after melt processing including a 100 hr anneal at 980°C in O_2, highly oriented surface layers developed on both type of samples.

INTRODUCTION

Critical current densities of polycrystalline $YBa_2Cu_3O_{7-x}$ are several orders of magnitude lower than those found in single crystals or thin films of the material. This is a major impediment to the use of high Tc superconductor in practical applications . Several theories have been proposed to explain this phenomenon, usually related to microstructural features such as grain misalignment and poor conduction through grain boundaries[1].

There is little agreement on the processing conditions needed for producing bulk ceramic bodies which exhibit a high critical current density. However, it is widely recognized that melt processing can be utilized to control the crystallographic alignment of grains[2,3]. In addition, this process can reduce the porosity and microcracking due to anisotropy in thermal expansion. The high density of these materials, however, causes a decrease in the oxygen diffusion rate, often leading to a depressed Tc in poorly oxygenated material[4]. A balance between these two parameters must be obtained.

In this paper, we concentrate on the effect of particle size and shape 2on the grain alignment in bulk forms of green pressed and sintered $YBa_2Cu_3O_{7-x}$ material.

EXPERIMENTAL PROCEDURE

Materials were synthesized from the solid state reaction of reagent grade Y_2O_3, $BaCO_3$ and CuO powders. These were ballmilled in a plastic bottle with distilled water and zirconia balls for 24 h, dried at 90 °C, granulated using a 40 mesh sieve, calcined in air at 925 °C for 18 h, cooled, then reground with mortar and pestle. The powder then was screened through a 40 mesh sieve, recalcined at 925 °C for 18 h, cooled and divided into two parts.

The first part shown in Fig. 1 was reground with mortar and pestle and consists of 50% -200 mesh(75 μm) and 50% -325 mesh(45 μm). The other part shown in Fig.2 was shaker milled in a plastic bottle with toluene and zirconia balls for 12 h, dried, and granulated through 70 mesh. The shaker milled powder was analyzed using a centrifugal sedimentation particle size analyzer. The average diameter of this powder was 1.10 μm. Both powders were then pressed at 92 MPa to make bars. These bars were heated to 1050°C, held for 2.5 h, cooled to 980°C, held for 100 h, cooled to 550°C for 36 h and cooled to room temperature.

The orientation of green and sintered samples were analyzed using x-ray diffraction.

RESULT AND DISCUSSION

Powder Characterization

The SEM micrograph in Fig.1 shows the shape of hand ground particles. The particles clearly are platelike in shape. Fig.2 shows the shape of particles obtained by shaker milling. These particle tend to agglomerate due to surface energy, but the shape of some individual particles are discernibly more isotropic in nature than the hand ground powder.

X-ray diffraction patterns of powders prepared by both methods are shown in Fig.3. The coarser hand ground powder with platy grains align in the process of being loaded into the edge drifted sample holder for x-ray analysis. This preferred orientation is shown by enhancement of (00l) peaks compared to random powder pattern.

The shaker milled powder, being much finer and less anisotropic did not align during loading. The x-ray pattern shows the expected intensity for the (00l) reflection and considerable peak broadening due to strain introduced by the fine particle size.

Orientation in the green samples.

X-ray diffraction patterns of the surfaces of a green pressed sample made of handground powder are shown in Fig. 4. The effect of alignment caused by uniaxial pressing can be seen on all surfaces of the bar. The strongest orientation is seen in the top surface which receives more pressure as compared to other surfaces. These patterns show that the c-axis of the YBa$_2$Cu$_3$O$_x$ crystals are perpendicular to the sample surfaces. However, the effect may depend on the magnitude of the pressure applied to form the bar.

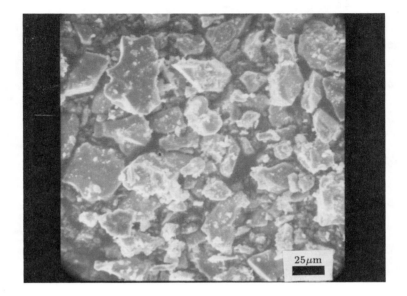

Figure 1: SEM micrograph of powder prepared by hand grinding

Figure 2: SEM micrograph of powder prepared by shaker milling

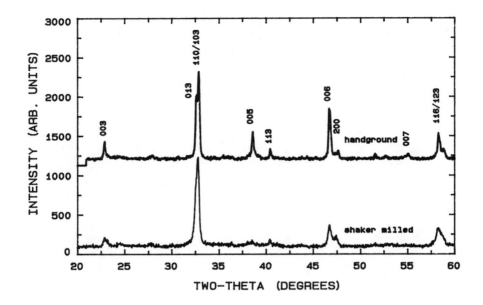

Figure 3: X-ray diffraction pattern of powder prepared by hand grinding and by shaker milling

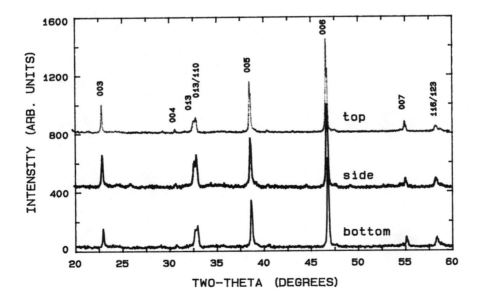

Figure 4: X-ray diffraction pattern of bar made of powder prepared by hand grinding

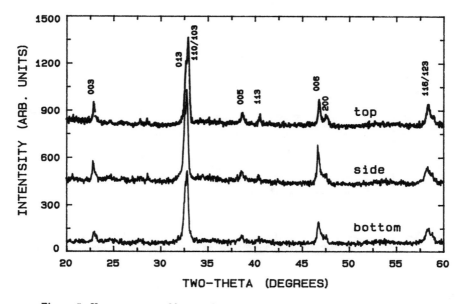

Figure 5: X-ray pattern of bar made of powder prepared by shaker milling

The x-ray diffraction patterns of a green pressed sample made of shaker milled powder are shown in Fig.5. No grain alignment occurs during pressing of the fine particles. The shape factor is believed to cause the differences in behavior of each powder during pressing. During calcination the growth of crystals in the direction perpendicular to c-axis is greater than that in the c-axis direction which results in a platelike particle[5,6]. When pressed, the particles, especially on the surface,   will orient with the largest dimension parallel to the surface. The smaller the particle the less will be the anistropic behavior of the powder which results in no orientation observed on the surface of the green pressed bar.

Orientation in the sintered sample.

The x-ray diffraction patterns of a handground sample after sintering is shown in Fig.6. Crystal alignment is clearly seen on the top and side surfaces, while no orientation is found on the bottom surface. The (013),(110) and (103) planes almost disappear from the X-ray diffraction pattern of the top surface of the sample showing a strong alignment of the grains. The sides also show the a-b axes aligned in the plane of the surface, in contrast to the result of work by A.F.Hepp et. al.[5] who sintered anistropic powder at 970° without utilizing melt processing. They reported that the sides of the sample showed a-b axes perpendicular to this surface. Apparently under proper conditions, sintering in the presence of a liquid phase results in a-b planes of $YBa_2Cu_3O_{7-x}$

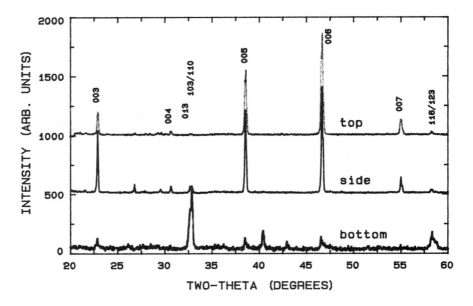

Figure 6: X-ray diffraction pattern of sintered sample made of powder prepared by hand grinding

crystals growing preferentially along the surface. Since crystal growth is a time dependent process, the longer the time the bigger the aligned crystals. The micrograph in Fig. 7 shows the fracture surface of a sintered, hand ground sample. The platelike layers perpendicular to the plane of fracture are clearly seen.

Fig. 8 shows the XRD-pattern of a sintered, shaker milled sample in the bottom spectrum and a sintered, hand ground sample in the top for comparison. Strong alignment occurred as compared to the green pressed sample. Despite a significant difference in the degree of alignment in the green pressed samples prepared by hand grinding and shaker milling (Fig. 4 and 5), melt processing with prolonged annealing produces similar textured surfaces on the samples. The specimens compacted from the handground powders may attain a higher degree of orientation after a shorter annealing time. More work must be done to define the optimum annealing time for the coarser, anistropic powder. Again, this finding is in contrast with the observation by Hepp et. al[5] who reported that isotropic grains do not align when sintered at 970°C. Referring to the bottom spectrum in Fig. 8, the shoulders on the peak at 32.70 and 46.68 two-theta indicate that the material is in the tetragonal form. This could be the results of inadequate oxygenation during sintering of the fine particles.

These results combine with previous work [7] indicate that melt processing

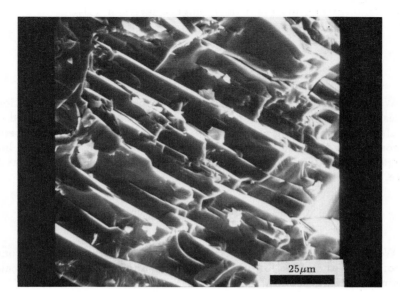

Figure 7: SEM micrograph of fracture surface of sintered samples made of powder prepared by hand grinding

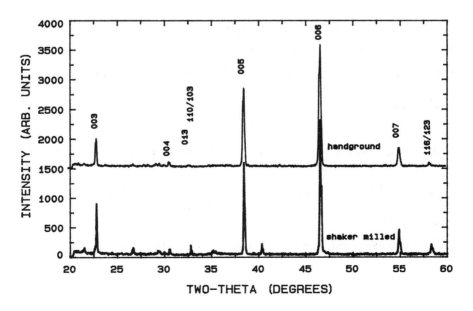

Figure 8: X-ray diffraction pattern of sintered sample made of powder prepared by shaker milling and hand grinding for comparison

is a powerful method to obtain thick, aligned layers of YB$_2$Cu$_3$O$_{7-x}$ on bulk forms.

CONCLUSION

We have shown that particle size and shape affects the orientation on surfaces of green pressed samples. The effect is less significant in sintered specimen after melt processing with prolonged annealing. Both samples prepared from coarse, hand ground and fine, shaker milled powder develop thick, highly textured layers on top and sides with the c-axes perpendicular to the as fired surface.

ACKNOWLEDGEMENT

We would like to thank the New York State Institute on Superconductivity of SUNY at Buffalo for providing funding and J.T. Baker Chemicals for providing reagent grade raw materials.

REFERENCES

1. J.W. Ekin et al.,J. Appl. Phys. 62(12), 4821-4828 (1987)

2. N.McN.Alford et al., Nature 332,58-59 (1988).

3. J.A.T.Taylor, P. Sainamthip and D.F. Dockery, Mat.Res. Soc.Symp.Proc.
   99,663-666 (1988)

4. K.Salama et al. , App. Phys. Lett. 54(23),2352-2354(1989).

5. A.F.Hepp, Ceramic Superconductors II (The American Ceramic Society, Inc., Westerville., 1988), p. 356.

6. R.B. vanDover, Proceeding of The First Annual Cambridge Conference on Commercial Applications of Superconductivity (World Tech Press, Cambridge, 1988) p. 236.

7. S. Kuharuangrong and J.A.T. Taylor, submitted to J. Am. Cer. Soc. (1990).

Multitarget Magnetron Sputtering of Large Area $YBa_2Cu_3O_{7-\delta}$ Thin Films

G. Wagner and H.-U. Habermeier

Max-Planck-Institut für Festkörperforschung, P.O. Box 800665, 7000 Stuttgart 80, Germany

Abstract

One important requirement for large scale applications of high T_C superconducting films as well as for some very interesting experiments in the field of microwave properties is the successful deposition of high quality large area thin films. The utilization of a multitarget magnetron sputtering system, especially designed for preparing film sizes up to two inches in diameter, allows a variety of different film compositions in the YBCO system. To avoid problems arising from the chemical properties of many barium compounds, we use targets of the intermetallic compound BaCu and YCu respectively. Argon is used as sputtering gas. Beside this, we let a flow of oxygen direct to the growing film to form the superconducting oxide. The substrate holder can be heated up by quartz lamps over 900°C. We are using two different ways to prepare thin superconducting films. The first way is an 'ex-situ' process. The oxygen partial pressure during deposition is very low due to the metallic surface of the targets. The superconducting phase is formed during a short high temperature post annealing in the deposition chamber at a pressure of 10mbar of pure oxygen without breaking the vacuum. The other way uses oxidized targets, where higher oxygen partial pressures are possible. Utilization of a suitable substrate temperature during deposition and omission of the high temperature post annealing lead to an 'in-situ' process. In this work we discuss both ways of sputter deposition and compare the properties of 'ex-situ' and 'in-situ' grown films. Our best films so far show metallic behaviour and a critical temperature (R=0) above 85K.

1. Introduction

Large area deposition of high T_C superconducting thin films is one of the most important requirements for large scale applications of the new materials as well as for some very interesting experiments in the field of basic research. Especially the microwave characteristics should be mentioned in this context. By reason of their low surface resistance, even in comparison with very pure metals, high quality large area films have the potential to become the first application of high temperature superconductivity. Although the laser ablation technique is the most common way of preparing high quality superconducting films, the deposition of large area films was not very successful so far. This is due to the small ablation area and the narrow angle where the stoichiometry is transferred to the film[1]. Nevertheless, Wu et al. [2] deposited high quality films on a rotating $SrTiO_3$ substrate with one inch in diameter. Greer et al. [3] succeeded in growing superconducting films on three inch sapphire wafers by scanning the laser beam over a 3.5 inch $YBa_2Cu_3O_{7-\delta}$ target. A very promising technique is metal organic chemical vapour deposition because of its high operation pressure. The preparation of YBCO films using a commercial MOCVD reactor, designed for subtrates up to 50 mm in diameter was reported for the first time by Zawadzki et al. [4]. Different sputtering techniques like single target sputtering using high pressures [5] or inverted cylindrical magnetrons [6] lead to high quality films on small substrate areas. Newman et al. [7] reported 'in-situ' growth of superconducting films over an area of 15cm^2 using off-axis sputtering from a composition target. In this work, we discuss the possibilities of sputtering large area $YBa_2Cu_3O_{7-\delta}$ thin films using two metallic targets.

2. Experimental

A schematic diagram of the most important parts of our sputtering system is shown in Figure 1. To get maximum flexibility to the film stoichiometry, we use co-deposition from three sputtering guns, each independently driven by its own RF power supply and automatic matching network. The three inch planar magnetron sputtering guns are mounted into the top flange of our deposition chamber with an angle of 30°, so that the central axis of all three guns intersect accurately at a common focal point. The substrate lies in this plane, about 150 mm away from the targets. This geometric design allows substrate sizes up to two inches in diameter.

To avoid problems either coming from the chemical instability of Ba or BaO or from a post annealing in wet oxygen to remove the fluorine resulting from a BaF_2 target, we use the intermetallic compound BaCu as barium source. In the process described in this paper, we use the intermetallic compound YCu as second target. The third gun is not in use. A gate valve between targets and substrates protects the targets from atmospheric contamination when changing the sample. This way presputtering time can be reduced to a minimum.

For heating up the substrate we use four rod-shaped quartz lamps which are mounted between two 3 mm thick plates (about $60 \times 70mm^2$) of a NiCr alloy. The lower one has a gold coating at the topside and a cooling coil at the backside. It is used as a reflector. The upper one is the heating plate carrying the substrate. The maximum temperature is more than 950°C which is measured by a thermocouple fitted inside the heating plate.

The flow of argon and oxygen into the vacuum chamber during deposition is individually controlled by mass flow controllers in the range of 1 to 5 sccm. The argon is introduced into the chamber adjacent to the targets, where the oxygen flows directly to the growing film passing a ringlike bended quartz tube enclosing the substrate. An automatic throttle valve in the pumping main keeps the total gas pressure on a constant value between 10^{-1} to 10^{-2}mbar. The measurement of the partial pressures using a differentially pumped mass spectrometer is very important for a successful reactive

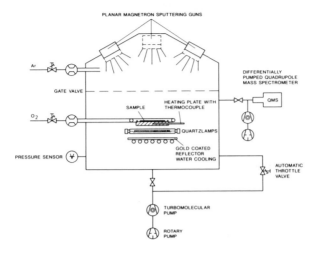

PLANAR MAGNETRON SPUTTERING GUNS

Ar

GATE VALVE

O₂

PRESSURE SENSOR

DIFFERENTIALLY
PUMPED QUADRUPOLE
MASS SPECTROMETER

QMS

SAMPLE

HEATING PLATE WITH
THERMOCOUPLE

QUARTZLAMPS

GOLD COATED
REFLECTOR
WATER COOLING

AUTOMATIC
THROTTLE
VALVE

TURBOMOLECULAR
PUMP

ROTARY
PUMP

Figure 1: Schematic diagram of the sputtering system

sputtering process. The stoichiometry of our films is routinely analyzed by atomic emission spectroscopy.

3. Results and Discussion

The oxygen which is mandatory for forming the superconducting phase, is introduced into the deposition chamber adjacent to the substrate. Nevertheless, the oxygen can react not only with the sputtered material but also with the targets. The oxidation of the targets' surface depends on the ratio of the absolute quantity of sputtered material to oxygen. As long as the sputtering rate is above a critical value for a given mass flow of oxygen, the targets will remain metallic. The BaCu target is shining in a bright green which is a simple indicator for its metallic surface, whereas the oxidized one causes a violet plasma colour. Another very reliable indicator is the DC self bias voltage. Figure 2 shows the behaviour of the self bias voltage with time at different values of the forward power which is a proportion to the sputtering rate. A constant mass flow of oxygen is switched on at the time t = 0. The lowest curve indicates a complete oxidation of the target. A short time after switching on the oxygen, the self bias voltage drops sharply. At this power the quantity of sputtered material is not high enough to absorb all the oxygen flowing into the system. The second curve representing a slightly increased sputtering rate shows only a small decrease of the self

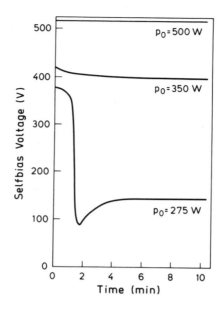

Figure 2: DC selfbias voltage after switching on a constant oxygen flow at different forward powers

bias voltage, due to a partial oxidation of the targets. At the highest power, the DC self bias voltage does not depend on the oxygen flow and the targets' surface remain completely metallic.

A mass spectrum of our sputtering atmosphere is shown in Figure 3. As long as the surface of the targets is metallic, the overall oxygen partial pressure (p_{O_2}) in the chamber is below 10^{-4}mbar, as it is shown in the upper curve (a). In this case p_{O_2} is found to be independent of the mixture ratio of oxygen and argon, because the sputtered material absorbs all the oxygen. An enhancement of the oxygen flow results in a very weak increase of p_{O_2}, while the targets' surface immediatly begins to get covered by an oxid layer which can be noticed by a decrease of the DC bias voltage. Oxygen partial pressures above 10^{-4}mbar are not possible before the whole targets' surface is oxidized. Then p_{O_2} mainly depends on the proportion of ingredients. Figure 3b shows the gas composition in the chamber using a 50 % oxygen / 50 % argon mixture with oxidized targets. The partial pressures are 5.3×10^{-2}mbar (O_2) and 4.3×10^{-2}mbar (Ar), respectively.

By this means, we have two different ways of preparing the films. The first one uses metallic targets, but the oxygen partial pressure is very low. This requires a substrate temperature below 560°C. Otherwise, the perovskite structure becomes thermodynam-

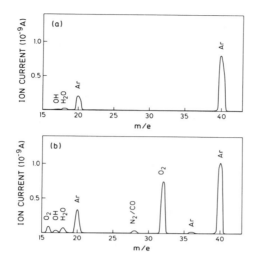

Figure 3: Mass spectrum of the atmosphere during sputtering
(a) using metallic targets' surface
(b) using oxidized targets' surface

ically unstable as reported by Bormann el al. [8,9]. Unfortunately, at such temperatures kinetic constraints were found to dominate over the thermodynamic aspects. It was impossible to form the superconducting phase under these conditions. Therefore, films deposited using metallic targets have to pass through a high temperature post annealing. For this reasons 'in-situ' growing of superconducting films is merely possible using the second way namely, oxidized targets allowing oxygen partial pressures above 10^{-2}mbar where the perovskite structure is stable up to higher temperatures [8,9].

In our preparation scheme developed so far, the stoichiometry of the films is controlled by tuning the forward power applied to the targets. The reduction of the targets' thickness with time, however, changes the sputtering rates. Therefore, we have to check the stoichiometry of our films routinely and adjust the ratio of the forward power every few runs. The temperature of the substrate holder during sputtering with metallic targets is 600°C. The films grow with a deposition rate of 150 to 200Å/min. After deposition, our films are annealed in the deposition chamber for 30 minutes at 850°C in oxygen at a pressure of 10 mbar without breaking the vacuum. These values are inside the stability range of the perovskite structure in Bormann's phase diagram [8,9]. While cooling down to room temperature, the chamber is floated with oxygen to atmo-

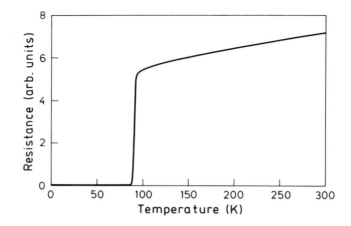

Figure 4: Resistance versus temperature diagram of an 'ex-situ' grown film

spheric pressure. We did not optimize our post annealing procedure so far. The result of a four probe DC resistance measurement plotted in Figure 4 indicates metallic behaviour above T_C and a narrow transition to superconductivity with $T_C(R = 0) = 86K$. The typical characteristics of the transition curves of epitaxially grown films, however, could not be observed. The X-ray diffraction pattern (Figure 5)points out a well crystallized but not textured polycristalline film. Moreover, a small amount of BaCuO$_2$ can be observed, due to a slight excess of barium and copper in the film resulting from the inaccurate control of the deposition rates of both targets. Nevertheless, the films have a very smooth surface and a high oxygen content as indicated by Raman spectroscopy [10].

Using the second variant of the process, the sputtered material is not enough to absorb all the oxygen, so that the chemical reaction with the oxygen also takes place at the targets' surface. The sputtering rate is now equal to the formation rate of the metal oxides, whereby the deposition rate is reduced to 15 to 20Å/min. Unfortunately, this form of the process is quite unstable. The reproducibility of the experiments is unsatisfactory. Therefore, the oxygen partial pressure must be regulated to a constant value in the range of 0.02 to 0.10 mbar. The stoichiometry of the films fluctuates and, thus, we have installed a crystal rate monitoring system. Many of our films produced by using oxidized targets, show a deficit in copper up to 40% but some films have excess in copper. Therefore, a third target seems to be necessary to control the copper content independently. We did not succeed in sputtering superconducting films with a T_C above 20 K using an 'in-situ' process and two intermetallic compound targets.

The homogeneity of the thickness of our films which is typically around 0.5μm is very good. It varies less than 20% over an area of 50mm in diameter. Up to now we

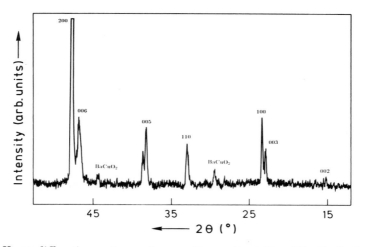

Figure 5: X-ray diffraction pattern of an ex-situ sputterd film. The $BaCuO_2$ is due to the film composition $YBa_{2.2}Cu_{3.5}O_x$.

did not rotate the substrate. Nevertheless, the film composition varies less then 10% along a strip with a length of 50 mm. But the edge of the substrate which is very close to one of the two targets shows an excess up to 40% of YCu or BaCu, respectively. Therefore a rotation of the substrate is required to get homogeneous film stoichiometry over the whole area of 50 mm in diameter.

4. Conclusions

In this work, we describe the deposition of YBCO thin films by co-sputtering from two metallic compound targets. We have shown that there are two completely different ways of growing the films. Using the first one, the oxygen partial pressure is very low during deposition due to the metallic surface of the targets. Therefore, a high temperature post annealing is mandatory to form the superconducting phase. The resulting films are polycrystalline, but very smooth and homogeneous even over a large area. An increase of the oxygen partial pressure which will make possible the 'in-situ' growth of superconducting films requires an oxidized targets' surface. Under these conditions, however, an accurate control of the oxygen partial pressure and the sputtering rates seems to be necessary.

Acknowledgements

The authors would like to thank O. Buresch, B. Leibold, and L. Viczian for character-ising the films by different techniques.

References

[1] R. SINGH; J. NARAYAN: Phys.Rev. B, **41** (1990) 8843.

[2] X. WU; R. MUENCHHAUSEN; S. FOLTYN; R. ESTLER; R. DYE; C. FLAMME; N. NOGAR; A. GARCIA; J. MARTIN; J. TESMER: Appl.Phys.Lett., **56** (1990) 1481.

[3] J. GREER; J. VAN HOOK: Proc. MRS Fall Meeting 1989 (in press).

[4] P. ZAWADZKI; G. TOMPA; P. NORRIS: Journal of Electronic Materials (in press).

[5] U. POPPE; J. SCHUBERT; R. ARONS; W. EVERS; C. FREIBURG; W. REICHERT; K. SCHMIDT; W. SYBERTZ; K. URBAN: Solid State Commun., **66** (1988) 661.

[6] X. XI; J. GEERK; Q. LINKER; O. MEYER: Appl.Phys.Lett., **54** (1989) 2367.

[7] N. NEWMAN; K. CHAR; S. M. GARRISON; R. O. BARTON; R. C. TABER; C. B. ECOM; T. H. GEBALLE; B. WILKENS: Appl.Phys.Lett., **57** (1990) 520.

[8] R. BORMANN; J. NÖLTING: Appl.Phys.Lett., **54** (1989) 2148.

[9] R. BORMANN: Proc. ICMC 90 (to be published).

[10] G. WAGNER; H.-U. HABERMEIER: Proc. E-MRS 90 (in press).

COMMERCIAL-SCALE PRODUCTION OF
HIGH TEMPERATURE SUPERCONDUCTING THIN FILMS

P. H. Ballentine, J.P. Allen
CVC Products, Inc., Rochester, NY 14603

A.M. Kadin, D.S. Mallory
Department of Electrical Engineering
University of Rochester, Rochester, NY 14627

ABSTRACT

The development of a commercial-scale sputtering process for the deposition of superconducting $YBa_2Cu_3O_7$ (YBCO) thin films is described. YBCO films are sputter deposited from a single 20 cm diameter powder target onto 2 inch substrates of MgO and $LaAlO_3$ using *in-situ* heating to produce high quality, c-axis oriented films without post annealing. By locating the substrate above the center of the magnetron erosion ring and by placing a "negative ion shield" between the target and the substrate, negative ion resputtering is avoided. Small shifts in composition are corrected by using a slightly Y-and Cu-rich target to produce films within 1% of stoichiometric composition. Films on $LaAlO_3$ and MgO have critical temperatures up to 88 K and critical currents above 5×10^5 A/cm^2 at 77 K. Uniformity is ±5% in thickness and ±7% in resistivity across a 2-inch diameter substrate. Microwave surface resistance measurements at 1.2 GHz and 4 K show $R_s \approx 0.1$ mΩ for YBCO ground planes and ≈ 0.3 mΩ for patterned YBCO striplines on MgO.

INTRODUCTION

As the quality of high T_c thin films continues to improve, more effort is being devoted to applications such as passive microwave components[1], high speed digital interconnects[2], and SQUID magnetometers[3]. Several key thin film requirements must be met for these applications to be successful. Most important the quality of the films must be maintained. It has been well established that even small changes in film composition or deposition temperature lower the critical current or raise the microwave surface resistance to the point where many applications are no longer practical. In addition, the film surface must be smooth on the submicron scale for patterning and application of subsequent thin films. It is generally accepted that the *in-situ* method of depositing crystalline films onto heated substrates is preferred over the conventional post-deposition annealing method because the maximum temperature is lower and the surfaces tend to be smoother with *in-situ* films.

Substrate compatibility is another key issue. High quality thin films have been grown on several different substrate materials, including $SrTiO_3$, $LaAlO_3$, MgO, and yttria stabilized ZrO_2. Although the best films to-date have been on $SrTiO_3$, the dielectric properties of this material make it unsuitable for high frequency applications. $LaAlO_3$ as well as well as a few more recently introduced perovskite-based materials[4] provide for epitaxial growth and have acceptable microwave properties. For applications which require the lowest microwave losses, Al_2O_3 will be required. And of course deposition of high T_c films on Si would greatly increase the range of practical applications. These last two materials, however, tend to react with the high T_c material, and either buffer layers will be required or growth temperatures will need to be reduced, probably below 500 °C.

Another issue is that of scale up. Most research efforts have used substrate sizes 1 cm^2 or smaller, and patterned devices will in most cases require larger areas. There are several factors which complicate scale-up of high T_c deposition processes. First, the newer substrate materials such as $LaAlO_3$ are only now becoming available in 2-inch diameter sizes. Second, precise composition control must be maintained uniformly over the entire substrate area. In this respect, each deposition method has its own inherent advantages and disadvantages. For example, sputtering of $YBa_2Cu_3O_7$ is generally accompanied by resputtering of the growing film by energetic oxygen atoms which originate at the target as negative ions[5]. This causes a shift in film composition and creates defects in the crystal structure. To overcome this, researchers have used off-axis sputtering[6] and/or operated at relatively high pressures, in excess of 100 mTorr.[7] Both of these options tend to lower the deposition rate and the off-axis geometry reduces thickness uniformity. Laser ablation has been used very successfully to deposit small-area films, but has yet to be scaled up above 1-inch diameters. Because laser ablation is a point source method, large-area deposition will require movement of the laser beam, the substrate, or both. In addition, the ablation process often results in the ejection of micron-sized particles which deposit on the film surface. Perhaps more problematic may be the presence of an evaporated component at wide ejection angles[8], which is off-stoichiometric and would have to be shielded while the ablated flux is allowed to deposit on the substrate. Another aspect of high T_c film deposition complicating scale-up is the need to heat the substrates uniformly to high temperatures during deposition. While the optimum deposition temperature appears to be a function of the deposition method used as well as operating conditions and substrate type, it has been reported, for a given process, that the substrate temperature must be tightly controlled to give films with the highest critical currents and critical temperatures. For example, Xi et al.[7] reported a drop in $T_c(R=0)$ to 50 K when the deposition temperature was lowered from 800 to 700 °C. Silver paint is often used to to thermally sink the substrate to the heater block and provide better temperature control and uniformity. This approach is likely to be incompatible with large area substrate processing or double-sided coverage, both of which are important for some microwave applications. In this paper we describe the current status in the development of a sputtering process for the *in-situ* growth of YBCO films from a 20 cm diameter target. This technique provides uniform deposition over 2-inch substrates and is compatible with two-sided coverage. Data on film resistivity, critical temperature, critical current, and microwave surface resistance are presented.

THIN FILM DEPOSITION

Two deposition systems are used in this work, a single target, single substrate research system (CVC SC-4000) for high T_c process development and a four-target system (CVC 601) for high T_c film and buffer layer deposition. A schematic of the sputtering geometry is shown in Fig. 1. $YBa_2Cu_3O_7$ films are deposited by rf magnetron sputtering from a single oxide target composed of loose powder spread over a 20 cm diameter water-cooled copper backing plate.[9] We use a loose powder YBCO target (as opposed to the more standard solid sintered target) because it permits easy change in composition and is less prone to cracking under thermal stress. We have also used a solid sintered YBCO target with cross-cut grooves to relieve thermal stresses and have obtained similar results. Substrates are placed 2.5 cm above the center of the target inside the magnetron erosion track, which is approximately 7.5 cm I.D. x 17.5 cm O.D., in order to avoid direct bombardment by energetic oxygen ions (or neutrals). In addition, we find it necessary to place a "negative ion shield" approximately 3 mm (within the cathode dark space) above the center of the target to further reduce energetic particle bombardment.[10] This approach is analogous to off-axis sputtering but has the potential for coating larger areas with better uniformity and at a higher rate. A 12.5 cm diameter inconel heater is used to radiatively heat the substrates up to 850 °C for *in-situ* growth and anneal the films in high partial pressures of oxygen. This approach permits substrates to be sequentially coated on both sides, and is consistent with substrate rotation for better uniformity. The substrate temperature is measured directly by clamping a type-K thermocouple between two sapphire substrates placed next to the sample substrate. It should be noted that the ion shield also serves as a radiation shield helping to maintain a uniform substrate temperature and reducing the heat load on the target.

After pumping down to below 10^{-7} Torr, the substrate is brought to the desired temperature and gas flow is set to establish 20 mTorr Ar and 10 mTorr O_2. RF power is ramped slowly to 450 W to avoid rapidly outgasing and disturbing the powder target. After deposition, the chamber is backfilled with 100 Torr of O_2 and the heater is turned off, allowing the substrate to cool to 100 °C in two hours.

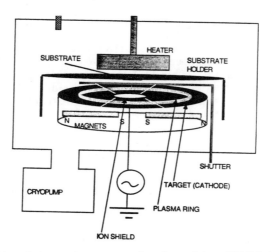

Fig. 1. Schematic of sputtering set up for depositing YBCO films using 8-inch target in rf magnetron mode with a central grounded ion shield between the target and the substrate

FILM PROPERTIES

Film thickness is measured using a stylus profilometer with either mechanically masked or photolithographically patterned steps and room temperature sheet resistance is measured using a four point probe. Surface morphology is observed using an SEM and film composition is measured by Energy Dispersive X-ray Analysis (EDXA) with occasional calibration by Inductively Coupled Plasma atomic emission spectroscopy (ICP). Crystalline structure is characterized by x-ray diffraction using Θ-2Θ scans. Superconducting properties are measured via two computer controlled cryogenic test stations, one with liquid cryogen and one with a closed cycle refrigerator. The dc R vs. T curves are taken using a 4 point measurement with pressed indium contacts. Critical currents are measured on 5 - 10 μm wide lines patterned by photolithography with a wet etch in 1% H_3PO_4 and contacted by evaporated Ag. Microwave measurements at 1 - 3 GHz are made using a stripline resonator with YBCO as the stripline, the ground plane(s), or both.[11]

As we have previously reported, large shifts in film composition due to negative ion bombardment could be avoided by placing the substrate inside the magnetron racetrack of a 20 cm target.[9] Further improvements in superconducting properties are obtained by placing a grounded shield within the dark space above the target center. Without the negative ion shield, films on MgO have ρ(300 K) $\approx$ 1700 μΩ-cm, and T_c $\approx$ 80 K. Furthermore, the c-axis lattice parameter is found to be extended from the ideal 11.67 Å to 11.86 Å. These properties are usually associated with oxygen deficient YBCO. However, post deposition anneals at 500 - 600 °C in oxygen do not increase T_c or contract the c-axis. Other researchers have also observed similar effects in sputtered YBCO and have attributed them to structural defects that are not the standard oxygen deficiency observed in bulk YBCO.[6,12] We conclude there is some residual high energy particle bombardment of the films, coming from the center of the target, which is not enough to cause large shifts in composition, but introduces defects into the film.   With the shield in place, $\rho(300 \text{ K}) \approx 550 \text{ } \mu\Omega\text{-cm}$, $T_c \approx 85$ K, and the c-axis lattice parameter is decreased to 11.67 Å.

Although large shifts in composition between target and film are not observed in the present geometry, finer compositional measurement does reveal a slight shift in film composition. Table I lists the compositions of the four targets used and some properties of the resulting films. Films deposited from a stoichiometric target (#1; $Y_{.17}Ba_{.33}Cu_{.50}$) are Ba-rich ($Y_{.14}Ba_{.36}Cu_{.50}$). These films have smooth surfaces but with depressed T_cs (50 - 60 K) and extended c-axis lattice parameters of 11.8 Å. Films deposited from a Cu-rich target (#2; $Y_{.15}Ba_{.28}Cu_{.57}$) are further enriched in Cu ($Y_{.14}Ba_{.24}Cu_{.63}$) and have very rough surfaces with granular Cu-rich material (probably CuO) on top of a highly c-axis oriented underlying film. The T_cs of these films, however, are significantly higher, around 85-86 K, and the c-axis spacing is 11.68 Å. Films deposited from a Y-rich target (#3, $Y_{.20}Ba_{.30}Cu_{.50}$) are slightly Y-rich ($Y_{.18}Ba_{.33}Cu_{.49}$) with T_c 83 K and a c-axis lattice parameter of 11.70 Å. Although the films are smooth and featureless to 0.1 µm under SEM inspection, there was some difficulty in making metallic contact to these films and we speculate that Y-rich material may be accumulating at the free surface. Interestingly, even though these films have slightly depressed T_c, the resistive portion of the curve has a large slope and often extrapolates below the origin at T=0. Films deposited from a very slightly Y-rich target (#4; $Y_{.18}Ba_{.31}Cu_{.51}$) are within 1% of stoichiometric metallic content. Figure 2 shows the Θ-2Θ x-ray diffraction curve for one of these films (run # 79) deposited on MgO at 740 °C. The film is primarily c-axis oriented with a lattice spacing of 11.67 Å and a rocking curve of 0.2° full width at half maximum on the (005) peak and there is some a-axis material present. Figure 3 plots R vs. T and J_c vs. T for run #79, showing $T_c = 88$ K and $J_c = 5 \times 10^5$ A/cm^2 at 77 K.

Table I Target compositions and resulting film properties

Target	Target composition	Film composition	T_c (K)	$\rho(300)$ ($\mu\Omega$-cm)	$\rho(300)/\rho(100)$
1	stoichiometric	$Y_{.14}Ba_{.36}Cu_{.50}$	50	1700	1.5
2	$Y_{.15}Ba_{.28}Cu_{.57}$	$Y_{.14}Ba_{.24}Cu_{.63}$	86	550	2.2
3	$Y_{.20}Ba_{.30}Cu_{.50}$	$Y_{.18}Ba_{.33}Cu_{.49}$	83	650	3.5
4	$Y_{.18}Ba_{.31}Cu_{.51}$	$Y_{.17}Ba_{.33}Cu_{.50}$	88	350	2.4

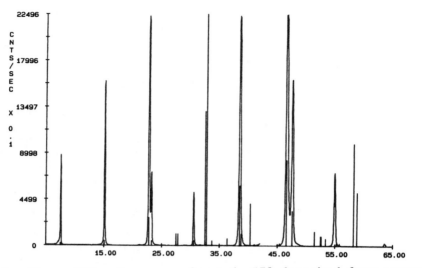

Fig. 2. X-ray diffraction scan of sample #79 deposited from target #4 (slightly Y-rich) on MgO at 740 °C. The (005) rocking curve is 0.2 ° full width at half maximum.

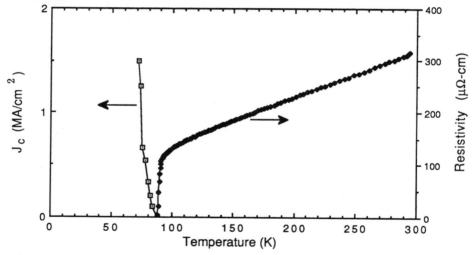

Fig. 3. R vs. T and J_c vs. T for sample #79 on MgO. The T_c is 88 K and J_c = 500,00 A/cm^2 at 77 K.

The small shifts in composition that we do observe tend to make the films Ba or Cu rich. This trend is opposite to what most groups have reported for sputtered YBCO, but can be understood in terms of the geometries involved. In conventional geometries, the substrates are located opposite the active region in the target Ba and Cu are resputtered from the film.   In the present case, the substrate holder is opposite the magnetron racetrack.   Ba and Cu that are resputtered from these regions are scattered back to the target and/or onto the substrate.   The relative composition shift was not the same for all targets.   In fact, the Ba/Cu ratio increased for target #1 but decreased for target #2.   This complex behavior may be due to a stratification in the resputtered material surrounding the substrate holder.

Films which are only 3% Ba rich (based on total metallic content) had significantly lower T_cs than films 13% Cu rich.   This result is consistent with Chew *et at.* [13] who have observed sharp decreases in T_c and J_c when the Y/Ba content of YBCO films was systematically decreased. Although target #4 gave films that were within 1% stoichiometric, we believe further refinements in target composition will give even better results.

We find the optimum deposition temperature for YBCO on MgO is in the 740 - 760 °C range, where the T_c is 84 - 88 K.   Increasing the temperature to 820 °C results in little or no a-axis material but lowers T_c to around 80 K, presumably the result of interaction with the substrate. Lowering the deposition temperature to 700 °C results in a T_c of 82 K.   For $LaAlO_3$ substrates, the optimum deposition temperature is somewhat higher, around 780 - 800 K, and the T_c is more consistently close to 88 K.

Figure 4 shows the deposition rate as a function of position across the substrate plane for three different total pressures (the Ar:O_2 ratio is held constant at 2:1).   These data were taken on unheated substrates with 450 W rf power and a target-to-substrate (T-S) spacing of 4.4 cm.   At 30 mTorr total pressure, the rate is 65 Å/min ±5% over a 6 cm diameter area with the highest rate at the outer edge of the substrate.   When the substrates are heated to 700 °C, the rate drops to 30 Å/sec, and when the T-S spacing is decreased to 2.5 cm and 700 °C heat applied, the rate drops to 12 Å/sec.

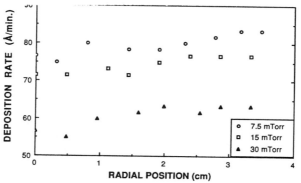

Fig. 4.   Deposition rate vs. position for YBCO showing a uniformity of ±5% across a 5 cm diameter area.

The variation in resistivity at room temperature for a 3500 Å film on a 2-inch dia. $LaAlO_3$ substrate, shown in Fig. 5, is ±7% across the entire substrate. The uniformity of superconducting properties across these areas have not yet been systematically measured. However, 1 cm x 1 cm substrates placed 2.5 cm from the center shows no decrease in T_c while at 3.8 cm, above the edge of the negative ion shield, the T_c drops by 5 K.

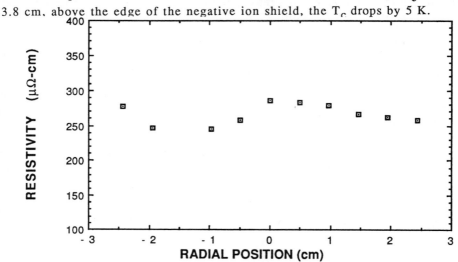

Fig. 5. Room temperature resistivity vs. position for a YBCO film on a 5 cm diameter $LaAlO_3$ substrate.

The microwave surface resistance for films on 2" x 0.5" MgO substrates deposited from target #4 was measured at 1.2 GHz using a stripline resonator technique. Figure 6 shows the calculated surface resistance as a function of temperature for an all-YBCO resonator at three power levels. At 4 K, $R_s \approx 0.3$ mΩ at -30 dBm, increasing to 0.8 mΩ at 0 dBm. When YBCO is used as a ground plane with a superconducting Nb center strip, the surface resistance of YBCO at 4.2 K was 0.1 mΩ or lower. This measurement is limited by experimental resolution and there does not appear to be any difference in Q between this and an all Nb resonator at 4 K. Preliminary measurements at 35 GHz and 77 K with an unpatterned film on $LaAlO_3$ were taken by substituting the YBCO film as the end wall of a Cu cavity. These results give a surface resistance of 11 m , which is a factor of 3.5 better than Cu at these conditions. We speculate the increased loss and strong power dependence with the YBCO strip line is due to local field concentrations near the edges of the center strip. We note there is only a weak power dependence for the Nb center strip with YBCO ground planes.

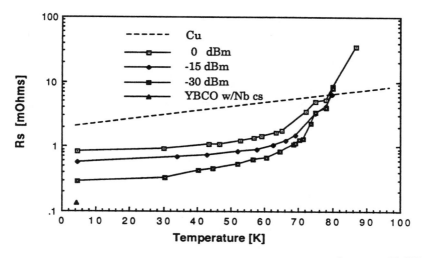

Fig. 6.   Microwave surface resistance vs. temperature for an all-YBCO resonator on MgO substrates.   The triangle is for YBCO ground planes with a Nb center strip.

SUMMARY

Thin films of $YBa_2Cu_3O_7$ were deposited by rf magnetron sputtering onto heated MgO and $LaAlO_3$ substrates.   Sputtering from a single relatively large (8 inch) oxide target allows for good control and reproducibility of composition, structure, and superconducting properties across a 2 inch diameter area.   The inclusion of a negative ion shield between the substrate and the target further improved film properties.   Slight shifts in composition between target and film were compensated by adjusting target composition.   Composition control is crucial; in particular, a slight excess of Ba in the films has a rather deleterious effect on superconducting properties.   Critical temperatures of stoichiometric films on MgO and $LaAlO_3$ were 86 and 88 K and critical currents were above 5 x 10^5 A/cm^2 at 77 K.   The microwave surface resistance of unpatterned films on MgO was 0.1 mΩ at 1.2 GHz and 4 K.

ACKNOWLEDGMENTS

The authors would like to acknowledge D. Hodge and R. Jammy at the University of Rochester, and R. Rath at CVC Products, Inc. for assistance in thin film deposition, Dr. P.K. Watson at Xerox Webster Research Lab for ICP analysis, T. Blanton and C. Barnes at Eastman Kodak Research Lab for x-ray diffraction, and Dr. E. Track of Hypres, Inc.   for preliminary microwave measurements at 35 GHz.   This research was supported in part by NASA Jet Propulsion Lab. and the New York State Institute on Superconductivity at Buffalo.

REFERENCES

1. J.D. Adam and G.R. Wagner, "High T_c Superconducting Thin Films: Processing, Characterization, and Applications", ed. Gerald Lucovsky, AIP Conf. Proc. No. 200, (AIP, New York, 1990) p. 227.

2. D.R. Dykaar, R. Sobolewski, J.M. Chwalek, J.F. Whitaker, T.Y. Hsiang, G.A. Mourou, D.K. Lathrop, S.E. Russek, and R.A. Buhrman, Appl. Phys. Lett. **52**, 1444 (1988).

3. J. Clarke and R. Koch, Science **242**, 217 (1988).

4. M. Berkowski, *et. al*, Appl. Phys. Lett. **57**, 632 (1990).

5. S. M. Rossnagel and J.J. Cuomo, in *Thin Film Processing and Characterization of High-Temperature Superconductors*, ed. J.M.E. Harper, AIP Conf. Proc. No. 164 (AIP, New York, 1988) p. 106.

6. C.B. Eom, J.Z. Sun, K. Yamamoto, A.F. Marshall, K.E. Luther, T.H. Geballe, and S.S. Laderman, Appl. Phys. Lett, **55**, 595 (1989).

7. X.X. Xi, G. Linker, O. Meyer, E. Nold, B. Obst, R. Ratzel, R. Smithey, B. Strehlau, F. Weschenfelder, and J. Geerk, Z. Phys. B - Condensed Matter **74**, 13(1989).

8. T. Venkatesan, X.D. Wu, A. Inam, and J.B. Wachtman, Appl. Phys. Lett. **52**, 1193 (1988) .

9. A.M. Kadin, P.H. Ballentine, J. Argana, and R.C. Rath, IEEE Trans. Magn. **25,** 2437 (1989) .

10. P.H. Ballentine, J. Archer, J. Allen, and A.M. Kadin, Soc. of Vacuum Coaters 33rd Technical Conf. Proc. , pp. 303-307 (1990).

11. A.M. Kadin, D.S. Mallory, and P.H. Ballentine, in *Superconductivity and Applications*, ed. H.S. Kwok, D.T. Shaw, and Y.H. Kao, pp. 757-766, (Plenum, NY, 1990).

12. J.A. Kittl, C.W. Nieh, D.S. Lee, and W.L. Johnson, Appl. Phys. Lett. **56**, 2468 (1990).

13. N.G. Chew, S.W. Goodyear, J.A. Edwards, J.S. Satchell, S.E. Blenkinsop, and R.G. Humphreys, Submitted to Appl. Phys. Lett. 1990.

DEGRADATION OF PROPERTIES OF YBa$_2$Cu$_3$O$_x$ SUPERCONDUCTORS SINTERED IN CO$_2$-CONTAINING ATMOSPHERES*

U. Balachandran, C. Zhang, D. Xu, Y. Gao,+ K. L. Merkle,+
J. N. Mundy,+ B. W. Veal,+ R. B. Poeppel
Materials and Components Technology Division
+Materials Science Division
Argonne National Laboratory
Argonne, IL 60439

G. Selvaduray
Materials Engineering Department
San Jose State University
San Jose, CA 95192

and

T. O. Mason
Department of Materials Science and Engineering
Northwestern University
Evanston, IL 60208

ABSTRACT

Stability of the YBa$_2$Cu$_3$O$_x$ (YBCO) superconductor in reacting with CO$_2$ in CO$_2$/O$_2$ gas mixtures during sintering was investigated as a function of the partial pressure of CO$_2$ and temperature. The transport critical current density, J$_c$, of the superconductor decreased drastically with increasing partial pressure of CO$_2$ in the gas mixture. As the partial pressure of CO$_2$ was increased, J$_c$ became zero (at 77 K) even though the major phase of the sample was still a superconductor according to magnetic susceptibility measurements. Microstructures and compositions of the samples were investigated by transmission electron microscopy in conjunction with energy-dispersive X-ray spectroscopy. Two types of grain boundaries were observed: ≈10% of the grain boundaries contained a second phase, and the regions near the remaining grain boundaries were tetragonal. At high partial pressures of CO$_2$, the YBCO completely decomposed to BaCO$_3$, Y$_2$BaCuO$_5$, and CuO.

---

*Work supported by the U.S. Department of Energy, Offices of Utility Technologies, Conservation and Renewable Energy, and Basic Energy Sciences-Materials Science, under Contract W-31-109-Eng-38, and by the National Science Foundation through the Science and Technology Center for Superconductivity, under Contract DMR88-09854.

INTRODUCTION

The transition temperature, critical current density, and width of superconducting transition of YBCO superconductors are influenced by atmospheric contaminants such as CO_2 and H_2O.[1-6] Jahan et al. indicated the formation of insulating phases when YBCO reacts with water vapor.[7] Reaction of YBCO with CO_2 has been reported by several researchers.[1,5,6,8] Gallagher et al. reported that at 1000°C in a 1% CO_2/O_2 mixture the YBCO phase was not decomposed, while in a 10% CO_2/O_2 mixture the YBCO phase was completely decomposed, forming $BaCO_3$, $Y_2Cu_2O_5$, and CuO.[1] Fjellvag et al. concluded that the reaction occurred in two steps: below 730°C the reaction products are $BaCO_3$, Y_2O_3, and CuO, while above this temperature the products are $BaCO_3$, $Y_2Cu_2O_5$, and CuO.[8] Because of limitations of the X-ray diffraction technique, neither Gallagher et al. nor Fjellvag et al., were able to study the spatial origin of the reaction between YBCO and CO_2, which may be very important with regard to the low value of the critical current density found in ceramic superconductors. Recently Cooper et al. used in-situ electrical conductivity measurements to study the kinetics of YBCO decomposition in a flowing 5% CO_2/95% O_2 atmosphere at 815°C.[9]

In this paper, we report on the degradation of properties (critical temperature, T_c, and critical current density, J_c) of YBCO superconductors sintered in CO_2-containing atmospheres. The microstructures and compositions of the samples were investigated by transmission electron microscopy, analytical electron microscopy, and secondary ion mass spectroscopy. The relationships between the properties and the partial pressure of CO_2 are discussed in terms of microstructural changes.

EXPERIMENTAL

Phase-pure orthorhombic powders of YBCO were prepared by mixing and grinding stoichiometric amounts of Y_2O_3, $BaCO_3$, and CuO and then calcining in flowing oxygen at a reduced total pressure of about 2 mm Hg at about 850°C; this was followed by low–temperature annealing in ambient–pressure oxygen.[10] The reduced pressure used in this process ensured the efficient removal of the CO_2 gas generated during the formation of the YBCO phase and resulted in the production of phase-pure orthorhombic powders. These powders were pressed into small pellets and sintered in the temperature range 900-1000°C for about 5 h in flowing (≈1 atm) O_2/CO_2 gas mixtures. The portion of CO_2 in the mixtures was 0-5%. The samples were cooled slowly to room temperature; a 12–h hold at 450°C was incorporated in the cooling schedule for oxygenation of the ceramics. J_c was determined by a standard four-probe resistivity measurement in liquid nitrogen.

A criterion of 1 μV/cm was used for J_c measurement. T_c values were obtained by resistivity and magnetization techniques. A low–field RF SQUID magnetometer was used for the magnetization measurements. Transmission electron microscope (TEM) specimen discs (3 mm in diameter) were cut from the sintered bulk samples, polished, and dimpled on both sides until a thin area at the center was obtained. The final TEM specimens were argon-ion–thinned at liquid nitrogen temperature.

RESULTS AND DISCUSSION

The J_c (at 77 K) values decreased as the CO_2 partial pressure in the sintering atmosphere was increased, finally becoming zero at a CO_2 partial pressure that is dependent on the sintering temperature. Resistivity measurements showed that the materials with $J_c = 0$ were semiconductive. Figure 1 shows the stability region for transport superconductivity with respect to the partial pressure of CO_2 at the four sintering temperatures (910, 940, 970, and 1000°C). Magnetization measurements for samples with $J_c = 0$ indicated that the majority phases of these materials were still superconducting. The onset temperature of superconductivity, about 90 K, was almost the same for all of the samples. Resistivity and magnetization measurements of samples with high and zero J_c suggested a strong blockage of superconducting current at the grain boundaries in the latter samples. A possible cause of the blockage could be a thin layer of a second phase at grain boundaries, formed during sintering due to the reaction of YBCO with CO_2 in the gas atmosphere.

TEM observations showed the presence of secondary phases at some grain boundaries. An example of one such secondary phase is shown in Fig. 2 for the sample sintered at 970°C in a 0.5% CO_2/O_2 gas mixture. The secondary phases were determined by X–ray energy–dispersive spectroscopy (XEDS) to be $BaCuO_2$ and $BaCO_3$. The widths of these grain boundary phases were much greater than the coherence length in YBCO superconductor. Such secondary phases therefore can obstruct the superconducting current passing across the grain boundaries so that the overall critical current density may decrease. These phases accounted for only about 10% of the observed grain boundaries, while the majority of the grain boundaries appeared quite sharp and had no obvious evidence of a second phase.

Because of the multitude of possible percolation paths, J_c would not become zero if only 10% of the grain boundaries were coated with a second phase. Therefore, a majority of the grain boundaries must be resistive enough to block the flow of superconducting current across the boundaries. By careful study of high-resolution electron microscopy (HREM) images, we found that the

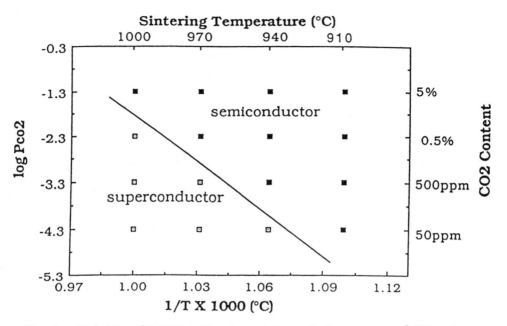

Fig. 1. Stability of YBCO with respect to partial pressure of CO$_2$ at various temperatures.

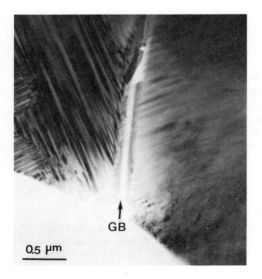

Fig. 2. TEM picture of grain boundary (GB) in YBCO sample sintered at 970°C in 0.5% CO$_2$/O$_2$ atmosphere; thick second–phase layer at GB is identified as BaCuO$_2$.

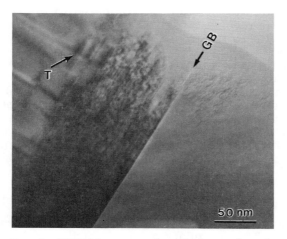

Fig. 3. HREM image of grain boundary in
YBCO sample sintered at 970°C in 0.5%
CO_2/O_2 atmosphere.

structure near the sharp grain boundaries was not orthorhombic,
but tetragonal. Figure 3 shows an HREM image of the lattice
fringes of (001) planes. By careful measurement of the interplanar
spacing, we determined that the spacing was about 1.19 nm at the
region near the grain boundary, while the spacing was approximately 1.17 nm in the region far from the grain boundary. From
neutron diffraction data,[11] it is known that the structure with c =
1.19 nm is tetragonal. Another indication of the presence of the
tetragonal structure is the termination of the twin structure, as
marked by T in Fig. 3. This can be taken as the signature of the
demarcation line between the orthorhombic and tetragonal
structures because the tetragonal structure has no twins.

A possible cause of the phase transformation from orthorhombic to tetragonal is the incorporation of carbon into the lattice
due to the CO_2 in the sintering atmosphere.[5,6] Carbon diffuses into
the lattice and expells the oxygen in the orthorhombic structure,
thus forming a tetragonal structure. Secondary ion mass spectroscopy (SIMS) was used to detect the carbon signal. The SIMS
results showed carbon segregation at grain boundaries and at
regions near grain boundaries.[6] Figure 3 illustrates a special
example that has a particularly large tetragonal region. We
observed that in most cases, such tetragonal regions near grain
boundaries varied from a few nanometers to several tens of nanometers. They were therefore very difficult to detect and were not
readily identifiable in most boundaries. The widths of these

tetragonal regions were quite large compared to the coherence length, and these regions block the superconducting current; therefore, the J_c goes to zero.

Our previous work indicates that the reaction rate with CO_2 depends on grain boundary orientation and structure.[5] Therefore, at low partial pressures of CO_2 (up to few hundred ppm), only a fraction of the grain boundaries may be modified sufficiently by such reactions to interrupt the superconducting current. For the samples sintered in 5% CO_2/O_2, no superconducting transition was observed. At this high level of CO_2, the YBCO phase completely decomposed into three different phases. The XEDS analyses revealed that these phases were $BaCO_3$, CuO, and Y_2BaCuO_5.

CONCLUSIONS

YBCO reacts strongly with CO_2 at high temperatures, leading to a decrease in J_c. The secondary phases, which formed at the grain boundaries as a result of reaction of YBCO with CO_2, block the flow of superconducting current. The secondary phases are identified as $BaCuO_2$ and $BaCO_3$ and appear as a thin layer coating the grain boundaries. The carbon segregation results in a phase transformation of the orthorhombic phase to the nonsuperconducting tetragonal phase. At CO_2 levels as high as 5%, the YBCO completely decomposes to $BaCO_3$, Y_2BaCuO_5, and CuO.

REFERENCES

1. P. K. Gallagher, G. S. Grader, and H. M. O'Bryan, Mat. Res. Bull., 23, 1491 (1988).

2. E. K. Chang, E. F. Ezell, and M. J. Kirschner, Supercond. Sci. Technol. J., 8, 391 (1990).

3. T. B. Lindemer, C. R. Hubbard, and J. Brynestad, Physica C., 167, 312 (1990).

4. T. M. Shaw, D. Dimos, P. E. Batson, A. G. Schrott, D. R. Clarke, and P. R. Duncombe, J. Mater. Res., 5, 1176 (1990).

5. Y. Gao, K. L. Merkle, C. Zhang, U. Balachandran, and R. B. Poeppel, J. Mater. Res., 5, 1363 (1990).

6. Y. Gao, Y. Li, K. L. Merkle, J. N. Mundy, C. Zhang, U. Balachandran, and R. B. Poeppel, Mater. Lett., 9, 347 (1990).

7. M. S. Jahan, D. W. Cooke, H. Sheinberg, J. L. Smith, and D. P. Lianos, J. Mater. Res., 4, 759 (1989).

8. H. Fjellvag, P. Karen, A. Kjekshus, P. Kofstad, and T. Norby, Acta Chem. Scand., $\underline{A42}$, 178 (1988).

9. E. A. Cooper, A. K. Bangopadhyay, T. O. Mason, and U. Balachandran, submitted to J. Mater. Res., September 1990.

10. U. Balachandran, R. B. Poeppel, J. E. Emerson, S. A. Johnson, M. T. Lanagan, C. A. Youngdahl, D. Shi, K. C. Goretta, and N. G. Eror, Mater. Lett., $\underline{8}$, 454 (1989).

11. J. D. Jorgensen, M. A. Beno, D. G. Hinks, L. Soderholm, K. J. Volin, R. L. Hitterman, J. D. Grace, I. K. Schuller, C. U. Segre, K. Zhang, and M. S. Kleefisch, Phys. Rev., $\underline{B36}$, 3608 (1987).

TEXTURING OF EPITAXIAL Y-Ba-Cu-O THIN FILMS ON CRYSTALLINE AND POLYCRYSTALLINE SUBSTRATES

H.S. Kwok, J.P. Zheng, S.Y. Dong, E. Narumi, F. Yang and D.T. Shaw
Institute on Superconductivity
State University of New York at Buffalo
Bonner Hall, Amherst, NY 14260

ABSTRACT

The degree of texturing of the grains in epitaxial Y-Ba-Cu-O films was studied by x-ray measurements. These high quality laser deposited films were all aligned with the c-axis perpendicular to the substrate surface, with a spread angle ranging from 0.3° to 4°, depending on the substrate quality. It was found that films on SrTiO3(100) and LaAlO3(100) were totally aligned with the a-axis either parallel or perpendicular to the [100] direction of the substrate. On MgO(100) there was a small fraction of the grains in the $<110>$ direction. For yttria-stabilized-zirconia, the a-axes of the YBCO grains were distributed in the $<100>$ and $<110>$ directions of the substrate. These observations are consistent with the degree of lattice mismatch in the various directions. For a polycrystalline substrate such as hastelloy, the in-plane texturing was found to be completely random, consistent with the random nature of the substrate. The correlation of these texturing results with the J_c values was found to be excellent.

1. INTRODUCTION

It is well-established that c-axis oriented Y-Ba-Cu-O (YBCO) films can be deposited on various substrates by a variety of techniques. Among the more sucessful ones are laser deposition, sputtering and evaporation[1-3], where superconducting films can be obtained in-situ without any post-deposition treatment. However, the texturing of these films has not been too well characterized, although it is believed to be quite good[4-6]. In this paper, a study will be reported on measuring the in-plane a- and b- axes orientations of the YBCO films on various substrates.

Our results showed that quantitative data on the in-plane texturing of the laser deposited films could be obtained using x-ray diffraction techniques. Different substrates were found to produce different degrees of texturing. Such texturing, of course, affects the critical current J_c and other electrical characteristics of the films. In particular, our results on polycrystalline substrates showed that there was no texturing on the substrate surface even though all the grains have parallel c-axes. This result is important in providing a baseline for improving the J_c of such films which are used in making superconducting tapes.

2. SAMPLE PREPARATION

The YBCO films to be discussed in this paper were all deposited using an ArF excimer laser with in-situ processing. The details of the deposition system and technique had been documented thoroughly[1]. The substrates used included SrTiO3, LaAlO3, MgO, yttria-stabilized-zirconia (YSZ), all cut with a (100) surface, and polycrystalline metals such as Hastelloy. Because of the lattice mismatch between YBCO and the substrates, true hetero-epitaxy in the sense of uniform layer-by-layer growth is impossible. Microstructural studies have shown that the film consists of many small grains with sizes ranging from 50 to 300 nm[4,7]. Within the grains, the crystallinity is excellent. For the case of SrTiO3, there were reports of excellent epitaxy because of the small lattice mismatch (0.4%). However, for most cases, there is a highly strained interface with the substrate where the lattice mismatch is compensated by dislocations. In all substrates, the Cu-O plane is always parallel to the substrate surfaces.

Using a θ-2θ x-ray measurement, it was found that only the (00L) peaks are present for all substrates, including the case of Hastelloy. Fig. 1 shows the example of MgO. The films were patterned into strips for J_c mesurements using the 4-probe method. J_c was defined using the 1 μV/cm criteria. A summary of the results is shown in Table I. In general the J_c(77K) is highest in SrTiO$_3$ and lowest in Hastelloy. Note that in all cases the T_c's are approximately the same. However J_c(77K) is over 4×10^6 A/cm^2 for MgO and only 4×10^4 A/cm^2 for Hastelloy. The subject of this paper is to explain this difference in J_c despite the similar T_c by texture analysis.

3. TEXTURE MEASUREMENT

Actually the subject of grain orientation is a matter of concern and has been studied by microstructural analysis[4-6]. For the case of MgO, a recent study showed that the a-axis of the YBCO grains was either parallel to the MgO[100] or MgO[110] axes[5]. The percentage of each orientation, however, could not be quantified because of the fine resolution of the electron microscope. Reference 5 reports a careful pole-figure texture analysis of post-anneal YBCO films. For high quality in-situ laser deposited films, such a study is lacking.

Fig. 2 shows the two different kinds of texturing that can be measured with x-ray analysis. The spread in the c-axis of the large number of grains is defined as Δ. It can be measured by fixing θ and rocking the film surface. Secondly, the alignment as well as the spread in the distribution of the YBCO[100] axis in the plane of the film can be measured in a Field-Merchant configuration as discussed below.

The result of the rocking curve measurement showed that there was always a tilt in the collective c-axis as shown in Fig.2. This tilt ranged from 0.3° to 3° and is dependent on the crystal supplier. We believe that this is due to improper polishing of the substrate so that the surface is not exactly (100). This was confirmed by performing the rocking curve measurement with the substrate only. The spread angle Δ of the c-axis ranges from 0.3° for SrTiO$_3$ and MgO, to 3° for Hastelloy[8]. This is in agreement with the polycrystalline nature of the latter substrate.

Fig. 3 shows the geometry used in the in-plane texturing study. Ordinarily, in a θ-2θ scan of the film, only the (00L) peaks can be observed. However, if the sample surface is tilted relative to the x-ray beam, as shown in Fig. 3, then other peaks can also be detected. In particular, if the film is tilted by $\alpha = 18.5°$, then the (109) peak can be detected at $2\theta = 77.24°$, assuming that the [100] axis is in the plane of incidence. Now if the film is rotated in its own plane (angle φ), the x-ray diffraction signal will be detectable only if there is a grain with an a-axis parallel to the plane of incidence. Thus a plot of the x-ray signal versus φ will provide a measure of the distribution of grain orientations on the film surface. This is essentially the Field-Merchant method of texture measurement (φ-scan).

Actually, the case of YBCO is slightly more complicated. It is because the crystalline structure is orthorhombic with a b. Hence the (109) and the (019) peaks occur at slightly different angles. For the (019) peak, the 2θ is 77.42°, a difference of almost 0.2° to (109). In the experiment, the angle θ had to be adjusted for each angle φ. This was done by peforming a rocking curve for each angle φ. Then the peak of the rocking curve was plotted against φ to provide the texturing

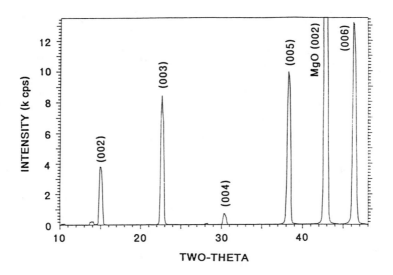

Fig. 1 θ-2θ scan of YBCO/MgO(100). The film thickness is 0.15 μm, deposited at 650°C by the ArF laser.

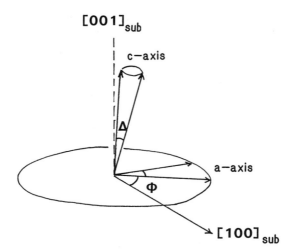

Fig. 2   Schematic showing the c-axis and a-axis texturing to be characterized.

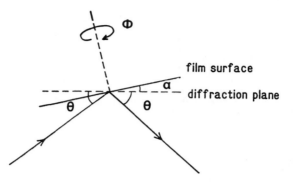

Fig. 3 Field–Merchant geometry for in-plane texturing analysis

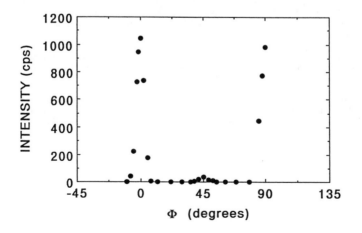

Fig. 4 φ-scan of the YBCO(109)/(019) peak for MgO.

data. This procedure also compensate for the slight nonparallelism of the c-axis between the grains. The drawback is that 90° twins cannot be distinguished.

4. EXPERIMENTAL RESULTS

The results are quite inteesting. For MgO(100) substrate, it was found that there were two equally strong peaks as φ was varied. They occurred at $\varphi = 0^\circ$ and $\varphi = 90^\circ$. Fig. 4 shows the results. It is believed that the FWHM of 7° is due to the instrument resolution. This was confirmed by performing the same measurement with the single crystalline substrate. Since the (109) and (019) peaks differ by only 0.2° in 2θ, they cannot be distinguished by our instrument. There can be both (109) and (019) components in the two observed maxima in the φ-scan. The conclusion is that the YBCO[100] axis is either parallel or perpendicular to the MgO[100] axis. As seen in Fig. 4, there is also a small peak at $\varphi = 45^\circ$. This corresponds to the a-axis parallel to the MgO[110] direction. From the peak values of the x-ray signals, the maxima at $\varphi = 45^\circ$ was about 4% of the value at $\varphi = 0$.

For MgO, therefore, it can be concluded that the grains are aligned with the a-axis either parallel or perpendicular to the [100] axis of the substrate. Only a small percentage is aligned with the MgO[110] axis. This is reasonable because the lattice mismatch in the $<100>$ direction is only $\sim 8\%$, while it is $\sim 50\%$ in the $<110>$ direction. The lattice mismatch is 1.8% if the YBCO[300] axis overlaps with the MgO[220] axis. Apparently such small lattice mismatch is not enough to overcome the large interface energy to encourage growth in this direction. In reference 5 where both the YBCO[100]//MgO[100] and YBCO[100]//MgO[110] are studied, the relative percentages are not known. From the present study, we can see that the latter type is insignificant. It is also concilded that the "near coincident site" theory is not important in determining the growth orientation of most of the grains[7].

For YSZ(100), the situation is quite different. A φ-scan of the (109) peak is shown in Fig. 5. It can be seen that there is a large signal at $\varphi = 45^\circ$. As a matter of fact, the ratio of the YBCO[100]//YSZ[110] signal to the YBCO[100]//YSZ[100] signal is about 1.9. In other words, the grains are mostly grown with the a-axis aligned in the $<110>$ direction of the substrate. This is again reasonable because the lattice mismatch in the $<100>$ direction is 11%, while it is only 4.5% in the $<110>$ direction. It should be noted that in determining the relative signals in the $<100>$ and $<110>$ directions, we also performed a φ-scan with just the substrate itself to locate the [100] axis.

Actually, it is surprising that there is such a large percentage of grains in the $<100>$ direction despite the much larger lattice mismatch as compared to the $<110>$ direction. It is possible that in determining the growth orientation, symmetry matching also plays an important role in addition to lattice matching considerations. This may be why so many grains are aligned in the $<100>$ direction despite the large lattice mismatch.

For SrTiO3 and LaAlO3, the results are as expected. There were only peaks at $\varphi = 0^\circ$ and $\varphi = 90^\circ$. There was absolutely no signal at $\varphi = 45^\circ$. This is reasonable because of the small lattice mismatches (0.4% and 0.8% respectively).

For the polycrystalline substrate, the φ-scan shows a complete random distribution of the a-axis in the plane of the substrate. Fig. 6 shows the result for YBCO on Hastelloy with a YSZ buffer layer[8]. This result is reasonable because the buffer layer is polycrystalline with no preferred orientation also.

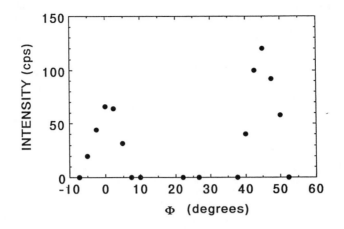

Fig. 5 φ-scan of the YBCO(109)/(109) peak for YSZ.

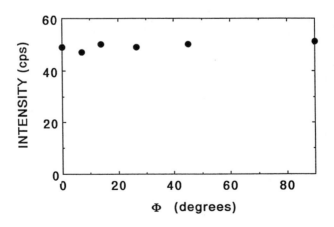

Fig. 6 φ-scan of the YBCO(109)/(019) peak for Hastelloy.

5. DISCUSSIONS AND CONCLUSION

From the above results, it can be inferred that in the formation of thin films on crystalline substrates, nucleation occurs at various local sites. These grains grow epitaxially in the direction of least lattice mismatch and symmetry mismatch. Eventually, the grains merge to form a complete film. Even for the case of complete wetting in YSZ[1], it is impossible to form just a single grain on the entire substrate surface.

All the experimental results are summarized in Table I. The texturing results explains why SrTiO3, MgO and LaAlO3 are good substrates to produce high J_c films of YBCO. However, it is interesting to note that the J_c for the YSZ film is as high as it is, since it is well-known that 45° misoriented grains will lead to a factor of 70 reduction in J_c10. Perhaps it is because of the small grain structure so that there is always a percolation path for the aligned grains to conduct the current in YSZ. In other words, one can at most disregard one half of the film as "bad". Hence J_c is reduced by about a factor of two.

For the case of Hastelloy, J_c is about 100x less than the optimum value. Again since the misoriented grains will have a 70x reduction in J_c, this is not an unreasonable result. In this case, since the grains are completely random, it is impossible to find a percolation path of grains with the same orientation. It has been calculated that the reduction in J_c for such a random network is about 50x. This factor just about accounts for the observed factor of 100. This implies that the grain quality in the polycrystalline case is probably similar to those on crystalline substrates.

In summary it can be seen that x-ray texturing measurements can be applied to quantify the grain orientation in polycrystalline films of YBCO. Since most "epitaxial" thin films are polycrystalline, such grain orientation information is of extreme importance in characterizing their performance. It can also be used to provide a criteria for the optimization of the deposition parameters to obtain high J_c films, especially on polycrystalline substrates.

This research was supported by New York State Institute on Superconductivity through the New York State Energy Research and Development Authority.

6. REFERENCES

1. Q.Y. Ying and H.S. Kwok, Appl. Phys. Lett., 56, 1478 (1990).
2. S. Prakash, D.M. Umarjee, H.J. Doerr, C.V. Deshpandey and R.F. Bunshal, Appl. Phys. Lett., 55, 504 (1989).
3. Q.X. Jia and W.A. Anderson, Appl. Phys. Lett., 57, 304 (1990).
4. R. Ramesh, D. Hwang, T.S. Ravi, A. Inam, J.B. Barner, L. Nazar, S.W. Chan, C.Y. Chen, B. Dutta, T. Venkatesan and X.D. Wu, Appl. Phys. Lett., 56, 2243 (1990).
5. M.G. Norton, L.A. Tietz, C.B. Carter, S.E. Russek, B.H. Moeckley and R.A. Buhrman, in Matl. Res. Soc. Proc. 169, J.A. Narayan et. MRS Publ. Pittsburgh, PA, 1990.
6. J.D. Budai, R. Feenstra and L.A. Boatner, Phys. Rev. 39, 12355 (1989).
7. D.M. Huang, T. Venkatesan, C.C. Chang, L. Nazar, X.D. Wu, A. Inam and M.S. Hegde, Appl. Phys. Lett., 54, 1702 (1989).
8. E. Narumi, L.W. Song, F. Yang, S. Patel, Y.H. Kao and D.T. Shaw, Appl. Phys. Lett., 56, 2684 (1990).
9. T.S. Ravi, D.M. Hwang, R. Ramesh, S.W. Chan, L.Nazar, C.Y. Chen, A. Inam and T. Venkatesan, preprint.
10. D. Dimos, P. Chaudhari and J. Mannhart, Phys. Rev. B 41, 4038 (1990).

TEXTURING HIGH TEMPERATURE SUPERCONDUCTORS AND MAXIMIZING FLUX PINNING IN SUPERCONDUCTING COILS

N.X.Tan and A.J.Bourdillon

State University of New York at Stony Brook
Department of Materials Science and Engineering
Stony Brook, NY 11794-2275

ABSTRACT

Anisotropies in both critical current carrying capacity and in fluxoid pinning from single-crystal high temperature superconductors determine optimum orientations for directionally solidified (from the partial melt) microcrystals in material for coil applications. Optimization depends partly on the selected coil diameter. In melt texturing, crystal growth is accompanied by liquid flow which generally leads to the production of second phases by segregational effects except where careful procedures are adopted to prevent these [1]. These segregational effects tend to reduce potential current carrying capacities even though they can in principle be used to increase flux pinning. Texturing of coils imposes special difficulties because of the mechanical weakness of the partially melted region. Progress in overcoming these difficulties is of great importance to a significant range of applications requiring either higher magnetic fields than are attainable with conventional superconductors, or the economy of superconduction at higher temperatures.

INTRODUCTION

High temperature superconductors such as $YBa_2Cu_3O_{7-x}$ {123} and Bi-Sr-Ca-Cu-O {BSCCO} material have anisotropic electrical conduction and magnetic properties [2]. In polycrystalline material, grain boundary weak links such as resistive boundary layers, high angle boundaries, crystallographic defects and composition variations at boundaries, result in low critical current densities (J_c) [3]. These phenomena can be used to explain why sintered bulk specimens typically have much lower J_cs than single crystal thin films [4]. However, bulk superconducting materials, such as are used in wires and coils, have promising potential applications in power transmission, magnets, motors, magnetic images, energy storage etc. [5]. These applications require high critical current densities, and so there has been considerable interest in developing material with high J_c for superconducting wires and coils.

Some techniques have been employed to texture ceramic superconducting wires by aligning grains in magnetic fields and by partial melting-texture-growth with thermal gradients [1,6]. The latter takes advantage of features of the phase diagram [7], whereby at around 1010°C {123} transforms, by incongruent melting, into liquid and solid Y_2BaCuO_5. At just below this temperature, growth of {123} can occur preferentially parallel to the (001) crystallographic plane, i.e. in the plane of maximum conduction. If a negative temperature gradient lies parallel to this plane, elongated grain growth along the gradient results in texture. The partial melting texture technique with thermal gradients has also been developed by using vertical thermal gradients with gravity used to direct the molten phase to the area of re-crystallization. In this partial melting regime the specimen can be kept rigid. Crystal growth, under these conditions, not only preserves stoichiometry but excludes second phases by zone refinement. As described below, in gravity aided texture growth (GATEG), the specimen is drawn slowly downwards through the temperature gradient of a furnace with narrow hot zone. The specimen is subsequently annealed in oxygen. Several other methods have been reported to increase J_c such as introducing second phases [8] and radiation damage to enhance flux pinning effects [9].

Texturing coils is more complex for several reasons: the shape is non-linear and anisotropic transport properties impose demands above those described previously, involving fibre texture in wires. The a-b planes in high T_c (critical transition temperature) materials generally carry maxima currents while the flux pinning force is weakest when the magnetic field, **B** is parallel to c axes. In BSCCO materials these effects are particularly strong because they have platelet granular morphologies. In coils strong flux pinning is desirable, particularly because at the temperatures envisaged for the operation of high T_c materials, residual resistance occurs through energy dissipation of mobile flux vortices. In order to achieve maximum current carrying capacity and to generate the highest magnetic fields in textured coils, optimized orientations will be required. A novel technique is being applied to develop sheet-textured BSCCO by taking advantage of its platelet grains. This technique has important application in texturing BSCCO coils.

The aligned platelet BSCCO materials can be further textured by GATEG. GATEG can be generally used to texture ceramics with preferential growth planes and incongruent melting.

EXPERIMENTS

Powders of $Bi_2Sr_2CaCu_2O_8$ {2212} and $Bi_2Sr_2Cu_2Cu_3O_{10+y}$ {2223} and of {123} were prepared by normal ceramic processing routes [1]. The grain size was in the range 1-10 μm.

a) sheet-texturing BSCCO

After grinding the sintered {2212} pellets, the {2212} powders are mixed with slow-hardening epoxy with volume ratio of ~20:1. The mixture is manually compacted in a copper sheath, followed by mechanical rolling with a standard workshop rolling mill. After setting, the metal sheath was chemically removed by diluted HNO_3 (~ 0.3 N). The epoxy in the tapes was dissolved away in acetone before sintering to achieve dense material after sintering. The {2212} tapes were finally placed on a ZrO_2 slab and were sintered at 840°C for 12 hours.

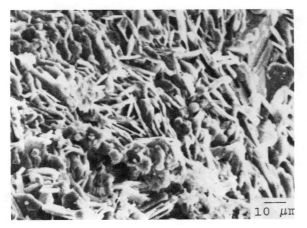

Figure 1 Secondary electron image (SEI) of a fractured cross-section of BSCCO superconducting tape, shows aligned morphology of BSCCO platelets parallel to the rolling direction.

The microstructure was characterized in an ISI SX 30 scanning electron microscope (SEM). Figure 1 shows a typical fractured surface morphology on the cross-section. The aligned morphology of {2212} platelets is observed parallel to the rolling direction. The a-b planes of {2212} are generally perpendicular to the cross-section so that the current can be carried on the a-b planes through the tapes. The aligned {2212} tape can be further textured by GATEG to promote texture growth. Experiments carried out by GATEG on {123} and {2223} are described below:

b) texturing {123} and {2223} by GATEG
The fabrication of {123} wires is described elsewhere [1,10]. Firstly specimens of {123} wires were textured at 1030°C in a conventional horizontal tube furnace with 150 mm long hot zone with edge gradients estimated to be about 0.3°C/mm. These specimens were subsequently compared with other specimens of {123} textured in a vertical tube furnace.

The microstructures of the specimens were examined in a JEOL JSM 840 SEM equipped with Link Systems energy dispersive spectrometer (EDS). Figure 2 shows that perfect stoichiometry was not retained: second phases of {211} and of CuO were observed together with majority {123}. The material had partial crystal alignment with high density except for the presence of a few large bubbles. Resolidified melt was found to be gathered at the base of the material, where some reaction with the crucible was observed.

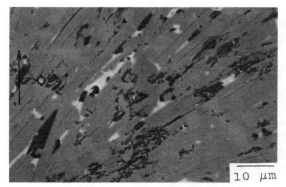

10 μm

Figure 2. Backscattered image (BEI) of polished YBCO, sintered in horizontal tube furnace. Arrow shows direction of fibre axis and temperature gradient.

Secondly, {123} was sintered in a vertical tube furnace. This furnace was used to avoid problems of melt segregation. The furnace was designed with a narrow hot zone (~3cm). A sintered wire specimen, about 20 mm long, was made to stand in a crucible containing alumina grog. The crucible was suspended on Kanthal wire which was drawn over a pulley and connected to a motor with gear. The speed could be adjusted between 10mm/s and 1 μm/min. The thermal gradient about the narrow hot zone was measured to be 1°C/mm at 900°C. Specimens were drawn upwards and downwards at various rates and temperatures. The optimal crystal growth was achieved by downward drawing of 3 mm/h at 1025°C. Thus single phase (except at the top, as described below) textured material was produced. On the cooling cycle, oxygen was passed through the furnace while the temperature was slowly cooled from 700 to 400 °C over 40 h. Figures 3 a & b demonstrates substantial grain growth without formation of second phases and a void from where a large 'single crystal' domain with volume >1mm^3 was extracted. By comparison, alignment was less perfect in specimens drawn upwards, and EDS showed evidence of considerable segregation. Moreover, with downward drawing, second phases of comparatively low melting point

were observed to be concentrated at the top end of the
specimen (figure 4) proving effects of phase purification
through zone refinement.   In a vertical furnace with
downward drawing, molten phases within the partial melting
regime, react with cooling solid so that crystal growth
parallel to the thermal gradient occurs. Since the
recrystallization occurs within the molten zone,
stoichiometry is preserved, while the specimen, being only
partially melted, remains rigid. Contamination occurs only
at the base by contact with supporting grog.   Furthermore
the observation of 'zone refining' leads to the expectation
that incidence of insulating grain boundary impurity phases
will be considerably reduced in this vertically textured
material.   Preliminary investigation in SEM supports this
expectation.

Figure 3 a.   SEI of fracture surface close to lower end of
specimen drawn <u>downwards</u> in vertical tube furnace.
b. SEI at higher magnification, showing perfectly aligned
microcrystals within a aligned domain.

The critical current density of a typical specimen was measured to be 5000 A/cm^2 at liquid helium temperature and 833 A/cm^2 at liquid nitrogen temperature.

100 μm

Figure 4. BEI near top end of specimen aligned in vertical furnace with downward drawing. Zone refinement is demonstrated by concentration of CuO (dark) at top.

Texture was also achieved in {2223} doped with Pb by GATEG at 910°C with process similar to that described above for {123} as observed in Figure 5. The dense morphology and second phases (black) CaCuO$_2$ and CuO are also shown in this figure. The formation of second phases can be due to the evaporation of Bi$_2$O$_3$ at 910°C and the excess CuO and CaO reacted and segregated out.

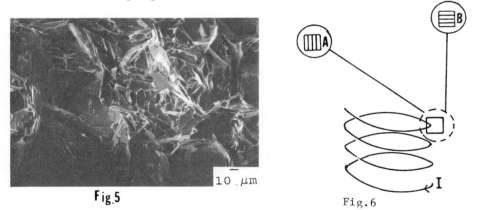

10 μm

Fig.5

Fig.6

Figure 5. SEI of nominal {2223} showing platellar grain growth on the specimen surface.

Fig. 6. Coil configuration and texture. In A, a-b planes are parallel to magnetic field vector, **B**; in B, a-b planes are perpendicular to **B**. Both A & B have a-b planes parallel to current vector.

DISCUSSION AND CONCLUSION

Sheet texture is the alignment of polycrystalline material in planar morphology (as in steel sheet-rolling) so that preferred crystallographic axes adopt distributions about orientations relative to the sheet normal and to the rolling direction. The texture develops partly through dislocation movement under stress and partly through grain growth [11]. The strength of rolled metals depends on the texture developed in rolling operations. Ceramics, being brittle, generally fracture under rolling stress, but rolling can in principle be used for compaction before sintering. Our preliminary results show that in this case plate-like grains can be made to take up preferential orientations with respect to the rolling direction and sheet normal (fig. 1). Fracture and disintegration can be avoided in a variety of ways two examples of which are as follows: (1) The plate-like grains can be formed into composite by suspension in an organic binder, such as polyethylene, before rolling. On subsequent sintering the binder is burnt away, but problems occur as the binder melts and surface tension forces the morphology to change. These problems can in principle be minimized by optimized choice of binder, including its concentration and removing it before sintering. (2) The plate-like grains are compressed inside a metallic sheath before rolling. The sheath is chemically removed before sintering. We have found that the resulting microstructure, while aligned, contains complexities due to the force vectors exerted on the ceramic powder by the combination of the external rollers and the metal sheath surrounding the powder.

Sheet texture can, in principle, be applied to fabrication of textured coils. Figure 6 shows the texture direction relative to the coil orientation. In the cases of both A and B, the aligned a-b planes are parallel to current vector i.e. for maximum current carrying ability. However, in case A, according to the magnetic properties of BSCCO [12], the flux pinning force is optimum since the Lorentz force is applied parallel to c-axis, where fluxon motion can be pinned by insulating layers between the Cu-O planes.

In conclusion, the development of material for superconducting coils above the level of liquid nitrogen requires the application of techniques for sheet texturing the anisotropic layered compounds. A careful choice of oriented crystal alignment can be used to maximize both current carrying capacity and flux pinning ability.

REFERENCES

1. A.J.Bourdillon, N.X.Tan, N.Savvides and J.Sharp, Mod. Phys. Lett. B 3, 1053 (1989)
2. T.R.Dinger, T.K.Worthington, W.J.Gallagher and R.L.Sandstrom, Phys. Rev. Lett., 58, 2678, (1988).
3. D.Dimos, P.Chaudhari, J.Mannhart and F.K.LeGouse, Phys. Rev. B 37, 1563 (1988)
4. A.M.Wolsky, R.F.Giese and E.J.Daniels, Scientific American, Feb., 45, (1989)
5. K.E.Easterling, C.C.Sorrell, A.J.Bourdillon, S.X.Dou, G.J.Sloggett and J.C.MacFarlane, Materials Forum 11, 30 (1988)
6. D.Farrell, B.Chandrasekhar, M.DeGuire, M.Fang, V.Kogan, J.Chem and D.Finnemore, Phys. Rev. B 36, 4025 (1987)
7. B.L.Lee, D.N.Lee, J. Am. Ceram. Soc., 72, 314, (1989)
8. M.Murakami, M.Morita, K.Doi, K.Miyamoto and H.Hamada, Jap. J. Appl. Phys. 28, L399 (1989)
9. R.B.Van Dover, E.M.Gyorgy, L.F.Schneemeyer, J.W.Mitchell, K.V.Rao, R.Puzniak and J.V.Waszczak, Nature 342, 55 (1989)
10. N.X.Tan and A.J.Bourdillon, to be submitted to Materials Lett. (1990)
11. C.Barrett and T.B.Massalski, Structure of Metals, 3rd revised ed. (Pergamon, N.Y. 1980), International Series on Materials Science and Technology, 35
12. B.D.Biggs, M.N.Kunchur, J.J.Lin, S.J.Poon, T.R.Askew, R.B.Flippen, M.A.Subramanian, J.Gopalakrishman and A.W.Sleight, 39, 7309 (1989)

ON THE FABRICATION, CHARACTERIZATION AND ENHANCEMENT OF YBCO FILMS ON LiNb0₃ SUBSTRATES DEPOSITED BY A SPUTTER S-GUN*

M. Eschwei, F.Lin, Q.G. He, S. Bielaczy and W.C. Wang
Electrical Engineering Department
Polytechnic University
Farmingdale, NY 11735

ABSTRACT

High quality YBCO films on LiNb0₃ substrate through in-situ deposition without post annealing have been reported. We have grown fine quality thin films with $T_c \approx 90°K$ on Y-cut LiNb0₃ substrate utilizing a sputter S-gun and a special post annealing process. The technique of thin film processing, interfacial studies, characterization by ultrasonics as well as film quality enhancement by ion implantation will be discussed in this paper.

INTRODUCTION

Epitaxial growth of superconducting YBaCuO film on LiNb0₃ has been reported.[1-3] Highly textured films reported thus far are through in-situ deposition without post annealing. Films through post annealing have been generally of lower quality. However, we have found that by adjusting the heating rate during annealing and finding the optimum maximum annealing temperature, both the transition temperature T_c and film orientation can be greatly improved. Furthermore, the micro-cracks induced by the mismatch in the thermal expansion coefficient between the film and the substrate can be mostly avoided. During the course of examining the lattice match between the YBCO film and LiNb0₃ substrate required for epitaxial growth. The formation of an interfacial layer has been found to exist between the film and the substrate. The newly formed layer may provide the necessary condition for the growth of the highly textured film. Some of the tetragonal, oxygen deficient YBCO film has been implanted with 0^+ ions at moderate dosage $\approx 10^{11}$ atoms/cm² in order to improve its stoichiometric composition. After implantation and low temperature annealing in air, the tetragonal phase film has been observed converting into the superconducting orthorhombic phase.

The paper will be presented as follows: First, the technique of thin film processing, including conditions of sputtering and post annealing, next, the interfacial studies and the characterization of the film by ultrasonics, and finally, film quality enhancement by ion implantation.

SPUTTERING

Films are deposited on LiNb0₃ substrates by r.f. sputtering utilizing a magnetron S-gun with a single sintered stoichiometric target. During sputtering additional magnets are employed to shape the plasma column so that the bombardment at the substrates by electrons and other ionized particles are at a minimum.[4] The general sputtering conditions are listed below.

---

*Research partially supported by N.Y. State Center for Advanced Technology in Telecommunications.

Target: Sintered YBa$_2$Cu$_3$O$_{7-x}$ superconductor
Substrate: Y-cut LiNbO$_3$
Substrate Temp: <150° C (No preheating)
Base Vacuum Pressure: 5x10^{-7}Torr
Sputtering gas (Argon) Pressure: 5x10^{-3}Torr
Oxygen Partial Pressure: 0
Deposition rate ≈45Å/min
Film Thickness: ≈1.5 μm
Source Power (r.f. sputtering): 150w

The as-deposited film, analyzed by x-ray diffraction, is amorphous and semi-insulating. Note that the optimum condition for obtaining high quality films on LiNbO$_3$ is to keep the oxygen partial pressure at zero.

ANNEALING

Due to the differential between the thermal expansion coefficient of the LiNbO$_3$ substrate and that of the YBaCuO compound, microcracks within the thin film are generally observed after post annealing. The annealed film is neither smooth nor with good adhesion to the substrate. It has been found, however, that by adopting different heating rates at various heating stages, the microcracks in the film can be mostly avoided and the superconducting properties improve as well. A typical annealing cycle for LiNbO$_3$ substrates is given below.

Heating:	From (°C)	To (°C)	Rate
	27	300	100°C/sec.
	300	600	40°C/min.
	600	875	20°C/min.
Flush with Helium at 875°c for 1.5 min.			
Cooling:	875	27	5°C/min
Oxygen Flow Rate during annealing is 3 Liters/min.			

The recorded resistance vs. temperature profile during annealing is shown in Fig. 1.[5] The optimum maximum annealing temperature, 875°C, is found to be critical, above which a non-superconducting phase prevails and below which T$_c$ reduces. Fig. 2 shows the superconducting properties of the annealed film with T$_c$ onset at 94°K and zero resistance at 87°K.

Fig. 3 is one of the x-ray diffraction pattern of the processed films, which highlights the (004) peak. Some of the processed films, however, do not possess such a strong (004) peak.

INTERFACIAL STUDIES

It is an experimental fact that high quality YBCO superconducting films can be grown on y-cut LiNbO$_3$ substrates. X-ray diffraction has shown that the c-axis of YBCO film is perpendicular to the substrate surface and the diagonal of the a-b planes (5.45Å) of the orthorhombic film is in alignment with the shorter side (5.15Å) of the rectangular cell in the a-c plane of the y-cut LiNbO$_3$ substrate.[2,3] It has been quite

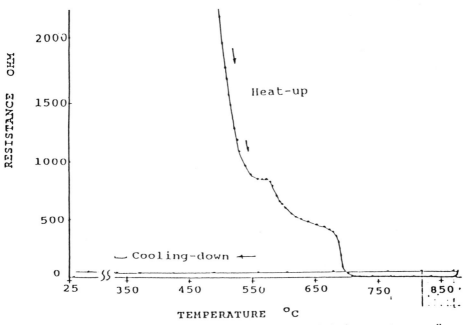

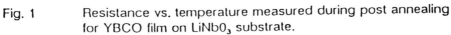

Fig. 1   Resistance vs. temperature measured during post annealing for YBCO film on LiNbO$_3$ substrate.

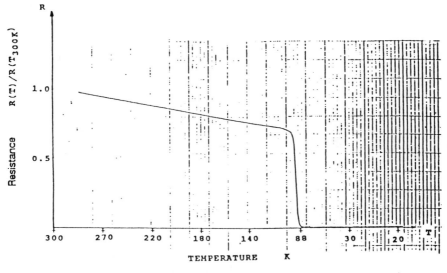

Fig. 2   Resistance vs. temperature for YBCO film on LiNbO$_3$ substrate.

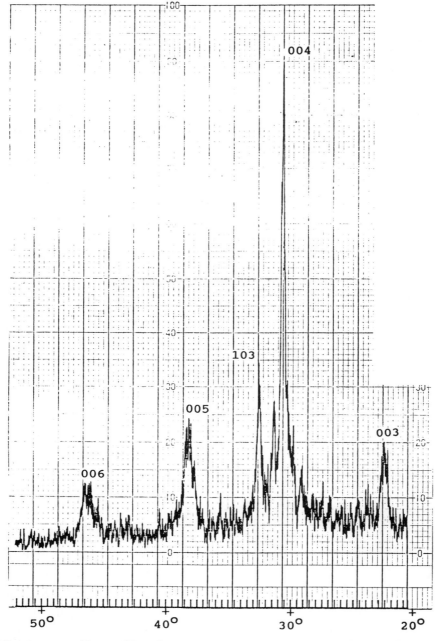

Fig. 3   X-ray diffraction pattern of a superconducting YBCO film on LiNbO$_3$ with strong (004) peak.

mystifying that such a highly textured film can be grown on the substrate with a mismatch more than five percent.

    We have also searched for clues in order to understand the growth mechanism. It has been reported that $LiNbO_3$ can crystallize in a nonstoichiometric form, $(Li_2O)_v(Nb_2O_5)_{1-v}$ with $0.48<v<0.5$.[6] It has also been known that when $LiNbO_3$ is heated in a vacuum, both Li_2O and oxygen will diffuse out; i.e., the surface layer of the $LiNbO_3$ substrate will become somewhat distorted. To demonstrate this point, some Y-cut $LiNbO_3$ substrates were heated in vacuum for some length of time (similar to the processing conditions of YBCO film). Subsequently, the substrates were analyzed by x-ray diffraction. Fig. 4a and 4b compares the x-ray diffraction pattern before and after the heat treatment. We note that the texture of the surface layer has been greatly altered. By coincidence the newly appeared diffraction peak at $2\theta=30.9°$ coincides well with the [004] peak of the YBCO film. We conjecture that the new surface layer offers a better lattice match. In addition, it may be of value to point out that niobium films have been successfully used as buffer layers for the growth of high T_c YBCO film on the alumina-ceramic substrate.[4] The Niobium layer, of course, would be oxidized during post annealing.

CHARACTERIZATION BY ULTRASONICS

    A number of anomalies on ultrasonic measurements of sintered YBCO high T_c ceramics have been reported.[7,8] In order to better understand the mechanism associated with the observed anomalies, more refined experiments have to be performed. Some refinements can be achieved by performing ultrasonic measurements on superconducting thin films, since high quality thin films are better oriented single crystals. Thus, these measurements will eliminate most of the complexities associated with polycrystalline ceramics. We have made such an attempt by depositing surface acoustic wave (SAW) transducers on y-cut, z-propagating $LiNbO_3$ substrates. Fig. 5a shows the surface wave attenuation vs. temperature measurement of a $LiNbO_3$ SAW delay line with Ag film deposited on the substrate surface. The r.f. signal applied to the input transducer is held constant at all temperature settings. At each temperature setting the corresponding output signal amplitude is recorded as one vertical trace in the oscillogram. After the first trace is taken, for example, we decrease the temperature by 8°K and shift the output trace horizontally to the right by one fifth a division, then we take the second trace; i.e., there are more than 35 traces (exposures) in each of the oscillograms 5a and 5b. Fig. 5b represents the same attenuation vs. temperature measurement, but with the silver film replaced by the superconducting YBCO film.Silver film is used in Fig. 5a, because the Ag film, just as the YBCO film would short-out the electric field induced by the SAW. Comparing Fig. 5a and 5b, it appears that three structural phase transitions have taken place in the YBCO film at temperatures 74°K, 118°K and 220°K respectively. It should be pointed out, however, that the above attenuation measurement is not precise, since all the four transducers, two in Fig. 5a and two in 5b are assumed to have the same temperature dependence. Refined experiments should be designed to measure the propagation loss only.

QUALITY ENHANCEMENT BY ION IMPLANTATION

    The technique of ion implantation has been used to introduce Cu atoms into Cu-deficient YBCO sputtered films to achieve the desired stoichiometric composition. After implantation the T_c, especially the zero resistance temperature, has been greatly improved. Yttrium atom have been implanted into films formed by coevaporation of

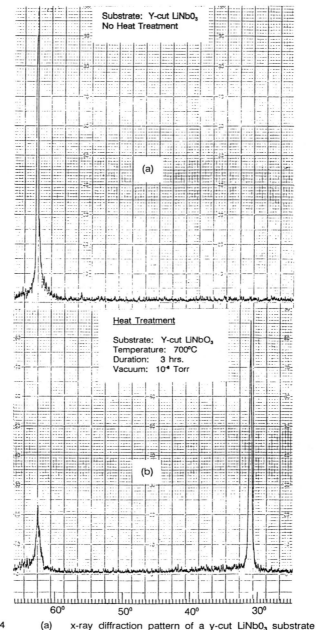

Fig. 4   (a)   x-ray diffraction pattern of a y-cut LiNbO₃ substrate
before heat treatment.

        (b)   after heat treatment.

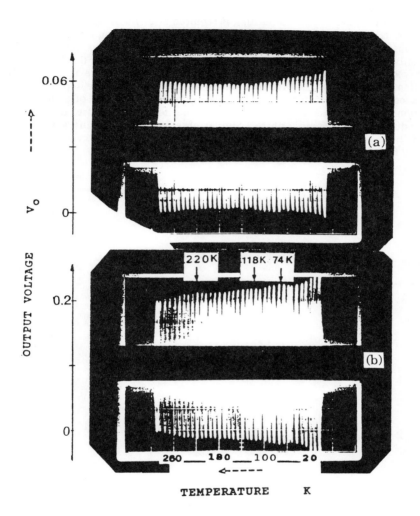

Fig. 5   (a)   Attenuation vs. temperature characteristics
                 of a Y-cut, z- propagating LiNbO$_3$ SAW delay line
                 with Ag film on LiNbO$_3$ surface. f=48.3MHz.
                 $V_{in(p \cdot p)}$=const. = 0.53 V at all temp.,
                 $V_{o(p \cdot p)}$=0.06 V at 294°K.

          (b)   Same as 5a, except Ag film is replaced by **a**
                 superconducting YBCO film.
                 $V_{in(p \cdot p)}$ = const. = 1.9 V, $V_{o(p \cdot p)}$=0.2 V at 295°K.

BaF$_2$ and Cu. The Y-implanted film was superconductiing at 85°K. In both cases, standard high temperature post annealing is required after implantation.[9,10]

Our laboratory experiments have been carried out in implanting 100 keV oxygen ions into the semiconducting YBCO tetragonal phase film at room temperature with relatively low dosage, ~5x10^{11} atoms/cm^2. After implantation, the intensity of the x-ray diffraction peaks are greatly enhanced. However, there is no shift in diffraction peak's position, i.e., the film remains tetragonal. When the implanted film is subjected to low temperature annealing (350°C) in air for a period of an hour, the diffraction peak is observed to shift from the tetragonal phase toward the superconducting orthorhombic phase. Similar observation on the diffraction peak shift has been reported in the process of plasma-reoxidation on oxygen deficient tetragonal films.[11,12].

REFERENCES

1. T. Venkatesan, C.C. Chang, D. Dijkkamp, S.B. Ogale, E.W. Chase, L.A. Farrow, D.M. Hwang, P.F. Micceli, S.A. Schwarz, J.M. Tarascon, X.D. W u and A. Inam, J. Appl. Phys., 63, 4591 (1988).

2. A. Hohler, D. Guggi, H. Neeb and C.Heiden, Appl. Phys. Lett. 54, 1066 (1989).

3. S.G. Lee, G. Koren, A. Gupta, Armin Segmuller and C.C. Chi, Appl. Phys. Lett. 55(12), 1261, 1989.

4. S. Onishi, H-M Shieh and A. Elshabini-Raid, Japanese J. of Appl. Phys., 28, (10), 542, 1989.

5. A. Davidson, A. Palevski, M.J. Brady, R.B. Laibowitz, R. Koch, M. Scheuermann and C.C. Chi, Appl. Phys. Lett. 52 (2), 157, 1988.

6. I.P. Kaminow and J.R. Carruthers, Appl. Phys. Lett., 22 (7), 326, 1973.

7. S. Bhattacharya, M.J. Higgins, D.C. Johnson, A.J. Jacobsen, J.P. Stokes, J.T. Lewandowski, and D.P. Goshorn, Phys. Rev. B 37, No. 10, 5901, 1988.

8. C. Duran, P. Esquinazi, C. Fainstein, and M. Nunez, Solid State Commun. 65, No. 9, 957, 1988.

9. M. Rubin, I.G. Brown, E.Yin and D. Wruck, J. Appl. Phys. 66 (8), 3940, 1989.

10. M. Nastasi, J.R. Tesmer, M.G. Hollander, J..F. Smith and C.J. Maggiore, Appl. Phys. Lett. 52 (20), 1729, 1988.

11. B.G. Bagley, L.H. Greene, J.M. Tarascon and G.W. Hull, Appl. Phys. Lett. 51 (8), 622, 1987.

12. A. Yoshida, H. Tamura, S. Morohashi and S. Hasuo, Appl. Phys. Lett. 53 (9), 811, 1988.

METAL-ORGANIC CHEMICAL VAPOR DEPOSITION (MOCVD) OF HIGH TEMPERATURE SUPERCONDUCTORS WITH ENHANCED CRITICAL CURRENT

Alain E. Kaloyeros, Aiguo Feng, and Elke Jahn
Physics Department, State University of New York at Albany, Albany,
NY 12222

Kenneth C. Brooks,
Chemistry Department, University of Illinois at Urbana-Champaign,
Urbana, IL 61801

ABSTRACT

High quality Y-Ba-Cu-O thin films were produced on single-crystal MgO and $SrTiO_3$ by metal-organic chemical vapor deposition (MOCVD) using metal chelates of β-diketonate ligands. The films were grown in a cold-wall CVD reactor at a reactor pressure of 10 torr and substrate temperature in the range 700-900^{0}C. Characterization studies were performed using Rutherford Backscattering (RBS), Auger electron spectroscopy (AES), scanning electron microscopy (SEM), and energy-dispersive x-ray spectroscopy (EDXS). The studies showed that the films were uniform, continuous, adherent and highly pure. Four point resistivity measurements showed that films had a sharp superconducting transition at 92K and exhibited critical current densities of ~$5x10^6$ A/cm^2 (B=0, T=77K).

INTRODUCTION

MOCVD is a thin film deposition process in which growth is achieved by transporting individual components, by means of volatile organometallic sources, to a substrate heated to a suitable temperature where they react to form the desired material. The advantages of MOCVD include its relative simplicity and controllability, leading to good adherence and reduced susceptibility to interfacial mixing and cross-contamination effects. In addition, it can produce a wide variety of materials at near-theoretical density, or controlled lower density. Reactions are irreversible, with no evidence of autodoping during epitaxial growth. This characteristic allows highly abrupt changes in doping and/or composition of epitaxial layers, with transitions within a few monolayers of growth.[1,2] These unique features have made MOCVD the technique of choice for the deposition of ceramic and electronic materials and have led, in recent years, to extensive studies of its applicability to the growth of high temperature superconductors.[3,4,5,6,7,8]

More importantly, MOCVD has many unique features which makes it especially attractive for the growth of high T_c superconductors for large-scale industrial applications, such as power transmission lines. These include the ability to provide high uniformity over a large area and extremely high growth rates. MOCVD, in contrast to physical vapor deposition techniques, does not require a vacuum environment and, accordingly, has the advantage of relative ease in scale-up to long lengths of tape. Also, it has the ability to coat substrates of complex shape and form, a feature which is limited only

470

by the crystal orientation of the substrate when epitaxial films are desirable.

Consequently, a research program has been initiated by the present investigators, in collaboration with the Superconducting Materials Group at Intermagnetics general Corporation (IGC), to deliver an MOCVD process for the growth of high-temperature superconductor (HTS) tapes for energy-related applications. The purpose of the initial research was to probe the suitability of MOCVD as a technique for the growth of high-quality HTS films with enhanced critical current.

The present paper is one in a series of reports by the present authors[91011] on the successful growth of superconductor thin films, showing a sharp transition at 92K and a current density of $5 \times 10^6 A/cm^2$ (B=0, 77K), by MOCVD from β-diketonate precursors, followed by in-situ post-deposition annealing in dry oxygen. In particular, it is shown that MOCVD-produced Y-Ba-Cu-O films exhibit quality equal to that of films produced by laser ablation and rf sputtered films.

EXPERIMENTAL TECHNIQUE

Three metal chelate precursors of the β-diketonate ligand 6,6,7,7,8,8,8-heptafluoro-2,2-dimethyl-3,5-octanedione, abbreviated as fod, were chosen as the source compounds for the Y, Ba and Cu. A cold-wall-type CVD apparatus, capable of operation in the range 10^{-9}-10^{-8} torr, was used for MOCVD of the 123 thin films. The experimental assembly is shown in figure 1 in the vertical cold-wall reactor configuration. It consists of two major components: a reaction chamber and a high-speed vacuum system.

The vacuum system is diffusion pump based. Type 304, stainless-steel, UHV Conflat flanges with copper gaskets are used for all feedthroughs, couplings, and connections. The reaction chamber is a stainless steel reactor of cylindrical shape of 150 mm inner diameter. The reactor is equipped with special support blocks/sample holder that allows one to use substrates of different shapes at variable angles, and to perform cold-wall CVD on large area substrates,up to 127 mm in diameter. A special oxygen-resistant heater stage was employed to ensure uniform heating of the substrates while avoiding any direct heating of neighboring surfaces in the reactor. The gas handling system is designed to deliver precisely metered amounts of precursors without transients due to pressure or flow changes. For this purpose, electronic mass flow controllers with flow accuracy of ±1% of full scale are employed.

In a typical deposition run, the source compounds were heated, in separate bubblers, to 45°C, 125°C, and 175°C, respectively, to sublime the Cu(fod)2, Y(fod)3, and Ba(fod)2. Hydrogen and dry and wet oxygen were employed as carrier gases to transport the vapor sources into the reactor. Deposition was carried out on MgO and SrTiO3. The samples were heated to temperatures in the range 700-900°C and MOCVD was carried out at a total reactor pressure of 10 torr. The deposition was followed by in-situ annealing in oxygen.

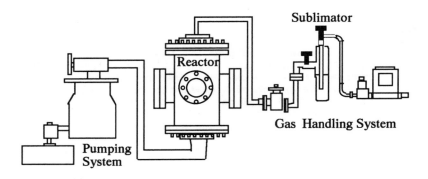

Figure 1.   Cold-wall CVD System.

METHODS OF CHARACTERIZATION

The films were characterized using four-point resistivity probe, Auger electron spectroscopy (AES), Rutherford backscattering (RBS), scanning electron microscopy (SEM) and energy-dispersive x-ray spectroscopy (EDXS). Auger spectra were recorded using a Physical Electronics PE595 instrument comprising a two-stage, retarding field/cylindrical mirror electron-energy analyzer, a coaxially contained electron gun with a beam normally incident on the sample surface and an electron multiplier. These components are contained in a demountable vacuum system evacuated to below a 10^{-10} torr. The analyzer transmission and resolution were 10% and ~0.6% respectively for all data. A beam energy of 5 keV was used with a beam current of 50nA, adjusted to provide convenient measurement conditions, but kept low enough to avoid specimen damage. Scanning electron microscopy measurements employing an 20 keV primary electron beam were performed using an ISI 2.2 scanning electron microscope, while energy dispersive x-ray spectroscopy (EDXS) analysis was carried out employing a LINK/Nucleus EX-2030 EDXS unit.

RESULTS AND DISCUSSION

Deposition Parameters

A typical deposition rate, for films produced at 700-900°C and a reactor pressure of 10 torr, was 15 Å/s. Typical film thickness, as measured by RBS, was 0.5-1 μm. The thickness variation across a typical substrate was found to be <5%/cm over the whole substrate.

Four-Point Probe Resistivity Measurements

A sharp transition temperature of ~92K was obtained for films grown at 800°C on single crystal MgO and SrTiO3 substrates (figure 2A). The films were annealed in-situ by a two-step cooling process: (1) a cooling down in oxygen atmosphere from 800°C to 450°C, where they were kept for thirty minutes, and then (2) a slow cooling down to room temperature. Subsequent critical current measurements gave ~5x10^6 A/cm^2 (B=0, T=77K).

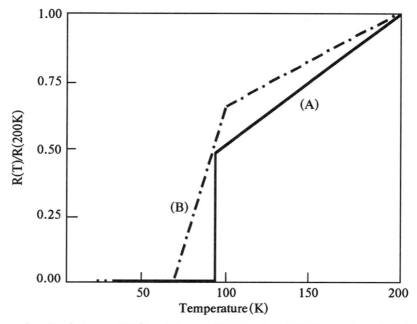

Figure 2. Resistance(T)/Resistance(200K) for MOCVD-produced Y-Ba-Cu-
O films on single-crystal MₒO and SrTiO₃ for two different
annealing regimes.

a relatively wide transition temperature around 75K (figure 2B).
Subsequent critical current measurements gave with critical current
densities of ~10^5 A/cm^2 (B=0, T=77K).

AES Results

Y-Ba-Cu-O films produced by MOCVD on single crystal MgO
substrates at 800°C and annealed according to the process described
above were studied by AES. The AES results were standardized using a
bulk Y-Ba-Cu-O sample with a T$_c$~92K.   All samples were sputter
cleaned with a xenon ion beam before Auger data collection to remove
surface contaminants.   The choice of a standard of composition and
chemical environment and bonding similar to that of the sputtered
films allowed us to achieve high accuracy in AES quantitative
analysis.   (Our results are based on the expectation that the
chemical and structural changes, if any, induced during the sputter
cleaning process are basically the same in the bulk sample and
sputtered films). In addition, all samples were analyzed under the
same experimental conditions, using a primary electron beam energy of
5 keV and an electron beam density of 50nA/(0.3mm)^2 to reduce sample
decomposition effects.

Figure 3 shows the Y, Ba, Cu and O Auger signals corresponding
to a bulk standard sample (fig. 3(A)) and to three different
locations on a Y-Ba-Cu-O thin film (figs. 3(B), 3(C) and 3(D)).

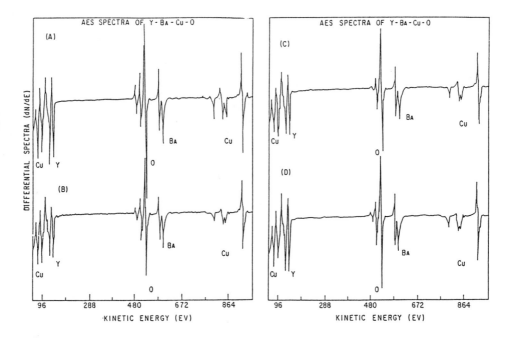

Figure 3. Differentiated Auger spectra of : (A) a bulk Y-Ba-Cu-O sample used as standard, and (B), (C), and (D) three different spots on the same Y-Ba-Cu-O thin film produced by MOCVD.

AES depth and lateral (x-y) quantitative analyses show that film composition, within the detection limits of AES, was practically constant over the entire film and independent of lateral or depth position. The composition can be expressed as $Y_{1.1}Ba_{1.9}Cu_{3.3}O_x$. The results are summarized in figure 4 which shows variations of Cu/Ba and Cu/Y ratios as a function of lateral position on the substrate. As can be seen, the Y-Ba-Cu-O films exhibited a uniform optimum composition of 123, regardless of position on substrate.

EDXS Results

To confirm the AES findings, energy dispersive x-ray spectroscopy (EDXS) was carried out using a LINK/Nucleus EX-2030 analyzer-minicomputer, equipped with acquisition and display facilities. The EDXS analysis, shown in figure 5, gave results in good agreement with the AES quantitative analysis and yielded a composition of $Y_{1.0}Ba_{2.1}Cu_{3.2}O_x$.

SEM Results

The scanning electron micrograph of an epitaxially-grown $Y_{1.0}Ba_{2.1}Cu_{3.2}O_x$ film, taken at a magnification of 2200X is shown in figure 5. The microstructure of the film consists of a typical

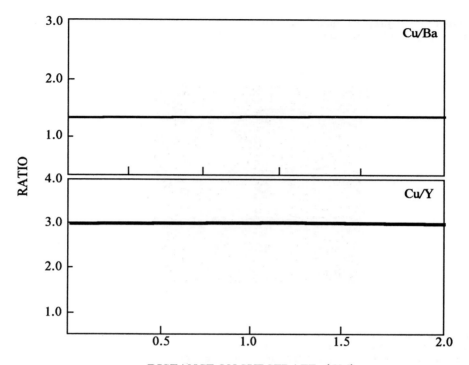

Figure 4(a) Variations of Cu/Ba and Cu/Y ratios as a function of lateral position for MOCVD-produced Y-Ba-Cu-O films.

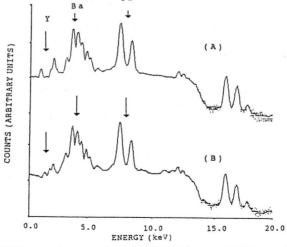

Figure 4(b) EDXS spectrum of a bulk and an MOCVD-produced 123 thin film.

pattern of highly oriented needles, with an extremely smooth surface and a very small concentration of voids.

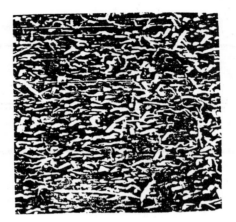

Figure 5   Scanning electron concentration of voids.microscope (SEM) micrograph using a magnification of 2200X, of a Y-Ba-Cu-O thin film produced by MOCVD.   A pattern of oriented needles is visible.

CONCLUSIONS

In conclusion, high-quality, high purity Y-Ba-Cu-O thin films were produced on single-crystal MgO and SrTiO$_3$ by metal-organic chemical vapor deposition (MOCVD) using metal chelates of β-diketonate ligands.   The films were grown in a cold-wall CVD reactor at a reactor pressure of 10 torr and substrate temperature in the range 700-900°C.   Characterization studies showed that the films were uniform, continuous, adherent and highly pure--carbon and fluorine contents were below the detection limits of the techniques used. Four point resistivity measurements showed that films annealed in a two-step annealing regime had a sharp superconducting transition at 92K and exhibited critical current densities of ~5x10^6 A/cm^2 (B=0, T=77K).   MOCVD-produced Y-Ba-Cu-O films thus exhibit quality equal to the best laser-ablation and rf sputtering produced films.

This "proof of concept" study thus indicates that MOCVD, because of its unique capabilities, that include the ability to produce films on large-area substrates of complex shape and form at high growth rates, is highly promising for energy-related applications, such as in transmission lines.

ACKNOWLEDGMENTS

This work was supported by the New York State Energy Resources and Development Authority grant#88F042A (AEK, KCB, AF, and EJ) and

SUNY-Albany grant#450300 (AEK, AF). The authors are greatly indebted to Michael Walker and Drew Hazelton, from the Intermagnetics General Corporation (IGC), for their technical advice and scientific discussions.

REFERENCES

[1] R. H. Moss and P. C. Spurdens, J. Crystal Growth 68, 96 (1984).

[2] C.R. Chen, and R. Moon, J. Cryst. Growth 77, 591 (1986).

[3] K. Watanabe, H. Yamane, H. Kurosawa, T. Hirai, N. Kobayashi, H. Iwasaki, K. Noto, and Y. Muto, Appl. Phys. Lett. 54, 575 (1989).

[4] Q.Y. Ma, T.J. Licata, X. Wu, E.S. Yang, and C.-A. Chang, Appl. Phys. Lett. 53, 2229 (1988).

[5] Y.L. Wang, W.T. Chou, T. Skotheim, R. Bhudani, M. Suenaga, and Y. Okamoto, Synthetic Metals 29, F621 (1989).

[6] A.J. Panson, R.G. Charles, D.N. Schmidt, J.R. Szedon, G.J. Machiko, and A.I. Braginski, Appl. Phys. Lett. 53, 1756 (1988).

[7] K. Zhang, B.S. Kwak, E.P. Boyd, A.C. Wright, and A. Erbil, Appl. Phys. Lett. 54, 380 (1989).

[8] P.H. Dickinson, T.H. Geballe, A. Sanjurjo, D. Hildenbrand, G. Craig, M. Zisk, J. Collman, S.A. Banning, and R.E. Sievers, J. Appl. Phys. 66, 444 (1989).

[9] A. E. Kaloyeros, A. Feng, J. Garhart, M. Holma, K. Brooks, and W. S. Williams, in Superconductivity and Applications, ed. H. Kwok (Plenum Press, New York, 1990).

[10] A. Feng, E. Jahn, and A.E. Kaloyeros, accepted for publication in the proceedings of the 1990 International Conference on Electronic Materials (MRS, European MRS, and the Japan Society of Applied Physics, Newark, NJ, 1990).

[11] A.E. Kaloyeros, K. C. Brooks, A. Feng, V. Tulchinsky, and J. Garhart, in High Temperature Superconductors: Fundamental Properties and Novel Materials Processing, eds. J. Narayan, P. Chu, L. Schneemeyer, and D. Christen (MRS, Pittsburgh, PA, 1989).

MAGNETIC ALIGNMENT OF CUPRATE SUPERCONDUCTORS CONTAINING RARE EARTH ELEMENTS (III)

F. Chen[*], J. Chen[*], J. Sigalovsky[**], R.S. Markiewicz[***], and B.C. Giessen[*]

[*]   Northeastern University, Dept. of Chemistry and Barnett Institute of
     Chemical Analysis and Materials Science, Boston, MA 02115;

[**]   Formerly Barnett Institute, Northeastern University, Boston, MA 02115;
     presently with Arthur D. Little Inc., Cambridge, MA 02140;

[***]   Northeastern University, Dept. of Physics and Barnett Institute of
     Chemical Analysis and Materials Science, Boston, MA 02115.

ABSTRACT

In ongoing research on the properties and fabrication of cuprate superconductors, we report here: experimental details on their magnetic alignment and its determination; a study of the "crossing over" from parallel to normal alignment (or failure to do so completely) in Gd-containing Bi-2212 and 123, respectively; and data on the complex magnetic field aligment of the non-superconducting 211 phase RE_2BaCuO_5 as a function of the rare earth (RE).

INTRODUCTION

As observed in the two previous papers[1,2], an intense magnetic field can be used to align crystalline grains of a material with magnetic anisotropy,[3] especially high-T_c superconductors (HTSC's), to prepare oriented materials,[4] analyze crystal fields,[5] study anisotropy effects[6] and critical currents[7,8] and, recently, carry out a novel double-alignment procedure[9,10]

As pointed out previously, in the rare earth (RE)-free superconductors, the magnetic anisotropy needed for alignment is associated with the two-dimensional CuO_2 plane conductivity, and the grains align with the c-axis parallel to the magnetic field, B. In RE-substituted materials, the grains also attempt to orient with the RE-moment parallel to B; however, while most RE moments are oriented parallel to the c-axis ("parallel" aligners), a few (Er, Eu, Gd, Yb, Tm) ("normal" aligners) lie normal to the c-axis, i.e., within the a-b plane, and, as shown before[1,2], point in specific directions in that plane. In the latter case, the relative magnitudes of the (normal) RE moment and the (parallel) plane conductivity moment (both of which vary with the RE

478

concentration) determine the direction and magnitude of the total moment.

Continuing an analysis of RE-substituted superconductors[1,2], we report here further information and observations on experimented methods; data on "cross-over" systems in which parallel alignment changes towards normal alignment as a function of rare earth concentration; and the magnetic alignment of the non-conducting "211" phase RE_2BaCuO_5 which is of importance in 123 superconductor preparation.

EXPERIMENTAL METHODS

The compositions of the materials studied are discussed in the respective sections. Details of the procedure used followed those presented in Refs. 1 and 2 and included mixing ceramics with epoxy (1:4) and holding in a magnetic field of 1 to 15 T at room temperature for ~20 minutes, followed by XRD analysis of the aligned, disc-shaped samples, with patterns taken from the face lying perpendicular to the vertical magnetic field used. In aligned Bi 2212(RE) and 123(RE) phases, (00ℓ) and (hk0) reflections are monitored for RE elements with c $\parallel$ B and c $\perp$ B, respectively.

An alignment percentage P is obtained either by a quantitative evaluation of the peak intensities, as discussed in Refs. 2,6, or by taking rocking curves. Rocking curves [normalized with respect to the intensity change solely due to the misalignment of the specimen plane normal with respect to the bisector of the source-sample-counter angle (180-2θ)] are used to obtain the

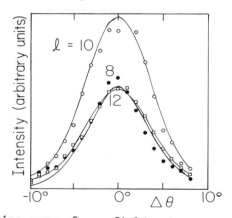

Fig. 1. Rocking curve for a field-oriented sample of Bi-2212:30% Ho. Solid lines are fit to Eq. 1 with $\theta_o = 5^o$; background corrected. [Neglecting background correction, $\theta_o = 4^o$ was found.]

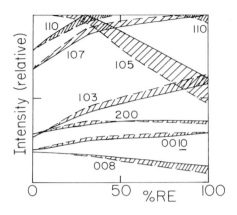

Fig. 2. Calculated X-ray intensities of RE-substituted Bi-2212 compound, as a function of RE concentration. Shaded region shows variation of intensity with RE species, varying from Ce(solid curve) to Lu (dashed curve).

angular spread of the (00ℓ) plane normals; an example is shown in Fig. 1.

In using relative peak intensities to assess texture, plots of the calculated intensities of important peaks for RE-doped superconductors as a function of RE content are useful. A plot of this type is given in Fig. 2 for the Bi-2212(RE) superconductors $Bi_2Sr_2Ca_{1-x}RE_xCu_2O_z$ as a function of x for the RE elements from Ce to Lu for the peak intensities indicated; in each case, peaks are normalized with respect to the intensity of the strongest peak in the pattern; in the case of 2212(Ce), that peak is (105) up to ~30% Ce, (110) from 30 to 95% Ce, and (107) from 95-100% Ce.

RESULTS AND DISCUSSION

Nature of Misorientation: In virtually all samples studied, we find a residual X-ray intensity at reflections which should not be present in perfectly oriented material, such as (110) or (107) in parallel-aligned Bi-2212. We have used these peaks to try to understand something about the nature of the misorientation. Thus, for mosaic spread about the c-axis, the intensities should vary as

$$I/I^r = e^{-(\theta/\theta_0)^2} \qquad (1)$$

where I^r is the intensity of that reflection in a randomly oriented powder, and θ is the angle between that reflection normal and the c-axis. Plotting log (I/I^r) vs θ^2, we generally find a

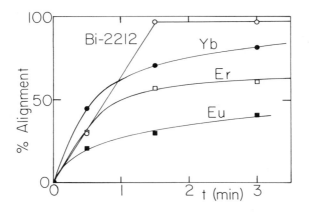

Fig. 3. Time dependence of magnetic alignment for Bi-2212 samples in epoxy. Data shown refer to pure Bi-2212 or with 30% RE substituted, RE = Yb, Er, Eu.

bimodal distribution: the (00ℓ)-reflections are relatively more intense, but the other reflections are all in nearly the same ratio as in unoriented material. That is, a certain fraction of the grains are well oriented, and the remainder are virtually unoriented. The percent of oriented grains is essentially equal to the P-factor introduced earlier[2,6]

Further evidence for this interpretation comes from Fig. 1. The solid curves are a fit to Eq. (1), assuming a common value $\theta_o = 5°$. Thus the (00ℓ)-reflections are consistent with simple gaussian broadening. However, Eq. 1 with $\theta_o = 5°$ would predict negligible intensities for all non-(00ℓ) reflections.

Effect of Curing Time and Applied Field: The characteristics of the epoxy used (Devcon "5 min" epoxy) as well as the strength of the applied field and the superconductor's magnetic moment affect the degree of alignment. This is shown in Fig. 3 for Bi-2212(RE) containing the "normal" aligners Eu, Er and Yb [a curve for undoped (parallel aligning) Bi-2212 is included for comparison.] It is seen that a strong field (here, B = 13T) is needed to obtain maximum and useful alignment for Yb and Er; for Eu, the normal moment is so weak that even a field as high as 13T will not produce adequate alignment. (At this time, the direction of the resultant moment for these cuprates is not known; it is assumed to be near <110> at the doping level of 30%.) A field of 1T (not shown) is insufficient in all cases. At room temperature, alignment times of ⩾3 min are seen to be required. After normalization for the HTSC magnetic moment, the plots would reflect solely the decreasing rate of viscosity-controlled

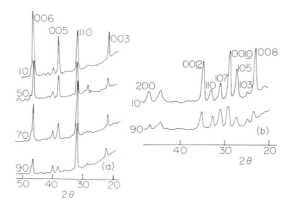

Fig. 4. X-ray reflection spectra of (a) 123:x% Gd and (b) Bi-
2212: x% Gd, showing partial crossover from parallel toward normal
aligner. In each case, x is given at the left side of the figure.

rotation with increasing alignment and the effect of increasing
viscosity of the epoxy with time.

Effect of a Weak Normal-Aligning RE (Gd):   In our previous
studies,[1,2] we found that Gd appeared to be a normal aligner in
123, but a parallel aligner in Bi-2212. This can be understood as
follows: the Gd moment always lies in the a-b plane, but is
relatively weak--hence comparable to the (c-directed) moment of
the CuO_2 planes. The magnetic field aligns the grains along the
net moment, and hence shifts from the parallel towards the normal

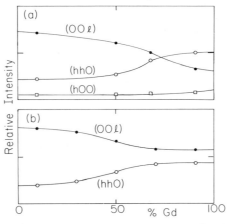

Fig. 5. Crossover from parallel toward normal alignment in (a)
123:x% Gd and (b) Bi-2212:x % Gd, showing relative intensities of
X-ray peaks vs. x.

direction as the Gd content is increased, as can be seen in the X-ray data (Fig. 4) and the plots of the (00ℓ) and (110) intensities derived from them (Fig. 5). It is seen that in $Y_{1-x}Gd_xBa_2Cu_3O_z$ the cross-over from predominantly "parallel" to predominantly "normal" alignment occurs only at x ~ .75, while for $Bi_2Sr_2Ca_{1-x}Gd_xCu_{2.2}O_z$ there is no cross-over point at all; the parallel alignment is weakened by Gd but is predominant at least to x ~ 0.9. Gd is thus not useful for normal alignment in either material. The actual direction of the magnetic moment of the sample is only approximated by this measurement of (110) and (00ℓ) intensities and is subject to a separate, more complete XRD study.[11] The change of moment magnitude and direction can also be observed by magnetic measurements.[11] Er and Yb are most suitable for normal alignment, and, for practical considerations, Yb is the element of choice.

Magnetic Alignment of RE_2BaCuO_5 (211) Semiconductor Phase: The 211 phase has a complex structure containing Cu in square pyramidal O-coordination.[12] It forms peritectically upon melting of most 123 superconductors and plays a major role in the preparation of useful bulk 123 material. It was thus of interest to study the magnetic field alignment of 211(RE).

Results are given in Table I and show the effects of the RE. Different RE's produce preferential alignment along each of the three crystallographic axes a, b and c. Thus, Sm aligns along [001], Er along [100] (and, to a lesser degree, along [010], and Yb aligns along [010]. Given the asymmetric, non-coplanar coordination of the RE atoms in two different sites having complex, 7-fold geometries in 211,[12] the complex and variable interaction of the f-electrons with the crystal field is not surprising.

It remains to be determined whether the 211 and 123 phases can both be aligned with respect to a common axis, if this is possible, melt grouwth with oriented 211 nuclei imposing their orientation on the 123 material may be possible.

Concluding Remarks: For the present, the important preference for specific axes within the a-b plane in magnetic field orientation discovered in the predecessor papers[1,2] is principally important for the double alignment process described in Refs. 9 and 10 The current results confirm that Yb is the normal-aligning RE additive of choice for both Bi-2212 and 123 and illustrate some the background of this work.

Table I: Magnetic alignment behavior of RE-211($RE_2Ba_1Cu_1O_5$)
vs RE-123

(211 lattice constants: a = 7.132Å, b = 12.181 Å,
c = 5.668 Å)

RE	Preferred Axis	
	211	123
Sm	a (very strong)	c
Eu	b (very strong)	b
Gd	a (weak)	b
Dy	a (weak)	c
Er	b (weak)	(a,b)
Yb	c (very strong)	b
Y	a (very weak)	c

ACKNOWLEDGEMENTS

This work was done in part at the superconducting magnet facility of the Francis Bitter National Magnet Laboratory, Cambridge, MA which is supported by the NSF at MIT. Our research was supported by DOE under subcontract from Intermagnetics General Corporation.

Publication No. 434 from the Barnett Institute.

REFERENCES

1.   F. Chen, R.S. Markiewicz and B.C. Giessen, in _Proc. 3rd Ann. Conf. on Superconductivity and Applications_, Buffalo, NY (1989), Plenum Publ., New York, NY, (1990), in press.

2.   F. Chen, R.S. Markiewicz and B.C. Giessen, in "High Temperature Superconductors, 1989, MRS Conf. Proc. Vol. (1990).

3.   D.E. Farrell, B.S. Chandrasekhar, M.R. De Guire, M.M. Fang, V.G. Kogan, J.R. Clem, and D.K. Finnemore, Phys. Rev. B 36, 4025 (1987).

4.   J.E. Tkaczyk and K.W. Lay, J. Mater. Res. 5, 1368 (1990).

5.   J.D. Livingston, H.R. Hart, Jr., and W.P. Wolf, J. Appl. Phys. 64, 5806 (1988).

6.   R.S. Markiewicz, K. Chen, B. Maheswaran, A.G. Swanson, and J.S. Brooks, J. Phys. Condensed Matt. 1, 8945 (1989).

7. R.H. Arendt, A.R. Gaddipati, M.F. Grabauskas, E.L. Hall, H.R. Hart, Jr., K.W. Lay, J.D. Livingston, F.E. Luborsky, and L.L. Schilling, in "High Temperature Superconductivity", ed. by M.B. Brodsky, R.C. Dynes, K. Kitazawa, and H.L. Tuller, Amsterdam, North-Holland (1988), p. 203.

8. K. Chen, B. Maheswaran, Y.P. Liu, B.C. Giessen, C. Chan and R.S. Markiewicz, Appl. Phys. Lett. 55, 289 (1989).

9. F. Chen, B. Zhang, R.S. Markiewicz, B.C. Giessen Appli. Phys. Lett. (1990) (submitted).

10. R.S. Markiewicz, F. Chen, B. Zhang, J. Zhang, J. Sigalovsky, S.Q. Wang, B.C. Giessen, this conference.

11. F. Feng, X.Y. Zhang, R.S. Markiewicz, B.C. Giessen, to be published.

12. C. Michel and B. Raveau, J. Solid State Chem. 43, 73 (1982).

MELTING POINTS OF RARE EARTH (RE) SUBSTITUTED 123 SUPERCONDUCTORS

Sheng-Qi Wang[*], Robert S. Markiewicz[**] and Bill C. Giessen[*]

[*] Department of Chemistry and Barnett Institute of Chemical Analysis and Materials Science, Northeastern University, Boston, MA;

[**]Department of Physics and Barnett Institute of Chemical Analysis and Materials Science, Northeastern University, Boston, MA.

INTRODUCTION

The substantial variation in the melting behavior of the 123 (RE) high-T_c superconductor (HTSC) phase $RE_1Ba_2Cu_3O_{6.5+\delta}$ with substitution of different RE's has been noted and partially studied before.[1-3] Because of the enormous practical importance of the melting points in the processing of this major HTSC compound (e.g. in melt growth or regrowth, use as crucibles or melt substrates, non-melting dispersoids,[4] etc.), we have undertaken an operational study of the softening and melting behavior for all accessible RE elements forming a superconducting 123 phase; we report this study here.

EXPERIMENTAL METHODS

Ceramic pellets were prepared by constituent oxide mixing and reacting, grinding and re-sintering, as usual. Samples were then heated in oxygen at atmospheric pressure ($\sim$ 1 atm O_2) in a well controlled furnace ($\pm$ 2°C) to various temperatures.

Subsequently, they were characterized by X-ray diffraction and were examined for indications of melting and softening by metallography and by inspection of shape change. Average grain sizes were determined and special attention was paid to the presence or absence of indications of eutectic melting or solidification.

RESULTS AND DISCUSSION

Metallographic data were derived from micrographs represented by the set given in Fig. 1 for $Nd_1Ba_2Cu_3O_x$ heated to 1090°C (a) and 1102°C (b); the change in grain structure and grain size between the two microsections is obvious and locates a melting reaction (identified in the literature[2] as peritectic: 123 -- →211 + L) between the two temperatures. Interestingly, we find this

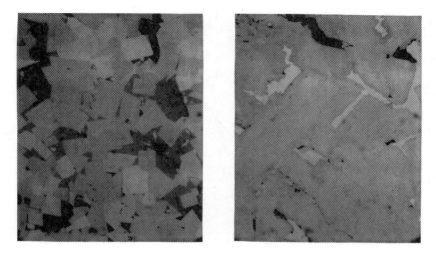

Fig. 1. Metallographs of Nd-123: left = heated to 1090°C (660 x magnification); right = heated to 1102°C. (330 x magnification).

peritectic reaction to be almost reversible on the time scale of this melting experiment: both samples in Fig. 1 contain predominantly the 123 phase, as seen by XRD. We therefore designate this melting mode (which we find for La, Sm and Nd) as quasi-congruent.

Our results are summarized and presented in Fig. 2 as a plot against the 123(RE) unit cell volume.[5] Alternatively, the RE^{+3} radius or the unit cell parameter $\underline{a}$ of the RE-123 phases could be used; these properties scale closely with each other.[5,6] However, the trend of melting point vs RE values is the opposite of that found for the pure RE metals or oxides.

In Fig. 2, we also include for comparison DTA data from two of the previous studies: early data from Katsui et al[1] and, very recently, Ullman et al.[2] (from the latter work, the 123 decomposition temperature m_1 is used). The agreement of these data sets with ours is quite satisfactory, the only substantial difference being in the data for RE = La, where we find a continuing increase in stability following the V_{RE} relation used, while Katsui et al.[1] find a small decrease from the value for 123-Nd which precedes it, departing from the linear relationship found here.

There is, however, no acceptable agreement between our data (and

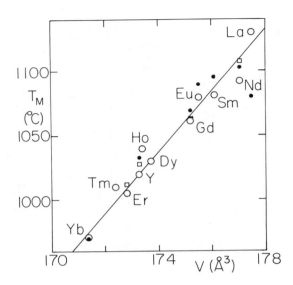

Fig. 2.   Melting point vs. unit cell volume for RE 123 compounds.
Open circles = results of present study;   filled circle = derived
from Ref. 1;   open squares = Ref. 2.

those in the other two studies) on the one side and those from a
study by Yoshiara et al.[7] which show much higher decomposition
temperatures for the late RE's (Dy-Yb) and much lower ones for the
early RE's (Sm-Gd), with a pronounced temperature maximum at Dy
and Ho, which is not apparent in our work.

CONCLUSIONS

We have obtained an empirical melting point curve for the RE-123
phases.   These melting (or decomposition) temperatures follow
closely a straight-line relationship when plotted vs unit cell
volume or RE ionic radius;   they extend over a rather large range
of $160^{\circ}C$, increasing from Yb to La.   We note a quasi-congruent
melting behavior for several of these phases (containing the early
Re's La, Sm and Nd).

ACKNOWLEDGEMENT

We are grateful for support of this work by SDI-O under
subcontract from Intermagnetics General Corporation.

Publication No. 435 from the Barnett Institute.

REFERENCES

1.   A. Katsui, Y. Hidaka, and H. Ohtsuka, Jap. J. Appl Phys. <u>26</u>, L 1521 (1987).

2.   J.E. Ullman, R.W. McCallum and J.D. Verhoeven, J. Mater. Res. <u>4</u>, 752 (1990).

3.   R.W. McCallum et al., Adv. Cer. Mat. <u>2</u> (38) 388 (1987).

4.   M. Murakami et al. Jap. J. Appl. Phys. <u>28</u>, 1189 (1989).

5.   J.M. Tarascan, W.R. McKinnon, L.H. Greene, G.W. Hull and E.M. Vogel, Phys. Rev. B<u>36</u>, 226 (1987).

6.   R. Beyers and T.M. Shaw, in <u>Solid State Physics</u> <u>42</u>, 135(1989).

7.   K. Yoshiara, F. Uchikawa, K. Yoshizaki and K. Nakahigashi, "High T_c Superconductivity", MRS Proc. Vol. (1990).

EFFECT OF GRINDING ON THE MORPHOLOGY AND SUPERCONDUCTING PROPERTIES OF $YBa_2Cu_3O_x$ POWDER

S.Rajendra Kumar, N.Ramadas, S.C.Mohan, S.Ravichandran
Corporate Research & Development Division,
Bharat Heavy Electricals Ltd., Hyderabad – 500593, INDIA

ABSTRACT

Preparation of $YBa_2Cu_3O_x$ by solid state reaction of constituent oxides or carbonates results in a powder that is normally coarse and has a wide size distribution. Preparation of the fine powder is of utmost importance in the processing of the material into wires and tapes. The grinding media used seem to exert considerable influence on the morphology of the powders. Generally long exposure to liquid media result in the degradation of the material.

The size, shape and distribution of powders ground in hexane, ethanol, xylene and deodorized kerosene for different time intervals up to 300 hours were studied. The degradation caused by exposure to the liquid media used for grinding and the effect of long term grinding on the crystal structure are also discussed.

INTRODUCTION

Several attempts are under way to produce high temperature superconductors in usable forms such as wires and tapes [1-4]. In order to be used in practical applications, these wires and tapes should be capable of carrying high currents. One of the important factors that has an effect on high critical current densities is the starting powder (precursor) size and distribution [5]. The high temperature superconductor $YBa_2Cu_3O_{7-x}$ (YBC) is commonly produced by the solid state reaction of the constituent oxides. This powder consists of coarse particles. In order to reduce the particle size, it is necessary to grind the reacted powder in a suitable liquid medium for sufficiently long periods. Long exposure to liquid media can lead to degradation of the powder [6]. This paper aims at studying the effect of long term grinding in different organic liquid media on the powder size, shape and distribution and to check for degradation of the powder, if any, by means of X-ray diffraction patterns.

EXPERIMENTAL PROCEDURE

The YBC powder was prepared by the solid state reaction of Y_2O_3, $BaCO_3$ and CuO in appropriate proportions. The raw material powders were mixed together in a planetary mill and calcined at $900^\circ C$ for 12 hours in air. The process was repeated thrice with intermediate grinding using pestle and mortar. The X-ray diffraction pattern of the powder so obtained showed the peaks corresponding to the orthorhombic YBC phase (Fig.1a).

The powder was divided into four groups. Each of the groups was loaded in agate pots containing different liquid media, viz. ethanol, hexane, xylene and deodorized kerosene. The pots were then loaded in planetary mill. Three agate balls placed in each of the pots served as the attrition media.The liquid media were replenished from time to time to make good the loss due to evaporation. Samples were removed from each of the pots at intervals of 50 hours. They were dried in an oven at 150°C. These samples were used for XRD studies, microstructural observation and SEM.

The XRD patterns were obtained using a Co-anode tube. Slides for optical microscopy were prepared by smearing a dispersion of the powder in ethanol on a glass slide. The SEM studies were carried out on green compacts of the powders. The particle size distribution studies were carried out using a Malvern Instruments Master Particle Sizer.

RESULTS AND DISCUSSIONS

Among the liquid media used for the study, hexane had to be replenished frequently because of rapid loss by evaporation. There were differences in the physical appearance of the powders ground in the different media. After evaporation of the liquids, the powder ground in kerosene and xylene did not show any macro agglomeration. The powder ground in ethanol agglomerated into flaky masses and the powder ground in hexane agglomerated into hard lumps.

XRD patterns of the powders ground in ethanol, xylene and kerosene did not show any extraneous peaks even after 300 hours of grinding. The powder ground in hexane started deteriorating after about 150 hours of grinding and turned brownish after 250 hours. The XRD pattern of the powder showed $BaCuO_2$, CuO, $BaCO_3$ and YBC peaks.

The XRD pattern of the powder ground in xylene for 250 hours (Fig.1b) showed that peaks corresponding to the (005), (014) and (104) planes of the YBC phase were suppressed. Only three sets of peaks indexed to combinations {(013), (103), (110)}, {(006), (200)} and {(123), (213), (116)} were observed. The same observation was made in the case of powders ground in kerosene and ethanol after 300 hours. In the case of the powder ground in xylene for 300 hours the relative intensities of the peaks corresponding to (006) and (123) have further reduced (Fig.1c). When the long term ground powder is heated to over 800°C the missing peaks reappear (Fig.1d). This seems to indicate that the YBC powder probably tends to the amorphous state by long term grinding.

SEM observation shows that the powders after 100 hours of grinding are less than 10 microns in size. The starting powder size was about 30 to 40 microns. The powder ground in kerosene for 100 hours showed a number of well defined elongated particles (Fig.2a). The powders ground in the other media showed irregular shaped particles with low aspect ratios after 100 hours of grinding (Fig.2b). All the powders showed a large variation in particle size after 100 hours. After 200 hours all the four powders showed highly irregular shaped particles with a narrow size distribution (Figs.3a,3b,3c,3d).

Fig.4 shows the particle size distributions of the powders ground in the various liquid media after 250 h. It can be seen that the powders ground in xylene and ethanol are less than about 10 microns in size. The powders ground in kerosene and hexane are coarser with the size ranging up to about 24 microns.

The following Table is a pointer to the effect of the liquid media on the particle size distribution:

Powder ground for 250 h in	Size distribution			Mean particle size, D50 (microns)	% Below 2 microns	Specific surface area (sq.m/cc)
	Size band (microns)	Band width (microns)	%			
Xylene	1.9-3.9	2.0	51.8	2.8	23.9	2.8634
Ethanol	3.0-6.4	3.4	51.9	4.3	3.7	1.6142
Hexane	3.0-8.2	5.2	53.3	6.0	1.3	1.1766
Deodorised kerosene	2.4-8.2	5.8	53.3	6.6	1.5	1.1481

It can be seen that the powder ground in xylene shows the narrowest size distribution, followed by ethanol, hexane and kerosene respectively. The fraction of particles of size less than 2 microns in the case of the powder ground in xylene (23.9%) is significantly higher than the corresponding values for the powders ground in other media. The powder ground in xylene also shows higher specific surface area and is expected to have better sinterability.

The optical micrographs did not give much information on the powder size and shape due to the fineness of the powder and excessive agglomeration. The micrographs of the powders ground for over 200 hours showed the presence of extraneous phases. Fig.5a is a typical micrograph showing the powder ground in xylene for 50 h. Fig.5b is the micrograph of the powder ground in xylene for 200 h showing additional phases.

CONCLUSIONS

Among the four liquid media used for grinding, viz. ethanol, hexane, xylene and kerosene, hexane is not suitable for long term grinding of YBC. Among the other three liquids the powder tends to be better dispersed in kerosene and xylene, than in ethanol.

There is some difference in the powder morphology depending on the liquid medium used for grinding for the first hundred hours of grinding. The powder ground in kerosene showed well defined elongated particles, whereas the powders ground in the other media showed irregular shaped particles. This difference could not be noticed when the grinding was continued for longer periods. The liquid media used for grinding affect both particle size and distribution. The finest particles and the narrowest size distribution were obtained when the powder was ground in xylene.

Long term grinding in ethanol, xylene and kerosene does not result in degradation detectable by XRD patterns. But the optical micrographs showed small quantities of extraneous phases after 200 hours of grinding.

YBC powder shows signs of tending towards amorphous state by long term grinding.

ACKNOWLEDGMENTS

The authors thank the BHEL management for permitting the publication of this paper and the DST, Govt. of India for funding our project.. We thank Mr.S.M.Gupta of DMRL, Hyderabad for assistance rendered in Scanning Elecron Microscopy. We thank the Chemical Engineering Dept. of I.I.T. Madras for carrying out the Particle Size Analysis. We also thank Dr.R.Somasundaram, Mr.G.Swaminathan and Mr.Kulvinder Singh for their constructive guidance and suggestions.

REFERENCES

1   R.W.McCallum, J.D.Verhoeven, E.D.Gibson, F.C.Laabs, D.K.Finnemore and A.R.Moodenbaugh, Adv. Ceram. Mater. 2B 388 (1987)

2   O.Kohno, Y.Ikeno, N.Sadakata, S.Aoki, M.Sugimoto and M.Nakagawa, Jpn. J. Appl. Phys. 26 1653 (1987)

3   R.Flukiger, T.Muller, W.Goldacker, T.Wolf, E.Seibt, I.Apfelstedt, H.Kupper and W.Schaver, Physica C 153-155 1574 (1988)

4   P.Regnier and L.Chaffron, Supercond. Sci. and Technol. 2 295 (1989)

5   T.E.Mitchell, D.R.Clarke, J.D.Embury, A.R.Cooper, J. Metals 6 (1989)

6   K.Imai. H.Matsuba, IEEE Trans. Magnetics 25 2045 (1989)

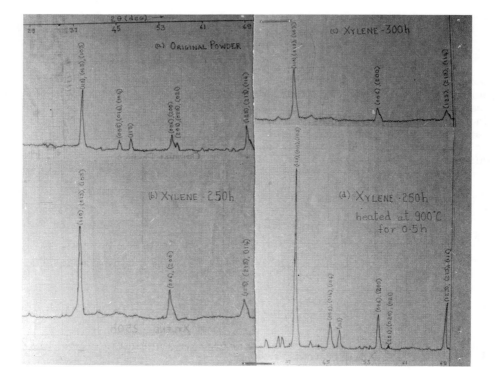

Fig.1 X-Ray diffraction patterns (Co K alpha radiation)
a)   Original powder
b)   Powder ground in xylene for 250 h
c)   Powder ground in xylene for 300 h
d)   Powder ground in xylene for 250 h followed by heating at
        900°C for 0.5 h

Fig.2a: Powder ground in deodorised kerosene for
100 h (2000 X) (SEM)

Fig.2b: Powder ground in xylene for 100 h (2000 X) (SEM)

Fig.3a: Powder ground in deodorised kerosene for 200 h (4000 X (SEM)

Fig.3b: Powder ground in xylene for 200 h (4000 X) (SEM)

Fig.3c: Powder ground in ethanol for 200 h (4000 x) (SEM)

Fig.3d: Powder ground in hexane for 200 h (4000 x) (SEM)

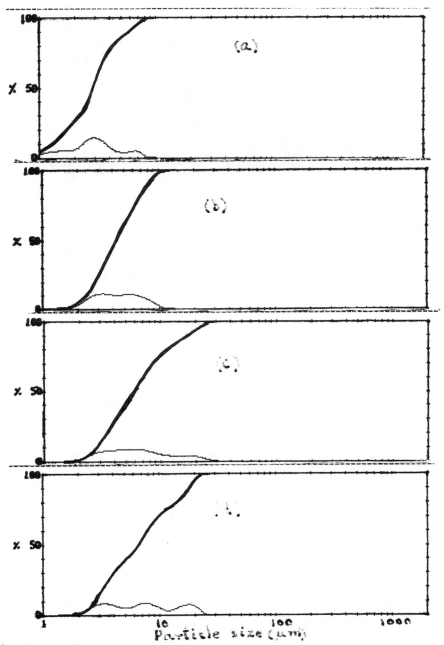

Fig.4:   Particle size distribution of powders ground for 250 h in
         (a) Xylene, (b) Ethanol, (c) Hexane, (d) Deodorised Kerosene

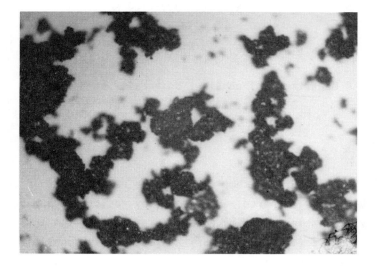

Fig.5a: Optical micrograph of powder ground in
xylene for 50 h (160 x)

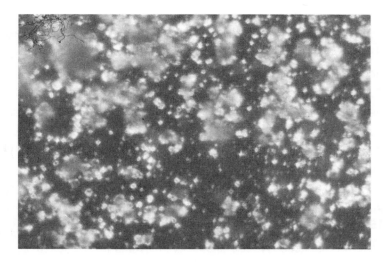

Fig.5b: Optical micrograph of powder ground in
xylene for 200 h (800 x)

NEW Tl-BASED SUPERCONDUCTOR TlPbSrRCuO WITHOUT Ca
WITH Tc ABOVE 100 K

Z. Z. Sheng, Y. Xin, and J. M. Meason

Department of physics, University of Arkansas,
Fayetteville, AR 72701

ABSTRACT

Ca-free TlPbSrRCuO samples (R = rare earths) with Tc up to above 100 K were prepared, and studied by resistance and ac susceptibility measurements and by powder x-ray diffraction analyses. A 1212-type phase $(Tl_{1-x}Pb_x)Sr_2(Sr_{1-y}R_y)Cu_2O_z$ is responsible for the observed superconductivity. A rare earth is required for the formation of the 1212 phase. Pb-doping is necessary to increase the Tc of the 1212 phase from 90 K to above 100 K.

INTRODUCTION

Discoveries of the 90 K TlBaCuO system[1,2] and 120 K TlBaCaCuO system[3,4] have stimulated numerous activities on the Tl-based superconductors. Two series of superconducting phases $Tl_2Ba_2Ca_{n-1}Cu_nO_z$ and $TlBa_2Ca_{n-1}Cu_nO_z$ were rapidly identified[5-8]. Subsequently, a number of new Tl-based superconducting systems were also discovered. Among them were the TlSrCaCuO system[9], the TlPbSrCaCuO system[10], the TlSrCaRCuO system (R = rare earths)[11], and the TlPbSrCaRCuO system[12]. Later, the Ca-free TlSrRCuO system was found to also be superconducting at about 40 K[13,14], and up to 82 K[13]. Further work showed that there is a 90 K phase in this Ca-free system[15,16]. Meanwhile, we found that when Tl is replaced partially by Pb, the TlPbSrRCuO system shows even higher Tc's up to above 100 K, and the phase responsible for the superconductivity is $(Tl_{1-x}Pb_x)Sr_2(Sr_{1-y}R_y)Cu_2O_z$ (1212 phase). In this paper, we report the preparation and characterization of TlSrRCuO samples (using Pr

and Tb as the representive of light rare earths and heavy rare
earths, respectively), and we show that a rare earth is required for
the formation of the 1212 phase and Pb-dopping is necessary for the
1212 phase to reach Tc above 100 K.

EXPERIMENTAL

TlPbSrRCuO samples were prepared using high-purity Tl_2O_3, PbO_2,
SrO ($SrCO_3$, or $Sr(NO_3)_2$), R_2O_3 (except CeO_2, Pr_6O_{11}, and Tb_4O_7), and
CuO. In a typical procedure, component oxides with certain metallic
atom ratios were completely mixed and ground. The powders were
pressed into pellets with a diameter of 15 mm and a thickness of
about 2 mm under a pressure of 7,000 kg/cm^2. The pellets were
placed in an alumina tube. The opening of the tube was covered with
an alumina plug. The alumina tube with contents was then put into a
preheated tube furnace, and heated at about 1000 ^oC in flowing
oxygen for 30 minutes, followed by furnace cooling to below 200 ^oC.
The resultant samples were examined by resistance and ac
susceptibility measurements and by powder x-ray diffraction
analyses. Resistance (ac, 27 Hz) was measured by the standard four-
probe technique with silver paste contacts. Ac (500 Hz) susceptibility was determined by the mutual inductance method. All measurements were made with computer control and processing, and were
carried out in a commercial APD closed cycle refrigerator. Powder
x-ray diffraction was carried out with Cu-Kα radiation with use of
a DIANO DIM 1057 diffractometer.

RESULTS AND DISCUSSION

Fig. 1 shows resistance-temperature curves for nominal samples
$Tl_{1.2}Sr_2(Sr_{0.6}Pr_{0.4})Cu_2O_z$ (A) and $(Tl_{0.9}Pb_{0.25})Sr_2(Sr_{0.6}Pr_{0.4})Cu_2O_z$
(B). These two samples were prepared in the same batch. Although
they exhibited very similar electrical transportation properties at
the normal state, they had different transition temperatures.
Sample A (without Pb) had an onset temperature of about 93 K and a

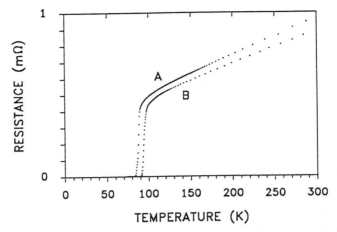

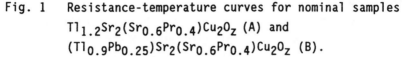

Fig. 1   Resistance-temperature curves for nominal samples
$Tl_{1.2}Sr_2(Sr_{0.6}Pr_{0.4})Cu_2O_z$ (A) and
$(Tl_{0.9}Pb_{0.25})Sr_2(Sr_{0.6}Pr_{0.4})Cu_2O_z$ (B).

zero resistance temperature of 84 K.   This result is similar to, but
much better than those we reported previously [15].   Sample B (with
Pb) had an onset temperature of about 100 K and a zero resistance
temperature of 91 K, about 7 K higher than those of the Pb-free
sample B.   Other experiments similarly showed that partial substitution of Pb for Tl increases Tc by 7 - 10 K.

Superconducting onset temperatures of the TlPbSrRCuO samples
could be even higher with the right preparation.   As an example,
Fig. 2 shows resistance-temperature curves for nominal samples
$(Tl_{0.8}Pb_{0.2})Sr_2Pr_{1.0}Cu_2O_z$ (C) and $(Tl_{1.5}Pb_{1.0})(Sr_{5.5}Tb_{0.5})Cu_3O_z$ (D).
Sample C showed clearly an onset temperature of 103 K although it
did not reach zero resistance down to 15 K, whereas sample D showed
an onset temperature about 105 K and a zero resistance temperature
of 80 K.

AC susceptibility measurements showed that TlPbSrRCuO samples
exhibited a strong diamagnetic signal around 90 K.   Above 90 K the
samples showed only a weak diamagnetic signal.   Fig. 3 shows, as a
typical example, ac susceptibility as a function of temperature for
a nominal sample $(Tl_{1.0}Pb_{0.2})Sr_2(Sr_{0.6}Pr_{0.4})Cu_2O_z$.   The resistance-

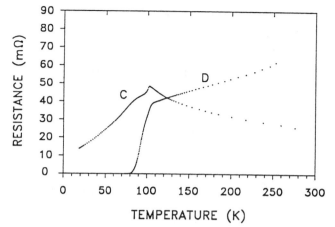

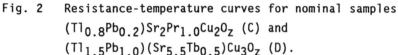

Fig. 2   Resistance-temperature curves for nominal samples
$(Tl_{0.8}Pb_{0.2})Sr_2Pr_{1.0}Cu_2O_z$ (C) and
$(Tl_{1.5}Pb_{1.0})(Sr_{5.5}Tb_{0.5})Cu_3O_z$ (D).

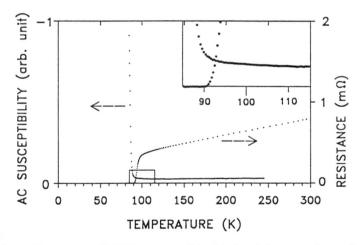

Fig. 3   AC susceptibility as a function of temperature for
$(Tl_{1.0}Pb_{0.2})Sr_2(Sr_{0.6}Pr_{0.4})Cu_2O_z$.

temperature curve for this sample is also shown in the same figure
for comparison. It can be seen that a strong diamagnetic signal
appears at about 90 K, which is roughly the same as the zero
resistance temperature of this sample. In the transition region,

the sample exhibited only a weak diamagnetic signal. The enlarged
transition region is shown in the insert, from which one can see a
weak diamagnetic signal up to about 100 K.

Powder x-ray diffraction analyses for a variety of TlPbSrRCuO
samples showed that most of the samples consist of a 1212 phase and
a 121 phase, and that the 1212 phase $(Tl_{1-x}Pb_x)Sr_2(Sr_{1-y}R_y)Cu_2O_z$ is
responsible for the observed 100 K superconductivity. As an
example, Fig. 4 shows powder x-ray diffraction pattern for the
sample $(Tl_{1.0}Pb_{0.2})Sr_2(Sr_{0.6}Pr_{0.4})Cu_2O_z$. This sample was nearly
pure 1212 phase. All peaks except three small peaks (marked by *

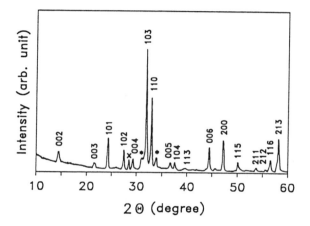

Fig. 4   Powder x-ray diffraction pattern for the sample
$(Tl_{1.0}Pb_{0.2})Sr_2(Sr_{0.6}Pr_{0.4})Cu_2O_z$.
*: 121 phase; x: unidentified impurity.

and x) can be assigned to a tetragonal unit cell of a = 3.8456(4) Å
and c = 12.204(2) Å. The peaks marked by * are the two strongest
peaks of the 121 phase, whereas the peak marked by x has not yet
been identified. $(Tl_{1.0}Pb_{0.2})Sr_2(Sr_{0.6}Pr_{0.4})Cu_2O_z$ was the starting
composition of the sample. We used Tl in excess by 0.2 mole of the
1212 stoichiometry. This excess was necessary to compensate the
loss of Tl during heating, so that the final composition of this
sample would be nearly $(Tl_{0.8}Pb_{0.2})Sr_2(Sr_{0.6}Pr_{0.4})Cu_2O_z$ (1212). The
indexed x-ray diffraction data are listed in Table I. By combining

Table I   Powder x-ray diffraction data for
$(Tl_{0.8}Pb_{0.2})Sr_2(Sr_{0.6}Pr_{0.4})Cu_2O_z$.

h	k	l	$d_{obs}(Å)$	$d_{cal}(Å)$	Int.
0	0	2	6.1047	6.1018	5
0	0	3	4.0659	4.0679	5
1	0	1	3.6688	3.6678	25
1	0	2	3.2539	3.2534	15
0	0	4	3.0472	3.0509	10
1	0	3	2.7960	2.7945	100
1	1	0	2.7176	2.7192	60
0	0	5	2.4417	2.4407	5
1	0	4	2.3916	2.3901	5
1	1	3	2.2616	2.2607	<5
0	0	6	2.0334	2.0339	20
2	0	0	1.9214	1.9228	25
1	1	5	1.8170	1.8164	5
2	1	1	1.7037	1.7030	<5
2	1	2	1.6554	1.6553	<5
1	1	6	1.6285	1.6287	10
2	1	3	1.5837	1.5840	25

composition, superconductivity, and powder x-ray diffraction data of
various samples, it is reasonable to conclude that a tetragonal 1212
phase with a general formula $(Tl_{1-x}Pb_x)Sr_2-(Sr_{1-y}R_y)Cu_2O_z$ is
responsible for the observed superconductivity of about 100 K.

For studying the effects of Pb and R on the formation of the
1212 phase and the superconducting behaviors of the samples, we
prepared two series of samples with starting compositions
$(Tl_{1.2(1-x)}Pb_x)Sr_2(Sr_{0.6}Pr_{0.4})Cu_2O_z$ and $(Tl_{1.0}Pb_{0.2})Sr_2(Sr_{1-y}Pr_y)-$
Cu_2O_z. Note that the Tl in the starting compositions was excessive
by about 20 % to their 1212 stoichiometry, which was used to
compensate for the Tl loss during heating. Powder x-ray diffraction

analyses showed that the diffraction lines of most of the samples
could be assigned to a 1212 phase and/or a 121 phase. By assuming
that the amounts of the 1212 phase and 121 phase are proportional to
the intensity of their strongest diffraction line (103 line for 1212
phase, and 102 line for 121 phase), we estimated percentage contents
of the 1212 phase and 121 phase in each sample. From the resistance
temperature curves of the samples, we obtained onset, midpoint, and
zero resistance temperatures for each sample. Since an onset
temperature is not well defined, we estimate the onset temperature
as the temperature at which the curvature in the resistance
temperature curve changes most rapidly. Table II lists the results
of resistance measurements and x-ray diffraction analyses for
samples $(Tl_{1.2(1-x)}Pb_x)Sr_2(Sr_{0.6}Pr_{0.4})Cu_2O_z$ with x = 0.0, 0.1, 0.15,
0.2, 0.25, 0.3, 0.4, 0.5, 0.7, and 1.0. The samples with x = 0-0.7
all contained a 1212 phase as the major phase and a 121 phase as
the minor phase. The sample with x = 0 had an onset temperature of

Table II   Onset (o), midpoint (m) and zero resistance (z)
temperatures and phase percentages for samples
$(Tl_{1.2(1-x)}Pb_x)Sr_2(Sr_{0.6}Pr_{0.4})Cu_2O_z$.

| | Tc (K) | | | phase (%) | |
x	o	m	z	1212	121
0.00	92	89	84	70	30
0.10	92	89	87	80	20
0.15	97	90	88	90	10
0.20	98	91	89	90	10
0.25	100	93	91	85	15
0.30	97	91	85	80	20
0.40	88	72	60	75	25
0.50	80	60	45	75	25
0.70	65	45	25	70	30
1.00	-	-	-	-	-

90 K and a zero resistance temperature of 84 K. When x = 0.1-0.3, the samples showed higher zero resistance temperatures and onset temperatures. The sample with x = 0.25 had a zero resistance temperature of 91 K and an onset temperature of about 100 K. These results suggested that Pb-dopping is necessary to increase the Tc of the 1212 phase from 90 K to about 100 K. When x = 0.4-0.7, the Tc's were depressed although these samples still consisted mainly of the 1212 phase. The sample with x = 0.7 contained a small amount of a cubic phase. When x = 1.0, the cubic phase became the main phase, and the sample was semiconducting.

Table III lists Tc's and phase percentages for samples $(Tl_{1.0}Pb_{0.2})Sr_2(Sr_{1-y}Pr_y)Cu_2O_z$ with y = 0.0, 0.1, 0.2, 0.3, 0.4, 0.5, 0.6, 0.7, 0.8, 0.9, and 1.0. The sample with y = 0.0 (i.e.

Table III   Onset (o), midpoint (m) and zero resistance (z) temperatures and phase percentages for samples $(Tl_{1.0}Pb_{0.2})Sr_2(Sr_{1-y}Pr_y)Cu_2O_z$.

y	Tc (K)			phase (%)	
	o	m	z	1212	121
0.0	-	-	-	-	99
0.1	93	-	-	10	90
0.2	98	90	70	25	75
0.3	98	90	88	55	45
0.4	100	92	90	90	10
0.5	100	91	85	98	-
0.6	100	90	-	98	-
0.7	101	88	-	98	-
0.8	100	82	-	98	-
0.9	101	70	-	95	-
1.0	53[*]	30	-	95	-

[*] A small signal of superconductivity was also observed at about 100 K.

without a rare earth) consisted essentially of the 121 phase, and
was not superconducting down to 15 K. The sample with y = 0.1
contained a 1212 phase as well as the 121 phase, and exhibited an
incomplete superconducting transition. With the increase of y (from
0.1 to 0.4), the 1212 phase increased and the 121 phase decreased,
and superconducting behavior of the samples improved. The sample
with y = 0.4 showed the best superconducting behavior with an onset
temperature of about 100 K and a zero resistance temperature of 90
K. When y > 0.5, the samples maintained a nearly pure 1212 phase,
but did not reach zero resistance down to 15 K. However, they
showed onsets of superconductivity around 100 K. The Tc of the y =
1.0 sample was significantly depressed. From the data in Table III,
we can make the following conclusions: (i) a 1212 phase is
responsible for the observed superconductivity, (ii) a rare earth is
required for the formation of the 1212 phase, and (iii) when y >
0.5, the zero resistance temperature of the 1212 phase is depressed.

In summary, a number of TlPbSrRCuO samples with Tc up to
above 100 K were prepared, and studied by resistance and ac
susceptibility measurements and by x-ray diffraction analyses. The
phase responsible for the 100 K superconductivity is a 1212 phase
with a general formula of $(Tl_{(1-x)}Pb_x)Sr_2(Sr_{1-y}R_y)Cu_2O_z$. A rare
earth is required for the formation of the 1212 phase. Pb-dopping
increases the Tc of the 1212 phase from 90 K to about 100 K. It is
believed that there is a new series of Ca-free superconductors
$(Tl_{1-x}Pb_x)Sr_2(Sr_{1-y}Pr_y)_{n-1}Cu_nO_z$ to be explored.

Acknowledgment We thank D.X.Gu, J.Bennett, D.Ford, and B.Xue
for their technical assistance, and Mrs. M.Megginson for her
assistance in the preparation of this manuscript. This work was
supported by Arkansas Energy Office grant AR/DNRG/AEO-UAF-88-005/S.

REFERENCES

1. Z.Z.Sheng and A.M.Hermann, Nature 332, 55 (1988).
2. Z.Z.Sheng, A.M.Hermann, A.El Ali, C.Almason, J.Estrada,
    T.Datta, and R.J.Matson, Phys.Rev.Lett. 60, 937 (1988).

3. Z.Z.Sheng and A.M.Hermann, Nature 332, 138 (1988).
4. Z.Z.Sheng, W.Kiehl, J.Bennett, A.El Ali, D.Marsh, G.D.Mooney, F.Arammash, J.Smith, D.Viar, and A.M.Hermann, Appl.Phys.Lett. 52, 1738 (1988).
5. R.M.Hazen, L.W.Finger, R.J.Angel, C.T.Prewitt, N.L.Ross, C.G.Hadidiacos, P.J.Heaney, D.R.Veblen, Z.Z.Sheng, A.El Ali, and A.M.Hermann, Phys.Rev.Lett. 60, 1657 (1988).
6. C.C.Torardi, M.A.Subramanian, J.C.Calabrese, J.Gopalakrishnan, K.J.Morrissey, T.R.Askew, R.B.Flippen, U.Chowdhry, and A.M.Sleight, Science 240, 631 (1988).
7. S.S.P.Parkin, V.Y.Lee, E.M.Engler, A.I.Nazzal, T.C.Huang, G.Gorman, R.Savoy, and R.Beyers, Phys.Rev.Lett. 60, 2539 (1988).
8. R.Beyers, S.S.P.Parkin, V.Y.Lee, A.I.Nazzal, R.Savoy, G.Gorman, T.C.Huang, and S.J.La Placa, Appl.Phys.Lett. 53, 432 (1988).
9. Z.Z.Sheng, A.M.Hermann, D.C.Vier, S.Schultz, S.B.Oseroff, D.J.George, and R.M.Hazen, Phys.Rev.B 38, 7074 (1988).
10. M.A.Subramanian, C.C.Torardi, J.Gopalakrishnan, P.L.Gai, J.C.Calabrese, T.R.Askew, R.B.Flippen, and A.M.Sleight Science 242, 249 (1988).
11. Z.Z.Sheng, L.Sheng, X.Fei, and A.M.Hermann, Phys.Rev.B 39, 2918 (1989).
12. R.S.Liu, J.M.Liang, Y.T.Huang, W.N.Wang, S.F.Wu, H.S.Koo, P.T.Wu, and L.J.Chen, Physica C 162-164, 869 (1989).
13. Z.Z.Sheng, C.Dong, Y.H.Lui, X.Fei, L.Sheng, J.H.Wang, and A.M.Hermann, Solid State Comm. 71, 739 (1989).
14. T.Itoch and H.Uchikawa, Phys.Rev.B 39, 4690 (1989).
15. Z.Z.Sheng, L.A.Burchfield, and J.M.Meason (to be published).
16. A.K.Ganguli, V.Manivannan, A.K.Sood, and C.N.R.Rao, Appl.Phys.Lett. 55, 2664 (1989).

EFFECT OF RF PLASMA DEPOSITION PARAMETERS ON $Y_1Ba_2Cu_3O_{7-x}$ THIN FILMS

A. Shah, E. Narumi, J. Schutkeker, S. Patel and D. T. Shaw
New York State Institute on Superconductivity, Amherst, NY 14260

ABSTRACT

Yttrium, barium and copper nitrates are dissolved in de-ionized water to form a 0.038M : 0.078M : 0.105M solution, which is used to generate, in-situ $Y_1Ba_2Cu_3O_{7-x}$ superconducting films in an argon-oxygen rf plasma. Rf power is operated at 4.5 kW, 13.56 MHz and 760 Torr pressure. The best zero resistance temperature, onset critical temperature and current density measured by the four probe transport method are 86K, 92K and 4×10^5 A/cm² at 77K and zero field. X-ray diffraction shows the films to be oriented with c-axis perpendicular to the substrate surface. Variations in the critical temperatures of the films and their microstructure depend on rf power, solution composition, solution concentration, plasma gas and aerosol carrier gas. It has been seen that the concentration of barium in the solution is more critical than that of copper. The concentration of the solution is related to the rf power; to prepare the best film, the concentration and the power levels have to be matched. While the variation of both aerosol carrier gas flow rate and the plasma gas flow rate causes a change in the structure and critical temperature of the film, the control of the former is more critical than the plasma gas flow rate.

INTRODUCTION

The technique of using rf plasma deposition has proved to be a viable process for the in-situ growth of superconducting $Y_1Ba_2Cu_3O_{7-x}$ thin films [1,2,3]. Using an evaporation-condensation process, this method has grown thin films by injecting an atomized aqueous solution of the nitrates of yttrium, barium and copper into an argon-oxygen induction plasma. This technique offers distinct advantages over many of the other deposition processes. High vacuum systems are unnecessary in the rf deposition process. The electrodeless plasma also provides an uncontaminated reaction zone. The use of a nitrate solution as a precursor is especially suitable because it allows the compound composition and the feed rates into the plasma to be accurately controlled.

A typical $Y_1Ba_2Cu_3O_{7-x}$ film on yttria stabilized zirconia (YSZ), under the present optimized conditions [1], has a critical temperature, T_c (R=0) of 86K; a T_c onset at 92K with a metallic behavior prior to onset and a critical current, J_c of 4×10^5 A/cm² at 77K and zero field. The film, which is deposited in 10-15 minutes, is usually 0.1-

0.2 µm thick. A typical film has a relatively smooth surface with large grains of 1-2 µm in size. X-ray diffraction of the film, using CuKα radiation, shows the film to be highly oriented with the c-axis perpendicular to the substrate. Only the 001 reflections are present. The c-axis length is consistently 11.67-11.68 Å. The full width half maximum (FWHM) of the rocking curve of this film is 0.55 degrees. An auger electron depth profile shows the composition to be homogeneous throughout the cross-section with a sharp interface at the YSZ substrate.

The rf plasma deposition technique has two main subsystems. One is the precursor nitrate solution and the nebulizer used to generate the the aerosol while the other is the plasma and plasma torch. Both the subsystems have an immense effect on the ultimate properties of the films. There is a narrow window in which the deposition parameters are just optimum to produce the highest T_c and J_c films. In the case of the precursor, the solution composition, solution concentration and the carrier gas flow rate are of importance. In the case of the plasma, the rf power and the plasma gas are of importance. The effect of these deposition parameters on the properties of the films has been studied in this paper.

SOLUTION COMPOSITION

The precursor solution is made by dissolving yttrium, barium and copper nitrates in de-ionized water to form a non-stoichiometric 0.038M : 0.078M : 0.105M solution (6g total by weight of the nitrates in 100 ml of de-ionized water, i.e. 6%). At this composition, films with the best T_c's are obtained. Decreasing the amount of copper in the solution, the T_c remains nearly constant up to a change of 1% in the copper content and a decrease by 1.75% depresses the T_c to only 75K. On the other hand, as soon as the copper content in the solution is increased by 0.25% the T_c decreases rapidly and ultimately looses its superconductivity when the copper content is increased by 1.5%. This is shown in Figure 1. An x-ray diffraction pattern of this film made with the copper rich solution shows the presence of CuO in them. This is also confirmed by the EDS back scattered image of this film. On the other hand, the amount of barium in the solution has to be carefully controlled because even a small increase or decrease of barium will cause a T_c degradation.

SOLUTION CONCENTRATION AND RF POWER

The films made by using the optimum 6% solution are usually smooth. The x-ray diffraction patterns show the presence of only the 001 reflection peaks. Now, if the solution concentration is lowered to 5g per 100 ml (5%) the film becomes patchy. The T_c drops to 62K even though the structure of the film, as shown by the x-ray diffraction pattern, is c-axis oriented. Prolonging the deposition time does not produce a smooth film, instead the structure of the film is destroyed by the extended exposure to the plasma. Now, as the solution concentration is increased to 8g per 100ml (8%) while keeping the other conditions the same, the deposition rate increases

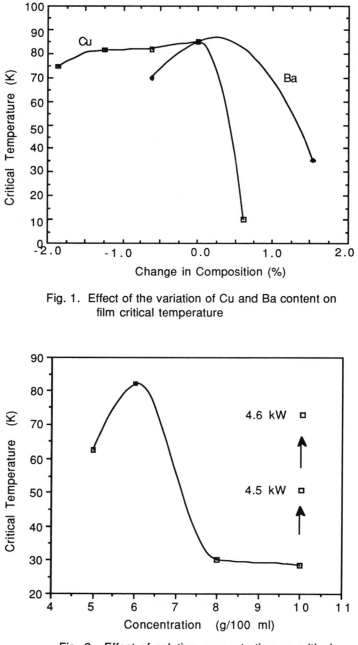

Fig. 1. Effect of the variation of Cu and Ba content on
film critical temperature

Fig. 2. Effect of solution concentration on critical
temperature of films made at 4.4 kW power

from 2 Å/sec to 4-4.5 Å/sec. The film is thicker for the same deposition time, but the T_c drops sharply to 30K. The x-ray diffraction pattern shows the presence of non-superconducting phases along with the 001 reflections from the superconducting phase. The film has lost its smooth microstructure. A further increase in the concentration to 10g per 100 ml (10%) again increases the deposition rate to 6 Å/sec but the structure degrades even further. The T_c drops to 27K. X-ray diffraction again shows the presence of the non-superconducting phases. By increasing the concentration, the dried aerosol particles are larger. This causes a problems as the particles do not all evaporate completely [4]. The reactions are not all complete, resulting in the formation of non-superconducting phases in the grains and grain boundaries. The technique turns into a spray-like process. Now, if the the rf power is increased to 4.5 kW, the T_c for the 10% concentration precursor increases to 50K and an even further increase to 73K is obtained by increasing the rf power to 4.6 kW. This is shown in Figure 2. The film begins to regain its structure by this increase in the power by 200 W. This also implies that for every precursor concentration there is a specific power level at which the plasma has to be operated. Finally, a further increase in the concentration to 13%, causes the film structure to change drastically. The film microstructure becomes granular and powder-like. The X-ray diffraction pattern shows that the film has a powder-like structure too. The film is still superconducting but at a low temperature of 15-20K. Figure 3 shows the film prepared at different solution concentrations.

The above effect is again observed when the rf power is varied, keeping all the other deposition conditions the same. As shown in Figure 4, the power versus the T_c curve peaks for a particular power level and the T_c drops off on both sides. Changing the power delivered to the plasma changes the temperature profile in the plasma and in its tail and this in turn changes the reaction zone temperature profile. Hence, the power level is directly related to the quality of the films.

AEROSOL CARRIER GAS AND PLASMA GAS

The aerosol is carried into the torch axially with oxygen as the carrier gas as shown in Figure 5. Now it is important that the aerosol is injected into the plasma. At low flow rates the aerosol particles tend to 'bounce' off the plasma while at higher flows the particles pierce through the plasma with a small residence time in the plasma[5]. In both these extreme cases unsuitable reaction conditions are present. This is corroborated by the adverse effect of these conditions on the films. At a flow rate of 3.6-3.8 lpm, the T_c of the film is high and at flow rates less than or more than this range the T_c drops off rapidly as shown in Figure 6. A similar effect is seen when the the plasma gas flow rate is changed. The plasma gas, oxygen, is injected radially as shown in Figure 5 and is surrounded by the swirling argon gas to keep it from coming in contact with the torch quartz tube. The change in T_c is not as drastic as that caused by the change in the carrier gas flow. This is because a small change in the plasma gas flow rate does not cause a drastic change in plasma parameters. On the other

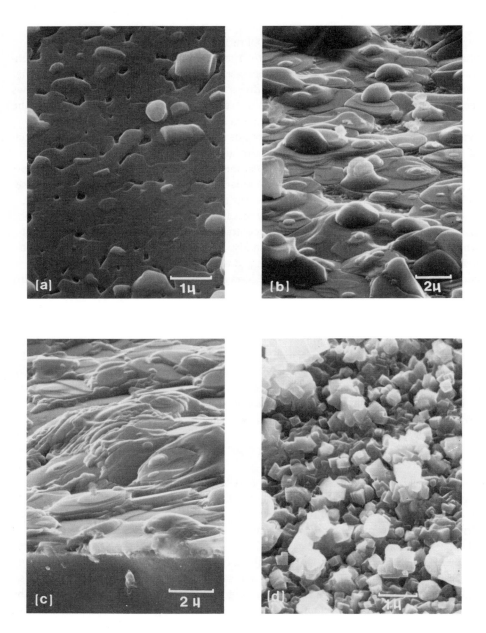

Fig. 3. Electron Micrographs of $Y_1Ba_2Cu_3O_{7-x}$ films made by using
nitrate solution concentrations of : (a) 6g/100 ml, (b) 8g/100 ml
(c) 10g/100ml and (d) 13g/100 ml.

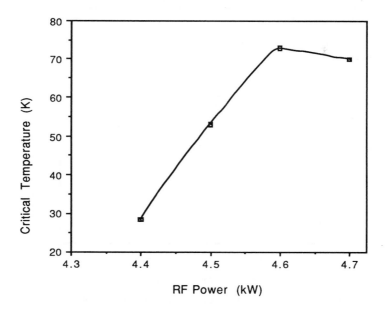

Fig. 4.   Effect of RF Power on the critical temperature of films
made by using the10% solution

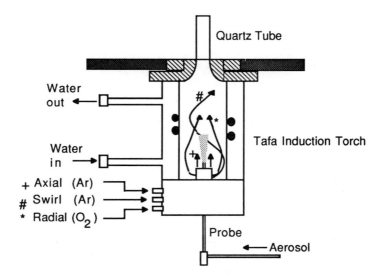

Fig. 5.   Direction of gas flow in Tafa Induction Torch

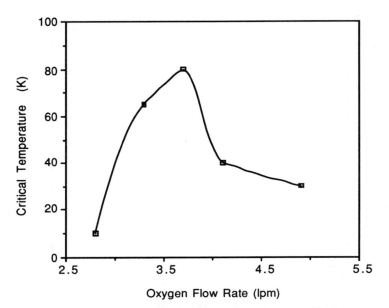

Fig. 6. Effect of aerosol carrier gas flow rate on critical temperature

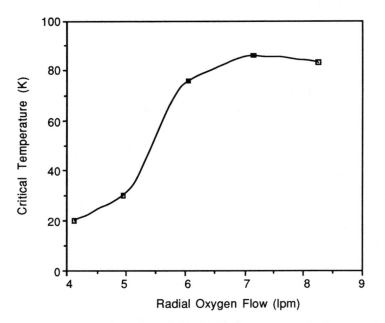

Fig. 7. Effect of radial plasma gas flow rate on critical temperature

hand, a small variation in the aerosol flow rate causes the plasma geometry and flow profiles to alter[5].

CONCLUSION

Film quality with regards to T_c and microstructure is highly dependent on the operating conditions of the plasma and the nitrate precursor solution. Extreme care is needed in preparing the solution. The solution has to remain uncontaminated and the solution concentration has to be maintained at the optimum level. The compositional ratios (especially, the barium ratio) have to be strictly maintained. The solution concentration to be used is tied to the rf power level and strict power control and a stable plasma is required. The carrier gas and the plasma gas have to be finely controlled as they control the plasma and hence, the structure of the films. Precise control of the aerosol carrier gas flow is necessary.

REFERENCES

1. A. Shah, E. Narumi, S. Patel and D.T. Shaw, Appl. Phys. Lett. **57**, 1452 (1990).
2. A. Shah, T. Haugan, S. Witanachchi, S. Patel and D.T. Shaw in **Proc. Mater. Res. Soc. Meeting,** Boston, December, 1989 (Material Research Society).
3. K. Terashima, K. Eguchi, T. Yoshida and K. Akashi, Appl. Phys. Lett. **52**, 1274 (1988).
4. T. Kodas, A. Datye, V. Lee and E. Engler, Appl. Phys. Lett. **65**, 2149 (1989).
5. M. I. Boulos, Pure & Appl. Chem., Vol. 57, No. 9, pp 1321-1352, 1985.

COMPARATIVE STUDIES OF OHMIC CONTACTS TO SUPERCONDUCTING YBCO THIN FILMS WITH DIFFERENT CONTACT MATERIALS

Q.X. Jia and W.A. Anderson, State University of New York at Buffalo
Department of Electrical and Computer Engineering
Institute on Superconductivity, Bonner Hall, Amherst, NY 14260

ABSTRACT

Electrical properties of the contacts with different contact materials on superconducting YBCO thin films were extensively investigated. The different contact materials including Au, Pd/Au, Ag, Yb/Cr/Al, RuO_2, RuO_2/Au, and RuO_2/Ag were chosen to explore the possibility of fabricating low resistance contacts on superconducting YBCO films with processes and materials compatible with semiconductor technology. Experimental results demonstrated the surface condition dependent contact properties regardless of the contact materials. Reproducible and controllable contacts were realized by using RuO_2 or the combination of RuO_2 with other metals.

INTRODUCTION

The formation of a very low resistance contact to high temperature superconducting YBCO thin films is one of the major problems in developing practical electronic and electrical applications. However, there are limited studies dealing with the ohmic contact to the YBCO thin films with a thickness in the range of 150nm used for electronic devices, although much work was done on bulk YBCO oxides. Experimentally, it is much more difficult to form low resistance contacts to YBCO thin films compared to that of YBCO bulk oxides because of the thinness of the films. The specific contact resistivity was less than 10^{-10} Ω-cm^2 for the Ag on bulk YBCO oxides [1,2] but as high as 10^{-3} Ω-cm^2 for Ag on YBCO thin films although a plasma oxidation was used to reoxidize a small amount of superconductor surface material which was exposed to air before electron beam evaporation of Ag [3]. The lowest specific contact resistivity was about 10^{-6} Ω-cm^2 when Ag was used as the contact for YBCO thin films, where the contact was thermally annealed at 500°C for four hours in an oxygen atmosphere [4]. The contact properties are also expected to be much more film surface dependent due to the higher ratio of the surface to volume of the films compared to that of the bulk materials.

In this paper, we present some experimental data concerning low resistance contacts on thin YBCO films with different contact materials including Au, Ag,

Pd/Au, Yb/Cr/Al, RuO_2, RuO_2/Ag, and RuO_2/Au. The advantages and disadvantages of using different contact materials for low resistance contacts on YBCO films will be discussed herein.

EXPERIMENTAL DETAILS

Superconducting YBCO thin films deposited either by laser ablation [5] or RF magnetron sputtering [6] were used in this study. The YBCO films deposited by laser had a thickness in the range of 150-200nm, a very good c-axis orientation, and a zero resistance temperature of around 85K. However, the YBCO film deposited by RF magnetron sputtering showed a trace of other phases but the same zero resistance temperature as a laser deposited film.

Contact pads with a diameter in the range of 0.8mm to 1mm were thermally evaporated onto YBCO thin film surfaces through a shadow mask. However, the RuO_2 was deposited using DC magnetron sputtering. Multilayer metal deposition, such as Yb/Cr/Al, was done sequentially without breaking the vacuum. The evaporation and sputtering were carefully done to prevent the thermal damage to YBCO films during evaporation or bombardment during sputtering. No intentional substrate heating was used during contact material deposition. Au wires with a diameter of 0.05mm, which were used as the interconnections for the later electrical measurement, were bonded to the contact pads using a glow discharge wire bonder.

The contact was thermally treated at 200 to $650^{o}C$ in oxygen at 1 atm to reduce the contact resistance or to enhance the oxygen content at the contact interface. This temperature region was chosen since it is the typical annealing temperature used for ohmic contacts in semiconductor technology. The time duration was controlled in the range of 15min although some of the samples were thermally treated at the desired temperature for several hours.

The contact resistance was measured both before and after thermal treatment using three and four terminal resistance vs temperature measurements as previously described [7]. The contact resistivity was defined as the product of contact resistance and contact area. The interface composition between contact materials and YBCO thin films was characterized by Auger electron spectroscopy (AES) depth profiling. Scanning electron microscopy (SEM) surface survey was also used to see the morphology changes after thermal treatment of the contacts.

RESULTS AND DISCUSSION

The contact resistance was YBCO film surface condition dependent regardless of the contact materials. Higher contact resistance was revealed for the contact on old YBCO films, where the YBCO film was exposed to air for several days. An even worse case was the appearance of semiconductor-like contact properties for the as-deposited contacts, such as Ag on old YBCO films [7]. Figure 1 shows the typical contact properties for the as- deposited Ag contacts on two different YBCO films. The semiconductor-like contact property could be removed by annealing the contact at a relatively low temperature of 450°C in oxygen ambient for 15 to 30min. High temperature annealing not only worsens the contact, but also degrades the YBCO film quality. This became very obvious both for the contacts of Ag and Pd/Au. The best contact with these two materials after approperiate thermal treatment gave a contact resistivity of about 10^{-4} to 10^{-3} Ω-cm^{2}at 77K, respectively. However, it should be pointed out that the Ag surface looks very rough after thermal treatment of the contact at a temperature of 400°C or above. Both SEM and AES surface survey showed the Ag island formation after this treatment [7]. The most obvious observation after thermal treatment of the Ag contact was the color changes of the Ag. The bright shiny Ag became milky and deep brown with increasing thermal treatment temperature.

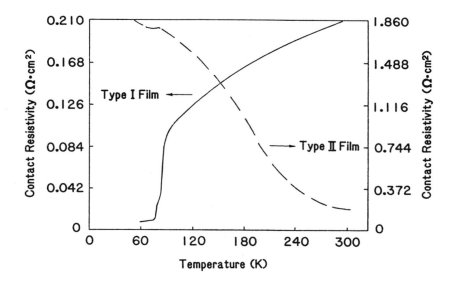

Figure 1.   Specific contact resistivity vs temperature for the as-deposited Ag contact on Type I and Type II YBCO thin films.

Au on YBCO thin films gave a better contact property than that of Ag. The degradation of zero resistance temperature after thermal treatment in the temperature range of 400°C to 650°C was not found although the contact resistivity was still relatively high. Specific contact resistivity less than 10^{-4} Ω-cm^2 at 77K was obtained with an Au contact on YBCO films. The degradation of the film quality including zero resistance temperature, critical current density, and resistance of the film at normal state was reported after thermal treatment of the Au contact above 350°C [8]. Specific contact resistivity as low as 10^{-8} Ω-cm^2 at 4.2K was also reported by using Au as a contact on the YBCO film with a thickness of 1 μm where the contact was treated at 850°C in oxygen for one hour [9].

Yb/Cr/Au showed very different contact properties on YBCO films compared to those of the Ag contact. There was no degradation of the film quality after thermal treatment of the contacts both for laser and RF sputtered films. Specific contact resistivity as low as 10^{-9} Ω-cm^2 at 77K was realized for RF sputtered films [10]. However, a relatively higher value for laser deposited thin films was found. It seems that the performance of the contact was film property dependent with this kind of structure.

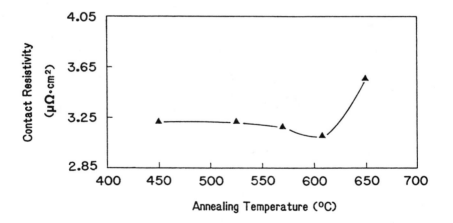

Figure 2.   Thermal treatment temperature-dependent contact property of RuO2 on YBCO films at 77K. The thermal treatment was done at the indicated temperature in oxygen for 15min.

A much better and reproducible contact was realized by using RuO_2 or a double layer structure. The highly conductive and thermally stable nature of RuO_2 makes it very attractive in metallization. Reproducible specific contact resistivity less than $5x10^{-6}$ Ω-cm^2 was achieved for a single layer RuO_2 contact. Figure 2 shows the specific contact resistivity at 77K vs thermal treatment temperature. The holding time at each temperature was 15min. Oxygen flow was maintained during this treatment. The best contact with a double layer structure showed contact resistivity as low as $5x10^{-8}$ Ω-cm^2 at 77K after an appropriate thermal treatment of the contact for a short time. The double layer structure provides other advantages compared to a single one. First, lower contact resistivity was demonstrated with a double layer structure. Secondly, the binding problem was solved due to the use of a double layer contact. It was very difficult to bind a gold wire directly to RuO_2 by a glow discharge wire binder.

CONCLUSIONS

The following points are concluded.

1) A low temperature process is necessary to maintain the intrinsic properties of the YBCO thin films.
2) A fresh film surface is critical to obtain low contact resistance. In situ contact electrode deposition is recommended to obtain much lower contact resistance.
3) Ag is not a very good contact material for very thin superconducting YBCO thin films although it shows very good contact property for bulk YBCO oxides.
4) RuO_2 might provide an alternative to realize a good contact for YBCO films due to it's good thermal stability and high conductivity.
5) It is much more difficult to realize a good contact on YBCO films than on bulk YBCO oxides.

ACKNOWLEDGMENT

We wish to mention the technical assistance from P.Bush and R.Barone. Q.Z.Zhu and J.P.Zheng of Dr. Shaw's and Dr. Kwok's groups, respectively, provided laser deposited YBCO thin films. This work was supported by the New York Research and Development Authority through the New York State Institute on Superconductivity.

REFERENCES

[1] J.W. Ekin, T.M. Larson, N.F. Bergren, A.J. Nelson, A.B. Swartzlander, L.L. Kazmerski, A.J. Panson and B.A. Blankenship, Appl. Phys. Lett. 52, 1819 (1988).

[2] S. Jin, M.E. Davis, T.H. Tiefel, R.B. van Dover, R.C. Sherwood, H.M. O'Bryan, G.W. Kammlott and R.A. Fastnacht, Appl. Phys. Lett. 54, 2605 (1989).

[3] A. Frenkel, M.A. Saifi, T. Venkatesan, P. Englandm, X.D. Wu and A. Inam, J. Appl. Phys. 67, 3054 (1990).

[4] T. Kobayashi, K. Hashimoto, U. Kabasawa and M. Tonouchi, IEEE Trans. Magn. 25, 927 (1989).

[5] S. Witanachchi, H.S. Kwok, X.W. Wang and D.T. Shaw, Appl. Phys. Lett. 53, 234 (1988).

[6] Q.X. Jia, D. Ren and W.A. Anderson, Int. J. Modern Phys. B3, 743 (1989).

[7] Q.X. Jia, W.A. Anderson, J.P. Zheng, Y.Z. Zhu, S. Patel, H.S. Kwok and D.T. Shaw, J. Appl. Phys., in press.

[8] M. Matsuda, A. Matachi, N. Hashimoto and S. Kuriki, Jpn. J. Appl. Phys. 29, L618 (1990).

[9] K. Mizushima, M. Sagoi, T. Miura and J. Yoshida, Appl. Phys. Lett. 52, 1101 (1988).

[10] Q.X. Jia and W.A. Anderson, J. Electronic Mater. 19, 443 (1990).

DEVELOPMENT OF HIGH AMPERAGE Bi-BASED
SUPERCONDUCTING TAPES ON SILVER

J. Ye, S. Hwa, S. Patel and D. T. Shaw
New York State Institute on Superconductivity
State University of New York at Buffalo
Amherst, New York 14260

ABSTRACT

Brush-on-method with partial-melt growth has been used to fabricate Bi-based 2212 superconducting thick films on silver. Transport properties show critical currents of 10-20 amperes (or equivalently J_c of 10^5 A/cm^2 below 20K and zero field), with a 50 percent degradation at 9 tesla field parallel to the film surface. Electron microscopy reveals the film to be layered, consisting of platelets with large aspect ratios and with the c-axis perpendicular to the substrate. Transport and microstructural properties of the films has been found to depend on the structural integrity of the substrate. Silver foil, 25 microns thick, has resulted in poor films with microcracks, whereas 75 micron thick silver foil have resulted in sturdy high quality superconducting films. Sputter deposited silver on polycrystalline MgO results in the total diffusion of the silver into the film with a corresponding degradation in film properties. Film processing and corresponding characterization results will be discussed.

INTRODUCTION

Heine[1] recently reported extraordinarily high critical current densities, J_c's at 4.2K in Bi-based silver wires (5.7×10^4 A/cm^2 at zero field and 1.5×10^4 A/cm^2 at 26T). Since then, Furakawa[2,3] has not only duplicated the results, but also showed that there is a remarkable degree of c-axis alignment (up to 95%) in similar Bi-based tapes with even higher J_c's. These wires and tapes exhibit disappointingly low J_c at 77K (10^3 A/cm^2 at zero field and 10^2 A/cm^2 at 20 mT). Although strong flux creep[4,5,6] can be used to explain the low J_c values at 77K, no reasonable explanation

524   © 1991 American Institute of Physics

is available at the present time for the high J_c values at 4.2K. Of course, flux creep is insignificant at this temperature and J_c is determined by flux pinning. For Bi-based (2212) compounds, a very high density of defects have been reported[7,8].

Most of the work on silver wires and tapes[1,2,3] has been based on the so-called "powder in tube" (PIT) technique, which involves filling of a silver tube with the BSCCO powder, sealing the ends, and densifying of the bulk BSCCO materials by repetitive extrusion and cold drawing. High temperature post-annealing is usually used, but the actual thermal cycle is vaguely discussed.

Tagano[9] has achieved highly oriented BSCCO tapes on silver strips without the densification used in the PIT process. Here, using the doctor-blade technique, a 50 μm thick green tape is laid on a 50 μm thick Ag foil and then heat treated. The resultant tape exhibits J_c values similar to those obtained with the PIT process.

Here we report typical results of the effect of several processing parameters on the transport and microstructural properties of 2212 BSCCO on Ag tapes fabricated by the brush-on technique followed by partial-melt growth.

TAPE FABRICATION

Figure 1 shows schematically the simple process of generating the BSCCO tape by the brush-on and partial melt growth technique. High quality 2212 BSCCO powder is suspended in butanol to form a paste. The paste is gently brushed on to the Ag foil (20 mm x 5 mm x 75 μm thick) mounted on a MgO support to form a 50 μm green tape. The tape is then placed into the furnace and goes through a heat treatment that consists of a fast ramp to the sintering temperature, which is slightly below the melting point, followed by a 20-30°C gradual cooling and finally through a quench process. The final superconducting BSCCO film is 10-15 μm.

EFFECT OF SINTERING TIME

The transport properties were found to be very sensitive to the sintering time. Short sintering times (<10 minutes) resulted routinely in T_c's above 85K and J_c's above 10^4 A/cm^2 at 77K and zero field. Figure 2 shows the variation of the critical current density with magnetic field parallel to the film across a 1 mm wide bridge.

Transport properties show critical currents of 10-20 amperes (or equivalently J_c of $\sim 10^5$ A/cm^2 below 20K and zero field), with a 50 percent degradation at 9 Tesla. Electron microscopy reveal the films to be layered, consisting of platelets with large aspect ratios and needle-like structures with typical size of 250 µm x 20 µm. The microstructure of such a film is shown in Figure 3. EDAX indicates the needle-like structure of Figure 3a to be other phases deficient in Bi, and the smooth layered structure shown in Figure 3b, under high magnification, to be 2212. Reduction in the number and size of the needle-like structures has been correlated to higher J_c's. This was accomplished by reducing the sintering time. A cross section of the film shown in Figure 4 reveals the fine layers, which are ~ 0.03 µm thick. The growth of these layers was found to be independent of the surface roughness of the substrate. XRD analysis reveals these films to be c-axis oriented normal to the substrate.

EFFECT OF SINTERING TEMPERATURE

The lowering of the sintering temperature from that of the optimum was found to be detrimental to the transport and microstructural properties. A decrease in the sintering temperature by 10ºC from that of the optimum resulted in the formation of Bi-rich phases and an increase in the number and size of the needle-like structures.

EFFECT OF SUBSTRATE THICKNESS

BSCCO films were grown under similar conditions on 25 µm and 75 µm thick Ag foils. The 25 µm thick substrate resulted in poor films primarily due to the handling difficulty arising from the removal of the Ag foil from the MgO support, or in the case without the MgO support, from the warping due to the high processing temperature. However, in either case, the bonding between the film and the substrate was sufficient to withstand peeling by mechanical bending of the films. Sputtered Ag on polycrystalline MgO was also used as a substrate. Total diffusion of the silver into the film with extensive degradation in film properties was observed.

CONCLUSION

The brush-on method with the partial-melt growth is capable of producing high amperage BSCCO tapes provided processing conditions are well controlled. The J_c's obtained by this technique are similar to the J_c's produced by other techniques such the powder in tube method. The simplicity of this process combined with the electrical properties of the Bi-based superconductors for high field applications, specifically in the 20-30K range, makes this attractive for rapid scale-up operations and development of actual devices.

ACKNOWLEDGEMENT

This work has been sponsored by the National Science Foundation. We would like to thank Dr. J. E. Tkaczyk of General Electric, Schnectady, New York, for the magnetic field measurements and P. Bush, T. Haugan and C. Li for providing the electron micrographs.

REFERENCES

1. K. Heine, J. Tenbrink and M. Thonner, Appl. Phys. Lett., 55, 2441 (1989).
2. Y. Tanaka, The September 1989 Meeting on Superconducting Materials and Magnets, page 30.
3. N. Uno, N. Enomoto, H. Kikuchi, K. Matsumoto, M. Miaura and M. Nakajima, The September 1989 Meeting on Superconducting Materials and Magnets, page 92.
4. D. Dew-Hughes, Cryogenics, 28, 674 (1988).
5. T.T.M. Palstra, B. Batlogg, R.B. van Dover, L.F. Schneemeyer and J.V. Waszczak, Appl. Phys. Lett., 54, 763 (1989).
6. J.Z. Sun, K. Char, M.R. Hahn, T.H. Geballe and A. Kapitulnik, Appl. Phys. Lett., 54, 663 (1989).
7. W.A. Fietz and W.W. Webb, Phys. Rev., 178, 657 (1969).
8. E.J. Kramer, J. Appl. Phys., 44, 1360 (1973).
9. K. Tagano, H. Kumakura, D.R. Dietderich, H. Maeda, J. Kase, T. Morimoto, B. Ullmann and H.C. Freyhardt, Proc. ICMC 90, Topical Conf. on HTSC, May 9, Garmisch-Partenkirchen, FRG.

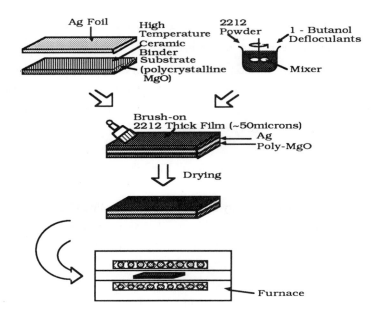

Fig. 1 Schematic diagram of the brush-on and partial melt growth on Ag foil.

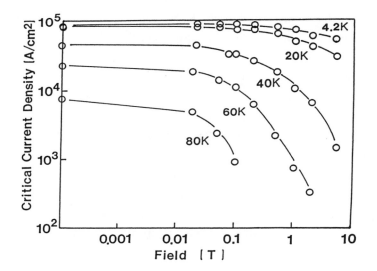

Fig. 2 Magnetic field dependence of the critical current density at various temperatures.

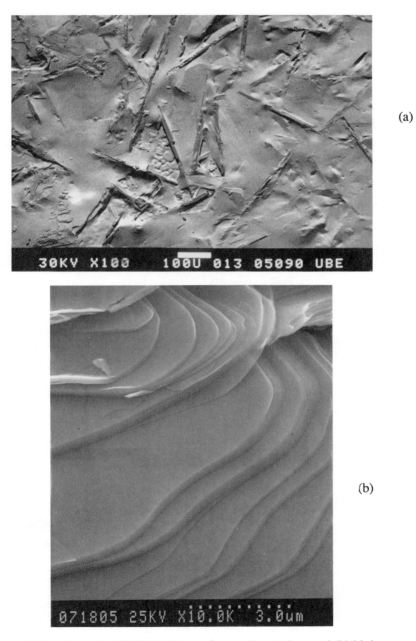

(a)

(b)

Fig. 3 SEM of a typical BSCCO film surface under (a) low and (b) high magnification.

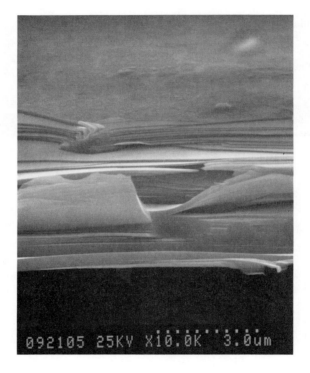

Fig. 4 Cross section of the BSCCO film showing the fine layered structure.

Deposition of Superconductive YBaCuO Films at Atmospheric Pressure by RF Plasma Aerosol Technique

H.H. Zhong, X.W. Wang, J. Hao and R.L. Snyder

Institute for Ceramic Superconductivity, NYSCC

Alfred University, Alfred, NY 14802.

ABSTRACT

Superconducting $Y_1Ba_2Cu_3O_{7-\delta}$ films were produced by a radio frequency (RF) plasma aerosol evaporation technique at atmospheric pressure without post-annealing. Aqueous solutions containing Y, Ba, and Cu were generated as an aerosol which was then injected into the plasma region. The ionized species were deposited onto substrates outside of both the plasma and flame regions. The substrate temperature was 400-600°C. The deposition rate is 0.01-100 μm/(min cm^2), and the film thickness is 1-200 μm. For an "as deposited" film on a single crystalline MgO substrate (100) with substrate temperature of 600°C, the onset temperature of the superconducting transition is 100K, with a transition width (10% - 90%) of 3K, and zero resistance at 93K. The critical current density of the film is 0.8 x 10^4 A/cm^2 at 77K. An optimum substrate temperature for this technique is discussed. The as-deposited films are compared with other post-annealed films (850°C, 1 hour). Since this technique does not require a vacuum environment it has potential for large scale production of thin films.

I. Introduction

Since the discovery of high T_c superconductivity[1], dozens of techniques have been developed to make films either inside of a vacuum chamber[2,3,4,5] or under normal atmosphere[6,7]. To-date, the best superconducting films are thin films with a thickness less than 10 μm, obtained by evaporation or sputtering methods in vacuum.[2,3] However, all of the vacuum technologies are limited in that they can produce only small film sizes and vacuum is ill suited for mass production procedures. Large scale production of films may rely on the development of non-vacuum techniques, such as tape casting[8], plasma spraying[6,7], or plasma vapor deposition. During the plasma vapor deposition of films[4,5], powder (or solution aerosol) is injected into the plasma (or flame) region, and the vapor is deposited onto a substrate to form a film thicker than 10 μm. Plasma spraying on the other hand, deposits the liquid or partially melted solid onto the substrate.

Our previous work[9] injected a fine powder into the plasma and, depending on conditions, sometimes produced plasma vapor deposited films but more often produced plasma sprayed films due to partial vaporization. We have modified this plasma technique, and obtained superconductive films with a solution mist injection method.[10]

II. Experiments

The starting powders were $Y(NO_3)_3$, $Ba(NO_3)_2$ and $Cu(NO_3)_2$ with purity of 99.9% or higher. These powders were mixed according to the stoichiometric ratio of Y:Ba:Cu in 1:2:3, and then dissolved in distilled (or deionized) water with a concentration of 150 g/l or less. The aqueous solution was then stirred thoroughly and poured into a plastic bowl, see Fig. 1.

A DeVilbiss Ultrasonic Nebulizer Model Ultra-Neb 99, normally used in hospital respiratory therapy, was used as an aerosol generator, to produce a mist in the space above the aqueous solution. Under the pressure of an argon or oxygen carrier gas, the mist is fed into the O_2/Ar plasma region. After travelling through the flame region, the ionized vapor is deposited onto a substrate. The substrate temperature is maintained at a fixed temperature during deposition. Between the plasma and flame regions, additional oxygen can be supplied through two auxiliary gas inlets. At the beginning of an experiment, the plasma oscillation and/or the mist injection may be unstable. A shutter is placed above the flame region to block unwanted vapor from the substrate. Depending on the concentration of the solution, the deposition rate varies from 0.01-$100\mu m$/(min cm^2). Depending on the deposition rate and the deposition time, the film thickness varies from 1 to $200\mu m$. Other experimental conditions are summarized in Table 1.

III. Results

Each film was formed on a substrate at a fixed substrate temperature during deposition. The substrate varied from 400°C to 600°C with the temperature accuracy of ±10°C. The "as deposited" films were black, and examined as described next.

1. Substrate temperature of 600°C.

Film resistances were measured by a standard 4-probe method. For an as-deposited film on a single crystalline MgO (100) substrate, the resistivity vs. temperature curve is shown in Fig. 2. The onset temperature is 100K, with a transition width (10%-90%) of 3K, and zero resistance at 93K. (The temperature accuracy is ±0.5K.) The critical current density of the film is 0.8 x 10^4 A/cm^2 at 77K. The thickness of this film is about 15 μm. Figure 3(A) is the x-ray diffraction pattern of an "as-deposited" superconductive film on a single crystalline MgO (100) substrate. As compared with a standard powder diffraction pattern of pure $Y_1Ba_2Cu_3O_{7-\delta}$ in Fig. 3(C), the film shows a pure 123 phase. The scanning electron microscope (SEM) reveals that grains of the "as-deposited" films are uniformly distributed, and grain sizes are smaller than 1 μm^3. Energy dispersive x-ray analysis (EDAX) shows uniform grain compositions containing the three desired metals.

Some other films were deposited at the substrate temperature of 600°C, and then post-annealed at 850 $\pm$ 10°C for 1 hour, with a flow of oxygen at a pressure of about 1 bar in the oven. The resistivity vs. temperature curve of a post-annealed film on MgO (100) substrate is shown in reference 10. The onset temperature is 105K, with a transition width of 6K, and zero resistance at 91K. The thickness of this film is about $20\mu m$. Figure 3(B) is the x-ray diffraction pattern of the post annealed superconductive film, and shows a pronounced (00ℓ) orientation. The SEM reveals that grains of the post-annealed films are uniformly distributed, and grain sizes are 1 x 1 x 10 μm^3. The surface morphology of the film is very similar to that of other post annealed films.[11] EDAX also shows uniform grain compositions containing the three desired metals.

2. Substrate temperatures of 500°C and 400°C

A film was formed at 500°C on a MgO (100) substrate. The X-ray diffraction pattern of the as-deposited film is plotted in Fig. 4(B). As compared with the standard 123 pattern shown in Fig. 4(C), the film shows a pure 123 phase. The film was then post-annealed at 850°C for 1 hour. The diffraction pattern of the post-annealed film is plotted in Fig. 4(A) and shows a pure 123 phase as compared with the standard 123 pattern of Fig. 4(C).

Another film was formed at 400°C on MgO (100). The diffraction pattern of the as-deposited film is plotted in Fig. 5(B). The major phase in the film is 123, as compared with the standard 123 pattern in Fig. 5(C). Another minor phase is barium copper oxide. A strong diffraction peak around 43 degrees is due to the substrate MgO. The film was post-annealed at 850°C for 1 hour. The pattern of the annealed film is plotted in Fig. 5(A). The dominant phase in the film is 123, and a minor phase is barium copper oxide.

3. Substrate temperature of 450°C

Films were formed at 450°C on MgO (100), or yttrium stabilized zirconia (YSZ) single crystal, i.e. ZrO_2 (100). These films were post-annealed at 850°C for 1 hour. The x-ray diffraction patterns of post-annealed films on MgO (100) and ZrO_2 (100) are shown in Fig. 6(A) and 6(B) respectively. The dominant phase in each film is 123 as compared with the standard pattern in Fig. 6(C). Another minor phase is barium copper oxide.

A graph of resistivity vs. temperature is shown in Fig. 7 for a film formed on a MgO (100) substrate. The onset temperature of superconductive transition is 102K, zero resistance temperature is 92K, and transition width is 3K. The critical current density is 1.3×10^3 A/cm^2 at 77K. The film thickness is about 15μm.

IV. Conclusions

It has been shown that the aerosol mist injection method can be used to produce oxide-superconductive films in ambient atmosphere by RF plasma evaporation. Besides the experimental set up shown in Fig. 1, we also have tried other configurations, i.e. mist or powder injection in the flame region, and powder injection in the RF plasma region. It is observed that films deposited by the mist injection methods are more uniform than that of the powder methods which are usually a hybrid plasma spray procedure. It is observed that films deposited by the ionized vapor from the plasma region are more homogeneous than that from the flame region. It is believed that oxygen is ionized into the O_2^+ state in the plasma region. The positive oxygen ion may be helpful in the formation of the superconductive $Y_1Ba_2Cu_3O_{7-\delta}$ structure[3]. It is also observed that when the substrate holder is ground electrically, the as-deposited film is superconductive. The optimum substrate temperature is 600°C for this non-vacuum technique.

Acknowledgements

This work was partially supported by the U.S. Air Force Rome Air Defense Center, The New York State Institute on Superconductivity, and The Center for Advanced Ceramic Technology at Alfred University. We wish to thank Prof. L. D. Pye, Prof. D. Shaw, Dr. S. Patel, Dr. J. J. Simmins, B. Chen, M. Pitsakis, M. Rodriguez and Dr. N. Burlingame for many valuable contributions.

References

[1] See, for example, A.W. Sleight, Science 242, 1519 (1988).

[2] B. Oh, M. Naito, S. Arnason, P. Rosenthal, R. Barton, M.R. Beasley, T.H. Geballe, R.H. Hammond, and A. Kapitulnik, Appl. Phys. Lett. 51, 852 (1987). M. Hong, S.H. Liou, J. Kwo, and B.A. Davidson, Appl. Phys. Lett. 51, 694(1987). D. Dijkkamp, T. Venkatesan, X.D. Wu, S.A. Shaheen, N. Jisrawi, Y.H. Min-Lee, W.L. McLean, and M. Croft, Appl. Phys. Lett. 51, 619 (1987).

[3] S. Witanachchi, H.S. Kwok, X.W. Wang, and D.T. Shaw, Appl. Lett. 53, 234 (1988).

[4] K. Terashima, K. Eguchi, T. Yoshida, and K. Akashi, Appl. Phys. Lett. 52, 1274 (1988).

[5] A. Koukitu, Y. Hasegawa, H. Seki, H. Kojima, I. Tanaka, and Y. Kamioka, Jpn. J. Appl. Phys. 28, L1212 (1989).

[6] J.J. Cuomo, C.R. Guarnieri, S.A. Shivashankar, R.A. Roy, D.S. Lee, R. Rosenberg, Adv. Ceramic Mat. 2. 442 (1987).

[7] W.T. Elam, J.P. Kirkland, R.A. Neiser, E.F. Skelton, S. Sampath, and H. Herman, Adv. Ceramic Mat. 2, 411 (1987).

[8] M. Ishii, T. Maeda, M. Matsuda, M. Takata, and T. Yamasunua, Jpn. J. Appl. Phys. 26, L1959 (1987).

[9] T.K. Vethanayagam, J.A.T. Taylor, and R.L. Snyder, Proc. of National Thermal Spray Conference, Oct. 1988, Cincinnati, OH.

[10] Preliminary results were reported, for example, X. Wang, H. H. Zhong, and R. L. Snyder, in Proc. of Conf. on Sci. & Tech. of Thin Film Superconductors, April, 1990, Denver, CO.

[11] H. S. Kwok, P. Mattocks, D. T. Shaw, L. Shi, X. W. Wang, S. Witanachchi, Q. Y. Ying, J. P. Zheng, "Laser deposition of YBaCuO Superconducting Thin Films", in Proc. of World Congress on Superconductivity; ed. C.G. Burnham, et al., (World Scientific, New Jersey, 1988).

Table 1 Deposition Conditions

Unit	Parameters	Value
Ultrasonic nebulizer	Power	70W
	Frequency	1.63MHz
Mist carrier gas, Ar or O_2	Flow rate	100–150 ml/min
Solution	Misting rate	2 ml/min
RF plasma generator	Power	30KW
	Frequency	4MHz
Plasma gas	Ar flow rate	15 l/min
	O_2 flow rate	40 l/min
Auxiliary gas, O_2	Flow rate	5 l/min
Film formation	Deposition rate thickness	0.01–100 μm/(min cm^2) 1–200 μm
Distance	Between substrate and top of plasma torch	7.5–12.5 cm
Film area		30–40 cm^2

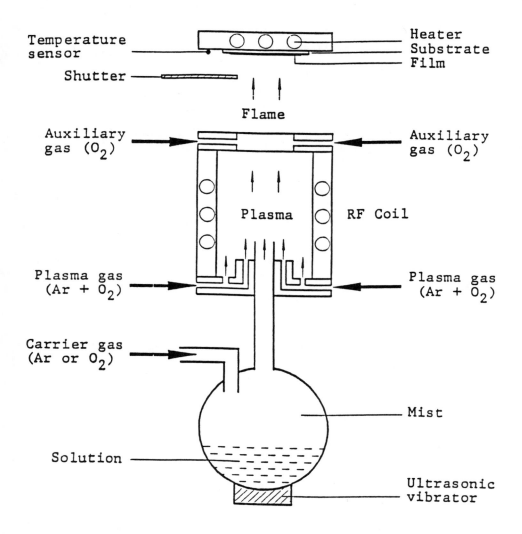

Temperature sensor

Shutter

Flame

Auxiliary gas (O_2)

Plasma

RF Coil

Plasma gas (Ar + O_2)

Carrier gas (Ar or O_2)

Solution

Heater
Substrate
Film

Auxiliary gas (O_2)

Plasma gas (Ar + O_2)

Mist

Ultrasonic vibrator

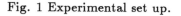

Fig. 1 Experimental set up.

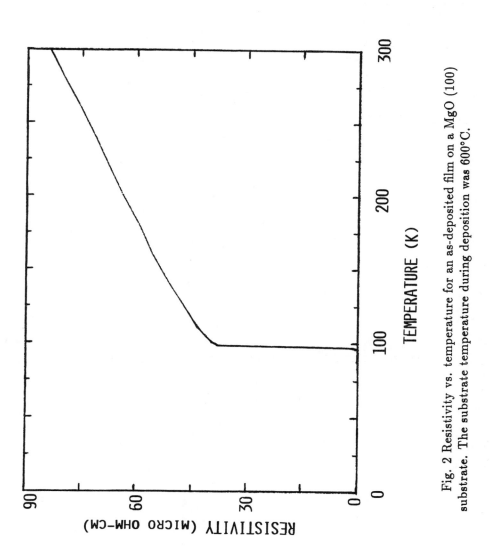

Fig. 2 Resistivity vs. temperature for an as-deposited film on a MgO (100) substrate. The substrate temperature during deposition was 600°C.

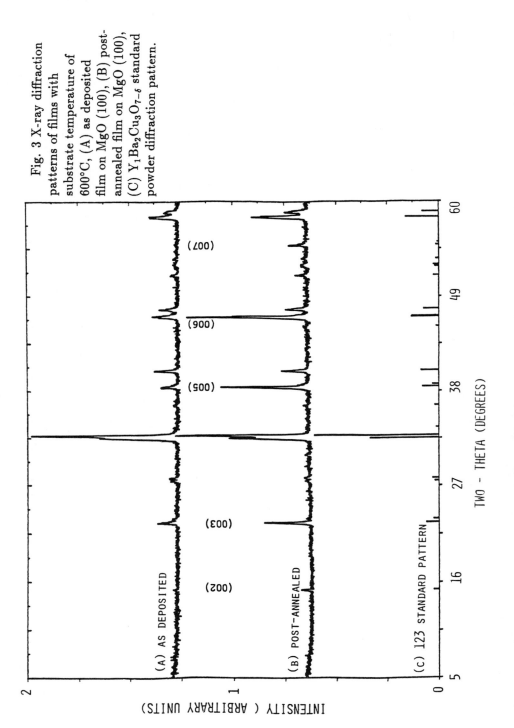

Fig. 3 X-ray diffraction patterns of films with substrate temperature of 600°C, (A) as deposited film on MgO (100), (B) post-annealed film on MgO (100), (C) $Y_1Ba_2Cu_3O_{7-\delta}$ standard powder diffraction pattern.

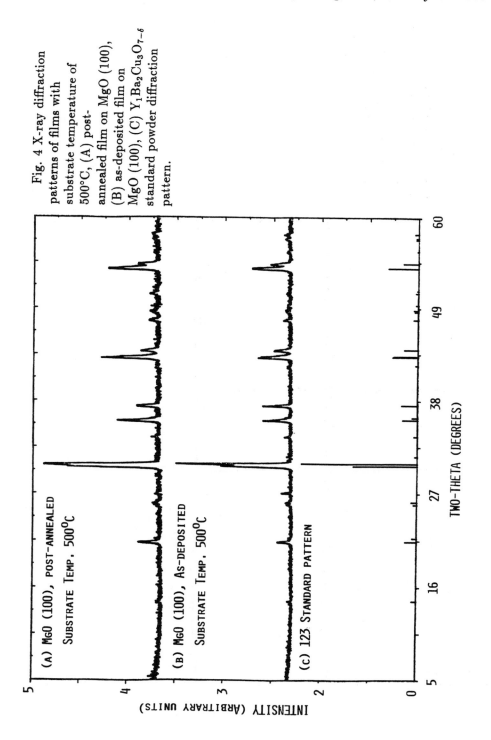

Fig. 4 X-ray diffraction patterns of films with substrate temperature of 500°C, (A) post-annealed film on MgO (100), (B) as-deposited film on MgO (100), (C) $Y_1Ba_2Cu_3O_{7-\delta}$ standard powder diffraction pattern.

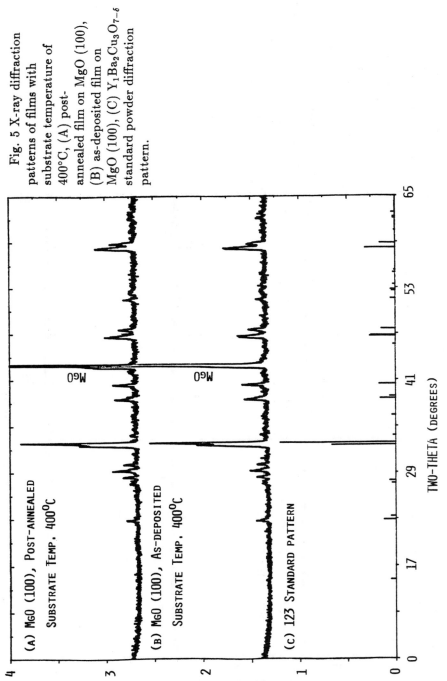

Fig. 5 X-ray diffraction patterns of films with substrate temperature of 400°C, (A) post-annealed film on MgO (100), (B) as-deposited film on MgO (100), (C) $Y_1Ba_2Cu_3O_{7-\delta}$ standard powder diffraction pattern.

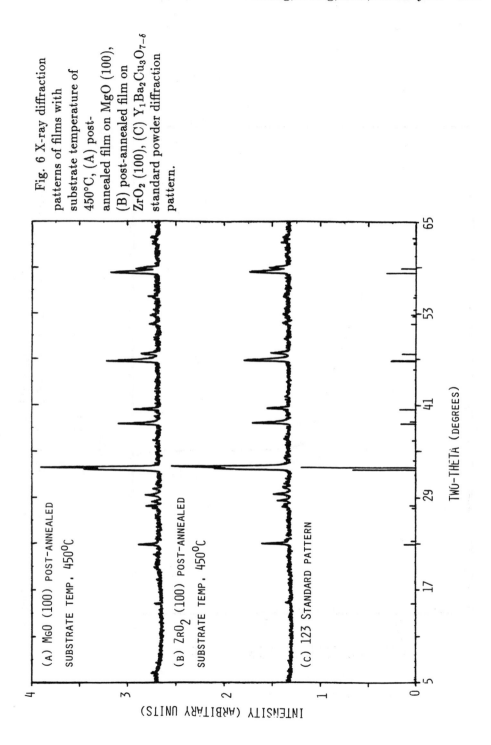

Fig. 6 X-ray diffraction patterns of films with substrate temperature of 450°C, (A) post-annealed film on MgO (100), (B) post-annealed film on ZrO_2 (100), (C) $Y_1Ba_2Cu_3O_{7-\delta}$ standard powder diffraction pattern.

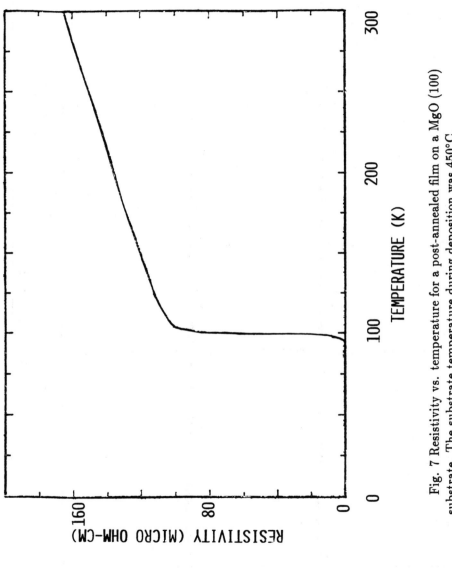

Fig. 7 Resistivity vs. temperature for a post-annealed film on a MgO (100) substrate. The substrate temperature during deposition was 450°C.

TEXTURED Bi-SYSTEM CRYSTALLINE STRUCTURES OBTAINED BY THE CZOCHRALSKI METHOD AND LPE

M.B.Kosmyna,A.B.Levin,B.P.Nazarenko,V.P.Seminizhenko

Institute for Single Crystals, Lenin av.60, 310141, Kharkov, USSR

Some problems of the controlled growth of the HTSC layered crystals and textured Bi-Sr-Ca-Cu-O films by the Czochralski technique and liquid-phase epitaxy are considered in the present paper. Layered crystals of the diameter 30 mm, thickness 5 mm and epitaxial films on $YAlO_3$ substrate were obtained.

Along with well-known and rather well-described methods for the preparation of the textured HTSC materials on the basis of Bi-Sr-Ca-Cu-O (BSCCO)[1,2,3] less considered is the problem of obtaining such materials by the Czochralski method and liquid-phase epitaxy (LPE).Some information on this problem is presented in [4,5,6,7].

The main idea of the present paper is to discuss the possibility to obtain materials the texture of which is set not by a simple temperature gradient and crystallization proceeding spontaneously but by a controlled way: growth of layered crystals on a seed using the Czochralski method and peculiarities of the liquid-phase epitaxy on single crystalline substrates.

Growth of a layered crystal was carried out on a high frequency device "Kristall-607" (p=20 kW, f=8 KHz) A platinum crucible, diameter 70 - 100 mm, was utilized as a container which was heated with high frequency current. Vertical temperature gradient over the melt did not exceed 20 K/cm and was created by means of ceramic Al_2O_3 screens, positioned above the crucible.

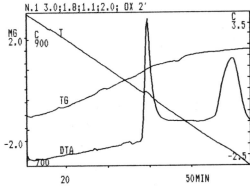

Fig.1. DTA and TG curves. Crystallization of the melt Bi:Sr:Ca:Cu=3.0:1.8:1.1:2.0

The powders Bi_2O_3, $SrCO_3$, $CaCO_3$, CuO were taken as an initial material in the ratio Bi:Sr:Ca:Cu:O = 3.1:1.8:1.1:2. The DTA curve built when cooling the melt of this composition at a rate of 5 K/min has only two exothermal peaks (see Fig.1), one of them at T 700 C below the temperature of complete crystallization of the melt

The seed rod was cut of the polycrystalline material, obtained by a slow cooling of the BSCCO system.

With the seed growing without pulling, polycrystalline ingots with chaotically arranged small crystallites were obtained. However at a pulling rate of 0.1 0.5 mm/h and frequency of rotation 5 min^{-1} we received layered crystals, with the diameter 10 -30 mm and thickness 5-10 mm. In such a crystal the single crystalline layers are parallel to the growth axis and have a crystallographic orientation (001), see Fig.2. The angle disorientation of the planes (001) of the single crystalline layers does not exceed 4 angular degrees. The results of the X-ray diffraction analysis showed that 70% of the ingot is represented by the phase 2212, about 30% - 2201 and a negligible amount of nonidentified phases. The temperature dependence of electric resistivity was measured on the samples cut

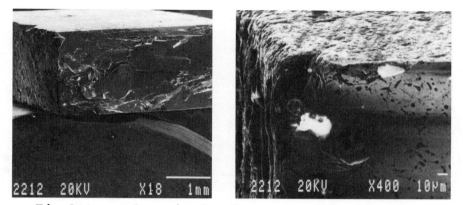

Fig.2 A sample,cut of a layered crystal,grown by the Czochralski method

from the crystal using the four-probe method (direct current 20 mA, contacts of the metallic In).

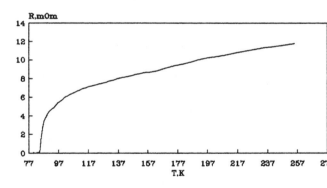

Fig.3. The resistivity curve vs temperature

The transition temperature Tc=82.8 K at the current direction in plane (001) (see Fig.3). At the current directed normal to the plane (001) the temperature dependence of resistivity has a semiconducting character up to the LH temperatures. The temperature dependence of the critical current in the range close to the transition temperature is shown in Fig.4. The critical current density at T=77 K and H=O is 300 A/cm.

Growth of thick textured films was performed according to the method described in[5,6,7].The initial material for this purpose was prepared by mixing Bi_2O_3 and CuO oxides in the ratios corresponding to the

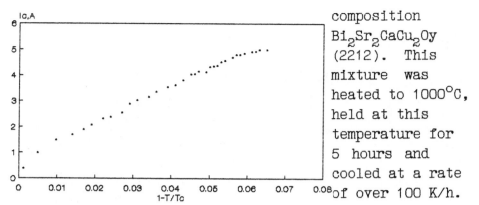

composition $Bi_2Sr_2CaCu_2Oy$ (2212). This mixture was heated to $1000^\circ C$, held at this temperature for 5 hours and cooled at a rate of over 100 K/h.

Fig.4. Critical current vs temperature   The obtained material BSCCO was mixed with KCl powder in the weight ratio BSCCO:KCl=1:5 and loaded into a platinium cricible of the 100 cm capacity.

This load (BSCCO+KCl) together with the crucible was placed into the furnace and heated to $900-1000^\circ$ C. Substrates of various single crystalline materials with surface treatment purity Ra 001 μm and typical size 10x15x0.5 mm were deeped into the solution-melt. Accidentally the temperature varied in a wide range: $800-1000^\circ$ C. On having deeped a substrate into the solution-melt the temperature of the latter was decreased at a rate of 1-20 K/h, which ensured, the growth of the epitaxial film. After finishing the growth process the substrates were taken out and cooled in air at a rate of 50-100 K/h.

The results of the study showed that out of the variety of the   tested   substrates:$LiNbO_3$, $LiTaO_3$, $Ba_xSr_{1-x}Nb_2O_6$, $Sr_2Nb_2O_7$, $SrTiO_3$, $Y_3Al_5O_{12}$, Al_2O_3, $La_2Ti_2O_7$, ZrO_2,   MgO, $YAlO_3$, $LaAlO_3$, $La_3Ga_5SiO_{14}$, $Gd_3Se_xGa_{5-x}O_{12}$ only some of them can be recommended for the epitaxial growth of films; they are $YAlO_3$, $LaAlO_3$, MgO.

Those are not only most resistant to the aggressive solution-melt but are rather close in the parameters of a unit cell (see Table 1).

Table 1 Parameters of the BSCCO lattice and substrates

Lattice parameters	BSCCO $\overset{8}{A}$	YAlO3 $\overset{9}{A}$	LaAlO3 $\overset{10}{A}$	MgO $\overset{10}{A}$
a	5.40	5.176	3.788	4.203
b	5.41	5.307		
c	31.0	7.355		

At the same time films valid for the study were grown only on the substrates of YAlO$_3$, with the orientations (010) and (110). Typical morphology of such films is shown in Fig.5.

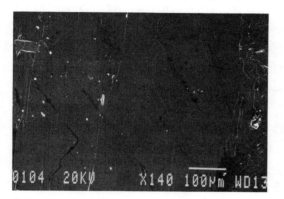

Fig.5.Morphology of the film Bi$_2$Sr$_2$CaCu$_2$O$_y$on YAlO$_3$(010) substrate

Unlike[2] we could not obtain continuous films on MgO substrates, though in chemical resistance MgO is a most suitable material. Morphology of a film on MgO, orientation [001] is shown in Fig.6. We consider that a discrete film is formed due to a low quality of the substrate surface and nonoptimal temperature conditions during experiments.

X-ray structural analysis of the BSCCO films obtained on YAlO$_3$, orientation (010), showed that in all the experiments easily formed is the superconducting phase, corresponding to the composition Bi$_2$Sr$_2$CaCu$_2$O$_y$ (2212). In the majority of cases the

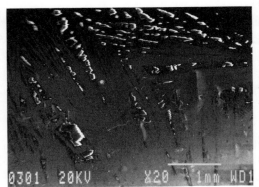

Fig.6. Morphology of the films
$Bi_2Sr_2CaCu_2O_y$ on MgO (001)
substrate

superconductivity was observed in the temperature range 80 – 86 K. In some samples (2223) phase was also found. A part from the phases (2212) and (2223) some samples had inclusions of a low temperature one (2201).

Samely as the autors[5], we observed variation of the C-parameter of the film in the course of experiments (see Table 2).

Table 2   Phase composition lattice parameters and transition temperatures of BSCCO films on $YAlO_3$ (010).

Experiment N	Phase presence	Lattice parameters, Å			Transition temperature Tc (R=0)
		a	b	c	
1	2212 (basis)	5.38	5.42	30.8	86 K
	2223 (<10%)	5.409	5.43	36.8	
2	2212 (basis)	5.39	5.409	30.8	78 K
	2212 (~20%)	5.39	5.409	30.39	
3	2212 (>20%)	5.38	5.409	30.899	83 K
4	2212 (>95%)	5.39	5.409	30.369	not HTSC 77K

As it is seen from the table, the superconducting property of films depends to a great extent not only on the presence of phases but also on the parameters of their lattice. The lattice parameters variation in BSCCO can be explained by oxygen deviation of the stoi-

Fig.7. Typical layered structure of the films $Bi_2Sr_2CaCu_2O_y$, parallel to the plane of $YAlO_3$ substrate (010)

chiometry and impurity entering the film from the solvent and substrate ions (primarily Y^{3+}). It should be also noted that the obtained films were of high structural perfection because the halfwidth of the rocking curves was 1.5 angular degrees (see Fig.7).

Using secondary ion mass spectroscopy it was shown that over the profile of the film the molar concentration of strontium ions 86-87 remained constant and the corresponding value for the strontium ions 88 revealed a tendency to the decrease. A similar decrease was also observed for copper ions. The main elements: Bi, Sr, Ca, Cu in different samples were in comparable amounts and the content of isotopes of the corresponding elements was determined according to their natural standards.

REFERENCES

1.J.S.Zhang,J.G.Huang,M.Jiang,Y.L.Ge,Y.Z.Wang, G.W.Qiao, Z.Q.Hu. Materials Letters, 8, 46-48 (1989).
2. W.Carrillo-Cabrera,W.Gopel,G.F.de la Fuente, H.R.Verdun. Appl. Phys. Lett., 55, 1032 (1989).
3. .Matsubara,H.Tanigawa,T.Ogura,H.Yamashita, M.Kinoshita,T.Kawai. Jap. J. Appl.Phys.,28, L1358 (1989).

4. A.Kurosaka,M.Aoyagi,H.Tominaga,O.Fukuda,H.Osanai. Appl. Phys. Lett., 55, 390 (1989).

5. G.Balestrino,R.Di Leo,M.Marinel et al. Preprint, MMS-HTSC conference Stanford, July 23-28,1989.

6. H.Takeya, H.Takei. Jap. J. Appl. Phys.,28 L1571 (1989).

7. M.B.Kosmyna,B.P.Nazarenco,G.Kh.Rozenberg et al. J.Low Temp.Phys.,16, 812 (1989)

8. I.E.Graboi,A.P.Kaule,Yu.G.Metlin. Chemistry and technology of the high-Tc superconductors, 6, -M: VINITI, 1988.(in Russian)

9. A.A.Kaminski. Laser crystals, "Nauka",M.,189 (1975).

10. R. Ramesh, A. Inam, W.A. Bonner et. al Appl.Phys.Lett., 55, L1138 (1989).

YBaCuO SUPERCONDUCTING THIN FILMS
PREPARED BY PLASMA-ENHANCED CHEMICAL VAPOR DEPOSITION

Kenji EBIHARA , Tomoaki IKEGAMI , Kazumi SUGA
and Tomoyuki FUJISHIMA

Department of Electrical Engineering and Computer Science,
Kumamoto University, Kurokami 2-39-1, Kumamoto 860,Japan.

ABSTRACT

YBaCuO superconducting thin films are prepared by using a plasma-enhanced metalorganic chemical vapor deposition. An rf discharge plasma is used to decompose the metalorganic chelate precursors. Optimization of the film deposition conditions is studied using optical emission spectroscopy. The rf process is also applied to decompose silicon compounds for preparation a SiO_2 insulating film on the $YBa_2Cu_3O_7$ bulk superconductor.

1. INTRODUCTION

The new materials deposition processes for superconductivity applications have been required to prepare the excellent uniform,compositional controlled and stoichiometric superconducting films at the low substrate temperature with high deposition rate. Metalorganic chemical vapor deposition (MOCVD) may be an appropriate process to obtain high Tc superconducting thin films for electronic applications. In the preparation of YBaCuO thin films using this MOCVD process,the substrates were heated to high temperatures such as 700° C, 650° C and 600° C [1-3] during the deposition. Plasma-enhanced metalorganic chemical vapor deposition (PEMOCVD) aims at lowering the process temperature and accelerating decomposition of source materials in the presence of plasma.

In this paper, we report the preparation of superconducting YBaCuO thin films by using an improved type of PEMOCVD process. An rf discharge was applied to decompose precursors by electronic excitation. Optimum deposition conditions for superconducting thin film structure were investigated by correlating the film properties with specified spectral intensities emitted from the process plasmas.

Moreover, in order to develop microfabrication techniques for superconductivity applications, the multilayer film of superconductor/insulator/superconductor (S-I-S) structure is preliminarily formed.

2. EXPERIMENT

Figure 1 shows a schematic diagram of the plasma-enhanced MOCVD system. An rf plasma was generated in a quartz discharge tube which was surrounded by an inductive coil connected to an rf (13.56 MHz) generator. A microwave power from a 2.45 GHz oscillator was also available for decomposition of the source materials in the previous work. [4,5] Chelates of Y, Ba,and Cu were used as precursors. Metalorganic compounds of $Y(dpm)_3$, $Ba(dpm)_2$ and $Cu(dpm)_2$ (dpm= dipivaloylmethane , $C_{11}H_{19}O_2$) were sublimated at the temperatures of 190° C, 300° C, 160° C, respectively. Each vapor was introduced at an outlet of the quartz tube and decomposed by coming in contact with the Ar/O_2 plasma flame. In order to make clear the optimum deposition conditions for each organic source, the chelates of Y, Ba, and Cu were separately decomposed at different deposition conditions. Each thin film was deposited on the substrate (MgO) heated at 400 - 500° C. The plasma state during the film deposition was monitored by using a spectroscopic technique. Optical spectral emission was studied to estimate the correlation between the film properties and plasma state.

In order to realize the thin-film superconducting devices, it is necessary to form a high-quality dielectric layer between the superconducting thin films. Silicon oxide films have been widely used for microelectronic applications including interlayer insulators. Recently, the use of SiO_2 as passivation layers for superconducting thin films has been studied . [6,7] SiO_2 films have been made by plasma deposition using SiH_4 as the silicon source and by using electron-beam evaporation of SiO_2.

We deposit the dielectric films from silicon compounds

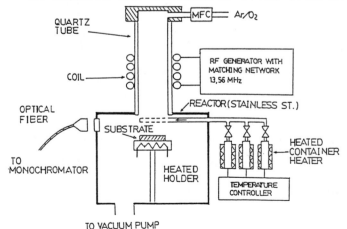

FIG.1.
rf plasma-enhanced metalorganic chemical vapor deposition system

such as tetraethoxysilane [TEOS, $Si(OC_2H_5)_4$] and hexame-
thyldisiloxane [HMDSO, $Si_2O(CH_3)_6$]. The deposition
system is a simple capacitively coupled arrangement operating at 13.56 MHz. The rf Ar/O_2 plasma is produced between internal electrodes separated by a distance of 2 cm.
Vapors of the TEOS and the HMDSO were introduced between
the electrodes. The typical working pressure is 200 mTorr
and the rf power of 100-200 W is supplied.

3. RESULTS AND DISCUSSION

The films were prepared under various deposition conditions. In order to find out the optimum preparation
parameters, coordination compounds of Y, Ba, and Cu were
separately vaporized at different conditions. The first
attempt to prepare superconducting thin films was to study
the deposition conditions suitable to each precursor.
$Y(dpm)_3$ was sublimated at the temperature of 190° C under
the base pressure of 3 mTorr. The vapor of the yttrium
precursor was introduced to the bottom of the rf discharge
tube. Two spectra from the plasmas are shown in Fig.2 for
the conditions of (a) 23 mTorr without Ar/O_2 gas and 150
mTorr with Ar/O_2 gas mixture. The rf power of 100 W was
supplied to the inductive coil for both plasmas. It is
clearly shown that there are remarkable difference between
the spectra for the (a) and (b) plasmas. Reactive gas of
Ar/O_2 mixture has strong effects on dissociation and
ionization of the precursor. The emission lines of
YI450.595, YII549.740 and YO585.900 nm were detected for
the 23 mTorr source plasma and the observed lines for the
150 mTorr plasma were YII395.036 and YO585.900 nm. Strong
$H\beta$ line (486.133) was also observed when Ar/O_2 gas mixture was introduced.
The $Ba(dpm)_2$ was heated to the temperature of 300° C.
The 30 mTorr source vapor was decomposed by supplying the
50 W rf power. In this case the potential lines due to Ba
species are impossible to distinguish at the resolution of
this experiment. Figure 2(c) shows the spectrum of the
$Ba(dpm)_2$ plasma at the total pressure of 150 mTorr when Ar
gas was introduced. BaI 597.170 nm was identified. In the
case of $Cu(dpm)_2$ where the plasma was produced with 100 W
rf power at the vapor pressure of 30 mTorr, CuI427.513 nm
line was observed as shown in Fig.2(d). On the other hand,
when Ar/O_2 gas was supplied, the existence of strong
emission from CuI and CuO species was shown : CuI522.007,
CuO472.100, CuO583.300 (in Fig.2(e)). Three source vapors
were transported to the reaction chamber and argon gas was
introduced. A typical spectrum is shown in Fig.2(f) for
the YBaCuO plasma which was generated with 100 W rf power
at the 150 mTorr. The emission lines arising from atomic
and molecular species appeared in the visible region
:YII395.036, YII549.740, CuI515.323, CuO527.900,

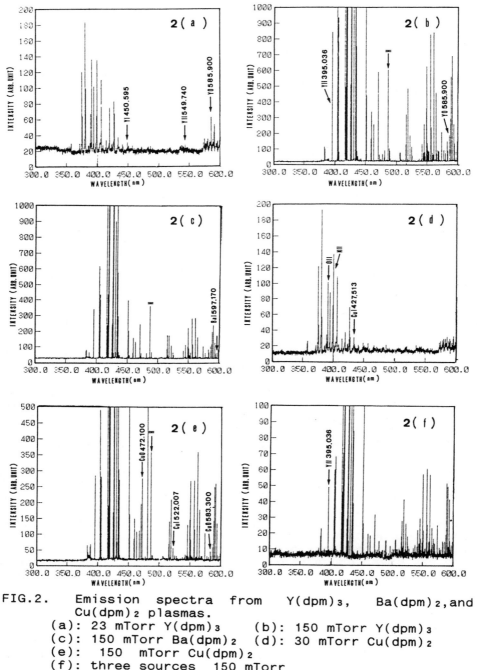

FIG.2.   Emission spectra from   Y(dpm)₃,   Ba(dpm)₂,and
         Cu(dpm)₂ plasmas.
         (a): 23 mTorr Y(dpm)₃   (b): 150 mTorr Y(dpm)₃
         (c): 150 mTorr Ba(dpm)₂ (d): 30 mTorr Cu(dpm)₂
         (e):   150 mTorr Cu(dpm)₂
         (f): three sources 150 mTorr

BaI553.555.

    After half an four of deposition , the films were cooled to room temperature under Ar/O_2 gas flow. The films deposited with each source precursor had smooth and shining surface and showed high resistivity. X-ray diffraction indicates that the as-deposited films are amorphous. The film deposition rate was about 0.2 nm/s. The deposited films annealed by heating in air 1 hour at 750° C did not show any structural transition. On the other hand, the films prepared with Ar/O_2 gas flow at 150 mTorr had very high growth rate of 5-10nm/s which is one order higher than that in OMCVD process.[1] Figure 3 shows the SEM of the $Cu(dpm)_2$ film annealed at 890° C for 20 min. Cubic-like grains appeared on the deposited surface. The $Ba(dpm)_2$ film annealed at the same condition as the $Cu(dpm)_2$ film showed a rough textured surface with grains 3-5 μ m.

    YBaCuO thin films were prepared simultaneously by decomposing $Y(dpm)_3$,$Ba(dpm)_2$ and $Cu(dpm)_2$. Each precursor was introduced to the reaction chamber through a heated stainless steel pipe. The partial pressures of Y, Ba and Cu sources were about 0.5 Torr, 1.5 Torr and 0.5 Torr, respectively. The rf power of 80 W was supplied to decompose each source vapor without Ar/O_2 gas mixture. The as-deposited films with brown color surface were electrically conducting. Thermal annealing at 750° C made the surface color change into dull gray. Figure 4 shows the surface

5 μm   1 μm

FIG.3. SEM micrograph of   FIG.4. SEM micrograph of
      $Cu(dpm)_2$ film   YBaCuO film
      annealed at 890° C   annealed at 750° C

morphology.   The SEM picture shows that phase transition
from amorphous structure to crystalline grains occurs by
annealing at 750° C. It is suggested that superconducting
YBaCuO films might be obtained by reduction of the film
thickness. Moreover, the electron probe microanalyzer
(EPMA) measurement shows that the YBaCuO films deposited
simultaneously with three sources are rich in copper
component. The result is supported by the fact that intensity of spectral lines originating in Cu atom and molecules is strongly enhanced under the above deposition
conditions.
    Fabrication of the S-I-S structure is a basic supercon-
ductivity technology for future superconductivity electronics. The plasma-enhanced CVD process may be a promising technique for in-situ fabrication of the superconducting devices. We preliminarily attempted to deposit silicon
dioxide insulator (SiO_2) on a high temperature superconductor $YBa_2Cu_3O_7$. The SEM photo of the plasma-enhanced
TEOS film indicates that the TEOS oxide covers the superconductor with high adhesion(in Fig(5)).

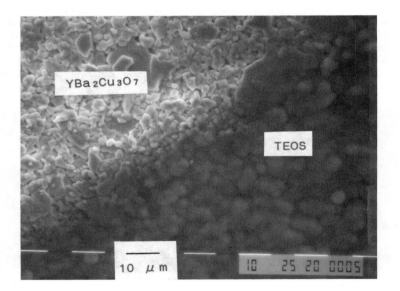

FIG.5. SEM micrograph of $SiO_2/YBa_2Cu_3O_7$ structure.
        TEOS was decomposed by rf plasma at total
        pressure of 200 mTorr.

4. CONCLUSION

A plasma enhanced metalorganic chemical vapor deposition was used to decompose metalorganic compounds of Y(dpm)$_3$, Ba(dpm)$_2$ and Cu(dpm)$_2$. Spectroscopic analysis for the decomposed plasmas showed that Y, Y$^+$, YO, Ba, Cu and CuO species are produced. After annealing at 750° C, the YBaCuO films deposited on MgO substrate demonstrated phase transition to crystalline grains.

REFERENCES

1. J.Zhao et al.,Appl.Phys.Lett.53,1750(1988).
2. K.Zang, B.S.Kwak, E.P.Boyd, A.C.Wright and A.Erbil, Appl.Phys.Lett. 54, 380(1989).
3. T.Nakamori, H.Abe, T.Nakamori, and S.Shibata, Jpn.J.Appl.Phys. 27,L1265(1988).
4. K.Ebihara, S.Kanazawa, T.Ikegami, and M.Shiga, Proceedings of 9th International Symposium on Plasma Chemistry(1989, Italy), Vol.3,p.1509.
5. K.Ebihara,S.Kanazawa, T.Ikegami and M.Shiga, J.Appl.Phys.68, 1151(1990).
6. O.X.Jia and W.A.Anderson,J.Appl.Phys.66, 452(1989).
7. T.S.Kalkur, R.Y.Kwok and D.Byrne, J.Appl.Phys. 67,918(1990).

GROWTH AND CHARACTERIZATION OF $Y_{1-x}Pr_xBa_2Cu_3O_{7-x}$ SINGLE CRYSTALS

E.Pollert, S.Durčok, J.Hejtmánek, L.Matějková, M.Nevřiva, M.Šimečková
Institute of Physics Czechoslovak Academy of Sciences, Na Slovance 2, 180 40 Prague 8, Czechoslovakia

Introduction

A sufficiently large single crystals of HTS materials of a good quality are needed if either the basic properties or the origin of limitations to the technological performances of this class of materials could be understood. It represents, however, a difficult task, particularly because of the incongruent melting of the YBCO low chemical stability and corrosion of the crucibles during the growth. From several approaches already applied the growth from nonstoichiometric melts seems to be meanwhile the most promising and an effort was devoted to improve this method.

Phase equilibria in Y-Ba-Cu-O system

The succesful crystal growth from a nonstoichiometric melt requires informations about the phase equilibria in the part of Y-Ba-Cu-O system including primary crystallization field of the YBCO phase. They were studied by a number of authors, e.g. the pseudobinary system $Y_2BaCuO_5-Ba_3Cu_5O_8$ [1] CuO - rich corner of the ternary system $YO_{1.5}-BaO-CuO$ [2], mostly DTA and X-ray analysis of the quenched samples were employed. This methods, however are not suitable for the study of solid-liquid equilibria. Thus a new procedure based on the separation of the melt from an equilibrated sample and the subsequent analysis of the solid residue, so caled soak method was developed

and used for the determination of the phase relationships
in the $CuO-BaCuO_2-YBCO$ ternary system [3]. The isothermal
cut at $950^{\circ}C$ in air was used as a starting point in our
systematic study of the growth of YBCO single crystals.
The found field of optimum compositions is substantially
narower than that reported by others authors, e.g. [4,5],
see Fig.1.

Growing procedure.

The upper temperature limit for the growth of crystals
is given by the incongruent melting point of YBCO phase
at $T \sim 1000^{\circ}C$, lower limit by the ternary eutectic temperature at $T \sim 905^{\circ}C$. In order to provide the temperature
range for the growth as large as possible it is suitable
to use a starting composition just below the melting
point of $YBa_2Cu_3O_{7-x}$. Long-time experiments are not
favourable because of the corrosion of crucibles. Con-
sequent contamination of crystals by the respective
cations can be seen from our experiments where the cooling
rates of 0.2÷0.3K/hour were applied. The crystals were
separated from the liquid by tilting of the crucible.

Heat treatment of crystals.

Most of the studied crystals were post-annealed in
order to assure their oxygen stoichiometry close to ideal
one. Two procedures were tested:
- reduction of the samples under nitrogen atmosphere
  at $590^{\circ}C$ for 2 hours followed by an reoxidation process
  under oxygen atmosphere, 2 hours at $590^{\circ}C$ and subsequently 100 hours at $450^{\circ}C$,
- oxidation of the samples under oxygen atmosphere 1460
  hours at $450^{\circ}C$.

Characterization.

Compositions of the prepared crystals were determined
by electron microprobe analysis and their c-lattice parameters by X-ray diffraction. A standard dc four-probe

method was used for the conductivity measurement of the crystals, in (ab) plane and along c axis in the temperature range of 4-300 K.

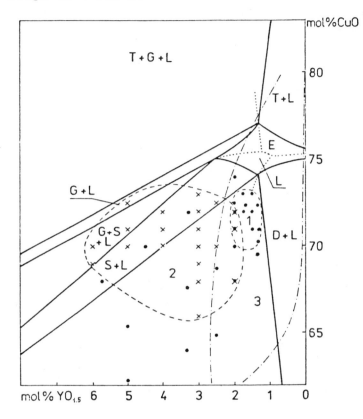

Fig.1. Isothermal cut in the Y-Ba-Cu-(O) system in air at 950°C. Projection of primary crystallization fields boundaries are denoted by dotted lines. Capitals denote coexisting phases: S-$YBa_2Cu_3O_x$, D-$BaCuO_2$, G-Y_2BaCuO_5, T-CuO, L-melt,•- our growing experiments, x-experiments of [4], 1-optimum compositional region according to our experiments, 2-region recommended by [4], 3-primary crystallization field according to [5].

Table I.

Prepared single crystals and their characteristic

Sample	starting composition /mol %/				average composition of the studied crystals	c /Å/	T_c /K/	(T=300K) $\rho_c/\rho_{a,b}$
	$YO_{1.5}$	$PrO_{1.5}$	BaO_2	CuO				
96	1.37		27.63	71.00	$Y_{0.90}Ba_{2.06}Sr_{0.04}Cu_{2.71}Al_{0.13}O_z$	11.731^x / 11.690^{xx} / 11.701		
97	1.37		28.47	70.16	$Y_{0.96}Ba_{1.99}Sr_{0.05}Cu_{2.72}Al_{0.15}O_z$	11.689	50	10.7
106	1.23	0.14	28.47	70.16	$Y_{0.89}Pr_{0.08}Ba_{1.98}Sr_{0.04}Cu_{2.79}Al_{0.14}O_z$	11.686	40	1.07
76	1.00	0.27	28.45	70.20	$Y_{0.81}Pr_{0.20}Ba_{1.94}Sr_{0.05}Cu_{2.64}Al_{0.21}O_z$	11.687	25	9
107	0.96	0.41	28.47	70.16	$Y_{0.67}Pr_{0.35}Ba_{1.93}Sr_{0.05}Cu_{2.52}Al_{0.25}O_z$	11.688	-	4.2
108	0.68	0.69	28.47	70.16	$Y_{0.54}Pr_{0.47}Ba_{1.94}Sr_{0.05}Cu_{2.67}Al_{0.24}O_z$	11.686	-	1.5

x as grown sample

xx oxygen gradient in the sample see Fig.4

Results and discussion

The results obtained on the studied crystals plates
of the dimensions up to 8x8 mm^2 are summarized in Table I.
Comparison of the starting composition and shows composition of the crystals that a gradual replacing of
ytrium ions for praseodymium, with the distribution
coefficient of praseodymium ions between the melt and
solid phase 1.5. is possible. Concentration fluctuations
detected in the crystals were found to be smooth as it
is illustrated for the end members of the studied series
see Figs. 2.and 3. and a similar conclusion can be made
from the comparison of the individual crystals coming
from the same batch, see growing run 97, Table I. Presence of strontium ions in the crystals is due to impurities in BaO$_2$ used as a starting material.

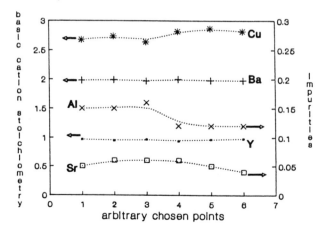

Fig.2. Concentration distribution in the sample 97

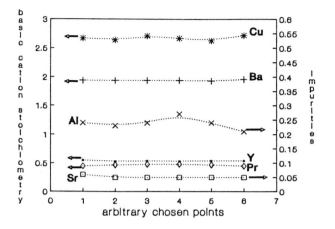

Fig.3. Concentration distribution in the sample 108

An important problem already mentioned is the contamination of crystals by Al^{3+} ions from the partially dissolved crucible which increases if the praseodymium ions concentration in the crystals increases too. It seems to be connected with the formation of vacancies in the YBCO structure. An occupation of Cu (I) sites by Al^{3+} ions gives rise to a local charge which can be compensated by two ways:

- by intercalation of oxygen anions into Cu(I) layers leading to an increase of the oxygen stoichiometry over the value of 7

- by formation of vacancies on Cu(I) sites in a neighbourhood of Al^{3+} ions.

The results show that the latter possibility is favoured.

The determined values of the **c** lattice parameters were used for a rough estimation of the oxygen stoichiometry in the studied crystals employing the relation

$$z = 7-x = 78.77 - 6.153 \ c$$

originally derived for the pure $YBa_2Cu_3O_{7-x}$ [6].
Nevertheless this approach seems to be justifiable also
for our crystals with regard to a negligible influence
of Y-Pr and Cu-Al replacement on the value of c lattice
parameter[7,8,9] The yielded values and from those deduced charge distributions are summarized in Table II.

In contrary to the long-time annealing experiments
which can be considered as equilibrium X-ray analysis of
the sample 96, treated according to the procedure combining reduction and subsequent reoxidation of the crystal,
revealed broadened difraction peaks with two maxima
corresponding to c= 11.702 and c=11.691, see Fig4.It indicates clearly existence of an oxygen concentration gradient varying in the range of z=6.59÷ 6.83.
In spite of constant concentration of holes increase of
the content of praseodymium and aluminium ions lead to
the effects already reported   i.e. gradual decrease of
the critical temperature, diminishing of the superconducting properties and substantial increase of the electrical resistivity. Nevertheless the anisotropic character
is preserved in the whole investigated concentration region, see Fig.5.

The role of praseodymium ions is in centre of interest of various authors and several influences were discussed. Let us note that the trapping of holes could be probably the most significant. Effect of aluminium ions as
well as vacancies seems to be more evident, they cut Cu(I)-O
chains and thus restrict motion of holes.

Table II.

Oxygen stoichiometry and charge distribution

sample	z	charge distribution	number of holes per formula unit
96	6.59	$Y^{3+}_{0.9}Ba^{2+}_{2.06}Sr^{2+}_{0.04}\left[Cu^{2+}(II)O_2^{2-}\right]_2\left[Cu(I)O_3\right]_{0.71}^{3.30-}Al^{3+}(I)_{0.14}\square_{0.15}O^{2-}_{0.46}$	0.22
97	6.85	$Y^{3+}_{0.96}Ba^{2+}_{1.90}Sr^{2+}_{0.05}\left[Cu^{2+}(II)O_2^{2-}\right]_2\left[Cu(I)O_3\right]_{0.76}^{2.92-}Al^{3+}(I)_{0.14}\square_{0.1}O^{2-}_{0.57}$	0.41
106	6.87	$Y^{3+}_{0.89}Pr^{(3+m)+}_{0.08}Ba^{2+}_{1.90}Sr^{2+}_{0.04}\left[Cu^{2+}(II)O_2^{2-}\right]_2\left[Cu(I)O_3\right]_{0.79}^{(3+m)}Al^{3+}(I)_{0.14}\square_{0.07}O^{2-}_{0.5}$	0.41
76	6.86	$Y^{3+}_{0.81}Pr^{(3+m)+}_{0.2}Ba^{2+}_{1.94}Sr^{2+}_{0.05}\left[Cu^{2+}(II)O_2^{2-}\right]_2\left[Cu(I)O_3\right]_{0.64}^{(2.75+m)-}Al^{3+}(I)_{0.21}\square_{0.15}O^{2-}_{0.94}$	0.4
107	6.05	$Y^{3+}_{0.67}Pr^{(3+m)+}_{0.35}Ba^{2+}_{1.93}Sr^{2+}_{0.05}\left[Cu^{2+}(II)O_2^{2-}\right]_2\left[Cu(I)O_3\right]_{0.52}^{(2,29+m)-}Al^{3+}(I)_{0.25}\square_{0.23}O^{2-}_{1.29}$	0.44
108	6.07	$Y^{3+}_{0.54}Pr^{(3+m)+}_{0.47}Ba^{2+}_{1.94}Sr^{2+}_{0.05}\left[Cu^{2+}(II)O_2^{2-}\right]_2\left[Cu(I)O_3\right]_{0.67}^{(3+m)-}Al^{3}(I)_{0.24}\square_{0.09}O^{2-}_{0.86}$	0.34

x as grown

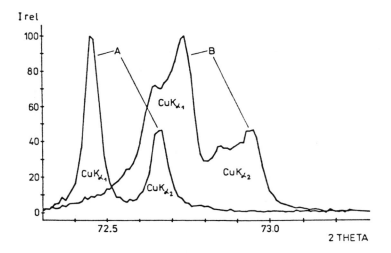

Fig.4. Comparison of the (009) Bragg reflections of the sample 96, A - as grown, B - annealed according to the procedure 1.

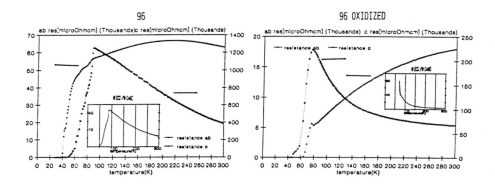

Fig.5. Temperature dependence of the electrical resistivity of the studied samples

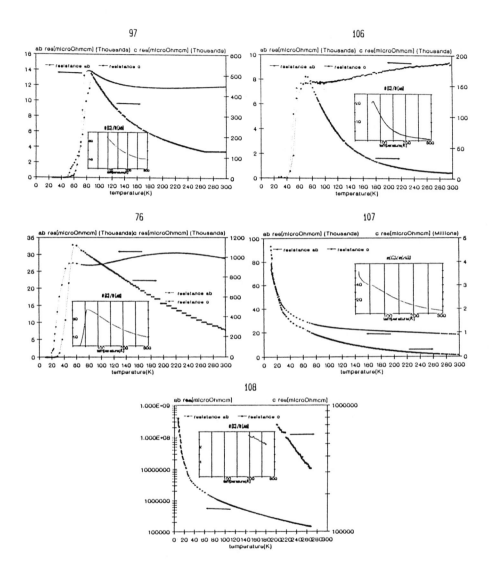

Fig.5. Continuation, temperature dependence of the
electrical resistivity of the studied samples

References

/1/ K.Oka, K.Nakane, M.Ito, M.Saito, H.Unoki,
Jap.Appl.Phys.27,L 1065(1988)

/2/ T.Aselage, K.Keefer, J.Mater.Res.3,1279(1988)

/3/ M.Nevřiva, P.Holba, S.Durčok,D.Zemanová,E.Pollert,
A.Tříska, Physica C 157,334(1989)

/4/ Th.Wolf, W.Goldacker, B.Obst, G.Roth,R.Flükiger,
J.Cryst.Growth 96,1010(1989)

/5/ K.Watanabe, J.Cryst.Growth 100,293(1990)

/6/ Z.Jirák, J.Hejtmánek,E.Pollert,A.Tříska,P.Vašek,
Physica C 156,750(1988)

/7/ T.Siegrist,L.F.Schneemeyer,J.V.Waszczak, N.P.Singh,
R.L.Opila,B.Batlogg,L.W.Rupp,D.W.Murphy, Phys.Rev.B,
36,1987,16,s.8365-8

/8/ J.L.Peng,P.Klavins,R.N.Shelton,H.B.Radonsky,
P.A.Hahn,L.Bernardez,Phys.Rev.B 40,4517 (1989)

/9/ B.Okai,M.Kosuge,H.Nozaki,K.Takahashi,M.Ohto, Jap.J.
Appl.Phys.27,L41(1988)

PROPERTIES OF LASER ABLATED SUPERCONDUCTING FILMS OF Y-Ba-Cu-O

S. Mahajan, R. L. Cappelletti, R. W. Rollins, and D. C.Ingram
Department of Physics and Astronomy
J. A. Butcher, Jr.
Department of Chemistry

Condensed Matter and Surface Sciences Program
Ohio University, Athens, OH 45701

ABSTRACT

We report successful preparation of thin films of $Y_1Ba_2Cu_3O_{7-\delta}$ (YBC) using the technique of laser ablation on single crystal (001) $SrTiO_3$. X-ray diffraction (XRD) pattern of the as deposited film showed the c-axis to be perpendicular to the substrate. The critical current was estimated to be 9×10^6 A/cm^2 ($H\|c$) and 4×10^8 A/cm^2 ($H\perp c$) at 12K as determined from magnetic hysteresis measurements. Initial base pressure in addition to oxygen pressure and substrate temperature is found to be important to achieve correct stoichiometry and high critical currents.

INTRODUCTION

The technique of laser ablation has proved to be a very versatile tool in the continuing endeavor of understanding the properties of high T_c materials. The technique has been used to produce some of the highest quality thin films of YBC, Bi and Tl superconductors[1-3] reported in the literature so far. We have used laser ablation to make films of YBC on (001) $SrTiO_3$ single crystals. The laser used for the purpose was a 248nm KrF excimer laser. The pulses were fired at the rate of 2-4 Hz. The ejected plume from the target was collected on (001)$SrTiO_3$ single crystal mounted on a stainless steel block welded to a resistively heated inconel strip. The substrate temperature was kept at 730°±15°C. The temperature was measured using a thermocouple mounted at the back of the substrate. Thermometry was checked at higher temperatures in situ by optical pyrometry. The deposition time was between 35-45 min for a 2000-3000Å thick film, the thickness being estimated from measurements of Rutherford backscattering (RBS). The base pressure achieved prior to deposition was ~ 2×10^{-5} Torr and the actual deposition was carried out at a 100-200 mTorr of oxygen pressure. After the deposition the temperature was slowly lowered to 400°C in about half an hour and held there for one hour at 1 atm of O_2 pressure and then slowly cooled to room temperature. We find that correct temperature and pressure are important to get c-axis oriented films and high critical currents.

EXPERIMENT AND DISCUSSION

The single crystals of $SrTiO_3$ were cleaned with
trichloroethylene, methanol and reagent grade acetone.   After
cleaning they were mounted on the stainless steel block using
silver paint.   RBS measurements show that the silver paint does not
diffuse through the substrate to the film.

The depositions were carried out at different base pressures
in the chamber, i.e. pressure before O_2 injection, obtained by a
cold-trapped oil diffusion pump.   The base pressures achieved prior
to the deposition were 70×10^{-3}, 10×10^{-3} and 2×10^{-5} Torr for three
different depositions.   After this, O_2 was injected into the
chamber and the deposition was carried at an oxygen pressure
between 1.5-1.8×10^{-1} Torr.   The substrate temperature was kept at
730°C in all the three depositions.

X-Ray Diffraction (XRD) pattern of the film grown at a base
pressure of $70 \mu m$ Hg (Fig. 1) showed some preferred orientation in
addition to the presence of a multiphase compound and it was non-
superconducting.   The superconductivity and the required
crystallinity was achieved in this film after a post-annealing
treatment at 900°C in flowing oxygen for one hour.   The critical

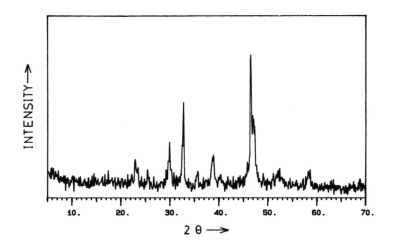

Fig. 1   XRD pattern of the film deposited at an initial
base pressure of $70 \mu m$ Hg. The pattern indicates the presence of mixed phases.

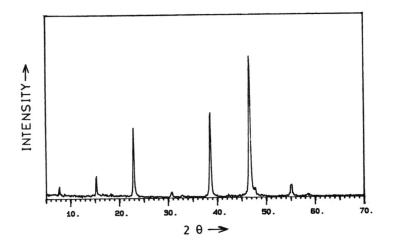

INTENSITY →

2 θ →

Fig. 2   XRD pattern of the film deposited at 2×10^{-5} Torr
initial pressure. The pattern shows a c-axis oriented
film.

current as determined by magnetic hysteresis measurements was 10^3
A/cm² at 15K.

The XRD pattern of the film grown at 10μm Hg base pressure
showed the presence of single phase compound and it was
superconducting but the J_c was still very low ($\sim 10^3$A/cm²) at 15K.
For the third film, the XRD pattern (Fig. 2) showed only multiples
of (001) peaks and it was found to be superconducting at 81K with a
critical current density of 9×10^6 A/cm² at 15K(H$\|$c).

The stoichiometry of the films was confirmed by RBS
measurements. The films made at base pressures of 70μm Hg and 10μm
Hg were found to be Cu and Ba rich while the cation (Y:Ba:Cu) ratio
in the third film was very close to 1:2:3.

The magnetic measurements were made on the PAR 150 vibrating
sample magnetometer. We present only the results of the
measurements made on the film deposited at a base pressure of 2×10^{-5}
Torr. Measurements were made for both field parallel and
perpendicular to c-axis. The sample film was 2600Å thick and
rectangular in shape (0.635cm x 0.635cm). The critical current
(J_c) was estimated, following Stollman et al[4], for H$\|$c(H$^\perp$film plane)
from:

$$J_c = \frac{60M_o}{d} \qquad (i)$$

For H$^\perp$c(H$\|$film plane) geometry we use the Bean formula:

$$J_c = \frac{40M_o}{t} \qquad (ii)$$

M_o in expression (i) and (ii) is the remanant magnetization obtained by applying a high field (11 kOe) to the sample and then reducing the field to zero. "d" is the diameter of the film area and "t" is its thickness. The critical currents measured this way are represented in (Fig. 3) as a function of temperature. For $H^{\perp}c$ the measured J_c at 12K is estimated to be ~ $4 \times 10^8 A/cm^2$. The inset in Fig. 3 shows the variation of J_c applied field ($H \| c$) at 12K. The high critical currents even at fields as high as 10 kOe indicate high pinning in these laser ablated films.

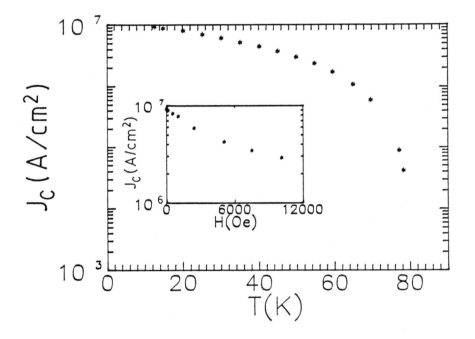

Fig. 3 Plot of critical current density vs temperature. The inset is a plot of J_c vs the applied field at 12K.

   In addition, for H∥c geometry we have studied the relaxation of magnetization process with time at various temperatures and we analyze our data using the expression

$$\frac{1}{M_o} \frac{dM}{d\ln t} = -\frac{k_B T}{U},$$

where U is the activation energy. The initial condition was obtained in the following way. The sample was first zero field cooled to 12K and then a high field of 11 kOe was applied and shut off to a very small remanant field of 35 Oe. The results are shown in Fig. 4.

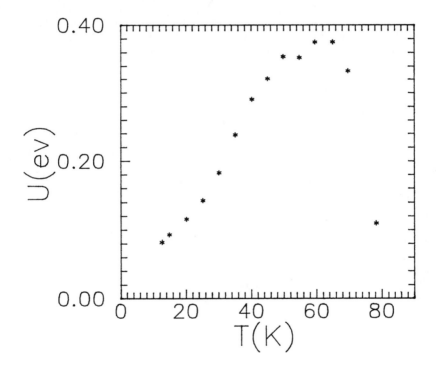

Fig. 4   Activation energy U (eV) vs temperature T (K).

The U vs T behavior shown here resembles that of Sun et al[5] who interpret their data in terms of a model involving a distribution of pinning energies and collective pinning. We would also like to mention the result of a relaxation experiment conducted for $H^{\perp}c$ geometry over a period of one hour. In this case no substantial relaxation was observed indicating that the pinning energies are much higher than $H \| c$ case.

CONCLUSION

We have found that substrate temperature and chamber pressure (initial base pressure as well as oxygen pressure during ablation) are important to achieve high orientation and high critical currents in thin films of YBC on $SrTiO_3$ by laser ablation. The most well-oriented film has $J_c(H^{\perp}c) = 4 \times 10^8$ A/cm^2 and $J_c(H \| c) = 9 \times 10^6$ a/cm^2 at 12K. Activation energy determined by magnetisation relaxation exhibits a strong temperature dependence peaking at 60K.

REFERENCES

1. X. D. Wu, A. Inam, T. Venkatesan, C. C. Chang, E. W. Chase, P. Barboux, J. M. Tarascon and B. Wilkens, Appl. Phys. Lett. 52, 754 (1988).
2. D. K. Fork, J. B. Boyce, F. A. Ponce, R. I. Johnson, G. B. Anderson, G. A. N. Connell, C. B. Eom and T. H. Geballe, Appl. Phys. Lett. 53, 337 (1988).
3. S. H. Liou, K. D. Ayelsworth, N. J. Ianno, B. Johs, D. Thompson, D. Meyer, J. A . Woollam and C. Barry, Appl. Phys. Lett. 54, 760 (1989).
4. G. M. Stollman, B. Dam, J.H.P.M. Emmen and J. Pankert, Physica C, 159, 854(1989).
5. J. Z. Sun, C. B. Eom, B. Lairson, J. C. Bravman, Y. H. Geballe and A. Kapiltunik, Physica C 162-164, 687 (1989).

COLD-WORKED, ANNEALED ALIGNMENT OF Ag CLAD Bi-2212 TAPES

F. Chen, R. Hidalgo, S.Q. Wang, X.Y. Zhang, R.S. Markiewicz and B.C. Giessen, Barnett Institute, Northeastern University, Boston, MA, 02115.

L.R. Motowidlo, IGC Advanced Superconductors Inc., Waterbury, CT 06704.

J.A. Rice and P. Haldar, Intermagnetics General Corporation, Guilderland, NY 12084.

ABSTRACT

Ag sheathed high T_c superconducting tapes of Bi-2212 were fabricated using the powder in tube method. X-ray diffraction studies of the as cold-worked (drawn and rolled) samples show some preferential (00l) alignment, but very broad peaks indicative of severe damage. Annealing removes the damage and greatly enhances the degree of orientation. Superconducting properties were also found to be extremely sensitive to cold-work and heat treatment conditions.

INTRODUCTION

Interest in the "powder in tube" method, to process high T_c superconducting wires and tapes, has been revived recently with the report of large critical current densities (J_c) at high fields in bismuth based systems.[1-4] Values as high as 8×10^4 A/cm^2 in fields up to 20 Tesla at 4.2K has been achieved.[3] At higher temperatures, 77K, J_c's of 1×10 at zero field[4] has also been realized in bismuth based superconducting composite tapes with a silver sheath. In addition, the "powder in tube" approach provides tapes, with sufficient mechanical strength and flexibility as well as offers the possibility of producing long lengths of technologically useful conductors. In fact, a similar process is currently in use at IGC to manufacture conventional multifilamentary low T_c superconductors. This technology has been used to develop a process to produce long lengths of metal oxide based composites to make practical material for device applications.[5]

Although the "powder in tube" method has been shown to yield superconductors with excellent properties, little has been reported on the optimum processing schedules. It is well known that the bismuth based superconductors are highly sensitive to preparation conditions. In this report, we address the effects of cold work and heat treatment on crystal structure, morphology and physical properties on Bi-2212 tapes clad with Ag. Conditions to obtain a high degree of orientation are also discussed.

EXPERIMENTAL

Tapes were prepared at IGC as follows: pre-reacted Bi-2212 powder was packed into silver tubes. To ensure complete filling of the tube space the powder was introduced gradually while tapping the tube. The ends were closed with silver plugs and lightly swaged. These billets were then cold-drawn on a laboratory draw bench. Additional processing involved rolling by providing successive passes to achieve a final thickness of 0.15 - 0.20 mm over the tape. A cross section of the rolled tape is shown in Figure 1. By carefully controlling the cold working conditions necking in the high T_c core could be reduced to a minimum. Lengths of 2-5 cm of the tape were cut from the long length and subjected to various heat treatment schedules. X-ray diffraction, AC susceptibility, scanning electron microscopy and transport J_c measurements were carried out at different stages of the processing. Critical current density results will be reported elsewhere.[6]

RESULTS AND DISCUSSION

X-ray diffraction studies were performed on the drawn and rolled tapes as well as on those that were annealed. After rolling, annealing was performed on samples in the Ag cladding using the following procedures. The tapes were supported on Bi-2212 powder, heated at 820°C for four days in air, then slowly cooled by turning off the oven and letting it cool to near room temperature (approximately 8 hours). Figure 2 compares x-ray diffraction spectra of (a) the starting powder, (b) a drawn and rolled sample and (c) a cold worked and annealed sample. In both cases (b) and (c), the silver cladding was carefully removed from one side of the sample and the powder left attached to the other side. The reflection spectra are then representative of the plane into which the tape was rolled. Both samples (b) and (c) show preferred orientation, with enhancement of the (00l) reflections. However, the unannealed sample shows considerably less alignment. Moreover, the peaks are excessively broad, indicative of considerable damage introduced in the cold working process. The precise nature of the damage (particle size, strains, stacking disorder) is not known at this time. By contrast, the annealed sample shows a very high degree of orientation, and most of the damage has been annealed out.

By detailed analysis of the x-ray intensities, we estimate that in the cold worked sample (b) about half of the grains are well oriented, and half remain randomly oriented. For the extruded and annealed sample, approximately 95% of the grains are well aligned.

Similar results were obtained on AC susceptibility data of the above samples.

Figure 3a. is the data for the starting Bi-2212 powder indicating a good development of the lower T_c phase (75K). Some excess CuO was present as evidenced in the XRD patterns. After packing in tubes and cold working AC susceptibility shows a complete degradation of the superconducting phases (Figure 3b.). This is in line with the XRD spectra which reveal broadening of the peaks resulting from severe damage, such as defects that may be introduced through cold working. Such degradation has also been observed by mechanical grinding of the bismuth based superconducting phases into powder.[7] However, recovery of the superconducting phase is possible by annealing as evidenced in Figure 3c. and Figure 2c. In fact, the annealing allows the Bi-2212 phase to grow with a strong texture with the c-axis oriented perpendicular to the rolling direction thus favoring the high current carrying planes along the length of the tape. The annealing step is, therefore, important to obtain grain growth and coupling that is not possible by cold working alone, to provide for high J_c's in these tapes. A repeated rolling-annealing routine densifies and textures the high T_c core further, enhancing the superconducting properties, provided the irrecoverable decomposition of the high T_c phase does not occur with excessive temperature and time of annealing.

We acknowledge the encouragement of Richard Blaugher and Bruce Zeitlin. This work was supported by DOE contract #DE-AC02-88-ER80607.

REFERENCES

1. K. Heine, J. Tenbrink and M. Thoner, Appl. Phys. Lett. 55, (1989) 2441.

2. H. Sekine, K. Inoue, H. Maeda and K. Numata, J. Appl. Phys., 66 (1989) 2762.

3. K. Sato, T. Hikata and Y. Iwasa, preprint.

4. S.X. Dou, H.K. Liu, S.J. Guo, J. Wang, M.H. Apperley, B. Loberg, C.C. Sorrell and K.G. Easterling, preprint.

5. L.R. Motowidlo, R.D. Blaugher, D.W. Hazelton, M.S. Walker and J.A. Rice, to be published in proceedings of Fall 1989 MRS Conference, Boston, MA.

6. L.R. Motowidlo and P. Haldar, to be published.

7. T. Kanai, T. Kamo and S.P. Matsuda, Japn. J. Appl. Phys., 29 (1990) L412.

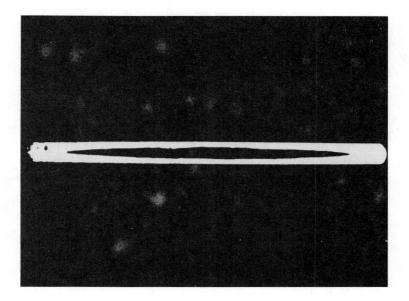

Figure 1. Cross section of drawn and rolled
tape high T_C conductor in Ag sheath.

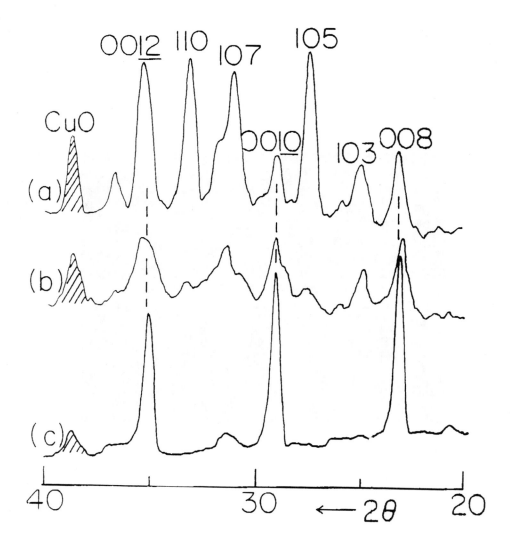

Figure 2. X-ray diffraction of Bi-2212 tape samples.
a.) Starting material, b.) drawn and rolled tape,
c.) cold-worked and annealed tape.

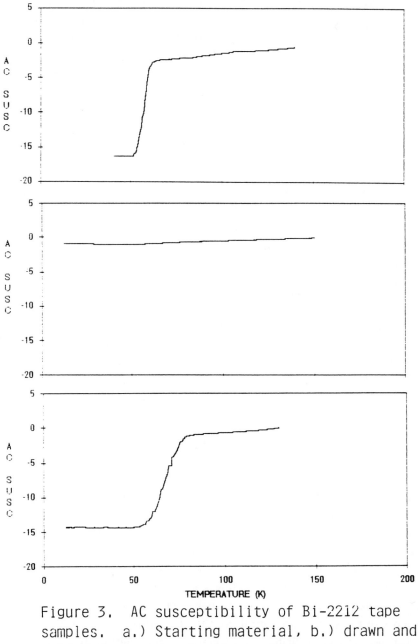

Figure 3. AC susceptibility of Bi-2212 tape
samples. a.) Starting material, b.) drawn and
rolled tape and c.) cold worked and annealed
tape.

HIGH TEMPERATURE REACTIONS IN THE
PROCESSING OF CERAMIC SUPERCONDUCTORS
VIA THE GLASS CERAMIC METHOD

D. P. Matheis and R. L. Snyder
Institute for Ceramic Superconductivity
NYS College of Ceramics
Alfred, NY 14802
C. R. Hubbard
High Temperature Materials Laboratory
Oak Ridge National Laboratory
Oak Ridge, Tn. 37831

Abstract

This paper contains a discussion of the glass ceramic processing methods and the important parameters which must be understood to control microstructure through the use of this process. Initial results of thermal analysis and high temperature diffraction analysis of glass crystallization, as well as high temperature diffraction analysis of the peritectic melting of the Bi-2212 phase are included.

INTRODUCTION

The glass ceramic process is an important method for the production of bulk bismuth superconductors which has been used by a number of authors. [1,2,3] Of the main methods for bulk sample preparation the glass ceramic method is the quickest, most convenient and least expensive. The solid state method is not well suited for the production of bismuth superconductor materials due to the fact that several long firings and regrinding steps are necessary to approach phase purity. The chemical processing techniques are not useful on a large scale due to the expense of the starting materials and the long involved process of dissolving, precipitating, ashing and firing. In the glass ceramic method the batch of oxides and carbonates is melted and quenched to form a glass. This glass is then ground and fired for varying periods of time (about 8 hours) to crystallize the

582

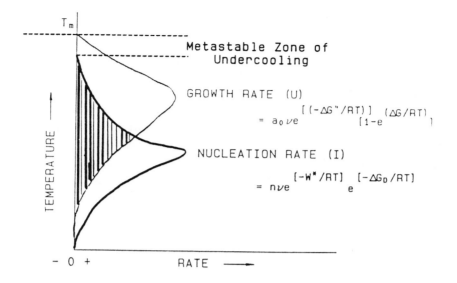

Figure 1: Nucleation and growth rates of glass crystallization.

superconducting phase. The glass ceramic method produces phase pure samples and the short firing times and decreases the chance of volatile loss of the bismuth. A more complete discussion of the advantages of the glass ceramic method and the use of this method in the bismuth system has been given by Bhargava et. al [4] and [3]

In order to use this method effectively to control microstructure the nucleation and grcwth of crystalline phases in the Bi-Sr-Ca-Cu-O system must be fully understood. In addition the peritectic melting sequence of the phases must also be understood if melt texturing techniques are to be used. The nucleation and growth of crystalline phases from a glass has been discussed by Turnbull and Cohen [5,6]. Figure 1 illustrates the nucleation and growth rates of glass crystallization discussed by Turnbull and Cohen. It is necessary to determine these curves for the phases in the bismuth system in order to carry out the microstructural tailoring necessary to produce good superconductors.

Since nucleation and crystallization curves are best determined dynamically the hot stage X-ray diffractometer was employed in this study. Some high temperature X-ray diffraction has been done to investigate crystallization temperatures and the phases present in partially melted states by Oka et. al [7,8].

EXPERIMENTAL PROCEDURE

Two samples were prepared for this study by the melt quench technique. The first sample contained a Bi-Sr-Ca-Cu stoichiometry of 2-2-1-2 (hereafter referred to as the stoichiometric sample) and the other sample with stoichiometry 2.16-1.53-0.95-2.0 (hereafter referred to as the non-stoichiometric sample) The stoichiometry of the non-stoichiometric sample is that suggested by Golden et. al [9] to give the highest T_c. Both samples were prepared by mixing the constituent oxides and carbonates in a mortar and pestle. These batches were then fired in a platinum crucible at 800°C for one hour to begin carbonate decomposition and then melted at 1100°C for ten minutes. The melt was quickly poured onto a liquid nitrogen cooled metal plate and then pressed from above with a liquid nitrogen cooled metal disk. The resulting, essentially amorphous material was then ground into a powder. A portion of the stoichiometric 2212 sample was heated to 870°C for eight hours to crystallize the Bi-2212 phase for the peritectic melting experiment.

Thermal analysis and high temperature diffraction were performed both at Alfred and at the HTML at Oak Ridge National Lab. Differential scanning calorimetry was performed at the HTML using an Omnitherm 1500S DSC. High temperature X-ray diffraction was performed at the HTML using a Scintag $\Theta - \Theta$ diffractometer with a hot stage sample mount and a solid state detector. High temperature X-ray diffraction was also performed at Alfred using a Norelco diffractometer with a hot stage and a backgammon type position sensitive detector.

RESULTS AND DISCUSSION

DSC analysis of the stoichiometric and non-stoichiometric samples as quenched are shown in Figure 2. This analysis shows a broad endothermic dip around 400°C indicating the onset of the glass transition region. This is immediately followed by a series of exotherms in both samples and then an endotherm which appears to contain more than one reaction due to its breadth and asymmetry. The reactions in the stoichiometric sample are shifted about 15 °C higher in temperature than the non-stoichiometric sample reactions. The initial exother-

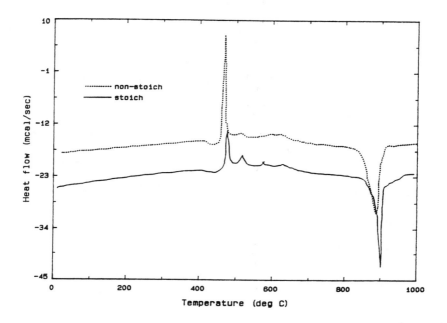

Figure 2: Differential scanning calorimetry analysis of as quenched stoichiometric and non-stoichiometric samples in air.

mic reaction is stronger in the non-stoichiometric sample than the stoichiometric sample and the opposite is true of the final endothermic reaction.

Figure 3 shows high temperature X-ray profiles of the non-stoichiometric sample from 300°C to 700°C. These profiles show the onset of crystallization of the Bi-2201 phase at around 465°C followed by slow crystal growth as seen in the sharpening of the diffraction peaks over time as the scans were run and temperature was raised.

Figure 4 shows high temperature X-ray profiles of the precrystallized stoichiometric Bi-2212 sample. These profiles show that the 2212 phase is stable through 870°C and that as heating continues to 900°C the sample undergoes a melting, leaving a phase indicated by Oka et. al [8] to be a strontium calcium cuprate. On cooling from 900°C the strontium calcium cuprate remains present and the Bi-2212 phase recrystallizes in a highly oriented manner, as evidenced by the strong 00l lines.

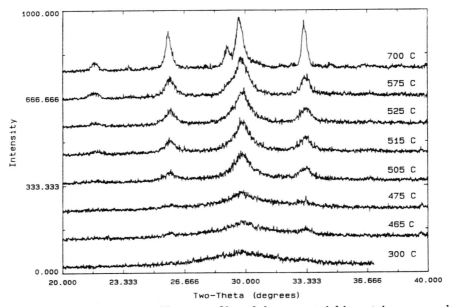

Figure 3: High temperature X-ray profiles of the non-stoichiometric as quenched sample performed at the HTML.

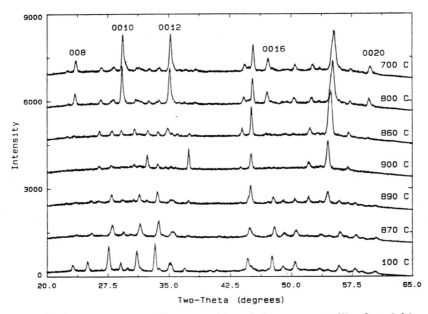

Figure 4: High temperature X-ray profiles of the precrystallized stoichiometric Bi-2212 sample performed at Alfred.

CONCLUSIONS

The onset of crystallization of the Bi-2201 phase corresponds to the initial exotherm in the DSC analysis. Since the exotherm takes place in about half a minute and the full crystallization takes places over several hours as seen in the X-ray patterns it is clear the crystallization of the Bi-2201 phase happens by a nucleation and growth process. The nucleation is very rapid as seen in the DSC while the crystal growth is very slow as evidenced by the high temperature X-ray. Thus although the nucleation of the 2201 phase cannot be avoided by a fast heating rate, growth of this phase can be minimized to facilitate more prominent growth of the 2212 phase at a higher temperature.

On recrystallizing from the melt the Bi-2212 phase is very highly oriented indicating that the melt texturing technique may be applicable to this material. It is likely that this technique would be best applied in a zone refining type procedure, very close to the melting point, trying to minimize the formation of the strontium calcium cuprate phase.

ACKNOWLEDGMENTS

We would like to thank the New York State Science and Technology Foundation and their Center for Advanced Ceramic Technology along with the New York State College of Ceramics for their sponsorship for this work. We are also indebted to the New York State Institute on Superconductivity and the Kodak corporation for their support. In addition we are grateful to the High Temperature Materials Laboratory of Oak Ridge National Laboratory which is supported by the office of Transportation Technologies of the US Department of Energy for use of their facilities.

References

[1] T. Kanai, T. Kumagai, A. Soeta, T. Suzuki, K. Aihara, T. Kamo and S. Matsuda, *Jpn. J. Appl. Phys.* **27** L1435 (1988)

[2] T. Komatsu, R. Sato, K. Imai, K. Matusita and T. Yamashita, *Jpn. J. Appl. Phys.* **27** L1839 (1988)

[3] A. Bhargava, A.K. Varshneya and R.L. Snyder, *Materials Letters* **Vol 8, No 10**, 245 (1989).

[4] A. Bhargava, A.K. Varshneya and R.L. Snyder, *Proceedings of the Conference on Superconductivity and Its Applications*, H.S. Kwok and D.T. Shaw editors, Elsevier New York (1988), 124-129.

[5] D. Turnbull, *Contemp. Physics* **10**, 473 (1969)

[6] D. Turnbull and M. Cohen, *J. Chem. Phys.* **29**, 1049 (1958)

[7] Y. Oka, N. Yamamoto, H. Kitaguchi, K. Oda and J. Takada, *Jpn. J. Appl. Phys.* **28** L213 (1989)

[8] Y. Oka, N. Yamamoto, Y. Tomii, H. Kitaguchi, K. Oda and J. Takada, *Jpn. J. Appl. Phys.* **28** L801 (1989)

[9] S. Golden, T. Bloomer, F. Lange, A. Segadaes, K. Vaidya and A. Cheetham, preprint (1990).

GLASS FORMATION AND TEXTURED CRYSTALLIZATION IN THE
Y₂O₃-BaO-CuO-B₂O₃ AND Y₂O₃-BaO-CuO-P₂O₅ SYSTEMS

B. J. Chen, M. A. Rodriguez and R. L. Snyder
Institute for Ceramic Superconductivity
NYS College of Ceramics
Alfred, NY 14802

Abstract

This paper presents a series of investigations of glass forming properties and crystallization behavior in both the Y_2O_3-BaO-CuO-B_2O_3
and Y_2O_3-BaO-CuO-P_2O_5 systems. The melting temperatures for the
glasses of varied compositions ranged from 1000°C to 1400°C. Different
heat treatment temperatures were used ranging from 700°C to 940°C.
A series of stoichiometries were studied to determine the glass forming
limits. The crystallized phases were analyzed by means of XRD and
SEM. The progressive development of 123 phase from the supersaturated glasses further proved the feasibility of obtaining superconducting
materials by using the glass-ceramic route. A study of the interface reaction between 123 and the glasses revealed that it is possible to crystallize the 123 phase from the glasses by a surface induced crystallization
technique.

INTRODUCTION

The feasibility and attractive aspects of obtaining superconducting materials by using the glass-ceramic route are currently the subject of a large
number of studies. Fabrication of glass-ceramics has been discussed extensively by Beall and Duke [1]. The attractive aspects and drawbacks of using
the glass-ceramic route were discussed in more detail by Bhargava *etal.* [2,3].
Technologically, the most important advantage is the possibility of using a
variety of continuous fabrication processes to make pore-free uniform products such as parts, sheets and wires. A major scientific advantage, often not

immediately obvious, is the possibility of arresting metastable phases into the structure. The glass-ceramic route, thus, give us an additional flexibility to examine yet undiscovered metastable phases. The most important disadvantage in glass-ceramic technology is that it may not always be possible to crystallize a specific phase.

This study focused on three approaches:

- A systematic study of the glass forming limits from binary to ternary in the system BaO-Y_2O_3 -CuO-B_2O_3 and the system BaO-Y_2O_3 -CuO-P_2O_5 .

- A series of studies on the crystallization properties of psudo-supersaturated glasses in the system BaO-Y_2O_3 -CuO-B_2O_3 .

- Surface induced crystallization studies using the $YBa_2Cu_3O_{7-\delta}$ "123" phase as the nucleating agent.

EXPERIMENTAL PROCEDURE

Glasses were formed from reagent grade $Ba(NO_3)_2$, CuO, Y_2O_3 , H_3BO_3[1] and $NH_4H_2PO_4$[2], which were melted in 10 g batches. The CuO-B_2O_3 glasses were melted in the range of 1000 to 1100°C for 30 minutes. The Y_2O_3 -B_2O_3 glasses were melted at 1500°C for 20 minutes. Both melts were performed using a platinum crucible. The $YBa_2Cu_3O_{7-\delta}$ -B_2O_3 glasses were melted in the range of 1200°C to 1300°C for 15 minutes in an MgO crucible. The $YBa_2Cu_3O_{7-\delta}$ -P_2O_5 glasses were melted in the range of 1300 to 1400°C for 30 minutes in an Al_2O_3 crucible. Melts were quenched between two steel plates to form about 2 mm thick samples. The crystallized samples were examined by powder diffraction (XRD) using a Siemens D500 diffractometer with Cu K_α radiation and a diffracted beam graphite monochromator. The SEM analysis was carried out using the ETEC instrument [3].

123 powder was prepared from reagent grade $Y(NO_3)_3$[4], $Ba(NO_3)_2$ and $Cu(NO_3)_2$[5] in stoichiometric proportions. The raw materials were mixed

[1] Fisher Scientific Company, Fair Lawn, NJ
[2] Aldrich Chemical Company, Inc., Milwaukee, WI
[3] ETEC autoscan, Hayward CA
[4] Alfa Products, Danvers, MA
[5] Fisher Scientific Company, Fair Lawn, NJ

and ground in a mortar and pestle, and heated at 800°C for 8 hours. After cooling the sample was ground to a fine particle size, heated at 940°C for 8 hours and then cooled to 450°C and held for 8 hours for oxidation.

Samples of both 123 and psudo-saturated base glasses were ground into fine powders. The 123 powder was pressed into pellets. Then the base glass powder was placed on top of the pellets and they were pressed again to form a double-layer. Carbowax 400[6] was used as a binder. The pellets were then heated to different temperatures for 5 hours to examine the interface reaction between 123 and the base glasses.

RESULTS AND DISCUSSION

Table 1 shows the compositions of the glass forming limits for the binary oxide-borate glasses. Two values are given for the $Y_2O_3:B_2O_3$ system representing both sides of a spinodal insolubility dome. Each value along with the other solubility limiting values were combined to produce compositions of two borate psudo-saturated in Y, Ba and Cu called A1 and B1. They formed very stable glasses (T_x - T_g is larger than 200°C) but are not completely saturated due to the modification of each ions activity in the presence of the other ions. A second approach to glass forming involved adding varying amounts of glass former to stoichiometric 123 mixtures. Experiments showed that when the concentration of glass former was reduced below four moles the glasses began to crystallize, so the $YBa_2Cu_3O_{7-\delta}$ -P_2O_5 and $YBa_2Cu_3O_{7-\delta}$ -B_2O_3 , compositions shown in Table 1 were used.

TABLE 1.Compositions of the saturated glasses

Binary	$Y_2O_3 :B_2O_3$	5:95, 30:70
glass forming	$BaO:B_2O_3$	45:55 [4]
limits	$CuO:B_2O_3$	55:45
Base glass A1	Y:Ba:Cu:B	1:5.8:8.6:44
Base glass B1	Y:Ba:Cu:B	30:29:43:198
123-P_2O_5 glass	Y:Ba:Cu:P	1:2:3:4
123-B_2O_3 glass	Y:Ba:Cu:B	1:2:3:4

[6]Union Carbide Corp., New York, NY

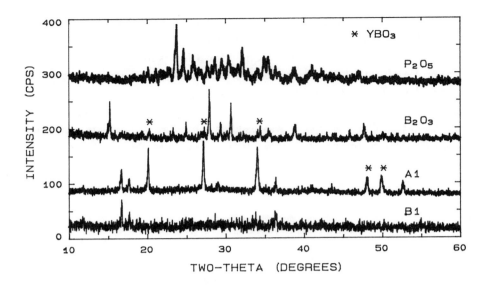

Figure 1: XRD of A1, B1, 123-P_2O_5 and 123-B_2O_3 glasses after heat treated at 850°C for 5 hours.

Figure 1 shows XRD patterns of the 123-P_2O_5, 123-B_2O_3, A1 and B1 glasses after heating to 850°C for 5 hours. The B1 composition is the most stable against devitrification with only a small amount of an unidentified phase crystallizing. The A1 and 123-B_2O_3 glasses precipitate YBO$_3$, CuO and the unidentified phase found in the B1 glass. In addition the 123-B_2O_3 shows additional lines probably due to the precipitation of borate phases as is clearly happening in the 123-P_2O_5 glass. We so no evidence of the crystallization of $YBa_2Cu_3O_{7-\delta}$ in any of these glasses.

Table 2 lists the compositions, heating temperatures and crystallized phases identified after heat treatment of three supersaturated glasses.

Table 2. Compositions and crystallized phases of B14, B15 and B16
supersaturated glasses.

	Glass composition Y:Ba:Cu:B	Heating temperature °C	Heating time (hours)	Phases identified after heat treatment
B14	1:5:8:7	750	5	CuO, unknown
		800	5	CuO, unknown
		850	4	123, CuO, YBO$_3$, unknown
		900	16	123, CuO, YBO$_3$, unknown
B15	1:5:8:6	900	16	123, CuO, YBO$_3$, unknown
		940	16	123, CuO, unknown
B16	4:6:6:7	900	16	123, CuO, unknown
		940	16	123, CuO, unknown

Figure 2 shows the XRD patterns of the B14 glass composition heated at different temperatures. This Figure illustrates the progressive development of the 123 phase as the heating temperature increases from 750 to 900°C in the presence of O_2. Only a single peak characteristic of the 123 phase was obtained around 32.8°2θ. This single peak indicates a tetragonal like lattice structure[2] which may be due to low solubility of oxygen in the glass. The intensity of the characteristic 123 peaks in surface diffraction patterns of B14 glasses are much higher than those for the powder diffraction patterns (see Figure 3). This suggests that the crystallization of the 123 phase began from the surface. The formation of a surface layer was observed by SEM analysis as shown in Figure 4. Analysis showed that the oriented layer thickness of 123 was proportional to the length of heat treatment. Samples heated at 850°C for only 4 hours indicated only a small quantity of the 123 phase by XRD. When heated at 900°C for 16 hours, well resolved peaks in the X-ray powder diffraction pattern denote an increased content of the 123 phase. Figure 5 shows similar results from B15 and B16 glasses which were heated for 16 hours at 900°C and 940°C, respectively. The development of the 123 phase above 900°C in both of the glasses is apparent.

Figure 6 shows the SEM results of the interface reactions between 123 and base glasses at different temperatures. At low temperatures (750 and 800°C), the interface reaction is not observed. Some wetting effect of a glassy phase on the 123 powder can be seen in (b) but there was no evidence

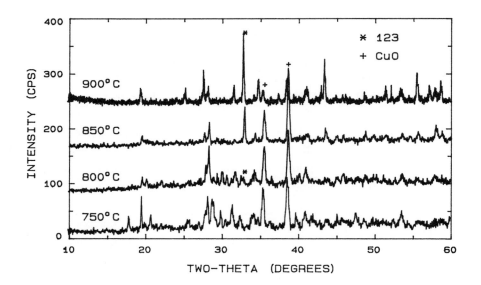

Figure 2: XRD patterns of B14 supersaturated glasses heated at different temperatures.

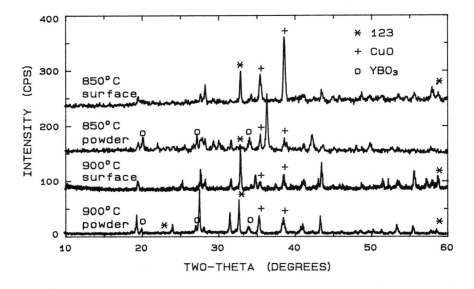

Figure 3: A comparison of surface and powder XRD patterns of sample B14.

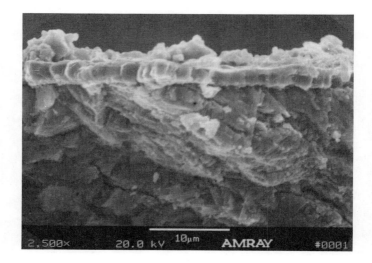

Figure 4: SEM of B14H3 showing the formation of a surface layer.

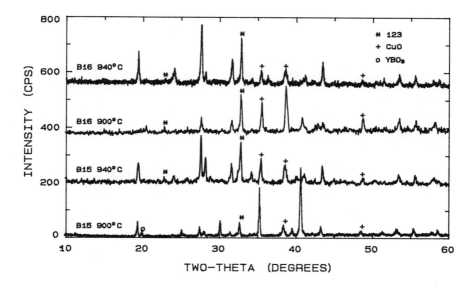

Figure 5: XRD of B15 and B16 supersaturated glasses heated at 900 and 940°C for 16 hours showing a significant development of the 123 phase from glass.

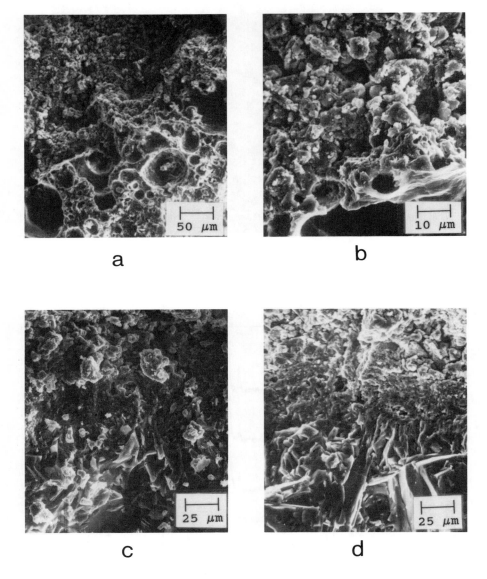

Figure 6: SEM of interface reactions between $YBa_2Cu_3O_{7-\delta}$ and base glass A1 at (a) 750°C , (b) 800°C , (c) 850°C and (d) 900°C for 5 hours

of surface induced crystallization on the base glasses. As the temperature increased to 850 and 900°C , heavy interface reactions were clearly seen (see (c) and (d) of Figure 6). Well aligned crystals at the interface of the 123 and base glass were evidence of surface induced crystallization.

CONCLUSIONS

- The approximate glass forming limit in molar ratio, is 1:2:3:4 for both Y:Ba:Cu:B and Y:Ba:Cu:P.

- A1 and B1 base glasses are stable glasses with T_x - T_g is larger than 200°C.

- 123-B_2O_3 glasses are more stable and crystallization is easier to control in comparison to 123-P_2O_5 glasses.

- The supersaturated glasses in an O_2 atmosphere showed progressive development of 123 phase with temperature.

- The 123 phase crystallizes from the surface and the amount of 123 phase increases with heating temperature and time.

- The 123 phase displays surface induced crystallization from the base glasses. Further studies will be focused on using supersaturated glass and choosing proper heating temperatures to control the crystallization and its texture.

ACKNOWLEDGMENTS

We would like to thank the New York State Science and Technology Foundation and their Center for Advanced Ceramic Technology along with the New York State College of Ceramics for their sponsorship for this work. We are also indebted to the New York State Institute on Superconductivity and the Kodak corporation for their support.

References

1. G.H. Beall and D.A. Duke, "Glass Science and Technology", **1**, D.R. Uhlmann and N.J. Kreidl editors, Academic Press, New York (1983).

2. A. Bhargava, A.K. Varshneya and R.L. Snyder, "Crystallization of Glasses in the System BaO-Y_2O_3 -CuO-B_2O_3 ", *Superconductivity and Its Applications*, H.S. Kwok and D.T. Shaw editors, Elsevier, New York 124-129 (1988).

3. A. Bhargava, A.K. Varshneya and R.L. Snyder, *Materials Letters*, **8**, [1,2] 41-45 (1989).

4. A. Bhargava, R.L. Snyder and R.A. Condrate, Sr, "The Raman and Infrared Spectra of the Glasses in the System BaO-TiO_2-B_2O_3 ," *Mat. Res. Bull.*, **22** 1603-1611 (1987).

THE USE OF GA$_2$O$_3$ AND IN$_2$O$_3$ IN FORMING SUPERCONDUCTING GLASS CERAMICS

B. J. Reardon and R. L. Snyder
Institute for Ceramic Superconductivity
New York State College of Ceramics
Alfred University
Alfred, NY 14802

Abstract

Glass-ceramic processing of Bi-system superconductors has been improved through the addition of intermediate glass formers. The intermediate glass formers used here are Ga and In. The Ga and Ga/In additions did produce $Bi_2Sr_2Cu_1O_6$ and $Bi_2Sr_2CaCu_2O_8$. The T_x -T_g difference for the Ga and Ga/In glasses were 73K and 89K respectively. The molar ratios used were 1:1 (2223:Ga) and 1:.9:.1 (2223:Ga:In).

INTRODUCTION

Glass-Ceramics are materials that are formed using typical glass forming techniques and are then heat treated to devitrify the desired crystal structure throughout the glass matrix. A goal in making a useful glass-ceramic is to first produce a glass whose difference between the glass transition temperature and the first onset crystallization is large enough to allow for good formability. The formed glass must then be heat treated to allow for nucleation and complete devitrification. The heat treatments must be sufficient to fully crystallize the desired phase but also avoid slumping and flowing of the product. The goal of this study is to find a glass that has an acceptable T_x -T_g difference, can maintain its form after long heat treatments, and that will promote the formation of the desired ceramic crystals.

Glass-Ceramics offer many advantages over typical solid state reactions. They produce a material which is typically 99.9% dense with higher strength as compared to typical ceramics[1]. These materials could have better connectivity of the superconducting crystals and the chemical compositions of the grain boundaries can be controlled so that inter-grain electrical conduction could possibly be enhanced[2]. All of these advantages should increase mechanical strength and critical current density. The fact that a glass is being used allows for the formation of almost any shaped product.

Bismuth is an intermediate glass former[3] and in the presence of other ions such as Ca and Sr, can form a glassy network. Glasses made from the appropriate Bi, Sr, Ca, and Cu stoichiometric ratios have been devitrifed to form superconducting crystals of $Bi_2Sr_2Cu_1O_6$ (2201), $Bi_2Sr_2CaCu_2O_8$ (2212)[4], and $Bi_2Sr_2Ca_2Cu_3O_{10}$ (2223)[5,6,7]. This technique has also been shown to work in the 1-2-3 system[8]. The purpose of this study is to improve upon these results by increasing the difference between the glass transition temperature and the first crystallization temperature.

The three most common superconductive phases in the Bi- system are the 2201, 2212 and 2223. The first phase to form in a typical reaction is the 2201. This phase has a T_c around 22K and usually crystallizes around 475°. The presence of 2201 is usually accompanied by CuO and a Ca-compound depending on the sintering conditions.

The next phase to form is the 2212 phase. This phase has a T_c at about 75K and forms from the combination of 2201, CuO and the Ca compound at about 830°C .

The 2223 phase is the most difficult to synthesize. This phase has a T_c around 110K[9] and therefore is most desirable. The formation of 2223 depends on the starting compositions (an excess of CuO appears to be most effective)[10,11], low oxygen partial pressure (lower than that present in air)[12], specific heat treatment temperature[13], long heat treatment times (100 hrs.)[14], and PbO doping[15,16]. The fact that the formation of 2223 is an ordering reaction[17] explains the need for long heat treatments. The requirement of a very narrow (+5K) temperature window for the stability of this phase can be understood by viewing the phase diagram of this system[18] and realizing that some liquid phase must be present in order to partially dissolve 2212 and precipitate 2223[19]. The need for excess CuO and low oxygen partial pressure is not completely understood[7,12]. Pb doping helps to stabilize and accelerate the formation of 2223[20].

EXPERIMENTAL PROCEDURE

This study attempts to crystallize the 2201, 2212 and 2223 phases in the Bi-system using additions of glass formers and intermediate glass formers. The additives used were Ga and a combination of Ga and In.

- Gallium

  The Ga study involved adding Ga to a 2223 cation stoichiometric batch composition to help increase the T_x -T_g difference. This method was first attempted by Gunthier et al.[21]. The basic starting materials, Bi_2O_3, CuO, CaO, $Sr(NO_3)_2$, Pb_3O_4, and Ga_2O_33 were weighed out it the stoichiometric proportions of 1:0 (1 mole 2223: 0 mole Ga), 1:.5, 1:1, 1:1.5, 1:2, 1:2.5, and 1:3. 2223 refers to 1.6 mole Bi, .4 Pb, 2 Sr, 2 Ca, and 3 Cu. These 10 gram batches were mixed by hand in a mortar and pestle for 40 minutes. Melting occurred at 1100°C in a Lindberg 1500 series resistance furnace in a Pt or MgO crucible for 7 minutes. All samples were poured on to an aluminum plate and quenched on top with a steel plate. The glasses produced, underwent X-ray diffraction (XRD), Differential Scanning Calorimetry (DSC) (Du Pont 1600 system), and Differential Thermal Analysis (DTA) (also Du Pont). The glasses were then heat treated in a Eurotherm Resistance furnace on Pt foil in air to determine the phases associated with the peaks found in the DTA and DSC runs. In a further effort to produce 2223 the composition of the glass was altered to contain 2224 and Ga in a 1:1 ratio. This glass was tested in the same way the others. The resulting glass was then heat treated at 830°C for 48 hr. in a nitrogen atmosphere to devitrify the 2223 phase.

- Gallium/Indium

  The Ga/In study involves taking the best 2223:Ga glass and partially replacing the Ga with In. This method was chosen because of the trend displayed in the work by Gunthier et al. who studied the effects of adding Group III oxides to the batch to help improve the T_x -T_g difference. Generally speaking they found that as the ion size increased the T_x -T_g difference also increased. Gunthier et al. did not study the effects of Indium. Since Indium is too ionic to be a an intermediate glass former it can not be used alone. Therefore it was used to partially replace Gallium in the 1:1 batch. The methods of study used in this experiment are exactly the same as that present in the Ga study. The Base composition used was a 1:1 where the 1 Ga was broken down into .9:.1 (Ga:In), .8:.2, .7:.3, .6:.4, .5:.5. In other words, 1:.9:.1, 1:.8:.2, etc. The In was obtained in the form of In_2O_3.

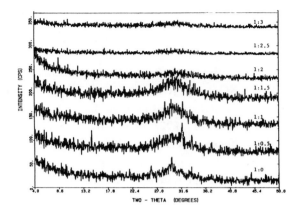

Figure 1: XRD patterns of Ga glasses with 2223 in increasing molar ratios.

RESULTS

- Gallium
  Figure 1 shows the XRD graphs of the glasses produced with Ga, each showing their amorphous character. Upon heat treatments at various temperatures as specified by the peaks in the DSC and DTA scans it was found that the first phase to devitrify after T_g was usually the 2201 phase. CuO also comes out at or around the same temperature. Table 1 presents the T_g 's and T_x 's of the 2201, 2212, and 2223 in the samples. The temperatures listed for the 2223 phase are an approximation since even the presence of a peak on the DTA scans were usually in question. Heat treatments at 850°C in air were not successful in producing 2223. The largest T_x -T_g difference (111K) occurred with the 1:1.5 sample. The 1:1 sample (the sample predominately studied by Gunthier et al.) displayed a T_x -T_g difference of 73K.

  The samples were then heat treated for 1 and 24 hours to help to identify the crystallization peaks. Table 2 shows the phases present at various temperatures. The 2212 phases were not present until the samples were heat treated at 775°C . The 0-1.5 mole compositions of Ga displayed the 2212 phase but as the amount of Ga increased the concentration of 2201 and CuO did also. This trend can be seen in Figure 2. Though the 1:1.5 composition had higher T_x -T_g differences

Table 1: This table presents the approximate glass transition temperatures
(°C) and crystallization temperatures of 2201, 2212, and 2223 in the Ga
addition samples.

Sample	T_g	2201	2212	2223
1:0	369	436	750	—
1:.5	405	474	775	850
1:1	433	506	766	825
1:1.5	425	536	750	837
1:2	430	559	743	850
1:2.5	475	609	750	800
1:3	450	609	738	—

it also contained 2201. Therefore the 1:1, which only formed the 2212
phase, is the best composition for obtaining 2212 in a glass-ceramic.
It should be noted that none of the samples slumped while being heat
treated.

- Gallium/Indium
  The glasses formed from the In addition can be seen in the XRD patterns in Figure 3. Heat treatments for 1 and 24 hours at various temperatures to ascertain the phases present at each DTA and DSC peak
  were carried out. Table 3 shows the results of the heat treatments. The
  1:.9:.1 had a T_x -T_g difference of 89K. This is larger than the original
  composition of 1:1 which was 73K. XRD analysis of the samples shows
  that 2201 and CuO generally crystallize first at low temperatures. Table 4 shows the XRD data obtained. At 775°C , the 2212 phase was
  able to devitrify. Figure 4 shows the XRD patterns of the heat treated
  samples at 775°C . All of these samples remained stable during heat
  treatment at 775°C .

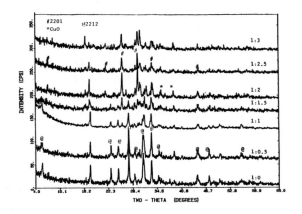

Figure 2: XRD patterns of heat treated Ga glasses with 2223 in increasing molar ratios at 775°C for 24 hrs.

Table 2: This table presents the phases present at each temperature (°C), or temperature range, or after a DSC scan for the Ga addition samples as determined by XRD.

Sample	DSC	400-500	530	775	850
1:0	2201	2201	—	2212	2212
1:.5	2201/CuO	2201/CuO	2201/CuO	2212	2212
1:1	2201/cuO	—	2201/CuO	2212	2212
1:1.5	2201/CuO	—	2201/CuO	2212/2201/CuO	melted
1:2	2201/CuO	2201/CuO	2201/CuO	2201/CuO	melted
1:2.5	2201/CuO	—	2201/CuO	2201/CuO	melted
1:3	2201/CuO	—	2201/CuO	2201/CuO	melted

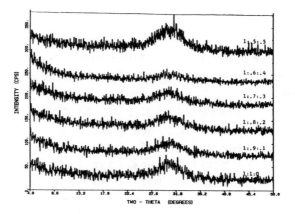

Figure 3: XRD patterns of In glasses with 2223 in increasing molar ratios.

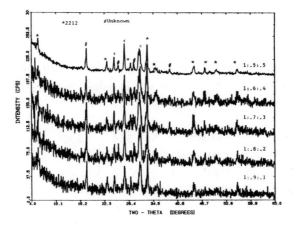

Figure 4: XRD patterns of heat treated In glasses with 2223 in increasing
molar ratios at 775°C for 24 hrs.

Table 3: This table presents the approximate glass transition temperatures (°C) and crystallization temperatures of 2201, 2212, and 2223 in the In addition samples.

Sample	T_g	2201	2212	2223
1:.9:.1	420	509	775	—
1:.8:.2	420	491	775	838
1:.7:.3	410	478	737	859
1:.8:.4	400	475	750	850
1:.5:.5	405	469	775	862

Table 4: This table presents the phases present at each temperature (°C), temperature range, or DSC scan for the In addition samples as determined by XRD.

Sample	DSC	400-500	500-550	600	775	850
1:.9:.1	2201/CuO	2201/CuO	2201/CuO	—	2212	melted
1:.8:.2	2201/CuO	2201/CuO	2201/CuO	—	2212	melted
1:.7:.3	2201/CuO	2201/CuO	—	—	2212	melted
1:.6:.4	2201/CuO	2201/CuO	2201/CuO	—	2212	2212
1:.5:.5	2201/CuO	2201/CuO	—	2201/CuO	2212	2212

CONCLUSIONS

As Table 1 shows adding 1 mole of Ga to 1 mole of 2223 does significantly improve the T_x -T_g difference. As the concentration of Ga increases the T_x - T_g difference also seems to increase. However, the position of T_g is not always clear and therefore not reliable. Furthermore, at higher concentrations of Ga it becomes more difficult to crystallize a superconductive phase.

The In addition (1:.9:.1) resulted in a T_x -T_g difference, as seen in Table 3, that was higher than that of the Ga sample (1:1). As the In concentration increased the T_x -T_g difference decreased because of the inability of In to form a glass.

ACKNOWLEDGMENTS

We would like to thank the New York State Science and Technology Foundation and their Center for Advanced Ceramic Technology along with the New York State College of Ceramics for their sponsorship for this work. We are also indebted to the New York State Institute on Superconductivity and the Kodak corporation for their support.

References

[1] P.B. Adams, M.G. Britton, J.R. Lonergan, G.W. McLellan, all of Corning Glass Works, "All About Glass," Corning Glass Works, Corning, (1984), pp(32-33).

[2] H. Zheng and J.D. Mackenzie, "Bi4Sr3Ca3Cu4O16 glass and Superconducting glass-ceramics," Physical Review B condensed Matter, Vol. 38, No. 10, 1/10/88.

[3] H. Rawson, "Inorganic Glass Forming Systems," (Academic London, 1967).

[4] A. Bhargava, R. L. Snyder and A. K. Varshneya, "Preliminary Investigations of Superconducting Glass-ceramics in the Bi-Sr-Ca-Cu-O System", Matl. Ltrs. 8[10], 425-431 (1989).

[5] D. Shi, M. Blank, M. Patel, D. G. Hinks, A. W. Mitchell, K. Van der Voort, H. Claus, "Superconductive Phases in the Bi- Sr-Ca-Cu-O System," Physica C 156, 822 (1988).

[6] T. Komatsu, R. Sato, C. Hirose, K. Matusita, and T. Yamashita, "Preparation of High T_c Superconducting Bi-Pb-Sr- Ca-Cu-O Ceramics by the Melt Quenching Method," Japanese Journal of Applied Physics Letters, Vol. 27, L2293 (1988).

[7] H. Zheng, J.D. Mackenzie, "Bi1.84Pb0.34Ca2Sr2Cu4Oy Superconducting Tapes with Zero Resistance at 100K Prepared by the Glass-to-Ceramic Route," Journal of Non-Crystalline Solids, Vol. 113 (1989).

[8] A. Bhargava, A. K. Varshneya and R. L. Snyder, "On the Stability of Superconducting $YBa_2Cu_3O_{7-\delta}$ in a Borate Glass-ceramic Matrix", Materials Letters, 8[1-2], 41-45 (1989).

[9] J.M. Tarascon et al. "Crystal Structure and Physical Properties of the Superconducting Phase Bi4(Sr,Ca)6Cu4O16+x," Physical Review B, Vol. 37, (1988), 9382.

[10] A.Ono, "Synthesis of t 107K Superconducting Phase in the Bi-Sr-Ca-Cu-O System," Japanese Journal of Applied Physics Letters, Vol. 27, (1988), L1213.

[11] S.J. Golden, T.E. Bloomer, F.F. Lange, A.M.Segadaes, K.J. Vaidya, A.K.Cheetham, "Processing and Characterization of Thin Films of the Two-Layer Superconducting Phase in the Bi-Sr-Ca-Cu-O System; Evidence for Solid Solution," , preprint (1990).

[12] U. Endo, S. Koyama and T. Kawai, "Preparation of the High T_c Phase of Bi-Sr-Ca-Cu-O Superconductor," Japanese Journal of Applied Physics Letters, Vol. 27, (1988) L1476.

[13] N. Kijima, H.Endo, J. Tsuchiay A. Sumiyama, M. Mizuno and Y. Oguri, "Reaction Mechanisms of Forming the High T_c Superconductor in the Pb-Bi-Sr-Ca-Cu-O System," Japanese Journal of Applied Physics Letters, Vol. 27, (1988), L1852.

[14] H. Nobumasa,K Shimizu, Y Kitano and T Kawai, "High T_c Phase of Bi-Sr-Ca-Cu-O Superconductor," Japanese Journal of Applied Physics Letters, Vol. 27, (1988), L846.

[15] S.A. Sunshine, T. Siegrist, L.F. Schneemeyer et al., "Structure and Physical properties of Single Crystals of the 84K Superconductor $Bi_{2.2}Sr_2Ca_{0.8}Cu_2O_{8+x}$," Physical Review B, Vol. 38, (1988), 893.

[16] S.M. Green et al., "Zero Resistance at 107K in the (Bi,Pb)-Ca-Sr-Cu oxide System," Physical Review B, Vol. 38, (1988), 5016.

[17] D. P. Matheis, and R. L. Snyder, "The Crystal Structures and Powder Diffraction Patterns of the Bismuth and Thallium Ruddlesden-Popper Copper Oxide Superconductors," Powder Diffraction, Vol. 5, No. 1, 3/90.

[18] C-L. Lee, J-J. Chen, W-J. Wen, J-M. Wu, T-B. Wu, T-S. Chin, "Equilibrium phase relations in the Bi-Ca-Sr-Cu-O system at 850 and 900°C ," Journal of Materials Research, Vol. 5, No. 7, July 1990.

[19] T. Hatano, K. Aota. S. Ikeda, K. Nakamura, K. Ogawa, "Growth of the 2223 Phase in Leaded Bi-Sr-Ca-Cu-O System," Japanese Journal of Applies Physics, Vol. 27, No. 11, 11/88.

[20] J.J. Cavax et al., "The Effect of PbO on the Growth of the 107K Phase in the Bi-System of Superconductors," Physica C, Vol. 153-155, (1988), 506.

[21] H. Abmann, A. Gunthier, B. Steinmann, "Preparation and Properties of Superconducting Crystallized Glasses and Enamels bases on the Bi-Sr-Ca-Cu-Oxide system," Icmc' 90 Topical Conference "High-Temperature superconductors, Materials Aspects" May 9-11, 1990 Garmisch-Partenkirchen, FRG.

THE EFFECT OF BaCO₃ PRECURSOR ON THE FORMATION OF PHASES IN THE BaO-Y₂O₃ -CuO SYSTEM

M. A. Rodriguez and R. L. Snyder
Institute for Ceramic Superconductivity
NYS College of Ceramics
Alfred, NY 14802

Abstract

Samples near the stoichiometry $Y_1Ba_4Cu_2O_x$ were prepared with and without the presence of carbonate. XRD and wet chemical analysis were performed to determine carbonate content. A compound with the stoichiometry of $Y_1Ba_4Cu_2(CO_3)O_{7\pm\delta}$ was stabilized when the precursor $BaCO_3$ was used. The absence of carbonate yielded the $Y_1Ba_4Cu_3O_{8.5+\delta}$ compound. Electrical measurements indicated no superconductive properties in the 10-298 K temperature range for the 142 oxycarbonate and the 143.

INTRODUCTION

This paper will focus on the composition found near the $YBa_4Cu_2O_x$ stoichiometry and therefore it shall be referred to as "142". The correct stoichiometry has long eluded determination and there has been much controversy as to the correct composition of this phase. This black color phase was first reported as the "132" since the stoichiometry was close to these values[1,2,3]. Deleeuw *etal.*[4] reported this structure to have the stoichiometry "385" however it was observed that this phase was unobtainable when all oxide or nitrate precursors were used. These reports were followed by claims that this compound was in fact an oxycarbonate solid solution over the 132 - 142 - 152 range of stoichiometries[5] since it was not possible to form the compound without the carbonate in the precursor material. Use of oxides and nitrates often lead to the formation of the compound $Y_1Ba_4Cu_3O_{8.5+\delta}$ "143" which was reported by Deleeuw as a cubic 8.1Å perovskite[6].

This study was undertaken to learn more about the 142 compound with regards to the possibility of it being an oxycarbonate and to examine the effects of carbonate precursor in relation to the superconducting $YBa_2Cu_3O_{7\pm\delta}$ "123" phase.

Table 1: Sample Preparation

Sample	Stoichiometry	Atmosphere	Precursors
1	142	Air	$BaCO_3$, Y_2O_3, CuO
2	142	O_2	$Ba(NO_3)_2$, Y_2O_3, CuO
3	143	O_2	BaO_2, Y_2O_3, CuO
4	143	Air	$BaCO_3$, Y_2O_3, CuO

EXPERIMENTAL PROCEDURE

142 and 143 samples were prepared by a solid state reaction with and without the presence of carbonate by mixing stoichiometric proportions of Y_2O_3, CuO, and $BaCO_3$, BaO_2 or $Ba(NO_3)_2$ precursor materials in a mortar and pestle according to table 1. The sample powders were calcined in alumina crucibles at 980°C for 36 hours with intermittent grindings using a heating rate of 10°C/min, and a cooling rate of 4°C/min.

Room temperature X-ray powder diffraction patterns were collected using a Siemens D500 diffractometer with a diffracted beam graphite monochromator using Cu K_α radiation. Diffraction patterns were collected from 5-65°2θ with a step size of 0.050°2θ and a count time of 1 second. Powder samples were mounted on top loaded sample holders.

To make bulk property measurements on the 142 compound, powder from sample 1 was pressed into a 1x1x5 cm bar. A 1.25 cm diameter cylindrical pellet was also pressed to test for Meissner effect. These samples were sintered in air at 980°C on an alumina setter. The heating rate was 10°C/min, soak time was 12 hours, and the cooling rate was 4°C/min. The sintered 142 bar was electroded in typical four-point arrangement and mounted on a cold finger of a closed-cycle helium cryostat for resistive measurements.

RESULTS AND DISCUSSION

Figures 1, 2, 3, and 4 show the X-ray powder diffraction patterns for samples 1 through 4, respectively. Significant support for the postulate that 142 is an oxycarbonate comes from the fact that it was not possible to make without the presence of carbonate within the precursor materials[5]. The 142 structure has no difficulty forming with $BaCO_3$ as the precursor material, as

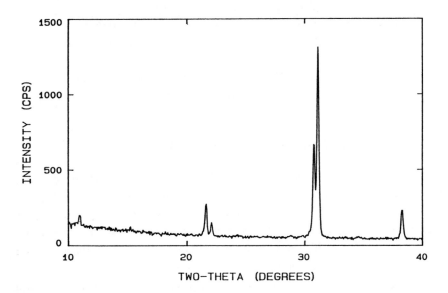

Figure 1: Powder X-ray diffraction pattern for sample 1; 142 stoichiometry prepared with BaCO$_3$ in Air. Pattern illustrates phase pure 142 phase.

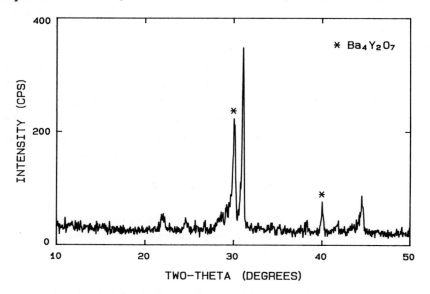

Figure 2: Powder X-ray diffraction pattern for sample 2; 142 stoichiometry prepared with Ba(NO$_3$)$_2$ in O$_2$. Pattern illustrates presence of cubic 143 and Ba$_4$Y$_2$O$_7$.

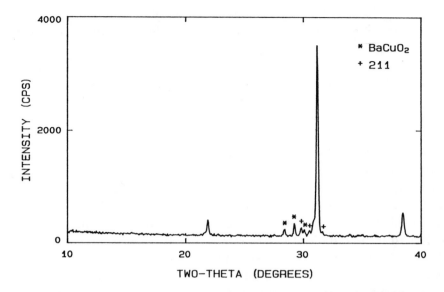

Figure 3: Powder X-ray diffraction pattern for sample 3; 143 stoichiometry prepared with BaO_2 in O_2. Pattern illustrates nearly phase pure cubic 143 with minor impurity phases of $BaCuO_2$ and 211 (the green phase).

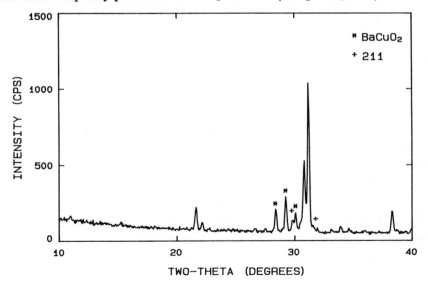

Figure 4: Powder X-ray diffraction pattern for sample 4; 143 stoichiometry prepared with $BaCO_3$ in Air. Pattern illustrates the presence of the 142 phase with impurity phases of $BaCuO_2$ and 211 (the green phase).

Table 2: Summary of Phases Observed in Samples 1-4

Sample	Stoichiometry	Carbonate	Major Phase	Impurity Phase(s)
1	142	Yes	142	-
2	142	No	143	Ba$_4$Y$_2$O$_7$
3	143	No	143	BaCuO$_2$, 211
4	143	Yes	142	BaCuO$_2$, 211

is demonstrated in Figure 1. This pattern illustrates a typical powder X-ray diffraction pattern for the 142 compound[7].

Sample 2 was of 142 stoichiometry made via reaction of Ba(NO$_3$)$_2$, Y$_2$O$_3$ and CuO in O$_2$. The diffraction pattern for sample 2 shown in Figure 2 indicated the formation of the cubic 143 phase and a second phase which appeared to be a Ba-Y-O binary compound. It is believed to be Ba$_4$Y$_2$O$_7$. To avoid atmospheric contamination the sample was left in the furnace under one atmosphere O$_2$ and only a small quantity of powder was used for X-ray analysis on a zero background holder. This accounts for the low intensity of the diffraction pattern.

The cubic 143 phase was isolated in phase pure form by de Leeuw *etal.*[4] using a spray-dried nitrate technique. A similar pattern for this compound was found for sample 3 which was synthesized using BaO$_2$ in O$_2$. See Figure 3. This diffraction pattern showed well resolved peaks for the cubic 143 structure as well as minor impurities of BaCuO$_2$ and Y$_2$Ba$_1$Cu$_1$O$_5$ "211" (the green phase). The presence of impurity phases in Figure 3 can be attributed to the slow kinetics of the solid state reaction.

Figure 4 displays the effect of carbonate on the ability for the cubic 143 phase to form. Sample 4 shows presence of the 142 compound with additional phases of BaCuO$_2$ and 211. It is apparent that the presence of carbonate facilitates the formation of the 142 phase. The results of phase analysis for samples 1-4 are summarized in table 2.

A very quick and empirical method of determining the presence of a carbonate was to perform a bubble test. If the 142 phase contained carbonate it would bubble when exposed to strong acid. This was indeed the case as 142 bubbled readily when 1 N nitric acid was dripped onto the powder. The black 142 powder was completely dissolved by the acid.

A wet chemistry analysis of carbon confirmed the presence of carbonate in the 142. Results showed the carbon presence in the 142 as 1.2 wt.%. Based on the stoichiometry, this works out to correspond to one carbon per unit cell within experimental error.

Measurement of the electrical properties of the 143 and 142 compounds showed the typical resistive behavior for a ceramic. Pellets of the 142 compound did not display Meissner effect at 77K. Bars of the 143 and 142 electroded for T_c measurement showed resistivity values in the $M\Omega \cdot cm$ range from 298 to 10 K.

The affinity of barium for carbonate is well known. Recent reports by Shaw *etal.*[8] of carbonate presence in the 123 compound show interesting correlations to the results presented here for the 142 phase. The presence of carbon detected in the 123 phase was roughly 0.5 to 1.2 wt% carbon which corresponds to approximately 1 carbon per 123 unit cell. Significant suppression of T_c was reported for 123 samples containing carbon and it is speculated that the location of carbonate group causes a lowering of T_c[8]. Observations of highly resistive behavior in the 142 compound correlates well with the observations of the 123 superconductor. The lack of electrical conductivity in the 142 could be the result of carbonate content.

CONCLUSIONS

- The presence of a carbonate precursor is required for the formation of the 142 phase.

- Wet chemistry indicates the 142 compound is an oxycarbonate compound with a stoichiometry of $Y_1Ba_4Cu_2(CO_3)O_{7\pm\delta}$.

- Analysis of electrical properties in the 10-298 K temperature range showed no superconductive behavior for the 142 compound.

ACKNOWLEDGMENTS

We would like to thank the New York State Science and Technology Foundation and their Center for Advanced Ceramic Technology along with the New York State College of Ceramics for their sponsorship for this work. We are also indebted to the New York State Institute on Superconductivity and the Kodak corporation for their support.

References

[1] K. G. Frase, E. G. Liniger, D. R. Clarke, J. Am. Ceram. Soc., **70** [9] C204-C205 (1987).

[2] G. Wang, S. J. Hwu, S. N. Song, J. B. Ketterson, L. D. Marks, K. R. Poeppelmeier, and T. O. Mason, Adv. Ceram. Mats. **2**, (3B) 313-325 (1987).

[3] R. S. Roth, K. L. Davis, and J.R. Dennis, Adv. Ceram. Mats. **2**, (3B) 303-312 (1987).

[4] D. M. DeLeeuw, C. A. H. A. Mutsaers, C. Langereis, H. C. A. Smoorenberg and P. J. Rommers, Physica C **152** (1) 152 (1988).

[5] R. S. Roth, C. J. Rawn, F. Beech, J. D. Whitler, J. O. Anderson, in *Ceram. Supercond. 2, Res. Update, 1988*, edited by M. F. Yan (Am. Ceram. Soc.: Westerville, Ohio) p. 13-26.

[6] D. M. DeLeeuw, C. A. H. A. Mutsaers, R. A. Streeman, E. Frikkee and H. W. Zandbergen, Physica C **158** 391-6 (1989).

[7] M. A. Rodriguez, J. J. Simmins, P. H. McCluskey, R. S. Zhou and R. L. Snyder, in *Adv. in X-ray Anal.*, edited by R. Jenkins, T. C. Huang, and P. K. Predecki (Plenum Publishing Corp.) **32** 497-505 (1989).

[8] T. M. Shaw, D. Dimos, P. E. Batson, A. G. Schrott, D. R. Clarke and P. R. Duncombe, J. Mater. Res., **5** [6] 1176-84 (1990).

RUDDLESDEN-POPPER TYPE PHASES IN THE La-Cu-O SYSTEM: A COMBINED THEORETICAL AND EXPERIMENTAL STUDY

Anurag Dwivedi, Mark A. Rodriguez, Yolande Berta, and A. N. Cormack
Institute for Ceramic Superconductivity
New York State College of Ceramics
Alfred University, Alfred, NY 14802.

ABSTRACT

Ruddlesden-Popper (R-P) type compounds in the parent La-Cu-O system have been reported to exist up to the $n=7$ series member of $La_{n+1}Cu_nO_{3n+1}$ [1]. La_2CuO_4 is the first member of the series and $LaCuO_3$, the $n = \infty$ member. The other members of the series can be regarded as combinations of these two end members.

The crystal structure of La_2CuO_4 has been well characterized but the structural information on the other members of this series is lacking in the literature. An atomistic computer simulation was performed to predict the equilibrium structures for the first five members of this series. XRD patterns were then calculated to aid in analyzing the experimentally prepared samples of higher order phases.

The ability to synthesize structurally stable bulk compounds of the $La_{n+1}Cu_nO_{3n+1}$ homologous series using a solid state and nitrate preparation technique has been investigated. Samples with starting stoichiometries corresponding to series members were reacted in both air and oxygen atmospheres. Results indicate bulk structural instability of series members higher than $n = 1$.

1   INTRODUCTION

Recent oxygen nonstoichiometry studies [2] suggest that the undoped, oxygen rich phase $La_2CuO_{4.08}$, shows superconductivity with T_c similar to that of doped La_2CuO_4 [3]. Stoichiometric La_2CuO_4 is, however, found to be nonsuperconducting [2]. This indicates that the excess oxygen provides the same level of doping of electronic defects (e.g. holes) as metal-ion substitution. Oxygen nonstoichiometry, thus, seems to play an important role in the superconducting phenomenon in this system. The nature of accommodation of excess oxygen in the La_2CuO_4 lattice is thus of immediate interest to crystal and defect chemists.

Some possible mechanisms for accommodating the excess oxygen in the structure are oxygen interstitial defects, metal vacancies, and the formation of Ruddlesden-Popper (R-P) type phases. Jorgensen *et al* have established the oxygen interstitial mechanism to be operative in La_2CuO_4 [4] and its nickelate analog La_2NiO_4 [5]; however, the existence of R-P type compounds has also been reported in La-Cu-O [1] and La-Ni-O systems [6].

The homologous series $La_{n+1}Cu_nO_{3n+1}$ in the La-Cu-O system is bounded by La_2CuO_4 ($n = 1$), and $LaCuO_3$ ($n = \infty$) members. These structures can be considered as a series of alternating perovskite and rock-salt layers. For higher order members in the $La_{n+1}Cu_nO_{3n+1}$ series perovskite blocks become richer in Cu-O layers whereas the size of the rock-salt blocks remains the same throughout the series.

Davies and Tilley [1] were first to report the existence of a Ruddlesden-Popper series $La_{n+1}Cu_nO_{3n+1}$ in the La-Cu-O system where they observed members as high as $n = 7$ by HRTEM. Lattice images showed intergrowths typical of higher order members but no powder X-ray diffraction patterns were reported illustrating bulk synthesis of $n > 1$ phases. Sekizawa et al [7] reported two phase mixtures for solid state samples having $n > 1$ compositions which were subjected to 1000°C in air. Picone et al [8] also observed two phase mixtures of La_2CuO_4 and CuO for solid state samples subjected to 1000°C in flowing oxygen. These results are not inconsistent since Davies and Tilley [1] also report a two phase mixture of La_2CuO_4 and CuO for their 950°C samples and only in their 900°C samples did they claim to observe higher series members.

Other reports have left some ambiguity concerning the ability to form bulk samples of $n > 1$ phases. The only published powder X-ray diffraction patterns for higher order members were reported by Torrance et al [9]. In this case no discussion of sample preparation technique or analysis was given. Also, atomistic computer simulation techniques [10,11] have predicted that the higher order members are energetically unstable with respect to La_2CuO_4, La_2O_3 and CuO.

In the present study, model structures for the first five members of the homologous series $La_{n+1}Cu_nO_{3n+1}$ have been developed, based on a simplified tetragonal unit cell, in a similar manner to the scheme used by Udaykumar and Cormack [12] for the Sr-Ti-O system. Atomistic computer simulation was employed and these model structures were allowed to relax to a minimum energy configuration using an appropriate potential model. The equilibrated structures of the first five compounds of the series were then used to calculate their powder XRD patterns. These calculated XRD patterns could serve as standards and may be used to aid in the identification of the synthesized higher order compounds since there is no other structural information available.

The structural stability of the $La_{n+1}Cu_nO_{3n+1}$ R-P type phases was studied by synthesizing the compounds using both a solid state and nitrate preparation technique. The chemical (nitrate) preparation was performed to achieve better reactivity as compared to conventional solid state methods which tend have very slow diffusion rates. This study was undertaken to determine if these phases could be synthesized in nearly ambient conditions as bulk compounds.

2   COMPUTATIONAL TECHNIQUES

The calculations are based on the Born model of the solid, which treats the solid as a collection of point ions with short range repulsive forces acting between them. This approach has enjoyed a wide range of success, although it has been found that the reliability of the simulations ultimately depends on the validity of the potential model used in the calculations. Detailed discussions on the simulation techniques have been presented by Catlow [13], and in the monograph edited by Catlow and Mackrodt [14]. Brief descriptions of interatomic potentials and lattice energy calculations are given below.

2.1   Potential Models

2.1.1   Pair Potentials

In the present study, the potentials used are described by a simple analytical function, the Born-Mayer central-force potential, supplemented by an attractive r^{-6} term, i. e.

$$V_{ij}(r_{ij}) = A_{ij}exp\left(-r_{ij}/\rho\right) - C_{ij}r_{ij}^{-6} \tag{1}$$

2.1.2   Shell Model

The polarizability of individual ions is included via the Shell Model of Dick and Overhauser [15] in which the outer valence electron cloud of the ion is simulated by a massless shell of charge Y and the nucleus and inner electrons by a core of charge X. The total charge of the ion is, thus, X+Y which equals the oxidation state of the ion. The interaction between core and shell of any ion is harmonic with a spring constant, k, given by:

$$V_i(r_i) = \frac{1}{2} k_i d_i^2 \tag{2}$$

where d is the relative displacement of core and shell of ion i.

For the Shell Model, the value of the free-ion electronic polarizability is given by:

$$\alpha_i = Y_i^2/k_i \tag{3}$$

2.1.3   Three Body Potentials

The three body potential form used in these calculations is given by:

$$V(r) = \frac{1}{2}K_3(\theta - \theta_0)^2 \tag{4}$$

where K_3 is a three body 'angle restoring' constant and $(\theta - \theta_0)$ is the deviation from the equilibrium bond angle θ_0. These potentials are used in order to include directionality of the bonds, especially those involving copper.

The potential parameters A, ρ, and C in Equation-1, the shell charges, Y and spring constant, k, associated with the shell-model description of polarizability and three body constant, K_3, have been derived by the procedure of 'empirical fitting', i.e. these parameters are adjusted by a least-squares fitting routine, so as to achieve the

best possible agreement between calculated and experimental crystal properties, in this
case, the structural features of the binary and superconducting compounds.

A three body potential model for Cu-O derived by Islam *et al* [16] was used in
preference to the pair potential model derived by us earlier [17]. Various potential sets
used in the present study are given in Tables I, II, and III. We had to modify the La-O
potentials for the change in coordination as will be described later. Lewis and Catlow
[18] discussed in greater detail the derivation of the potential parameters for oxides.

2.2   Lattice Energy Calculations

The lattice energy is the binding or cohesive energy of the perfect crystal (per unit cell
or per formula unit). It is of central importance in treating thermochemical properties
of solids and in assessing the relative stabilities of different structures. Moreover, its
derivatives with respect to elastic strain and displacement are related to dielectric,
piezoelectric and elastic constants and phonon dispersion curves.

The lattice energy is calculated in the Born model of the solid by the relation

$$U = \frac{1}{2} \sum_i \sum_j V_{ij} \tag{5}$$

where the total pairwise interatomic potential, V_{ij}, is given by:

$$V_{ij}(r_{ij}) = q_i q_j / r_{ij} + A_{ij}\ exp(-r_{ij}/\rho_{ij}) - C_{ij} r_{ij}^{-6} \tag{6}$$

with the first term representing the coulombic interactions between species i and j, and
the last two terms, the non-coulombic short range contributions discussed above. The
lattice energy is thus calculated exactly and the only limitations in the procedure arise
from a lack of precise knowledge of the interatomic potentials.

Calculation of the equilibrium atomic configuration involves adjusting the coordinates until the internal basis strains (i.e. the net forces acting on a species) are totally
removed. Details of the procedure have been outlined by Cormack [19] and Parker [20].

2.3   Calculation of XRD Patterns

Since no experimental structural information has been reported for higher series members, calculated X-ray powder diffraction patterns were determined using atomic positions based on computer simulated minimum energy structures. A local version of
the program POWD10 [21] was used to calculate the diffraction patterns. Anomalous dispersion corrections were applied to all atoms, and all calculations used Smith's
tabulation of the Cromer and Weber [22] atomic scattering factors. The calculations
used $K_{\alpha 1}$ radiation with a wavelength of 1.5405981 Å. The calculated intensities were
derived from a Cauchy profile with a width at half-height of 0.07° at $2\theta = 40°$. This
sharp half-width was chosen so that all theoretically resolvable lines would be seen.

Table I: Short Range Potential Parameters Used in the Present Work:

Potential form: $V(r) = A\ exp(-r/\rho) - C\ r^{-6}$

Interaction	A	ρ	C	Source
$Cu^{2+}-O^{2-}$	294.15	0.40023	0.000	[16]
$Cu^{3+}-O^{2-}$	583.93	0.35402	0.000	[16]
La[6]-O	1644.98	0.36196	0.00	[16]
La[9]-O	1575.0	0.36196	0.00	[16],Present work
La[12]-O	2066.0	0.36196	0.00	[16],Present work
$O^{2-}-O^{2-}$	22764.200	0.1490	43.00	[16,18]

Table II: Shell-Model Parameters Used in the Present Study:

Interaction	Shell Charge (Y)	Spring constant (k)	Source
Cu^{2+}(Core)-Cu^{2+}(Shell)	1.000	∞ (Rigid Ion)	[16]
Cu^{3+}(Core)-Cu^{3+}(Shell)	0.000	∞ (Rigid Ion)	[16]
La(Core)-La(Shell)	-0.25	145.0	[16]
O(Core)-O(Shell)	-2.389	42.0	[16]

Table III: Three-Body Potentials Used in the Present Study:

Bond	K_3	θ_0	Source
O(1)-Cu^{2+}-O(1)	2.1889	90	[16]
O(1)-O(2)-Cu^{2+}	3.2813	38	[16]
O(1)-Cu^{3+}-O(1)	2.1889	90	[16]
O(1)-O(2)-Cu^{3+}	3.2813	38	[16]

3   EXPERIMENTAL PROCEDURE

Conventional solid state samples were prepared by mixing stoichiometric ratios of binary oxides CuO and La_2O_3 using a mortar and pestle made of alumina. Samples were fired at 850°C and 900°C for 96 hours in air with intermittent grindings.

For chemical synthesis a nitrate preparation was performed. Solutions of lanthanum and copper nitrate were prepared and analyzed to determine cation concentration. Stoichiometric volumes were titrated, mixed and atomized onto a hotplate heated to approximately 400°C. The resulting green-brown compound was then placed in alumina crucibles and heated to 800°C. After the ashing step, samples fired in air were heated to 900°C for 72 hours. Samples fired in oxygen were heated to 900°C for 48 hours.

The X-ray data for the lanthanum cuprate compounds were collected using a Siemens D500 diffractometer with a diffracted beam graphite monochromator using Cu K_α radiation. Patterns were taken from 2-67°2Θ with a count time of 1 second and step size of 0.050°2Θ. Powder samples were mounted on top loaded sample holders.

4   RESULTS AND DISCUSSION

4.1   Reliability of the Interatomic Potentials

The first step in any atomistic simulation is to test the reliability of the potential model used in the study. In a systematic investigation of potentials for CuO, the pair potential model was found to be inappropriate and rather a three body form which takes the covalancy of the bonds into account, should be used. Also La-O potentials, which were derived for La_2O_3, could be transferred to a model for La_2CuO_4 only after appropriate modification in pre-exponential repulsive parameter 'A'. This modification is required due to the fact that La ions are in six-fold coordination in the hexagonal La_2O_3 whereas they enjoy a nine-fold coordination in La_2CuO_4. The higher coordination implies a higher La-O nearest neighbor distance and thus a larger 'A' potential parameter. The coordination of La increases further in the perovskite type $LaCuO_3$ where it is surrounded by 12 oxygen ions. It is important to mention here that in R-P type compounds corresponding to $n > 1$ there are two different types of La sites: one which is in perovskite type blocks (12-fold coordinated) and one at the interface of the rocksalt and perovskite block which is 9-fold coordinated (Fig-1).

We modified the 'A' parameter of La-O short range potential to account for this change in the La coordination (Table-I). We raised the La[6]-O 'A' value to match the structural properties of La_2CuO_4 in order to obtain an appropriate Potential set for La[9]-O interaction. Similarly a hypothetical cubic perovskite structure of $LaCuO_3$ is required to calculate suitable La[12]-O short range potential set. It is difficult to synthesize the compound $LaCuO_3$ in pure form unless excess oxygen is provided [23]. $LaCuO_3$ synthesized in excess oxygen has been reported to have a rhombohedral structure, whereas in R-P type compounds, it is in the form of cubic perovskite layers.

Crystal structure of $LaCuO_3$ reported in the literature [23], thus, could not be used for deriving La[12]-O repulsive parameter, 'A', accurately.

In the present study we derived the La[12]-O repulsive parameter by assuming that the lattice parameter of the cubic cell is the same as its rhombohedral counterpart. Excellent agreement of the computer simulated structure with XRD structures reported in the literature [24] for La_2CuO_4 (Table-V and Fig-2) is indicative of the reliability of the potentials used in this study. Due to the unavailibility of any structural information of members $n > 1$, the reliability of La[12]-O potentials could not be tested.

4.2   Model Structures of Compounds ($La_{n+1}Cu_nO_{3n+1}$)

Model structures for the first five members of the homologous series were generated. This was done by adding a Cu-O layer to the perovskite slab of each member to create the next highest n member. These series members have very similar structural trends as are seen in the system Sr-Ti-O [12]. But we observe that the perovskite and rocksalt blocks in the Sr-Ti-O compounds are electrically neutral, whereas in the La-Cu-O compounds they are not. These model structures were based on simplified tetragonal structures of La_2CuO_4. The stoichiometries and some structural features of these compounds are reported in Table-IV.

The location of holes was assumed to be on Cu sites (since we can not treat them otherwise) which converts some of the Cu^{2+} to Cu^{3+} (Table-IV). It is important to note here that the location of Cu^{3+} with respect to Cu^{2+} ions should first be derived in order to generate the realistic structures. There is absolutely no clue in the present literature about this. We equilibrated the structures of the first few members of the series using various possible Cu^{2+}/Cu^{3+} distributions. The energy for each of the combinations proved to be more or less the same. Trends in the calculated dielectric constants helped us determine the sites occupied by Cu^{3+} in the members $n > 1$. Model structures of n=1 and n=2 members are shown in the Figures 1a and 1b.

4.3   Equilibrated Structures and Their X-Ray Diffraction Patterns

These model structures were relaxed using an energy minimization technique to give a final equilibrated structure. These equilibrated structures contract along the 'c' direction with respect to their model structures. This contraction of model structure (c_{model} - $c_{equilibrated}$) increases with n. The model and equilibrated structures are reported for the first and second members of the series in Table V and VI, respectively. The results for the equilibrated structures for the higher order members will be presented elsewhere [25]. For the $n = 1$ compound, it is quite encouraging to note that atoms at sites La(4e) and O(4e) in the model structure relaxed appropriately during equilibration.

The $n = 1$ La_2CuO_4 calculated pattern (Fig-2) shows good agreement in comparison to the observed pattern with the exception that the computer calculated patterns did not predict an orthorhombic splitting observed in the experimental data; this is because computer simulated structures were simplified to tetragonal rather than the

Table IV: **Some members of R-P homologous series of La-Cu-O compounds:**
Generic Formula: $La_{n+1}Cu_nO_{3n+1}$

'n' in formula	Compound	Cu^{+3}/Cu^{+2} ratio	No. of atoms per unit cell	c (Model) (Å)
1	La_2CuO_4	None/1	14	13.2487
2	$La_3Cu_2O_7$	1/1	24	22.0812
3	$La_4Cu_3O_{10}$	2/1	34	30.9136
4	$La_5Cu_4O_{13}$	3/1	44	39.7461
5	$La_6Cu_5O_{16}$	4/1	54	48.5786
6	$La_7Cu_6O_{19}$	5/1	64	57.4110
7	$La_8Cu_7O_{22}$	6/1	74	66.2435
8	$La_9Cu_8O_{25}$	7/1	84	75.0760
9	$La_{10}Cu_9O_{28}$	8/1	94	83.9084
10	$La_{11}Cu_{10}O_{31}$	9/1	104	92.7409
∞	$LaCuO_3$	1/None	5	a=b=c=3.9

Table V: **Model, Equilibrated, and XRD Structures of La_2CuO_4.**
Space Group: I4/mmm (No. 139)

Parameter	Model	Equilibrated	XRD [24] (For $La_{1.85}Ba_{0.15}CuO_4$)
Lattice Parameters:			
$a (= b)$ Å	3.7817	3.78996	3.7817
c Å	13.2487	13.21226	13.2487
Co-ordinates: **(S.G. I4/mmm)**			
Cu^{2+}: 2a	z=0.0000	z=0.0000	z=0.0000
La^{3+}: 4e	z=0.3333	z=0.3619	z=0.3607
O^{-2}: 4c	z=0.0000	z=0.0000	z=0.0000
O^{-2}: 4e	z=0.1667	z=0.1811	z=0.1824

Table VI: Model and Equilibrated Structures of n=2 member ($La_3Cu_2O_7$).
Space Group: I4mm (No. 107)

Parameter	Model	Equilibrated
Lattice Parameters:		
$a (= b)$ Å	3.7817	3.7991
c Å	22.0812	21.6129
Co-ordinates: (S.G. I4mm)		
Cu^{2+}: 2a	z=0.9	z=0.8765
Cu^{3+}: 2a	z=0.1	z=0.0745
La^{3+}: 2a	z=0.3	z=0.2990
La^{3+}: 2a	z=0.5	z=0.4742
La^{3+}: 2a	z=0.7	z=0.6634
O^{-2}: 4b	z=0.1	z=0.0852
O^{-2}: 4b	z=0.9	z=0.8802
O^{-2}: 2a	z=0.0	z=0.0
O^{-2}: 2a	z=0.2	z=0.1849
O^{-2}: 2a	z=0.8	z=0.7748

true orthorhombic symmetries. One should note that this assumption is not too severe since this structure has very little orthorhombicity ($a \approx c$). In a $\sqrt{2} \times \sqrt{2}$ type unit cell, dimensions are reported to be [26] 5.406 Å× 5.370 Å× 13.15 Å.

Calculated XRD patterns of $n = 1, 2, 3, 4$, and 5 members are shown in Fig-3. For higher members, the structures have not been experimentally resolved yet. Therefore the synthesis of the $n = 2$ and higher order members was attempted and analysed using our calculated XRD patterns. The regions around 25° and 40°2Θ as well as superlattice peaks at low angle show characteristic differences in the patterns of the different members. These differences would aid in identification of particular members in experimental diffraction patterns.

One should keep in mind that any inaccuracy in interatomic potentials would reflect in the equilibrated structures and further in calculated XRD patterns. Reliability of La[12]-O short range potentials used in the present study is still to be proved as there is absolutely no structural information available for comparing the results of members $n > 1$.

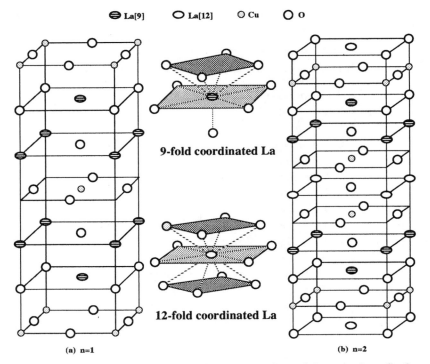

Figure 1: Model crystal structures of the first two members of the series $La_{n+1}Cu_nO_{3n+1}$

4.4   Synthesis of R-P type Compounds

The results for the solid state samples prepared at 850°C for 96 hours in air atmosphere appear in figure 4. The air fired $n = 2, 3, 4$ samples display well resolved patterns for the $n = 1$ La_2CuO_4 phase as well as CuO (tenorite) and a trace phase of La_2O_3. The peak intensity for the copper oxide phase increases with n.

The results for the solid state n=3 sample prepared at 900°C for 96 hours in air atmosphere appear in figure-5. This pattern displays multiple unidentified peaks near the highest intensity peak for La_2CuO_4. These peaks, initially, were speculated to be due to the coexistance of higher order R-P phases, however, the higher order phases could not be identified unambigously.

To confirm the presence of $n > 1$ phases in the above $n = 3$ sample, HRTEM was performed. We did not observe lattice images of any higher order phases. Images, however, showed high concentration of dislocations in this sample.

The results for the nitrate prepared samples at 900°C in air and oxygen atmospheres appear in figures-6 and 7, respectively. The air fired $n = 2, 3$ samples shown in figure 6 displayed well resolved patterns for the $n = 1$, La_2CuO_4, phase as well as CuO (tenorite) and a trace phase of La_2O_3. The peak intensity for the copper oxide phase increased with n. The oxygen fired nitrate prepared samples displayed in figure-7 also showed powder diffraction patterns containing the $n = 1$ La_2CuO_4 phase and the tenorite phase increasing in concentration with the higher copper containing stoichiometries.

Synthesis studies reveal that the higher order members $La_{n+1}Cu_nO_{3n+1}$ are not stable in ambient conditions. Energetics (reported elsewhere [11,25]) of these compounds also suggest thermodynamic instability of these phases with respect to La_2CuO_4, CuO and La_2O_3. Kinetic processes may, however, outweigh the thermodynamical driving force for decomposition and it may be possible to produce these homologous series compounds (e.g. as seen by Tilley [1]) if prepared in special conditions e.g. very high oxygen pressure. Calculations suggest that these compounds, if produced, will be in a metastable in ambient conditions form and hence may have limited technological application.

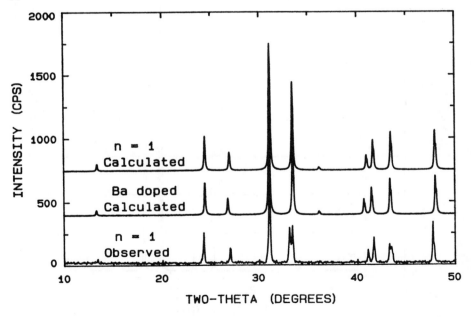

Figure 2: Comparision of Calculated and Experimental XRD patterns for La_2CuO_4.

5   CONCLUSIONS

The calculated XRD patterns show superlattice peaks and other obvious differences between the patterns of these compounds to differentiate between various R-P compounds. These patterns can be used to identify the experimentally prepared samples of higher order members.

X-ray diffraction patterns indicate the existence of La_2CuO_4 and CuO in 850°C and 900°C samples with $n > 1$ stoichiometries; the CuO concentration increases with n.

Based on our synthesis results, undoped $La_{n+1}Cu_nO_{3n+1}$ homologous series members higher than $n = 1$ (La_2CuO_4) do not appear to be structurally stable as bulk phases under conditions of air or 1 atm oxygen. This is consistent with the previous study on the energetics of the R-P compounds in the La-Cu-O system [11,25], which suggests that higher order members are not thermodynamically stable and will tend to decompose to the end members La_2CuO_4 and $LaCuO_3$. The $LaCuO_3$ phase, being unstable under ambient conditions, is likely to further decompose to the binary oxides which is consistent with our observation of excess CuO in $n > 1$ stoichiometry samples.

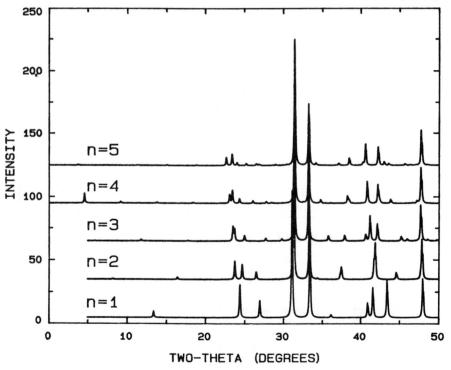

Figure 3: Calculated XRD patterns for first five members of series $La_{n+1}Cu_nO_{3n+1}$.

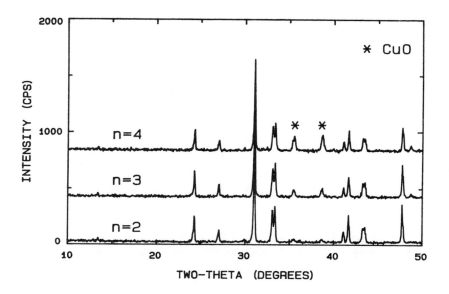

Figure 4: Powder X-ray diffraction patterns for $La_{n+1}Cu_nO_{3n+1}$ series compositions fired at 850°C for 96 hours in air prepared by solid state reaction.

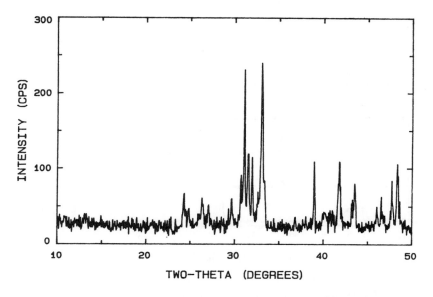

Figure 5: Powder X-ray diffraction patterns for n=3 series member fired at 900°C for 96 hours in air prepared by solid state reaction.

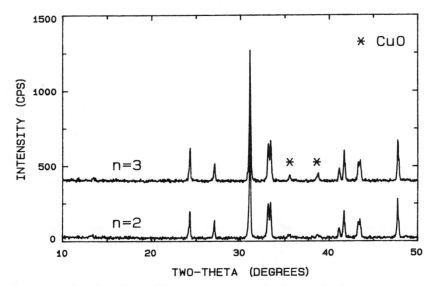

Figure 6: Powder X-ray diffraction patterns for $La_{n+1}Cu_nO_{3n+1}$ series compositions fired at 900°C for 72 hours in air. Samples prepared by nitrate method.

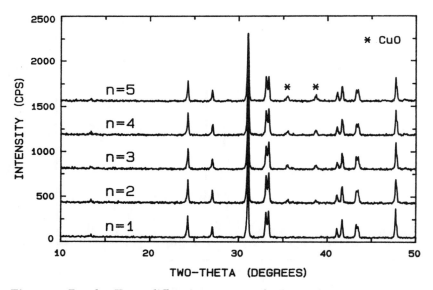

Figure 7: Powder X-ray diffraction patterns for $La_{n+1}Cu_nO_{3n+1}$ series compositions fired at 900°C for 48 hours in O_2. Samples prepared by nitrate method.

References

[1] Davies, A. H., and Tilley, R. J. D., Nature **326** 859-61(1987).

[2] Dabrowski, B., Hinks, D. G., Jorgensen, J. D., and Richards, D. R., Mater. Res. Soc. Symp. Proc. Vol. 156, p69.

[3] Bednorz, J. G., and Muller, K. A., Z. Phys. B**64**, 189-193 (1986).

[4] Jorgensen, J. D., Lightfoot, P., and Pei, Shiyou, invited manusccript submitted to the LT-19 Satellite Conference on HIGH TEMPERATURE SUPERCONDUC-TIVITY, Aug 13-15, 1990, Queens' college, Cambridge University, Cambridge, England. To be published in a special, supplementary issue of Superconductor Science and Technology.

[5] Jorgensen, J. D., Dabrowski, B., Pei, S., Richards, D. R., and Hinks, D. G., Phys. Rev. B, **40**, No-4, 2187(1989).

[6] Drennan, J., Tavares, C. P., and Steele, B. C. H., 1982, Mat. Res. Bulletin. **17**, 621(1982).

[7] Sekizawa, K., Takano, Y., Takigami, H., Tasaki, S., and Inaba, T., Jap. J. Appl. Phys., **26**(5), L840-1 (1987).

[8] Picone, P. J., Jenssen, H. P., and Gabbe, D. R., J. Cryst. Growth, **91**, 463-7 (1988).

[9] Torrance J. B., Tokura, Y., and Nazzal, A., Chemtronics, Vol-2, Sep., 120(1987).

[10] Dwivedi, A. and Cormack, A. N., The Proceedings of Ceramic Sc. and Tech. Congress in conjunction with the 22nd pacific coast regional meeting, Anaheim, CA. Oct 31 - Nov 3 (1989).

[11] Dwivedi, A. and Cormack, A. N., The Proceedings of the Intl. Conference on Superconductivity (to be published in a special issue of the Bulletin of Materials Science by Indian Academy of Sciences, Bangalore, India), Banglore, India, Jan 10-14 (1990).

[12] Udaykumar, K. R., and Cormack, A. N., J. Amer. Cer. Soc. **71**, C-469(1988).

[13] Catlow, C. R. A., in Mass Transport in Solids, edited by F. Beniere and C. R. A. Catlow, p1, Plenum (1983).

[14] Catlow, C. R. A. and Mackrodt, W. C., eds. Computer Simulation of Solids, Lecture Notes in Physics, 166, Springer, (1982).

[15] Dick, B. G. and Overhouser, A. W., Phys. Rev., 112[1], 90 (1958).

[16] Islam, M. S., Leslie, M., Tomlinson, S. M., and Catlow, C. R. A., J. Phys. C: Solid State Phys. **21**(1988), L109-L117.

[17] Dwivedi, A., and Cormack, A. N., in the Proceedings of third annual conference on superconductivity and application, Buffalo, NY, Sep 19-21, 1989. "Superconductivity and Applications", ed: H. S. Kwok, Plenum 1990.

[18] Lewis, G. V. and Catlow, C. R. A., 1985, J. Phys. C:Solid State Phys., **18**, 1149(1985).

[19] Cormack, A. N., Solid State Ionics, **8**, 187 (1983).

[20] Parker, S. C., Computer Modeling of Minerals, Ph. D. Thesis, Univ. of London, London, UK, (1983).

[21] D. K. Smith, M. C. Nichols, and M. E. Zolensky, *A Fortran IV Program for Calculating X-ray Powder Diffraction Patterns, version 10*, College of Earth and Mineral Sciences, Pennsylvania State University (1983).

[22] D. T. Cromer and J. B. Mann, Acta. Cryst. **A24** 321 (1968).

[23] Demazeau, G., Parent, C, Pouchard, M., and Hagenmuller, P., Mat. Res. Bull. Vol. 7, pp. 913-920, 1972.

[24] Jorgensen, J. D., *et al*, Phys. Rev. Letters, **58**, No. 10, 1024(1987).

[25] Dwivedi, A., Rodriguez, M. A., and Cormack, A. N., Manuscript in preparation.

[26] Grande, Von B., Muller-Buschbaum, Hk, and Schweizer, M., Z. anorg allg. Chem. **428**, 120(1977).

Applications

ENERGY APPLICATIONS OF SUPERCONDUCTIVITY

T.R. Schneider
Electric Power Research Institute, Palo Alto, CA 94303

S.J. Dale
Oak Ridge National Laboratories, Oak Ridge, TN 37831

S.M. Wolf
U.S. Department of Energy, Washington D.C. 20585

ABSTRACT

Recent progress in developing high-temperature superconductors has enhanced the economic viability of energy applications such as power systems, motors, material processing and handling, refrigeration, transportation, and power electronics. This paper discusses the technical and economic issues associated with these applications.

INTRODUCTION

The initial discovery by Bednorz and Muller of superconductivity at 35 K, followed by Chu and Wu's discovery of the so-called 1-2-3 compounds that superconduct at 95 K, caused great excitement in the scientific community. Great expectations of advances were reported in the popular press. Today, scientific excitement remains high, but the realists' view is that it may take many years to apply superconductivity in large-scale electrical equipment.

In 1989, tremendous scientific progress was achieved, including the discovery of whole new families of superconductors. Compounds based on bismuth and thallium, with transition temperatures near 100 K and 125 K respectively, were discovered in rapid succession in January and February 1989. Towards the end of the year, a new lead-based family with transition temperatures near 70 K was reported. Higher transition temperatures, some as high as room temperature (about 300 K), have been reported but not substantiated. The existence of several different families of materials has advanced scientific understanding of these new phenomena and offered the promise of yet higher operating temperatures.

While new materials have been discovered, and understanding of the structure and chemistry has been greatly advanced, a comprehensive theory has not yet emerged. However, a physical picture of the nature and behavior of the charge carriers in the new material is evolving.

Superconductors can be characterized by three important parameters: T_c, the critical transition temperature; H_c, the critical magnetic field; and J_c, the critical current density. Above specific values for each parameter, the superconductor reverts to its normal state. Below these critical values, the super-conductors exhibit the essential characteristics of persistent dc currents and perfect diamagnetism.

These new superconducting materials are able to superconduct above the boiling point of liquid nitrogen (77 K) and allow liquid nitrogen rather than liquid helium to be used as a coolant. This high critical temperature surprised the physics community and promised the possibility of broader use in everyday life. The second parameter, critical magnetic field, does not pose any serious technical obstacle because the measured values are higher than the levels required for currently conceivable applications.

The third parameter, critical current density, is the most challenging technological hurdle to clear for many energy applications of superconductivi-ty. Although the new superconductors have achieved very high -- and techno-logically exciting -- critical current density values ($> 10^6$ A/cm^2) in the form of small single crystals and single-crystal thin films (up to a few microns thick), the results for polycrystalline "bulk" samples have been less encouraging. Polycrystalline strands or tapes are the most logical route to fabricating a wire. Very recently, the highest performing superconductors in bulk form have achieved nearly 10^5 A/cm^2 under low-temperature (4.2 K) conditions, but far lower current densities at the higher liquid nitrogen temperatures.

INCENTIVES FOR DEVELOPING HIGH T_c SUPERCONDUCTORS

There are a number of sectors in the U.S. economy where potential energy savings through improved efficiency can be large. The total electric energy sales in the U.S. in 1988 were 2,566 x 10^9 kWh. Motors account for about 64% of the annual U.S. electric energy use, with motors in the size of 125 hp and above accounting for 47% of that consumption. The transporta-tion sector accounts for about 27% of the total U.S. energy consumption, and petroleum provides 97% of the energy used in this sector. Thus, there are significant potential energy savings if electric devices using high-temperature superconductors can be used in motor and transportation applications.

From an electrical energy conservation viewpoint, elimination of the resistive loads in the conductors of electrical equipment will have significant impact. Another major advantage from the use of superconductors is in the production of strong magnetic fields. The ability to apply magnetic fields above the normal 1-2 Tesla has the advantage that the iron core can be eliminated in devices such as motors, generators, transformers, and other magnet applications. This leads to reduced size and weight, in addition to increased efficiency due to the elimination of the iron core losses.

What may be the most important factor for evaluating the new high T_c superconductors for practical applications is that the critical temperature of these materials is above the boiling point of liquid nitrogen. Many of the potential applications of superconductors are ruled out as being impractical when cryogenic helium is required as a refrigerant. Liquid nitrogen will make some of these applications feasible. Both the availability and physical properties of nitrogen gives it a significant advantage over helium.

In the sections that follow, the particular issues relevant to each of the major study areas are discussed.

POWER SYSTEM APPLICATIONS

The electric utility industry may eventually become one of the largest users of superconductors. This industry has researched superconducting transmission cables, generators, and energy storage for more than two decades, and the latest developments have renewed interest in superconductors.

Transmission lines that can carry electricity enormous distances without resistive losses are one of the most promising potential applications for new superconductors. Losses from conventional transmission lines average about four percent, but by eliminating even those small losses, a superconducting line might pay for itself over its 40-year life. Transmission lines require relatively low current densities compared to other superconductor applications. And the flexibility requirement can be relaxed because transmission cable can be installed in underground rigid or semi-rigid individual sections. As for ac losses, a zero-resistance dc transmission line would completely eliminate the issue. However, only future research can determine if the enormous length and surface area of a superconducting transmission cable can be cooled and insulated economically.

Generators whose rotors are wound with superconducting wire of adequate current-carrying capacity can produce magnetic fields much higher than those produced by generators with copper windings. Superconducting generators would thus be perhaps only half the size of ordinary generators, and potentially less costly. Generators with low-temperature superconductor rotor windings have already been designed. However, the new superconductors may be too brittle for the rigors of generator operation and maintenance, or possibly too brittle to fabricate into rotor windings. The current density of the new superconductors must also be increased to levels of over 10^5 A/cm^2.

Energy storage is another exciting prospect for superconductors. Superconducting magnetic energy storage (SMES) systems would enable utilities to efficiently store electric energy for later use, thereby providing an alternative to pumped hydroelectric storage plants. Commercial-scale SMES plants would be based on a doughnut-shaped electromagnetic coil, perhaps hundreds of meters in diameter, wound with superconducting wire. Electricity flowing thorough this electromagnetic coil would generate a magnetic field that would store electric energy. Once a direct current is circulating in a superconducting electromagnetic coil, the current will persist indefinitely with no losses. The resulting magnetic field can be discharged for electric energy as needed at almost any power level without loss of efficiency. Overall SMES efficiency from ac input to ac output can be more than 90 percent, compared with pumped hydro's 70-75 percent. Although low-temperature superconductors have been shown to be technically feasible, high-temperature superconductors will require further advancement to be able to handle the very large current densities (>300,000 A/cm^2) required for energy storage.

MOTORS

None of the possible applications for superconductors would have more impact on energy productivity than electric motors, which account for about 64% of all electrical energy used in the United States. Superconducting windings could make it possible to significantly increase the magnetic field in the motor's air gap (and thus boost power) without necessarily adding to the motor's weight and volume. Overall, the reduction of losses from conventional windings could make superconducting motors extremely efficient.

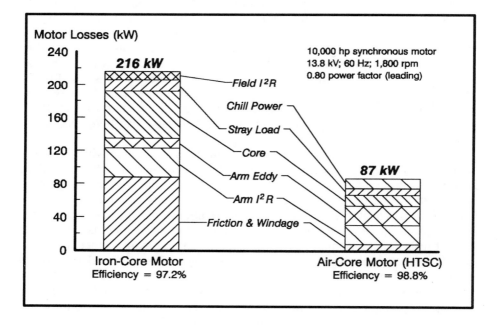

Figure 1: Motor Losses

In addition to eliminating iron losses in a motor, the elimination of iron from the motor design allows the use of substantially higher magnetic fields and a reduction in size and weight for a given performance rating. Also, the elimination of iron results in less variation of motor efficiency with load.

Earlier efforts indicate that there is no conceptual difficulty in applying superconducting technology to any standard motor configuration. The problems that limit the direct substitution of superconductor for conventional conductor in motor applications are associated with ac losses, mechanical integrity of superconductor and cooling systems, the need for helium cooling and associated high cost, and electro-mechanical stability in the motor during startup and load variation. With the exceptions of the largest dc homopolar and ac synchronous motors, ac losses and controller requirements effectively negate any energy savings resulting from the application of low-temperature superconductors.

The application of high-temperature superconductors in rotating machines would potentially provide benefits in the form of even higher efficiency, reduced size and weight, and reduced capital costs due to reduced refrigeration load and potentially higher operating fields.

Ac losses in the superconductor can be a major obstacle. However, an economic trade-off exists between ac losses and refrigeration capacity: Because less refrigeration capacity is required for the new superconductors, higher ac losses can be accepted without reducing the motor's cost-effectiveness.

ELECTROMAGNETIC PUMPING

Electromagnetic pumps operate by using the reaction between a magnetic field and a current passing through an electrically conducting fluid. Electromagnetic (EM) pumps have been employed as an alternative to mechanical pumps. Because EM pumps do not require seals and have no moving parts in contact with the fluid, they are more reliable than mechanical pumps, especially when handling liquid metals. The pump can also be oriented in any position.

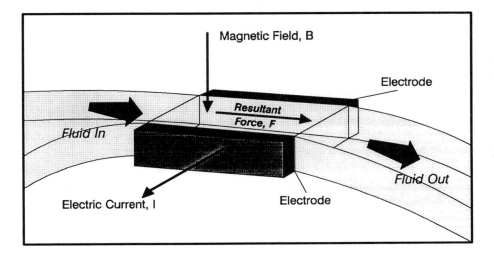

Figure 2: Principle of Electromagnetic Pump and Thruster

Current mechanical pumps generally have an efficiency in the 80-85% range. EM pumps have an efficiency in the 40-50% range. The current use of EM pumps is limited to pumping of electrically conducting fluids, such as aluminum, sodium, or lithium, which present special problems to conventional pumps. Many types of EM pumps have been designed and built, but none of them employ superconductors.

For pumping liquid metals, the use of superconducting windings results in a reduction of pump size (and mass) and an increase in pumping efficiency over an EM pump with conventional windings. The pump efficiency can be improved from about 44% for the conventional EM pump to 68% for the EM pump with a superconducting winding. Only the resistive losses were considered in this comparison and are set to zero for the superconducting winding. The analysis did not include the cryogenic refrigeration system for cooling of the rotor.

Superconducting EM pumps could find an economical application in transporting nonmetallic fluids that are difficult to handle because of their radioactivity, chemical reactivity, or corrosiveness and abrasiveness (e.g., abrasive slurries). In these cases, the improved reliability of an EM pump could more than offset its low efficiency relative to a mechanical pump. However, if such a fluid has an electrical resistivity of the same or greater order as that of sea water, the potential benefits gained by using superconducting EM pumps appear to be marginal.

MAGNETIC HEAT PUMPS AND REFRIGERATION

Magnetic heat pumps have been used routinely in the laboratory during the past fifty years to provide small amounts of refrigeration at extremely low temperatures (under 1 K). Around 1900, it was discovered that when certain paramagnetic or ferromagnetic materials were subjected to an intense magnetic field, alignment of their magnetic moments caused the material to warm. Conversely, removal of the magnetic field caused cooling. This behavior is called the magnetocaloric effect (MCE) and forms the basis of a class of Carnot-like heat engines called magnetic heat pumps (MHPs). The MCE is highest for certain rare-earth compounds operated around their Curie temperature, but is limited to changes in temperature of approximately 20 K (2 K/Tesla) which is achievable only for magnetic fields of 10 Tesla or more. Fields of this magnitude can best be obtained through the use of superconducting magnets.

The MHP is directly analogous to a conventional vapor-compression heat pump/refrigerator; the magnetic field corresponds to pressure and the magnetization corresponds to volume. Manipulation of the magnetic field can be accomplished in one of three ways: (1) raising or lowering the magnetic field, (2) moving the magnetic material relative to the magnet, or (3) moving a shield between the magnetic material and the magnet.

Most MHP development efforts to date are for cryogenic refrigeration at temperatures below 20 K. Cooling powers of watts are typical. For commercial interest, the MHP devices need to provide refrigeration and heat pumping around room temperature with capacities of 10s-100s of <u>kilo</u>watts. A number of experimental devices have been built using gadolinium as a working material because of its 293 K Curie temperature.

An economic comparison was made between a conventional vapor compression refrigeration and various MHPs to provide 50 kW of cooling at $0°F$ and 500 kw heating at $240°F$. The capital cost of the MHP is estimated to be 2-3 times the conventional refrigeration and 3-4 times for heating. However, due to the improved efficiency, the estimated operating costs (excluding cooling costs for the magnet) for the MHP are less.

Gadolinium represents a significant cost item for the MHP, and other materials should be investigated. Designs and prototypes of MHPs suitable for a large market should also be explored, and their feasibility should be determined using low-temperature superconductors.

MATERIALS PROCESSING

There are significant potential applications for superconductors in materials processing utilizing either modest magnetic fields of large volume or intense magnetic fields of small volume. Fields under 2 Tesla applied during the processing of polymers, ceramics, metals, and composites can strongly affect their properties. Intense fields of 10 Tesla or more can be used in gyrotrons to produce high frequency electromagnetic radiation that might be used to interact with matter in novel ways to enhance properties or improve processes.

Magnetic processing appears to be of potential value to a wide variety of material-driven industries: electronics, telecommunications, aerospace, automotive, etc. Magnetic fields can be used to manipulate the nature of growing crystals, orientation of components within a system (crystals, fibers, etc.), and the distribution of constituents within a system, such as dopants in semiconductors, conducting polymers, and glasses.

MAGNETIC SEPARATION

Magnetic separation is the use of magnetic forces to separate materials of various composition on the basis of their magnetic properties. Magnetic separation is important in terms of economy, efficiency, and energy productivity for the processing of ferrous ores/metals and kaolin clay, and the recovery of magnetic materials from a variety of types of solid waste and scrap. There are more uses of magnetic separation -- particularly those involving weakly magnetic materials -- just now being adopted for large-scale industrial use. These new applications include those for the removal of environmentally objectionable species from waste water and waste gases, and for the magnetic pretreatment of water to prevent carbonate scale formation in equipment such as dishwashers, heat exchangers, and steam boilers.

An important observation can be made that relates to the use of higher temperature superconducting materials for magnetic separation. Liquid helium superconducting magnetic separators have been found to be reliable and economically superior to resistive electromagnetic separators on a substantial industrial scale in the growing kaolin industry. It follows that if higher temperature superconducting separators (such as those that might be cooled with liquid nitrogen) can be made that are as reliable and comparable in capital cost, then they would automatically find a niche for a substantial industrial application. By analogy, higher temperature superconducting separators could displace resistive electromagnetic separators in other existing industrial applications and new applications that are being developed.

MATERIALS FABRICATION

The availability of practical superconductors at liquid nitrogen temperatures can make many new fabrication uses available through application of very high magnetic fields. Important information concerning the technical barriers can be obtained by experiments using low-temperature superconductors. Fields of 10-12 Tesla are readily available from such magnets.

This is sufficient pressure to deform a rod or tub of material of low-flow stress such as lead or tin. Experiments with these materials and with low-temperature superconducting solenoids could clarify the pressure/deformation relationships as a function of solenoid inner diameter/workpiece diameter. They could also clarify the relationship between workpiece velocity and the rate of penetration of flux lines into the workpiece surface. Studies of the reactions of the solenoid to the magnetic pressure variations would provide information in two areas: the reaction of the solenoid body itself to the imposed forces, and the electrical reaction to the increase magnetic field as the workpiece penetrates the flux lines.

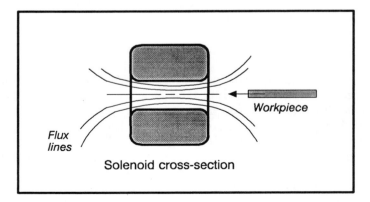

Figure 3: Schematic of Workpiece and Solenoid Relationship

TRANSPORTATION

The new MAGLEV vehicle technology is perhaps the most developed of all transportation technologies for superconductor applications. The Japanese repulsive-force-levitated vehicle originally developed under the auspices of the Japanese National Railway uses low-temperature superconductors. Therefore, high-temperature superconductors must be regarded as enhancing rather than enabling this technology. The use of superconductivity allows the MAGLEV vehicle to operate with a large space gap (10-15 cm) between the vehicle base and the guideway. With such a large gap, the vehicles can travel at speeds up to 500 km/h or more without excessively tight tolerances on guideway alignment.

Use of high-temperature superconductors would provide a considerable reduction in cost and complexity of the cryogenic system, as well as savings in operating energy and mass. Reliability would also significantly increase. For a typical vehicle route, the energy savings is potentially on the order of 5-15%. However, the superconducting magnets represent only a fraction of the overall MAGLEV cost.

Electromagnetic thruster systems for ship propulsion have been investigated previously and small models have been built to demonstrate the concept. In the EM thruster, an electrical current is passed directly through the sea water and interacts with an applied magnetic field to propel the vessel forward. The design of the EM thruster involves considerations similar to those associated with the construction of a dc conduction EM pump.

POWER ELECTRONICS

With the advent of modern power semiconductor devices, such as the transistors and thyristors introduced thirty years ago, power electronics have played an increasing role in the control of electric power devices. Two of the most common applications of power electronic devices are for motor drives and switching-mode power supplies.

The major potential energy savings with superconducting power electronics devices will be in motor controllers. Any superconducting motor will require power electronic controllers. Motor controllers that use semiconductor devices consume about 2% of the energy used by the motor, whereas motor controllers using superconducting "transistors" would, in principle, consume almost no energy. It is estimated that currently over 120×10^9 kWh of electric energy are consumed by motor controllers each year. This is likely to increase with the expanded use of power electronic controllers.

Among the concepts considered for a superconducting power electronic switch, the thin film superconducting switch offers the most promise because of its relatively high power and voltage capabilities and superior controllability. In its simplest conceptual form, it consists of a thin film of superconducting material capable of being switched from the superconducting (on) state to the normal (off) state and vice versa. The film cross section is governed by the current to be conducted, and the length of the device is governed by the resistance needed to minimize power losses in the normal state. The preferred method for turning such devices "on" and "off" involves the use of an electric field to modulate the superconductor free energy and, hence, its critical temperature, T_c.

Made of indium and indium oxide film, Ta-doped $SrTiO_3$ and $Ba_xSr_{1-x}TiO_3$ superconductors have achieved complete field-induced switching. However, the critical temperature of these materials is in the range of 1 K, making them impractical.

Although these materials are not quite adequate to make a superconducting motor controller economically viable, they should be used to develop switching techniques that will be directly applicable to new materials when they become available.

SUMMARY

From this survey and assessment of applications, it has become clear that significant opportunities for energy applications of superconductors exist today and that the range of applications will greatly increase when new superconductors become practical. Several applications are entering the marketplace today using conventional low-temperature superconductors. In particular, low-temperature superconductors in magnetic separation, magnetic levitated trains, and magnetic processing of materials are establishing a record of success in both lowering energy costs and achieving technical results not readily accomplished without superconductivity.

Further development of these technologies will occur even without the new superconductors. These applications do assure a ready market for the new materials once they become available. Successful development of the new superconductors will certainly expand the range of economic application.

The magnetic field required for many of these applications are readily achievable by the new superconductors. However, the desired critical current densities at liquid nitrogen temperatures are only achieved by thin films or single crystals. There is no consensus as to whether high values of J_c can be achieved in bulk samples of these materials at liquid nitrogen temperatures. Figure 4 identifies desirable targets and actual superconductor performance at 4.2 K (liquid helium temperature).

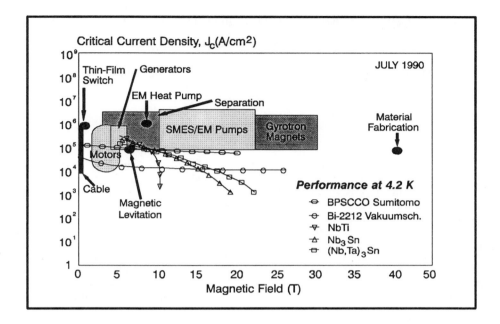

Figure 4: Performance Targets for Energy Applications

Brittleness and other poor mechanical properties of high-temperature superconductors also remain important obstacles to their use in practical devices. High-temperature superconductivity appears to call for layered copper oxide structures, and macroscopic samples of such crystals exhibit poor mechanical properties.

While it is still too early to make reliable market forecasts, the successful development of the new higher-temperature superconductors is certain to affect electric energy use in many different ways. Direct substitution in current electrical apparatus will reduce energy use. New applications, typically utilizing magnetic fields, will add further value to the use of electric energy. The magnitude of the impact of this remarkable discovery will depend upon our ability to reduce this scientific breakthrough to engineering practicality.

HIGH TEMPERATURE SUPERCONDUCTIVITY SPACE EXPERIMENT (HTSSE): PASSIVE MILLIMETER WAVE DEVICES

D. U. Gubser, M. Nisenoff, S. A. Wolf, J. C. Ritter and G. Price
Naval Research Laboratory
Washington DC 20375-5000

ABSTRACT

The Naval Research Laboratory (NRL) is exploring the feasibility of deploying high temperature superconductivity (HTS) devices and components in space. A variety of devices, primarily passive microwave and millimeter wave components, have been procured and will integrated with a cryogenic refrigerator system and data acquisition system to form the space package, which will be launched late in 1992. This Space Experiment will demonstrate that this technology is sufficiently robust to survive the space environment and has the potential to significantly improve space communications systems. The devices for the initial launch (HTSSE-I) have been received by NRL and evaluated electrically, thermally and mechanically and will be integrated into the final space package early in 1991. The performance of the devices will be summarized and some potential applications of HTS technology in space system will be outlined.

INTRODUCTION

The designers of satellite systems, both civilian and military, demand the maximum of system capabilities at a minimum of weight, volume and electrical input power. Superconductivity is an electronic and electrical technology which provides the ultimate in performance at a minimum of electrical input power. This latter fact has been known to spacecraft designers for more than 20 years. However, all superconducting materials known before 1986 had to be operated at temperatures in the vicinity of 4 K, the boiling point of liquid helium, This required a very sophisticated, heavy and bulky cryogenic enclosure for typical spacecraft mission of five to ten years. Therefore, the spacecraft designers had very little interest in superconductivity. Nevertheless, there have been a number of scientific space missions, such as the Infra-Red Astronomical Observatory (IRAS), which were flown despite the burden of liquid helium refrigeration since these missions required the ultimate in performance.

The advent of superconductivity at temperature near and above that of liquid nitrogen, 77 K has dramatically reduced the cryogenic refrigeration burden associated with the use of superconductivity in space systems. A plot of superconducting transition temperature as a function of time is shown in Fig. 1.

Using passive cooling techniques, under certain spacecraft conditions, temperatures down to about 70 K to 100 K can be obtained. Realizing the significance on this spectacular breakthrough in superconductivity on future spacecraft technologies, the U.S. Naval Research Laboratory (NRL) initiated a program in 1988 known as the High Temperature Superconductivity Space Experiment (HTSSE) which is an extremely focussed and coordinated program whose goal is to demonstrate the feasibility of incorporating this revolutionary technology into space systems.

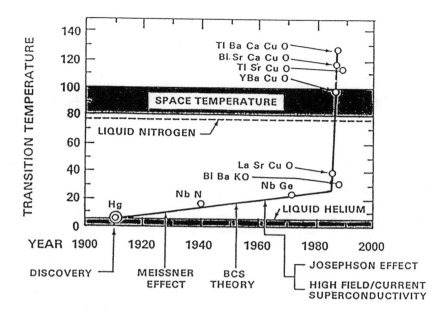

Fig. 1. A plot of the highest known transition temperature for superconductors as a function of time. Along the time axis several milestones in the history of low temperature superconductivity are noted.

The expected payoffs from this Space Experiment are threefold:
(1) a definite demonstration that high temperature superconductivity would make possible breakthroughs in spacecraft operational capability.
(2) prepare an experimental package containing HTS devices and components ready for flight testing in 1992, and,
(3) the results of the several Space Experiments (after the initial experiment in 1992, it is envisioned that there would be subsequent experiments, roughly every two years) would enable spacecraft designers to evaluate the benefits of using superconducting components in all of their systems thus revolutionizing spacecraft design.

In order to achieve these goals, and to speed the transition of HTS devices from the laboratory into space systems, the HTSSE program has attempted to focus the US superconductivity research community, both government-funded as well as industrial-funded, on development of devices and components for space. The customary approach for developing a new technology is to emphasize materials optimization first, then develop devices using these materials, followed by addressing the issues involved in packaging and integrating the new devices into a system. However, the philosophy behind the HTSSE program was a parallel approach where at the same time as materials and device optimization was done the contractors also addressed the issue of packaging these devices for space deployment. This approach is a very aggressive one with some risk as very close coordination has to be maintained throughout the entire development cycle. However, the approach adopted by HTSSE will drastically reduce the time required for insertion of this technology into space systems.

PROGRAM SCOPE

In order to accelerate the development of HTS devices for space applications, the HTSSE program has attempted to leverage the existing government programs in high temperature superconductivity (such as the DARPA, SDIO and the various services programs), and to focus these activities as well as industrial in-house programs toward space deployment. The approach followed in the HTSSE program is illustrated in Fig. 2.

2.1 Device Fabrication

Devices were fabricated by a number of industrial laboratories under contract with NRL, by NRL, and several National Aeronautical and Space Administration (NASA) laboratories and by, at least, one foreign research organization. The vendors were given a certain degree of flexibility in selecting the type of device they chose to provide. When the selection was made in the Spring of 1989, most of the vendors selected to provide microwave devices, both thin film as well as bulk. Each participant in HTSSE was requested to provide two "prototype" devices to NRL for testing near the end of calendar year 1989 for electrical testing and five "optimized" devices prior to 30 June 1990, for eventual integration into the Space Experiment.

Five "optimized" devices were requested to allow verification of their performance electrically, mechanically, thermally and for radiation survivability. Requiring the delivery of five devices also provided NRL and the entire superconductivity community some indication of how mature the technology had become. The electrical measurements were used to compare with those provided by the vendor and to establish a reference for the devices once they had been incorporated in to the Space Experiment. The thermal, vacuum, and shock and vibration testing was performed to verify that the devices and packaging were robust enough to withstand the rigors of launch and space operation. The devices also were given a preliminary radiation effects characterization to estimate whether they would be capable of surviving at least one year mission in the radiation fluences to which a typical spacecraft might be exposed.

Each of the devices had to pass these tests to be considered for inclusion in the Space Experiment. The other criterion, and the one that was essential for demonstrating the usefulness of this technology in space, was that the performance characteristics of the devices exceeded by an order of magnitude the capabilities provided by current technologies used in spacecraft systems.

2.2 Space Experiment Package

The objective of the HTSSE experiment is to place these devices in space, cool them to their operating temperature and monitor their electrical properties during the duration of the mission to ascertain failure modes, if any, of HTS devices in an space environment. Accordingly, the Space Experiment package had to be capable of providing the required cryogenic environment and also to monitor the electrical performance of the devices. Therefore, while the devices were being fabricated and evaluated, NRL scientists and engineers were addressing the spacecraft package itself.

The characteristics of the prototype devices received near the end of 1989 indicated that a cold stage temperature of 77 K and below was required. Some of the devices functioned well at 77 K but most of them had much improved characteristics as the temperature was lowered further. Therefore, the goal was set that the cryogenic

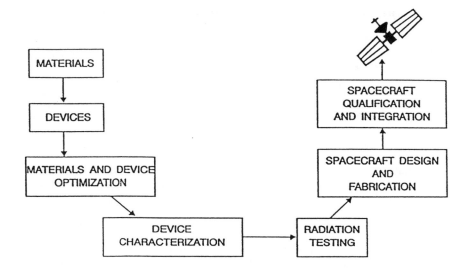

Fig. 2. The approach followed during the HTSSE program.

refrigeration had to provide a cold stage temperature of 77 K with a goal of achieving temperatures near 70 K. Also it was estimated that there might be as many as 15 thin film devices and 4 "bulk" devices (i.e., microwave cavities and IR detector assemblies).

2.3 Cryogenic Refrigeration System

A number of cooling techniques were evaluated for the HTSSE mission.
(1) <u>Dewar containing liquid nitrogen cryogen</u> Liquid cryogenic systems are inherently the most reliable means for producing stable cryogenic environments as they contain no moving parts. However, for the estimated heat load and the desired one-year mission duration, a very heavy and bulky dewar system would have to be built especially for the Space Experiment. The cost and delivery time for such a dewar system was not consistent with the milestones outlined for the Space Experiment.
(2) <u>Joule-Thompson Cryocooler</u> An open-cycle Joule-Thompson (J-T) cooler is inherently very reliable as there are no moving parts in the cold portion of this type of cooler which is based on the adiabatic expansion of high pressure gas through a narrow orifice. The heat load and the planned mission duration required a very large volume of gas and thus, this type of cryogenic refrigeration system was not considered suitable for HTSSE.
(3) <u>Stirling Cryocooler</u>. The third class of cryogenic refrigeration considered for HTSSE was the Oxford Stirling Cooler manufactured by the British Aerospace Company. This cryocooler provides 800 milliwatts of cooling capacity at 80 K and about 500 milliwatts at 70 K for an electrical input power (at 300 K) of about 50 watts. The weight of the cooler is about 8.4 kg. One cooler of this type has successfully been operated in the laboratory for a period of time exceeding 24,000 hours, about three years. Although this type of cooler was designed for a European Space Agency mission, it has not been flown so flight data is not available. Nevertheless, NRL has

purchased two of these coolers for the HTSSE mission. One will be used as the primary cooler, while the other will serve as a redundant back-up unit.

A preliminary design of the Space Experiment package has been made with the following estimates of the three dominant heat leaks from ambient (spacecraft) temperature into the cold stage:

Mechanical Supports (total of six)members)	70 milliwatts
Radiation Losses	250 milliwatts
Input-Output Leads Heat Leak (for 20 pairs	170 milliwatts
of coaxial cables)	---------------------
Total Estimated Heat Load	550 milliwatts

The major uncertainty in this estimate is the radiative lose which was based on an emissivity of 0.004 watts/cm^2/K, which was based on forty layers of multi-layered aluminum coated mylar sheet. This estimate indicated that single British Aerospace Stirling cooler will be capable of providing an operating temperature of less than 77 K for the cold portion of the experiment and, very likely will achieve temperatures close to 70 K.

As was mentioned earlier, the two Stirling coolers will be built into the Space Experiment package. Each cooler will be thermally connected to the cold stage via a gas-activated thermal switch. The operational cooler will be thermally shorted to the cold stage which the other, non-operating cooler will be thermally isolated from the cold stage. If the first cooler should fail, its thermal switch would be opened, the cooler turned off, the second cooler turned on, and once it is down to temperature, its thermal switch would be closed to connect that cooler to the cold stage. Thus redundant coolers would prolong the mission duration.

It should be noted that even lower temperatures - in the vicinity of 60 K - might be achieved IF both coolers could be operated simultaneously. However, this would require about 100 to 120 watts of input electrical power, much in excess of the present estimate of the total power that will be available for the Space Experiment, which is about 65 watts of average power.

2.4. Measurement Instrumentation

During the flight of the Space Experiment, on-board instrumentation will monitor the electrical characteristics of each of the devices about once each day and the results telemetered to a ground station.

Most of the devices procured for the HTSSE flight will operate at microwave frequencies, and the description of the on-board instrumentation will be limited to that system. Ideally, one would like to have on the satellite a complete microwave network analyzer, such as the HP-8510. Unfortunately, there is no space-qualified version of such an instrument. Therefore, NRL engineers are assembling a system which will only measure the amplitude of the transmission of the HTSSE microwave devices. See Fig. 3.

The basic signal source will be a frequency synthesizer which can be stepped in 10 kHz intervals from 4.250 GHz to 5.250 GHz. A second measurement band from 8.500 GHz through 10.500 GHz will also be available using a signal obtained by a frequency doubler from the synthesizer in the lower frequency band. In each band, the signal will be applied to the input connector of ten-position coaxial switch located at ambient (spacecraft) temperature. (The variation in the insertion loss of the switch from throw to throw has been measured to be less than +/- 0.03 dB). Each output terminal of the switch will be connected via a 0.040 inch diameter coaxial cable approximately one foot long to a device located at the cold stage of the space package. This very small diameter coaxial cable was selected to minimize the thermal heat load on the cold stage (each coaxial cable represents a 3 milliwatt heat load) but with an associated 12 dB attenuation loss for the electrical signal. (This loss can be tolerated during this experiment but certainly will not be acceptable in a operational system which might be deployed in the future.) The output from each device is connected to a receiver at ambient (spacecraft) temperature by another 0.040 inch diameter coaxial cable. One pair of coaxial cables will be used for a zero attenuation "through" connection to the verification of the calibration of the measurement system. Preliminary estimates are that the microwave transmission of the devices will be measured with an uncertainty of less than 0.5 dB.

2.5. Launch Vehicle

Negotiations for a launch opportunity have resulted in a commitment for a launch for the HTSSE package late in 1992. Since this would be a secondary mission, HTSSE will use the excess satellite budgets in terms of weight, volume and electrical power. The projected utilities available to HTSSE were about 100 kg, and about 85 watts of prime electrical power. Weight and volume are not very crucial issues. However, electrical power must be used for both the cryocooler, which will be functioning continuously, and the measurement instrumentation, which will be operated intermittently. The electrical power available to HTSSE of about 85 weatts is very marginal. It is this limited power budget that has restricted HTSSE to the use of only one Stirling cryocooler, and thus restricted the operating temperature to the vicinity of 77 K but certainly no lower than 70 K. (A budget of more than 120 watts would have allowed the simultaneous operation of both cryocoolers which would have insured operation at temperatures near 60 K, a more comfortable operating margin. The primary payload would provide the telemetry instrumentation and a frequency standard to which the HTSSE frequency synthesizer standard would be compared. The on-board computer would control the experiment and process the output data into a form suitable for telemetering to a ground station, probably, once per day.

DEVICE FABRICATION

Devices for the HTSSE were obtained by several different means. The majority of devices were procured under contract from seventeen different industrial research organizations. Additional devices were obtained from government laboratories such as the Naval Research Laboratory and the Jet Propulsion Laboratory. In addition, one foreign research laboratory, the University of Wuppertal working in collaboration with Interatom, a subsidiary of Siemens, provided one device at no cost to the program. The participants in the HTSSE program are summarized in Table I which lists the participants and a brief description of the device supplied. Note that the majority of devices are microwave components, most of them in thin film format although a number of bulk cavities were also included. The only non-microwave devices were the IR bolometer array from Honeywell and the thermal isolator from Lockheed. This reflects the opinion of the suppliers and the HTSSE program personnel that passive high frequency devices had the highest probability of success in time for the planned 1990 delivery and the 1992 launch date.

In order to illustrate the type of performance enhancement that can be achieved using HTS materials, a few examples will be cited. In Figure 4 contains a drawing of a

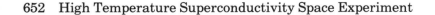

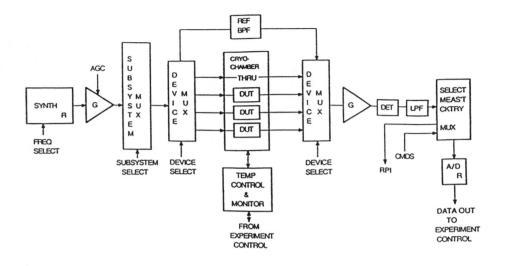

Fig. 3. Schematic of one channel of the measurement circuit to monitor the electrical
characteristics of the microwave devices during the HTSSE space flight.

four-pole edge-coupled filter fabricated at Lincoln Laboratories using a COMSAT
Laboratory design and AT&T Bell Labs doubled sided YBCO films on LaAlO3. A
normal metal version of this filter exhibited a 4 dB insertion loss while the HTS version
exhibited a insertion loss of less than 0.5 dB. Note the abrupt roll-off of the skirts of
the filter and the very low out-of-band rejection, in excess of 50 dB.

Another filter, designed at the Naval Research Laboratory, is shown in Fig. 5.
This filter, which was a five-pole, edge-coupled microstrip filter, had an insertion loss
of less than 0.5 dB at 77 K and even lower values at lower temperature. The normal
state equivalent of this filter was about 4 dB. This filter was fabricated using a normal
metal ground plane: even lower insertion loss values would be achievable if a HTS
ground plane had been used.

A third example of an HTS device fabricated for the HTSSE program is a ring
resonator fabricated at NRL using YBaCuO thin film on MgO substrate. See Fig. 6.
The Q-factor of this device is greater than 3,000 for temperatures below about 77 K.
this is to be compared to a value of 300 which is typical of a normal metal resonator of
the same design. HTS metallization can also improve the Q-factor of air-filled and
dielectrically loaded cavities.

IMPLICATIONS OF HTS TECHNOLOGY ON SPACECRAFT SYSTEMS

The NRL High Temperature Superconductivity Space Experiment has been
described and the performance of several HTSSE devices discussed. The question
remains: what impact might HTS technology have on future generations of spacecraft
systems. Several examples will now be presented.

Low insertion loss components The low insertion loss characteristics of passive HTS
components such as transmission lines, delay lines, couplers,. etc., will drastically

minimize losses between the antenna and the amplifier in a satellite receiver. This reduction in loss will improve the system signal-to-noise ration (S/N) which can result in either improved range, wide signal bandwidths or increased signal discrimination. Alternatively, this improved S/N for the receiver may translate into lower transmitter power which, on a satellite, could result in a reduction in weight, volume and electrical power. This latter advantage would impact communications satellites, both civilian as well as military.

Higher Q-value filters and resonators. Low insertion loss, very sharp roll-off filters that can be fabricated from HTS materials will provide more efficient utilization of frequency bands, adjacent channels can be closer together without interfering with each other. Furthermore, thin film HTS filters can provide characteristics previously possible only with waveguide technology. Thus, the introduction of HTS thin film technology into spacecraft systems, will drastically reduce weight and volume without sacrificing performance. Both of these features will impact commercial communication satellite design as well as military systems.

CONCLUSION.

The High Temperature Superconductivity Space Experiment (HTSSE) will develop HTS devices and components with characteristics superior to that achievable using conventional technology for introduction into spacecraft systems. The structure of the program is such as to leverage the US national program in high temperature superconductivity and to focus a major portion of this effort onto space applications. A variety of devices have been fabricated by a number of industrial and government laboratories, their characteristics measured both by the vendor and by NRL engineers. After space qualification of these devices in suitable packages, these devices will be integrated with a cryogenic refrigerator and a measurement systems so that the properties of these devices can be monitored during the planned one to one and one-half year space mission. The HTSSE is planned for a launch on a Navy rocket late in 1992.

ACKNOWLEDGMENTS

The High Temperature Superconductivity Space Experiment is managed by the Naval Research Laboratory and funded by the Space and Naval Warfare Systems Command of the U.S. Navy. The support of other government agencies such as the Defense Advanced Research Projects Agency (DARPA) and the Strategic Defense Initiative Organization (SDIO) which partially funded related research and development work at some of the HTSSE contractors is also acknowledged as is the Internal Research and Development (IR&D) funds of several of these organizations. The work at the Jet Propulsion Laboratory was funded by NASA while the devices supplied by the University of Wuppertal and Interatom, a subsidiary of Siemens, were provided at no cost to the HTSSE program.

In summary the naval Research Laboratory gratefully acknowledges the hard work and dedication of the scientists and engineers at NRL and the other program participants who made possible this outstanding progress which will insure that a very successful High Temperature Superconductivity Space Experiment will be ready for launch in 1991.

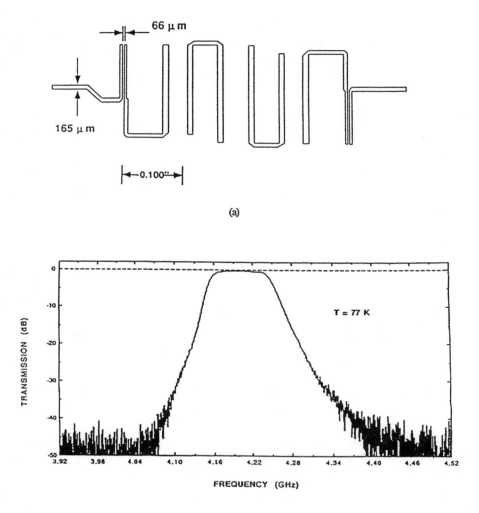

(a)

(b)

Fig. 4. (a) Diagram of four-pole Chebyshev microstrip filter fabricated by
            Lincoln Laboratories using YBCO film made by A T&T Bell
            Laboratories after a design by COMSAT Laboratories.
        (b) Transmission Coefficient of four-pole Chebyshev filter. Note the
            very low insertion loss in-band and the very steep roll-off
            characteristics of the filter. Measurement of the out-of-band
            rejection was limited by noise floor of the network analyzer.

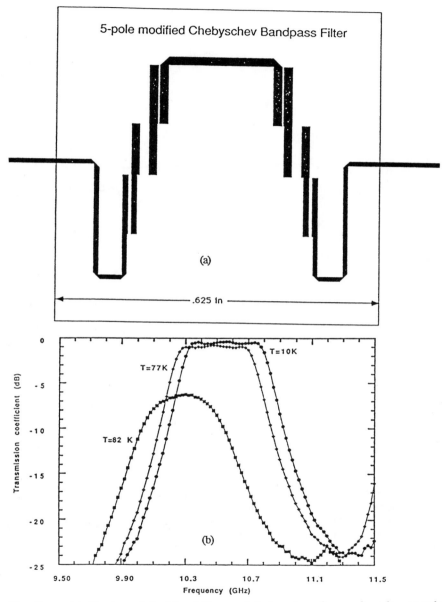

(a)

.625 in

(b)

Fig. 5.   (a) Design of the Naval Research Laboratory four-pole, edge-coupled
microstrip filter centered at 10.5 GHz The filter was fabricated
using YBCO thin films on magnesium oxide substrates.

(b)   Transmission coefficient of NRL four-pole, edge-coupled
microstrip filter. Data is shown for both a filter fabricated using
gold metallization measured at 300 K and 10 K for comparisons
to the HTS filter ,measured at 77 K and 10 K.

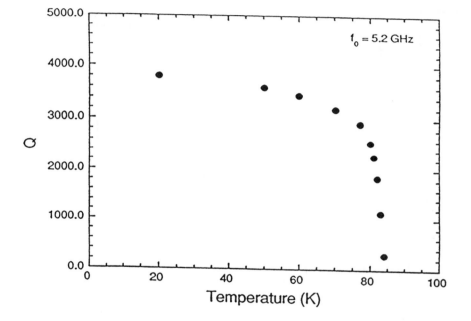

Fig. 6. Quality factor of thallium-based HTS end-coupled resonator at 4.8 GHz
fabricated by Superconducting Technology, Inc. as function of
operating temperature. A similar resonator fabricated using gold
metallization would have a Q-value of 300 in this temperature range.

TABLE I - List of Suppliers of Devices to HTSSE and Devices to be Delivered.

Participating Organization	Device Provided to HTSSE
A T&T Bell Laboratories Murray Hill NJ	YBCO ring resonators at 4.74 GHz and 9.5 GHz
DuPont Wilmington DE	Thallium based ring resonator at 4.9 GHz
Ford Aerospace , Palo Alto CA	Dielectric resonators with YBCO thin films at 10 GHz.
General Atomics, San Diego CA	Coated YBCO cavity at 10 GHz.
General Electric, Syracuse NY	Coplanar YBCO resonator at 4.8 GHz.
High Tc Superconco, Lambertsville NJ	Bulk YBCO cavity at 10 GHz Thick film YBCO microwave power limiter
Honeywell, Minneapolis MN	I x 12 linear array of YBCO transition edge Infra Red Bolometers.
Hughes Research Laboratory, Malibu CA	Six pole YBCO filter at 9.6 GHz
HYPRES, Inc., Elmsford NY	2 nanosec. YBCO delay line, 4 GHz wide.
ICI Composites, Wilmington DE	Thick film coated YBCO cavity at 5 GHz.
Lockheed Space and Missiles Palo Alto CA	Thallium based thermal isolator
MIT Lincoln Laboratories, Lexington MA	Six pole, 300 MHz bandwidth filter at 4.7 GHz
David Sarnoff Research Center,Princeton NJ	Four pole YBCO 50 MHz B.W. filter at 9.3 GHz.
Superconducting Technology, Inc. Santa Barbara CA	Thallium based ring resonator at 4.9 GHz.
TRW, El Segundo,CA	YBCO end-coupled resonator at 10.4 GHz
Westinghouse R&D Center, Pittsburgh,PA	Four-pole YBCO edge-coupled filter at 9.4 GHz.
College of William and Mary,Williamsburg VA	Electropheretically coated cavity at 1.42 GHz
Naval Research Laboratory Washington DC	Four-pole, edge-coupled YBCO filter at 9.1 GHz YBCO ring resonator at 9.4 GHz.
Jet Propulsion Laboratory, Pasadena, CA	YBCO low-pass filter with cut-off at 9.7 GHz.
University of Wuppertal Wuppertal, FRG	YBCO patch antenna resonant at 5 GHz

Engineered YBCO Microbridges

R. W. Simon and J. A. Luine
TRW Space & Technology Group
Redondo Beach, CA 90278

INTRODUCTION

The perovskite cuprates offer the possibility of a superconductive microelectronic technology that operates at temperatures above 50 K and perhaps as high as 77 K. The primary building block of such a technology is the Josephson junction. In conventional materials, Josephson junctions take the form of multilayer tunneling devices but in the cuprates such devices are still lacking. The absence of a true tunnel junction has necessitated the development of a variety of microbridge structures. This paper discusses the overall situation with respect to engineered cuprate microbridges and, in particular, focuses on our group's two successful techniques for microbridge fabrication.

THE MICROBRIDGE ALTERNATIVE

While true tunnel junctions have been notably lacking in the cuprates there have been many demonstrations of the Josephson effects. Very early on there were successful SQUIDs made by exploiting natural weak links occurring in both sintered pellets of YBCO[1] and in post-annealed thin films[2]. The improved quality of cuprate thin films has eliminated the presence of the grain boundaries that can be exploited for Josephson devices. Therefore, it has become necessary to artificially engineer weak links in thin films in order to fabricate functioning microbridges. Recent work in a number of laboratories has yielded encouraging results in such microbridge fabrication. The high-T_c cuprates may be the materials for which microbridges are finally pushed to their practical limits for applications.

HTS MICROBRIDGES

There have been many geometries applied to the problem of cuprate microbridges but only standard Josephson geometry that has yielded reasonable results has been the S-N-S microbridge which can be either a planar devices or a multilayer device. S-N-S planar junctions have been studied by a number of groups[3,4] and have demonstrated the features expected of Josephson devices. The characteristic voltages ($I_c R_n$) of these devices have been quite low (<50µV), which is to be expected from theoretical considerations of the proximity effect[5]. Such junctions also place unreasonable demands on lithography if they are to function well at elevated temperatures. It is unlikely that S-N-S planar structures containing noble metals will lead to practical high-temperature devices.

Studies of sandwich S-N-S junctions have yielded very useful information about cuprate Josephson devices. The high temperatures associated with cuprate growth eliminate the possibility of fabricating symmetric sandwich structures containing noble metals and cuprate counterelectrodes. However, studies of YBCO-Ag-CS (where "CS" is a conventional metallic superconductor) have demonstrated the necessity of transporting pair currents in the favorable a-b plane directions[1,2]. More recently there have been studies of S-N-S sandwich structures containing alternative normal layers composed of oxide conductor (or superconductor) films. Sandwiches composed of different members of the bismuth-based cuprates[6] fall into this category. Perhaps of greater interest is work on sandwiches containing a-axis YBCO films and doped $SrTiO_3$ normal layers[7]. As pointed out in Ref. 5, this system may provide the ideal implementation of the S-N-S geometry using the superconducting cuprates since it eliminates fundamental problems of low resistance and diminished critical current that are characteristic of structures incorporating noble metal layers. It remains to be seen whether reproducible results can be obtained from this junction technique.

GRAIN BOUNDARY JUNCTIONS

The simplest cuprate Josephson device to date is a microbridge formed by growing a film over a bicrystal (two different crystalline faces fused together) which creates a grain boundary at the crystal interface[8]. The resultant device (fabricated by IBM) behaves in accordance with the well-known RSJ model for microbridges but has a suprisingly low $I_c R_n$ product. While this microbridge is rather elegant and performs well, it is rather unsuitable for extended applications since it requires a bicrystal interface for every device. Similar results have been obtained by the Cornell group using "mosaic growth" films that contain a relative small number of clean, well-defined grain boundaries[9]. These microbridges have the drawback that the grain boundaries are located all over the films which reduces the critical currents in the films and increases the potential noise in SQUIDs. Despite these problems, both of these grain-boundary devices provide good model systems for studying the properties of cuprate Josephson junctions.

There are two other examples of multilayer junction structures that have demonstrates successful microbridge operation. One is a trilayer sandwich of YBCO films incorporating a layer of PrBCO, which is a semiconducting variant of the 1-2-3 compound. The second is an attempt to produce a traditional edge junction in which two YBCO films are connected through an edge contact incorporating a barrier layer. While both of these geometries have yielded functioning Josephson devices, it appears that neither one operates in the way they were intended.

The published data on the Bellcore trilayer[10] suggests that the PrBCO layer is not responsible for the current flowing in the device. The relatively low normal-state resistance is consistent with conduction through microfilaments, pinholes, or other defects in the Pr layer. The critical currents of the devices do scale with junction area, suggesting that there must be many current paths that are evenly dispersed. Improved device processing may yield very different results from these junctions in the future.

The IBM edge junctions[11] appear to behave much like the above trilayers. Both devices have IV characteristics that are not well-described by the RSJ model. These junctions may also be formed by microfilamentary connections between the two superconducting films rather than transport through an intended barrier layer. To date, the edge junctions have not demonstrated high device yields, but this situation may change with improved processing.

The most elementary form of microbridge is created by restricting the flow of current between two superconducting banks by lithographic patterning, thinning, ion damage, or some other process. In conventional superconductors, it is feasible to create regions of constricted current flow on the length scale of the coherence length and thereby produce true Josephson devices. In the cuprates, one cannot practically work on this length scale because of their extremely short coherence length. As a result, junction dynamics is more likely to be governed by flux motion with a characteristic length of the perpendicular penetration depth. Nevertheless, the resultant devices an exhibit essentially all the featured desired in true Josephson devices.

A variety of methods have been successfully employed to create current-concentrating microbridges in cuprate films. In general, what is required is a means of reducing the current-carrying area in a patterned films over the shortest length possible. This can be accomplished by physical modification of the film, chemical modification of the film, or both. We investigated an early example of such a process in our group by growing YBCO films across a thin aluminum stripe[12]. The aluminum poisoned the superconductor film and produced a region of greatly reduced critical current density which in turn allowed us to demonstrate functioning SQUIDs in high current density films. Another good example of physical modification is the use of pulsed electric fields to locally overheat constrictions in YBCO films[13] and create weak links. Whether this process creates grain boundaries or microfilaments is unclear. In any case, the resultant devices exhibited well-controlled (although small) $I_c R_n$ products.

Our own research has centered around the use of two different techniques to locally alter the properties of YBCO films: focused ion beam implantation and substrate step edges. While the microscopic details and device dynamics of these two processes are not yet well understood, both methods have yielded good results for the fabrication of both rf and dc SQUIDs.

FOCUSED ION BEAM MICROBRIDGES

There is a long history of using ion beam damage to alter the properties of superconducting films. In recent years, ion beam damage has been shown to controllably reduce the critical currents in YBCO films[14]. Following this work, we have make use of focused high-energy ion beams to engineer weak links into epitaxial ErBCO and YBCO films. Our initial work[15] demonstrated that this technique could be used to produce dc SQUIDs operating up to approximately 60 K with up to 50% critical current modulation by an applied magnetic field. We report on the details of the microbridge fabrication process elsewhere[16].

Fig. 1. Silicon ion tracks across two weak links
in focused ion beam dc SQUID.

We have performed a number of experiments investigating the effects of varying ion energy, ion dose, film thickness, and film morphology upon the fabrication of focused ion beam (FIB) microbridges. We have obtained our best results for devices made from 200 nm films irradiated with a flux of ~3 x 10^{13} 300 keV Si^{++} ions/cm^2 over a ~100 nm long bridge. Figure 1 shows ion beam tracks across two constrictions in a sputtered YBCO film that form the active devices in a dc SQUID. In general, we have demonstrated a process with a reasonable yield of functioning SQUIDs (typically 40%) but with a considerable scatter in critical currents.

Individual microbridges fabricated by the FIB process have IV characteristics comparable to those of a number of other devices, for example the Bellcore trilayers and the IBM edge junctions. Figure 2 shows the IV for one of these devices under microwave irradiation and features well-defined Shapiro steps. The critical currents of these bridges are typically in the 50-500 μA range, three orders of magnitude below that of unirradiated links on the same wafers. We believe that these microbridges are formed by destroying the superconductivity in the bulk of the film in the irradiated region, leaving only weak connections between the banks of undisturbed material.

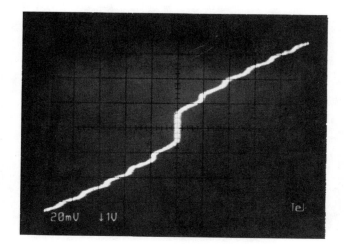

Fig. 2. IV characteristic of FIB microbridge
under 11 GHz irradiation (x=20μV/div, y=100μA/div)

These results are consistent with a model (based in part on output from the Transport of Ions in Matter computer model) that suggests the following picture for the formation of weak links. Most of the damage is caused by low energy ($\leq$ 1 keV) oxygen recoils within 5 nm of the primary ion tracks. Along the recoil track vacancies are created every few tenths of a nanometer. At the doses used, the mean distance between primary ion tracks is ~1.8 nm. Therefore, the recoil tracks overlap and create a dense web of lattice damage. Since these lengths are on the scale of the coherence length in the YBCO, there is significant disruption of the phase coherence within the irradiated regions. Given the randomness of the lattice damage, there may be large variations in the coupling through these regions, which may account for the appreciable scatter in critical currents observed in our experiments.

These variations in weak-link critical currents limit the performance of individual dc SQUIDs due to asymmetries between the two junctions comprising the SQUID, although we have observed relatively large critical current modulation in applied fields. It remains to be seen whether the I_c imbalance can be reduced by improved fabrication techniques.

Of greater concern is the inevitable reduction of critical temperature that accompanies the critical current reduction in the FIB microbridges. We have not observed any functioning devices above about 60 K and degrading modulation performance with increasing temperature generally restricts the useful operating range of the FIB SQUIDs to 40-50 K. Comparable results have been reported recently by the Julich group[17] using e-beam lithographic patterning of a PMMA mask to limit the spatial extent of a large-area 100 keV argon ion beam. The operating range of the resultant SQUIDs was the same as we observed. Despite the current drawbacks, small-area ion damage appears to provide a reasonably reliable technique for fabricating microbridges in epitaxial YBCO films.

STEP-EDGE MICROBRIDGES

The second technique that we have concentrated on is the substrate step-edge microbridge[17]. This microbridge fabrication process consists of etching a sharp step in a substrate prior to thin film growth and then depositing a film thinner than the step height across the step. The microbridge is formed by the disruption of film growth across the step. The result of this process is a 2-3 order of magnitude reduction in critical current across the step which in turn leads to microbridge behavior that is comparable to many of the devices mentioned above. The favorable aspects of this process are that the yield of working devices is quite high and that the critical current reduction across the step is not necessarily associated with a large critical temperature reduction. We have fabricated SQUIDs using the step-edge process that operate at temperatures up to 82 K and have measured a flux noise at 10 Hz of 1.5×10^{-4} $\Phi_0/\sqrt{Hz}$ at 77 K. We have operated rf SQUIDs at 27 MHz and at 10 GHz with good success. We report on the details of these devices elsewhere[18,19,20].

APPLICATIONS POTENTIAL OF MICROBRIDGES

The overriding question pertaining to the step-edge microbridges and all the engineered weak link devices is that of their potential utility for microelectronic applications. A number of specific criteria come to mind in evaluating the various microbridges now available. Of paramount concern for multidevice applications is the issue of process yield. With respect to this issue, there two types of yield to consider. First there is wafer-to-wafer yield. From the standpoint of manufacturability, this is quite important. At the current level of technology development, on the other hand, it is of little consequence. Of greater interest is the device-to-device yield. This directly determines our ability to fabricate multigate circuits using microbridge junctions. For simple SQUIDs, on the other hand, even this sort of yield is non-critical. In any event, the step-edge process and the FIB process have already demonstrated high enough yields to allow us to test simple circuits containing up to five SQUIDs.

The key issue for any Josephson technology is that of critical current control. Josephson circuits are designed around specific current levels and the real junctions must come close to the target values and must not vary significantly (perhaps no more than 5%) across a chip. It is in this respect that trilayer tunnel junction technology excels. To date, there is no HTS microbridge technology that provides an acceptable level of critical current control (with the possible exception of the Berkeley "shock" junctions that do not appear to thermally cycle well.) The step-edge junctions tested to date have exhibited 100-200% I_c variations in rf SQUIDs and order-of-magnitude I_c variations in dc SQUIDs. It may be the case that step-height and film-thickness variations can account for most of these variations. Improved processing techniques will shed light on this question but for now we do not have a process with tight controls on critical currents.

A third consideration for microbridge fabrication technology is the issue of extendability. A useful device process must be able to make large numbers of devices on a wafer scale. Thus techniques like the IBM bicrystal approach cannot really be extended in this sense. The step-edge technique appears to be extendable within the limits of the geometry control discussed above. Certainly any of the multilayer techniques under development are capable of wafer-scale fabrication.

The last consideration we will mention is that of the performance characteristics of the active devices themselves as expressed by such figures of merit as the $I_c R_n$ product and the SQUID noise. We have neither accurately measured these parameters nor, in the case of the I_cR_n, determined the significance of the measured values. The SQUID noise measured in a number of different engineered microbridges - particularly the intrinsic value inferred from the step-edge data - is low enough for most applications. If the I_cR_n products determined from step-edge IV curves are meaningful, then these values are quite acceptable as well. We summarize the properties of many of the existing microbridges in Table I, which speculates on the ability of the various techniques to meet the performance requirements outlined above.

Microbridge type	Yield	I_c Control	Extendability	I_cR_n	Noise
Bicrystal G. B.	++	++	---	--	+++
Mosaic G. B.	++	+	-	-	---
Shock junction	++	+++	--	-	++
Step-edge	++	+	+++	++	++
Foc. Ion Beam	+	-	+	+	+
"Edge" junction	--	-	++	++	++
PrBCO trilayer	+	+	+++	-	++
SNS planar	--	--	++	---	++

Symbols: +++ Excellent   ++ Very Good   + O.K.
          - Not good   -- Major problem --- Disaster

Table I - Summary of evaluation criteria for YBCO microbridges.

CONCLUSIONS

There has been substantial progress made in the development of YBCO Josephson devices. We now have several fairly reliable techniques by which microbridge junctions can be fabricated in high-quality, epitaxial thin films. Furthermore, additional improvements in the quality of superconductor thin films should not impair our ability to produce working microbridges. We can now proceed to optimize the properties of these microbridges and begin to use them to fabricate integrated SQUIDs, logic gates, and various components of signal processing circuits. While the sandwich tunnel junction made from cuprate superconductors continues to elude us, the prospects for microelectronic applications using microbridges have improved dramatically.

ACKNOWLEDGMENTS

The authors acknowledge the benefits of many fruitful discussions with A. H. Silver, M. R. Beasley, T. H. Geballe, P. A. Rosenthal, and G. Deutscher. The TRW work reported here is that of colleagues J. B. Bulman, J. F. Burch, S. B. Coons, K. P. Daly, W. D. Dozier, R. Hu, A. E. Lee, C. E. Platt, S. M. Schwarzbek. M. S. Wire, and M. J. Zani. Our work reported here was supported in part by ONR/DARPA under contract N00014-88-C-0747.

REFERENCES

1 J. E. Zimmerman, J. A. Beall, M. W. Cromar, R. H. Ono, *Appl Phys Lett*, **51**, 617 (1987).

2 R. H. Koch, C. P. Umach, G. J. Clark, P. Chaudhari, R. B. Laibowitz, *Appl Phys Lett*, **51**, 200 (1987).

3 D. B. Schwartz, P. M. Mankiewich, R. E. Howard, L. D. Jacke, B. L. Straughn, E. G. Burhart, A. H. Dayem, The observation of the ac Josephson effect in a $YBa_2Cu_3O_7$ junction, *IEEE Trans. Mag.*, **25**, 1298 (1989).

4 M. S. Wire, R. W. Simon, J. A. Luine, K. P. Daly, S. B. Coons, A. E. Lee, R. Hu, J. F. Burch, C. E. Platt, to be published in Proceedings of 1990 Applied Superconductivity Conference.

5 G. Deutscher, R. W. Simon, On the proximity effect between normal metals and cuprate superconductors, submitted to *Appl Phys Lett*.

6 K. Mizuno, K. Higashino, K. Setsune, K. Wasa, *Appl Phys Lett*, **56**, 1469 (1990).

7 D. Chin, T. Van Duzer; private communication.

8 R. Gross, P. Chaudhari, D. Dimos, A. Gupta, G. Koren, Thermally activated phase slippage in High-Tc grain-boundary Josephson junctions, *Phys Rev Lett*, **64**, 228 (1990).

9 D. H. Shin, J. Silcox, S. E. Russek, D. K. Lathrop, B. Moeckly, R. A. Buhrman, Clean grain boundaries and weak links in high T_c superconducting $YBa_2Cu_3O_7$ thin films, *Appl Phys Lett*, **57**, 508 (1990).

10 C. T. Rogers, A. Inam, M.S. Hegde, B. Dutta, X. D. Wu, T. Venkatesan, Fabrication of heteroepitaxial $YBa_2Cu_3O_7$-$PrBa_2Cu_3O_7$-$YBa_2Cu_3O_7$ Josephson devices grown by laser deposition, *Appl Phys Lett*, **55**, 2032 (1989).

11 R. B. Laibowitz, R. H. Koch, A. Gupta, G. Koren, W. J. Gallagher, V. Foglietti, B. Oh, J. M. Viggiano, <u>All high T$_c$ edge junctions and SQUIDs</u>, *Appl Phys Lett*, **56**, 686 (1990).

12 R. W. Simon, J. F. Burch, K. P. Daly, W. D. Dozier, R. Hu, A. E. Lee, J. A. Luine, H. M. Manasevit, C. E. Platt, S. M. Schwarzbek. D. St. John, M. S. Wire, M. J. Zani, <u>Progress towards a YBCO circuit process</u>, Proc. Conf. On Sci. & Tech. of Thin-film Superconductors, Denver, April,1990 (to appear.)

13 D. Robbes, A. H. Miklich, J. J. Kingston, Ph. Lerch, F. C. Wellstood, J. Clarke, <u>Josephson weak links in thin films of YBa$_2$Cu$_3$O$_7$ induced by electrical pulses</u>, *Appl Phys Lett*, **56**, 2240 (1990).

14 Alice E. White, K. T. Short, R. C. Dynes, A. F. J. Levi, M. Anzlowar, K. W. Baldwin, P. A. Polakos, T. A. Fulton, L. N. Dunkleberger, *Appl Phys Lett*, **53**, 1010 (1988).

15 M. J. Zani, J. A. Luine, R. W. Simon, R. A. Davidheiser, <u>Focused ion beam high-Tc superconductor dc SQUIDs</u>, submitted to *Appl Phys Lett*.

16 M. J. Zani et al; submitted to 1990 Applied Superconductivity conference.

17 G. Cui, Y. Zhang, K. Herrmann, Ch. Buchal, J. Schubert, W. Zander, A. E. Braginski, C. Heiden, <u>Properties of rf-SQUIDs fabricated from epitaxial YBa$_2$Cu$_3$O$_7$ films</u>, submitted to LT-19 Satellite Conference on HTS, Cambridge (1990).

18 K. P. Daly, W. D. Dozier, J. F. Burch, S. B. Coons, R. Hu, C. E. Platt, R. W. Simon, <u>Substrate step-edge YBa$_2$Cu$_3$O$_7$ rf SQUIDs</u>, submitted to *Appl Phys Lett*.

19 R. W. Simon, J. B. Bulman, J. F. Burch, S. B. Coons, K. P. Daly, W. D. Dozier, R. Hu, A. E. Lee, J. A. Luine, C. E. Platt, S. M. Schwarzbek, M. S. Wire, M. J. Zani, <u>Engineered HTS microbridges</u>, to be published in Proceedings of 1990 Applied Superconductivity Conference.

20   K. P. Daly, Charles M. Jackson, J. F. Burch, S. B. Coons, R. Hu, <u>YBCO step-edge rf SQUID biased at 10GHz</u>, to be published in Proceedings of 1990 Applied Superconductivity Conference.

CURRENT ACTIVITIES OF SUPERCONDUCTING WIRES AND CABLES

K.Matsumoto and Y. Tanaka

The Furukawa Electric Co., Ltd., Yokohama R&D Laboratories, 2-4-3, Okano, Nishi-ku, Yokohama, 220 Japan

ABSTRACT

The state of the art and recent progress in the development of superconducting wires and cables are reviewed and discussed. Excellent superconducting properties of the practic superconducting wires, are of special interest in the applications such as MRI, NMR spectrometer, partical accelerator, SMES and so on. The practical superconducting wire for DC applications have been also developed. In a well controlled condition, the critical current densities of Nb-Ti wire and bronze-processed Nb_3Sn wire have achieved over 3.8×10^5 A/cm^2 at 5T and 3.15×10^4 A/cm^2 at 16T, respectively. On the other hand, the Ag-sheathed oxide superconducting wires such as YBCO and BSCCO are developed actively. The transport critical current densities are reached over 3.5×10^4 A/cm^2 at 77K and zero magnetic field for a c-axis oriented BSCCO wire.

INTRODUCTION

In recent years, considerable attention has been given to both the enhancement of the critical current densities and the decrement of the ac losses in the practical superconducting wires and the new oxide superconducting wires. The critical current densities in Nb-Ti and Nb_3Sn multifilamentary superconducting wires with filament sizes of around few microns have been developed by various optimization efforts in the applications such as MRI, dipole magnets for high quality in magnetic fields and NMR spectrometer for high magnetic fields. The flux pinning mechanisums in these wires are developing actively on the basis of α-Ti precipitates and cold working or grain boundaries and the value of Tc or Bc_2. (1)-(3) But there are many unwanted phenomena in the high Tc, oxide superconducting wire such as the pinning mechnisum, the weak link, the flux motion and so on. An understanding of these electromagnetic phenomena is usefull for enhancing transport current. Espetially a controll in the c-axis orientation has realized drastically the higher transport critical current densitis of the Ag-sheathed BSCCO superconducting wires. (4)

PRACTICAL SUPERCONDUCTING WIRES

Nb-Ti ALLOY WIRES

The important factors which should be controlled in order to enhance the transport critical current densities in the Nb-Ti alloy wires are not only the intrinsic pinning force in the filaments but also keeping the uniformity of the filaments in the logitudinal direction of the wire during fabrication processes. However, it is difficult to realize these two factors simultaneously, in practice. Usually, the transport critical current density has been diominated by the uniformity of filaments related to intermetallic compounds formed around filament surfaces during heat treatments of the composites such as extrusion or thermomechanical process for α-Ti precipitates and the micro-structure. Consequently, they make a chance of sausasing in the filaments resulting in the degradation of the transport current densities. For discussing the intrinsic Jc of Nb-Ti wires, This non-uniformity of the filaments must be eliminated.

Four types of composites were used in this study.(6) One conductor was prepared by a single extrusion process. Seven Nb-Ti alloy rods were stacked into a Cu tube, co-extruded, and drawn down to appropriate diameters (sample A). The other two were made through a double extrusion process resulting 351-and 1057-filament composites were drawn down into various diameters(sample B and C). The last sample D was of a three component composed of Cu for stabilization isolated by a Cu-Ni matrix. These conductors have diffusion barriers around the filaments to prevent the formation of intermetallic compounds during heat-treatment processes. Figure 1 is a photomicrograph of sample D in cross section.

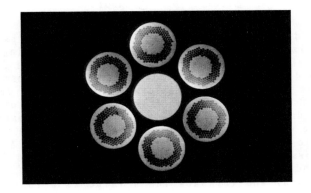

Figure 1   Cross seciton of AC superconducting wire using 0.5 μm Nb-Ti filaments

For these conductors, multiple heat-treatments to achieve the higher transport current were carried out mainly for 40 hours at 375 C. Critical current measurements were performed on specimens of more than 0.5 m length with 0.3 m voltage gap under the magnetic fields from 1 to 10T. The criterion of critical current was an overall composite resistivity of $10^{-14}\Omega$m. The n value related to the superconducting nomal state transition was determined over the voltage range of 10μV/m-100μV/m.

The relationship between the obtained Jc in 5T or 8T and NbTi filament diameter is shown in Figure 2. Sample C revealed the maximum Jc of 3830 A/mm^2(5T, df=2.5μm) and sample B has the maximum Jc of

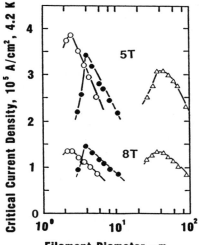

Figure 2   Effects of filament diameters on Jc
at 5T and 8T
△:Sample A,   ●:Sample B,   ○:Sample C

1450 A/mm^2 (8T, df=4 μm). Reduction of filament diameters is related to the initial size of Nb-Ti rods and the amount of drawing strain, ε = ln(Ao/A), which is significant factor to estimate the Jc of Nb-Ti wires made by different thermaomechnical processes. Figure 3 shows Jc (5T and 8T) as a function of final drawing strains, εf. Jc values at 5T of sample B and C increase linearly as the increase of εf, where εf means the drawing strain after last precipitation heat-treatment. The optimum Jc values are obtained around εf=4.5 to 5.0, and successively, the Jc values decrease steeply. While sample A with large filaments shows gradual increase of Jc against εf and its peak is obtained around the same εf.

Figure 4(a) and (b) show the n value as a function of the magnetic fields, which is the resistive transition index, E/Eo = $(I/Io)^n$. Sample A shows the ideal linear dependence against the magnetic fields, however, in the case of sample B, the values deviate from the ideal dependence in the lower field. Especially, heavy cold worked specimens show the large degradation of the n value in the low fields. At first, it is thought that the degradation may be brought about the filament sausaging and consequently the optimized Jc of sample B may be determined dominantly by extrinsic factors.

On the other hand, from Bc_2 and Tc measurements of sample A and B, the values of Bc_2 change slightly against the final strain in the range from 10.4 to 10.5T, and the values of middle point Tc of sample B decreases about 0.8K against the amount of the cold work of εf=5.6. But the relationship between the Jc values and the change in Bc_2 and Tc is unclear in this stage, because Matsushita et al predict

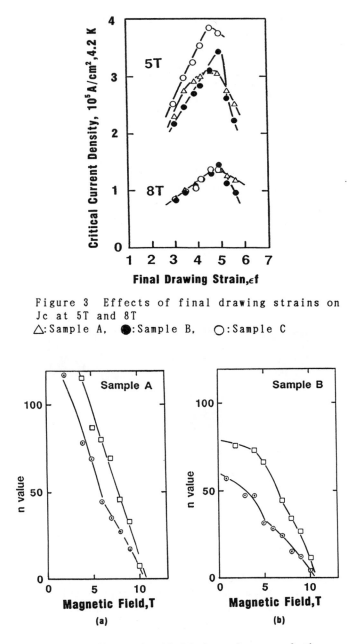

Figure 3   Effects of final drawing strains on
Jc at 5T and 8T
△:Sample A,   ●:Sample B,   ○:Sample C

Figure 4   Magnetic field dependences of the n
value of Nb-Ti, samle A (a) and sample B (b)
□: ε f=3.1,   ◉: ε f=5.3

the significant reduction of the elementary pinning force does not
occur through the decrease of Tc due to proximity effect on the
arrays of the Nb-Ti/α-Ti multilayers. (5)
    For the ac applications, sample D with 0.5μm filaments embed-
ded in high-resistance Cu-30 wt% Ni alloy has been performed, resulting a 100 kVA superconducting magnet at the power frequency. (7)
Figure 5 is a photograph of the completed 100 kVA magnet. The ac
losses of the wire will be in the area of 100 kW/m^3 at 50 Hz under
the magnetic field of 1T. In order to lower the ac losses, the submicron NbTi filament and the appropriate filament spacing are designed
because the main loss of the ac superconducting wire in the low
magnetic field is a hysteresis loss due to the filament size and/or
the filament spacing. Although the enhancement of the critical current density in the low field will be expected, the increase in the
effective filament diameter by the proximity effect due to the decrease of the filament spacing become the increase in the hysterisis
losses. These design concepts for the ac wire should be applied to
the other ac applications. (8)

Figure 5   100kVA superconducting magnet

Nb$_3$Sn COMPOUND WIRES

    In the area of technological uses of high field superconductor,
one of the most reliable and actual processes has been the so-called
bronze process. In particular, the most significant character in
multifilamentary bronze-processed Nb$_3$Sn wire is its current carrying
capacity, resulting in the big projects such as MFTF-B at LLNL or
CTC-CBC at JAERI. As the commercial high field superconductor, the
multifilamentary bronze-processed Nb$_3$Sn wires should be more developed in their superconducting properties such a transport current
density.

Three Nb₃Sn wires containing Ti and/or Ta were used in present study.(9) The wires were fabricated using 0.7 wt%Ti-alloyed Nb, 7.5wt% Ta-alloyed Nb and pure Nb filaments with a Cu-14.3 wt% Sn-0.2 wt%Ti alloy matrix. The wires used in this study were fabricated by a general extrusion and drawing techniques. Of the wires studied here, the 2.3μm and 1.5μm filament diameter were designed in order to achieve the high critical current density. The conditions for heat-treatments were those which were required by the filament sizes enough to form the Nb₃Sn layer. Critical current measurements were performed on specimens of rather short length, 30-50mm with 10 mm gap between the voltage taps under the magnetic fields from 10 to 23 T. The criterion of critical current was an overall composite resistivi-ty of $10^{-11} \Omega m$ - $10^{-13} \Omega m$.

The results of critical current measurements at 4.2K for the typical specimens are shown in Figure 6. The difference in Jc between pure Nb filaments and the alloyed filaments are clearly seen under the higher magnetic fields. As pointed out before, the effectiveness of enhancing Jc over that for pure Nb filaments by simultaneously alloying Nb₃Sn with Ti or Ta increases steeply as the magnetic field is rised beyond 14T. Also, it is interesting to note that the abso-lute values of Jc at 20 T for the Ta-alloyed sample is above the value(85 A/mm²). It is suggested that there exists a considerable opportunity in reducing the size of the magnet, thus the cost, if this value is taken into consideration for designing very-high-field magnet(>20 T).

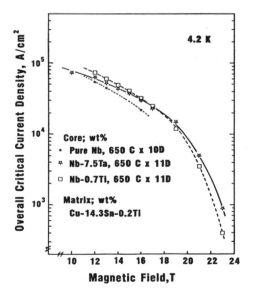

Figure 6 Effects of alloying to Nb cores on Jc at high magnetic fields

It is also interesting to note that there are considerable
variations in n values among alloying Nb_3Sn wires by increasing
magnetic field. The n-H plots for two filament series, $2.3\,\mu m$ and
$1.5\,\mu m$, are shown in Figure 7. All of these values are determined
from V-I curves between $10\,\mu V/m$ and $100\,\mu V/m$. The n values of alloying Nb_3Sn wire show the reduction by increasing magnetic fields. It
is seen that the field dependency of the n value is reflected in
change of superconducting properties due to flux flow in filaments.
Because, it is another reason why the n values in lower magnetic
fields deviate form $(1-H/Hc_2)$ relation as shown in the case of non-
uniform filaments such as sausaging. But, actually, it should be
concluded that the values of Jc in the wire with finer filaments,
$d_f=1.5\,\mu m$, at these high fields are primarily limited by the non-
uniform size of filaments rather than by their superconducting
properties such as the strain effect.

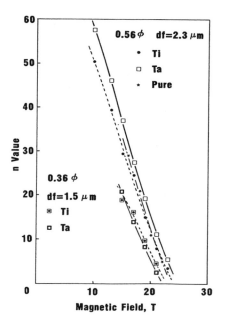

Figure 7   Effects of filament size and magne
tic fields on the n values

OXIDE SUPERCONDUCTOR

Oxide superconductors have the same characteristics of hardness
and brittleness as the intermetallic compound superconductors such as
Nb_3Sn and V_3Ga. Thus a direct application of the techniques elaborat-

ed for the intermetallic compounds is useful to attempt in making for oxide superconducting wires. Recently, sufficiently long oxide super-conducting wires and tapes have been made by the composite process. In the composite process, a metal tube such as silver was filled with oxide powder. The loaded tube was then processed by drawing and rolling into wire and tape, successively heat-treated at 800 C- 900 C for 50 hr under the controlled oxygen atmosphere. Figure 8 shows Jc-H characteristics of Ag-sheathed BSCCO superconducting wires at 77K. The Jc values remarkably depend on the degree of crystal orientation, F. For instance, Jc in a case of F=99% has reached $3.5 \times 10^4 A/cm^2$ under zero magnetic field at 77K. The Jc values at 77K could keep high level up to 0.01T. However Jc steeply decreased beyond 0.1T. (4)(10)On

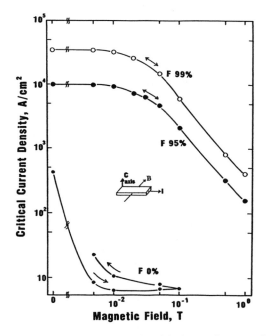

Figure 8   The magnetic field dependence of critical current density at 77K for Ag-sheath ed BSCCO tapes with various F values (F:a degree of c-axis orientation)

the other hand, in order to measure the transport current at helium temperature, two aligned wires with different F value were prepared. As shown in Figure 9 the Jc indicated F99 % decreased with the mag-netic field from 6.0×10^5 A/cm^2 at 1T to 2.5×10^5 A/cm^2 at 10T and then kept almost constant up to 23T. (10) With respect to the trans-port current of the oxide superconducting wires, these approaches are very promising to develop the high feild superconducting wires and ciols.

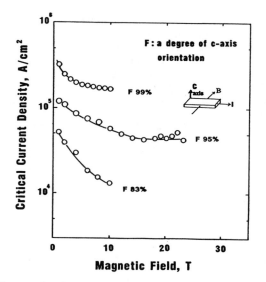

Figure 9 The magnetic field dependence of
the Jc at 4.2K for Ag-sheathed BSCCO tapes
with various F values. The magnetic fields
were applied parallel to the tape surface
and normal to the current flow direction.

CONCLUSION

(1) Superconducting transport critical current density, Jc, in field
up to 10 T and 23 T were measured for paractical NbTi and Nb_3Sn
wires with fine filaments.
(2) The best values of Jc (without Cu) at 5 T for NbTi wire and at
16T for Nb_3Sn wire were 3.8×10^5 and 3.2×10^4 A/cm^2, respectively.
(3) In order to achieve more high critical current density, it was
shown that not only non-uniformity of the filaments should be minimized, but also the intrinsic factor for pinning had to be optimized
much more.
(4) Transport critical current densities of the oxide superconducting wires were strongly affected by the degree of crystal orientation in the Ag-sheathed BSCCO system.

ACKNOWLEDGEMENTS

I am very grateful to T. Sano and K. Yamada for much careful
experimental works, and N. Enomoto and M. Mimura for helpful discussion. The high-field experimental data were obtained using the magnet
facilities of the Tohoku University.

REFERENCES

1. P.J.Lee and D.C.Larbalestier, J. Mater. Sci. 23(1988)2523-2536

2. D. C. Larbalestier, P. J. Lee, Li Chengren and W. H. Warnes, Cryogenics 24(1986) 13-18
3. M. Senaga, K. Tsuchiya, N. Higuchi and K. Tachikawa, Cryogenics
   25(1985) 123
4. N. Enomoto et al., 36th Spring Convention of the Japan Soceity of
   Appl. Phys., (1989)2a-PC-16
5. T. Matsushita, S. Otabe, and T. Matsuno, Adv. Cryog. Eng.,
   vol. 36(1990)263.
6. K. Matsumoto and Y. Tanaka, 11th Int. Conf. on Magnet Technology,
   Sep. (1989)
7. S. Akita, T. Ishikawa, T. Tanaka, K. Oishi, K. Matsumoto, K. Wada,
   Y. Tanaka and M. Ikeda, Furukawa Review, 7(1989)50
8. O. Tsukamoto, Y. Tanaka, K. Oishi, T. Kataoka, Y. Yoneyama, T. Takao
   and S. Torii, Proc. of 2nd Int. Symp. Superconductivity, Nov.
   (1989)26
9. H. Sakamoto, K. Yamada, T. Sano, Y. Tanaka, H. Ii and E. Yamaguchi,
   to be presented at 43th Meeting on Cryogenics and Superconductivity, Yokohama, Japan, May(1990)
10. M. Mimura et al, Int. Cry. Mat. Conf., Garmisch-Partenkirchen(1990)

MECHANICAL DAMAGE ASSESSMENT IN PROTOTYPE SUPERCONDUCTING CABLE COMPONENTS

Robert D. Hilty and Roger N. Wright
Materials Engineering Department
Rensselaer Polytechnic Institute
Troy, New York 12180

ABSTRACT

The use of high T_c superconductors, namely $YBa_2Cu_3O_{7-x}$, in the design of electrical componentry shows a great potential for energy savings in the transmission and generation of electrical power. Through characterization of the physical and mechanical properties of these materials, their usefulness in engineering practices can be predicted. The strain required to form and propagate cracks has been considered. The role of crystallographic texture in achieving optimum mechanical and electrical properties will also be discussed.

INTRODUCTION

The development of high temperature superconductors has created the possibility for ultrahigh efficiency energy transport and generation. The economic savings possible from this type of development have created an increased interest in the production of superconducting cable components. Although several configurations have been proposed for cable conductors, this study focuses on $YBa_2Cu_3O_{7-x}$ fabricated into superconducting composite tape. The tape was supplied, as developmental materials, by Intermagnetics General Corporation of Guilderland, NY.

The tape conductor is a three layer composite consisting of a layer of substrate, buffer and superconductor. The conductor is likely to see a variety of environments before it can be placed in service. Therefore, each of the three layers must work together to maintain mechanical and electrical integrity. A schematic of the composite conductor is shown in Figure 1. The Ni substrate is 0.25mm (0.010 in) thick, with a coating of MgO approximately 0.025mm (0.001 in) thick and 0.1mm (0.004 in) of $YBa_2Cu_3O_{7-x}$. Continuous layers of $YBa_2Cu_3O_{7-x}$ were applied to the substrate by an aerosol technique then sintered at 910 C, followed by an equilibrating heat treatment in oxygen at 450 C [1].

EXPERIMENTAL PROCEDURE

The objectives of this study were three-fold. First, perform a characterization of the crystalline texture of the superconductor as fabricated. Second, impose a mechanical strain upon the samples. Third, characterize the damage to the tape as a function of applied strain.

678

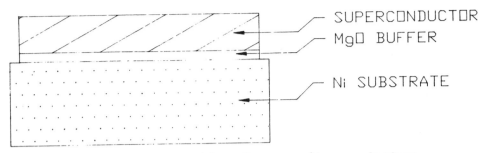

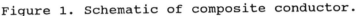

Figure 1. Schematic of composite conductor.

Current theory proposes that the Cu-O planes in $YBa_2Cu_3O_{7-x}$, which are normal to the c axis, are the primary sources of the superconducting phenomena [2]. Thus, the desired texture requires the c axis to be normal to the superconducting plane, with variations in the a and b axis orientations. The high thermal expansion coefficient in the c direction of $YBa_2Cu_3O_{7-x}$, requires the c axis to be oriented normal to the plane of the substrate in order to decrease thermal stresses due to expansion mismatch. The texture of the ceramic can therefore greatly influence its properties. The preferred grain orientation of the $YBa_2Cu_3O_{7-x}$ compound was evaluated using standard x-ray pole figure generation techniques.

A mechanical strain was imposed on the samples by way of bending. The samples were bent around mandrels of varying radii in a controlled manner such that the superconducting layer would not come in contact with the mandrel, see Figure 2. The bending creates a uniaxial tensile strain in the sample. The strain can be expressed by the equation:

$$\epsilon_b \approx \frac{T}{2R}$$

where T is the overall thickness of the sample, R is the radius of the mandrel and ϵ_b is the strain due to bending. The degree of strain varied from 0 to 0.2% in steps of 0.05%.

The final step was to evaluate the degree of damage in the superconductor due to processing and mechanical straining. This was performed in two ways. Scanning electron microscopy (SEM) was a primary source of information about the superconducting surface layer and its mechanical properties. Secondly, a comparison was made between the degree of mechanical strain and the resistivity at varying temperatures. Resistance in the samples was measured by a standard four point probe test in temperatures ranging from 298°K to 55°K.

Sintering is a common method used in the fabrication

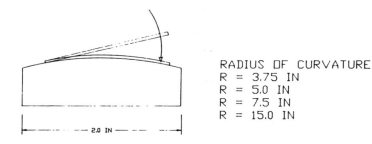

RADIUS OF CURVATURE
R = 3.75 IN
R = 5.0 IN
R = 7.5 IN
R = 15.0 IN

Figure 2. Deforming mandrel for inducing strain.

of high Tc superconductors into cable components. The
sintering process can however, lead to lower critical
current densities and lower transition temperatures as
well as a decrease in overall material density due to a
low particle density. At sintering temperatures below
940°C, significant porosity remains in sintered 123
compound [3]. The initial superconducting current
densities for this sample configuration are typical for
sintered material, 10 A/cm^2 at 70°K [4]. This low value
is primarily due to a low particle density which
generates weak links in the continuity of the Cu-O
conducting planes.

RESULTS

X-ray analysis has generated the pole figure shown
in Figure 3. The (113) peak was the only segregated
strong peak, with small MgO and Ni peaks present. The
pole figure shows a peak along the 60° circle. Thus, a
maximum intensity area lies in this angular orientation.
The indices of this point can be calculated knowing the
center pole (113) and the angular variation (about 60°).
The peak can be shown to be of the orientation (66$\bar{5}$).
This orientation is approximately a (11$\bar{1}$) orientation and
is not oriented in the c axis direction. Note that the
texture is not very strong in this sintered product. The
probability of any local structure being oriented in the
(66$\bar{5}$) direction is 1.41 times that of most other
directions. The departure from a c axis orientation is
less favorable for several reasons. The thermal
expansion coefficients are very anisotropic, varying by a
factor of 4, so the possibility of microcrack formation
at grain boundaries is much more intense [5]. This
texture also fails to strongly align the Cu-O planes
which promote the superconducting nature of $YBa_2Cu_3O_{7-x}$.
The superconductor has been found to display some degree
of cracking at all applied strains greater than 0.05%
Figures 4-6 show the cracking of the superconducting

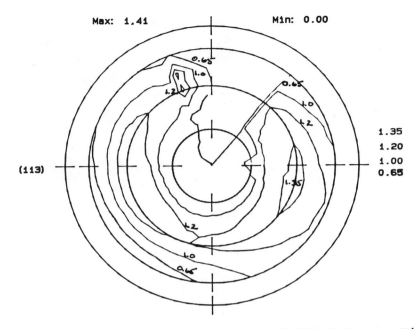

Figure 3. (113) pole figure of YBa$_2$Cu$_3$O$_{7-x}$ on Ni

layer for various levels of strain.   Macrocracks,
visible at 50x, are not seen in the unstrained sample,
but appear in any sample receiving even a 0.05% strain.
Microcracks, visible at 500x, are seen at seemingly
random locations in all of the samples, with an increase
in crack density as the degree of mechanical strain
increases.

The low particle density, which evolves from the
sintering process, creates a non-homogeneous layer of
material.   Local inhomogeneneities then become primary
sources for stress concentrations and thermal expansion
mismatch.   Figure 4 shows that crack formation in the
ceramic occurs at or below strains of 0.05%.   Fracture
with a small applied strain is most likely to occur on a
microscopic scale, in a localized region where the stress
is highly increased due to a local aberration in the
geometry.   The localized stress quickly reaches a
critical value, at which the material fractures.   At this
point the load is redistributed to the next area of
stress concentration.   The stress concentrations in the
composite conductor studied can be formed in several
ways.   These include: diffusion of nickel into the
superconducting layer, orthorhombic to tetragonal phase
transformations, voids in the sintered material and sharp
edge geometries formed when the tape is fabricated.

All of the cracks present in the conductor are

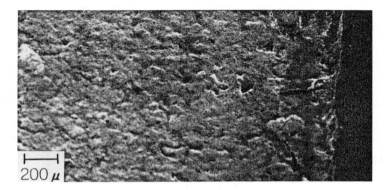

Figure 4. 123 Superconductor: strain = 0

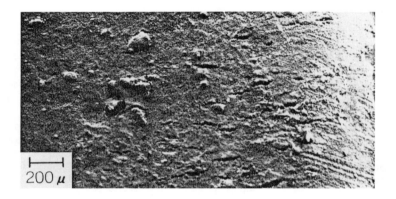

Figure 5. 123 Superconductor: strain = 0.05%

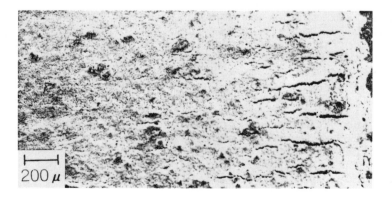

Figure 6. 123 Superconductor: strain = 0.20%

oriented perpendicular to the longitudinal direction of
the sample, which is also the direction of maximum
tension. Since some of the cracks existed previous to
the straining, some previous longitudinal tensile strain
must have been present. It is most likely that these
cracks were formed by mismatch in thermal expansion. The
thermal expansion coefficients for the materials used to
fabricate the conductor are given in Table 1. By using a
buffer, in

Material	Thermal Expansion Coefficient ppm/°C
$YBa_2Cu_3O_{7-x}$	a 12 b 10 c 40
Mgo	13.5
Ni	15.3

Table 1. Thermal expansion coefficients for
superconductor composite materials.[2,4]

this case MgO, the thermal mismatch between Ni and 123
compound can be eased, decreasing the thermal strain.
The thermal expansion coefficients for Ni and Mgo are
probably greater than the in plane coefficients for
$YBa_2Cu_3O_{7-x}$; this normally puts the cooled down
superconductor in a compressive stress state. However,
tension might occur on rapid reheating, if some low
temperature stress relief has occurred. Thermal strain
on the order of 0.10% is possible. In a sintered
material with low particle density, voids within the
conducting layer are bridged by joined particles. Since
the particle bridges have varying angles and voids into
which they may deflect, fracture may occur in
compression.
     Of course, if the c axis of $YBa_2Cu_3O_{7-x}$ is
substantially in the plane of the conductor, the
$YBa_2Cu_3O_{7-x}$ would expand more than the Ni and MgO, and
tension would occur on cooling.
     The degree of cracking and crack propagation has
been shown to be a function of the degree of strain in
the material. As the crack density increases in the
conductor, some of the current that would normally flow
through the superconductor is shunted across the Ni
substrate. Figure 7 shows the relationship of strain
(i.e. crack density) to conductor resistance and
temperature. The three strains used for this figure,
0.05%, 0.10% and 0.15%, show a fundamental relationship

between the degree of strain and the resistance.   These
samples did not become superconducting due to oxygen loss
and water absorption during the experimental procedure.

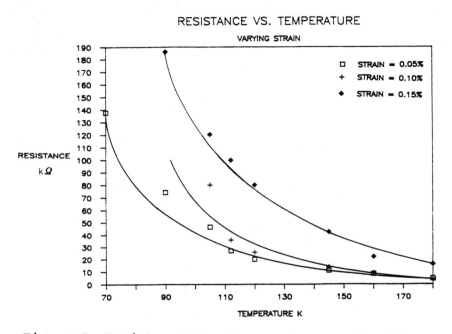

Figure 7. Resistance vs. Temperature for varying
degrees of mechanical strain

CONCLUSIONS

•   The texture of the composite conductor is shown to be
($66\bar{5}$).   This texture does not strongly promote alignment
of Cu-O planes for superconductivity; it also increases
thermal expansion mismatch between the substrate and
ceramic.

•   The existence of inhomogeneous material due to
substrate diffusion, edge fabrication and phase changes
in the superconducting layer facilitates crack formation
and propagation.

•   Crack propagation in a composite conductor can occur
at strains as low as 0.05%, with a rapid increase at
strains above 0.15%.

•   A fundamental relationship exists between the crack
density and the conductor resistance such that increased
cracking increases resistance.

ACKNOWLEDGEMENT

The authors would like to acknowledge the support of the New York State Institute on Superconductivity for their contributions to this endeavor. Specimen materials and resistance measurements were supplied by Intermagnetics General Corporation, Guilderland, NY.

REFERENCES

[1] Blaugher, R.D., Hazleton, D.W., Rice, J.A., "Development of a composite Tape Conductor of Y-Ba-Cu-O", unpublished paper, Intermagnetics General Corporation, Guilderland, NY, 1989.

[2] Poole, C.P., Datta, T., Farach, H.A., <u>Copper Oxide Superconductors</u>, John Wiley and Sons, New York, 1988.

[3] Lanagan et al, "Ceramic Fabrication Technology for High T_c Materials", <u>Superconductivity and Its Applications</u>, Elsevier Science Publ., New York, 1988.

[4] Hazelton, D.W., Blaugher, R.D., Walker, M.S., "Fabrication of a Technologically Useful Conductor Using Ceramic Superconductor", unpublished paper, Intermagnetics General Corporation, Guilderland, NY, 1989.

[5] Jin, S., Tiefel, T.H., Sherwood, R.C., Davis, M.E., van Dover, R.B., Kammlott, G.W., Fastnacht, R.A., Keith, H.D., "High Critical currents in Y-Ba-Cu-O Superconductors", <u>Appl. Phys. Lett</u>. 52 (24), 2074, 13 June 1988.

[6] O'Bryan, H.M., Gallagher, P.K., <u>Advanced Ceramic Materials</u>, 2, 640 (1987).

HIGH RESPONSIVITY RADIATION DETECTION BY HTS FILMS

Yeong Jeong* and Kenneth Rose

Center for Integrated Electronics
Rensselaer Polytechnic Institute
Troy, New York 12180-3590

ABSTRACT

High responsivity detection (>1 KV/W) of 9 GHz microwave radiation by high temperature superconducting films has been observed around 60K. Measurements were made on YBaCuO films deposited on MgO substrates. The responsivity is related to the strongly nonlinear current-voltage characteristics of these films and the films behave like an SNS junction. We discuss the characteristics of the observed detection mechanism, its physical origin and potential applications.

INTRODUCTION

Detection of microwave and/or infrared radiation is one of the most promising applications of superconducting films. Some time ago we discovered voltage responsivities approaching 10 KV/W at 9 GHz in granular tin films with high sheet resistances.[1,2] Time constants were less than 100 ps. Subsequent research[3] showed that such films could behave as coherent arrays of weak link detectors. Leung et al[4] have reported reactively sputtered NbN/BN granular films can act as fast far infrared detectors with a voltage responsivity of about 1V/W and a response time less than a nanosecond for pulsed 385 μm radiation. Competitive semiconductor detectors such as HgCdTe photoconductors have high responsivities around 20 KV/W.[5]

When high temperature oxide superconductors were formed it was natural to inquire whether they would exhibit a high responsivity detection mode. For weak links the voltage responsivity is related to the nonlinearity of the voltage-current characteristic by

$$r_v = \frac{1}{2}\frac{d^2V/dI^2}{dV/dI} \tag{1}$$

However, one expects a quantum limit to how large r_v can become, corresponding to the reciprocal of some minimum detectable current. We estimate this current by dividing the voltage quantum corresponding to single photon absorption breaking an electron pair, hf/2e, by the microbridge resistance R. This corresponds to an upper limit of

$$r_v \lesssim \frac{2e}{hf}\ R \tag{2}$$

(Similar, reciprocal relations are well established for SIS junctions and current responsivity.[6]) Rewriting (1) we have

$$r_v = \frac{1}{2}\frac{dR/dI}{R} \tag{3}$$

Plotting log R as a function of I using a semilog graph gives a direct indication of r_v. More exactly, we should use the dynamic resistance to evaluate r_v. However, a power series

---

* Now at Samsung Advanced Institute of Technology, P.O. Box Suwon 11, Kyung Ki-Do, Korea.

expansion of the characteristic shows that use of the dynamic resistance gives a conservative estimate of r_v.

We have used semilog plots of R versus I to search for regions of high responsivity detection by Y-Ba-Cu-O films.[7,8] Fig. 1 shows the dependence of resistance on current for a $YBa_2Cu_3O_{7-x}$ film at various temperatures. This film was 680 nm thick in a strip structure 1 mm wide with 3 mm between voltage terminals. Fig. 1 shows that this film is a poor high temperature superconductor. Although a transition to superconductivity begins at 85K, significant resistance is seen at higher currents down to 32K. Even at 32K there appears to be a critical current of about 500 μA, corresponding to $J_c \approx 70$ A/cm² on the basis of the film cross section. In the intermediate region at higher temperatures the resistance of the film is reduced substantially below the normal state but significant resistance remains.

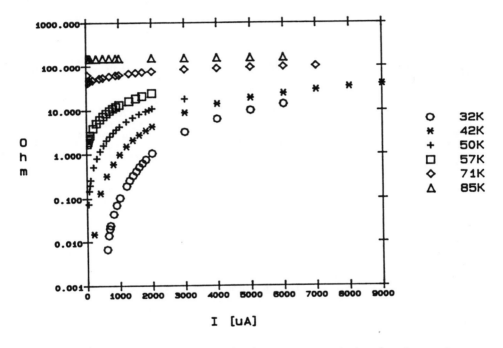

Fig. 1 Semilog plot of the dependence of resistance on current at various temperatures, $YBa_2Cu_3O_{7-x}$ film (90418) in a strip structure.

Regions where the resistance changes rapidly in Fig. 1 would be expected to have a high voltage responsivity on the basis of (3). Calculations based on the experimental results in Fig. 1 show that at 50 K r_v should rise rapidly at lower currents, exceeding 10 KV/W for I $\lesssim$ 800 μA. At higher temperatures or currents r_v drops to much lower values.

In this paper we report confirmation of high voltage responsivities by measurements at 9 GHz on similar films. We describe our experimental results, indicate their significance for applications, and explore the physical basis for this behavior.

EXPERIMENTAL MEASUREMENTS

Y-Ba-Cu-O films were deposited on (100) single crystal MgO substrates at ambient temperature by sequential electron beam evaporation. Our standard procedure has been to deposit the film in six layers Cu/Ba/Y/Ba/Y from substrate to surface with equal thicknesses of the same material. The as deposited film was then heated to 900°C in 20 minutes and sintered at 900°C for half an hour. After sintering the furnace was cooled to 450°C where it was annealed for seven hours. Sintering and annealing were performed in flowing oxygen.

The particular film on which we report microwave measurements was 680 nm thick and had an as-deposited composition Y:Ba:Cu = 1:2:5. The film was deposited through a shadow mask to produce a strip structure $L \times W = 2.4 \times 0.4$ mm^2. The onset of the superconductive transition occurred at 95K when the film had a normal state resistance of 120 Ω, corresponding to a sheet resistance of 20 Ω/sq. and an estimated resistivity of about 1.4 mΩ-cm. The average TCR above 95K was +81 mΩ/K. By 70K the film resistance had dropped to 20 Ω and was ohmic for measurement currents below 100 μA. The film became fully superconducting at low currents below 60K.

The DC I-V characteristics and voltage responsivity calculated using (3) are shown in Fig. 2a and 2b. The straight lines in Fig. 2a correspond to ohmic resistance. Note that the highest responsivities appear around 60K when the total film is just entering the superconducting state. We would expect lower responsivities at higher and lower temperatures. These measurements were made by mounting the film on a dip stick which could be lowered into a helium storage container. The temperature of the sample was determined by its height in the storage dewar.

To measure responsivity at microwave frequencies the film was mounted in a sample holder at the end of a silver plated stainless steel x-band waveguide. The brass sample holder formed a short circuit at the end of the waveguide. The sample was attached to the sample holder by four brass clamps which served as electrical contacts. A small amount of silver paint was applied at the corners of the sample to make better electrical contact between the clamps and the samples. A Lakeshore Cryogenics DT-470 temperature sensing diode was attached to the bottom of the sample holder. The waveguide was free to move in the dewar's top plate so that the sample holder could be positioned at varying heights above a liquid helium bath for measurements at varying temperatures. Although simple, this scheme used excessive amounts of liquid helium even when insulating baffles were placed above the sample holder to reduce vapor streaming. The same apparatus was used to measure responses above liquid nitrogen temperatures as the sample warmed. An HP 620A microwave generator tuned to 9 GHz was used as the signal source. The incident power was 1 mW (0 dBm).

The measured response of the sample at various temperatures is shown in Fig. 3. The voltage in Fig. 3 is the difference of voltages measured across the sample with and without microwave power. Note that the response at 57K is larger than the values at 51K or 68K, in agreement with our expectations from Fig. 2. Note also that much higher bias currents are required to observe appreciable responses at low temperatures.

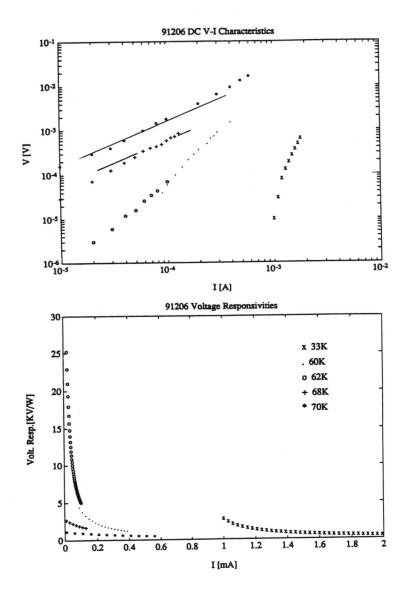

Fig. 2 Characteristics of Y-Ba-Cu-O film (91206) in a strip structure at various temperatures.

    a. DC V-I characteristics
    b. Calculated voltage responsivities in the classical limit

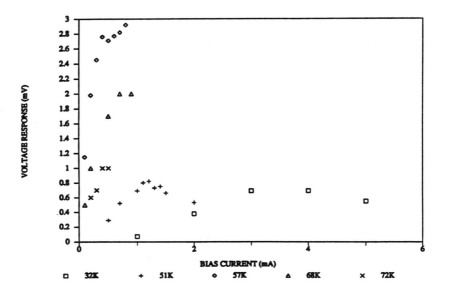

Fig. 3 Measured voltage response of Y-Ba-Cu-O film (91206) to incident microwave power. Dependence on bias current at various temperatures. Incident microwave power is 1 mW at 9 GHz.

To convert responses to responsivities we need to determine the power absorbed in the sample. Since our sample is conducting strip 0.4 mm wide backed a by close conducting wall, we can estimate the induced rf current flow by considering the current flow in a shorting end plate and calculating the fraction of flow in a 0.4 mm wide strip at the center of the waveguide. We estimate $I_{rf} = 140$ μA for 1 mW incident power. This is about 3% of the total induced current. $P_{abs} = I_{rf}^2 R_{ac}$ where $R_{ac} = dV/dI$ can be determined from the dc V-I characteristics in Fig. 2. I_{rf} is independent of temperature and bias current, so long as R_{ac} is small, but R_{ac} is not. At 57K R_{ac} varies from 2.2 to 9.4 ohms as bias current varies from 100 to 500 μA. Over this range of bias currents P_{abs} varies from 42 to 180 nW.

Fig. 4 shows the bias-current dependence of voltage responsivity obtained from Fig. 3 in this fashion. It also shows the calculated responsivity variations from Fig. 2b at corresponding temperatures. Although the magnitudes of observed and calculated responsivities are in good agreement, reaching 20 KV/W, they occur at higher bias currents. We believe the reason is that only a fraction of the bias current flows through the weak link responsible for microwave detection.

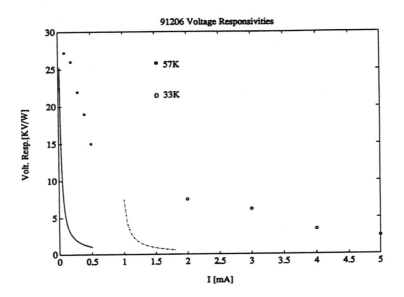

Fig. 4 Comparison of calculated responsivities (Fig. 2b) with responsivities estimated from measured voltage response (Fig. 3) at corresponding temperatures. *, o are responsivities estimated at 57 and 33K. —, — · — are responsivities calculated at 60 and 32K.

Similar measurements were attempted on a comparable film strip 5 mm wide immersed in liquid nitrogen where we would expect to see a bolometric response. For this film the room temperature resistance was 400 Ω and the slope of the resistance change $\gamma = dR/dT \approx 35 \, \Omega/K$. At a bias current of 10 μA, we observed a voltage response of 120 nV for 0.1 mV incident GHz radiation. The induced rf current estimated as above was 490 μA. Since the sample resistance was 90 Ω, $P_{abs} = 21 \, \mu W$, corresponding to a voltage responsivity of 5.7 mV/W! A low γ and a high thermal conductance substrate contribute to the low responsivity. The main point is that we are seeing high responsivity response to microwave radiation in a material whose bolometric responsivity is low.

An alternative way of estimating the voltage response is given by Kanter and Vernon.[9] They have derived an expression for the contribution of a small signal current $I_s \sin \omega t$ to the average voltage across a resistively shunted Josephson junction. For $\omega \ll \omega_o$, the Josephson frequency, they find only a contribution from the second order term

$$< V_2 > = \frac{I_s^2}{4} \frac{d^2 V}{dI^2} \tag{4}$$

This allows the voltage response to be expressed in terms of I_s and the curvature calculated from the DC V-I characteristic Using the value of $I_s = 140\ \mu A$ derived earlier, the calculated $<V_2>$ from (4) is compared with the observed output voltage in Fig. 5.

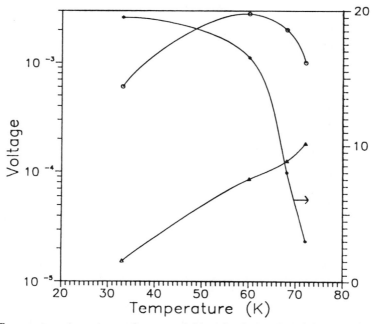

Fig. 5 Temperature dependence of measured (○) and calculated (△) voltage responses for Y-Ba-Cu-O film (91206). Division gives the voltage ratio on the rightside scale.

The observed voltage is much larger than the calculated voltage and their ratio is shown as well. The simplest exploration for this discrepancy is that our estimated value of I_s is too small. At 60K the ratio from Fig. 5 is 16. Since the rf current of 140 μA was determined previously as an rms current, we have used $I_s = \sqrt{2}\ (140\ \mu A) = 200\ \mu A$. Equal voltages correspond to $I_s = 800\ \mu A$. If the coupling of power to the film is greater than we estimated by a factor of 16 then the peak voltage responsivity we observe at 57K would be reduced from 27 KV/W to 1.7 KV/W. Above the transition to complete superconductivity around 60K, the ratio of observed to calculated voltage drops sharply. This would seem to indicate reduced coupling to our sample.

APPLICATION IMPLICATIONS

DC V-I characteristics indicate that our films behave like resistively shunted Josephson junctions (RSJ) in series with a parasitic inductance. Fig. 6 compares the V-I characteristic for Y-Ba-Cu-O film 91206 at 33K with the McCumber model[10] for a weak-link junction and series inductance. Fig. 6 shows the fit of the zero inductance characteristic, $\beta_L = 0$, to

series inductance. Fig. 6 shows the fit of the zero inductance characteristic, $\beta_L = 0$, to our experimental results. Comparing our characteristic with the normalized characteristics shown in Fig. 6 of McCumber[10], we find our characteristic corresponds to $\beta_L \approx 0.3$.

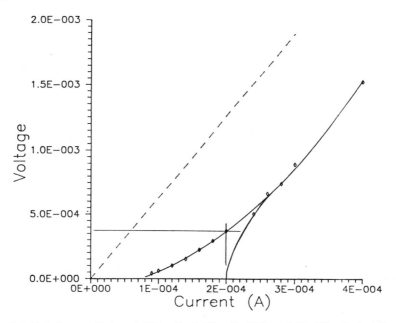

Fig. 6 DC V-I characteristics of Y-Ba-Cu-O film (91206) at 33K. The dashed line corresponds to the normal resistance of the film. The solid lines correspond to the McCumber models for $\beta_L = 0$ and 0.3.

Considerable information of value for ascertaining applications can be obtained from Fig. 6. The intercept of the $\beta_L = 0$ characteristic with the current axis determines i_m, the maximum Josephson current, in McCumber's theory. For our film, $i_m \approx 200$ μA. The bias current at which voltage first appears (5 μV) was 80 μA, corresponding to a normalized current $\alpha = 80/200 = 0.4$. This agrees well with McCumber's Fig. 5 for $\beta_L \approx 0.3$. Our parasitic inductance can be estimated from

$$\beta_L = \phi_o / 2\pi i_m L \tag{5}$$

where $\phi_o = 2.07$ fWb is the flux quantum. For $i_m = 200$ μA and $\beta_L \approx 0.3$ we find L ≈ 6 pH. It is interesting that the parasitic inductance for the niobium wire used in a niobium point contact was 600 pH. Since this was the point contact used in Kanter's[11] study of Josephson junction mixing at 95 GHz, it suggests that our films are quite compatible with microwave applications.

A particularly important number can be obtained by determining the (i_m, v) coordinates for our $\beta_L \approx 0.3$ characteristic. On McCumber's normalized scale (α, η) $\approx$ (1, 0.53) for $\beta_L \approx$

$$\eta = GV/i_m = V/i_m R_j \qquad (6)$$

we find $i_m R_j \approx 360/0.53 \approx 680 \ \mu V$. $i_m R_j$ is commonly used as a figure of merit for weak links. The value we get is comparable to that seen for good conventional SNS microbridges[12] and approaches the $i_m R_j \gtrsim 1$ mV value often used as a guideline for applications.

PHYSICAL MECHANISMS

We have observed a significant amount of residual resistance between 60K and 95K in these films. As reported earlier,[8] we believe that this residual resistance is related to proximity effects. The proximity effect induces superconductivity in the normal metal region of SNS structures. In the clean limit, the characteristic length $(1/K)$ of induced superconductivity in the normal metal is given by[13]

$$1/K = \frac{\hbar V_N}{2\pi k_B T} \qquad (7)$$

where V_N is the Fermi velocity of the normal metal.

Due to this proximity effect the length L of the normal metal is modulated and the resistance of the normal metal becomes

$$R_N = (\rho_N/A)(L - 2/K)$$
$$= R_{NO}(1 - 2\hbar V_N/\pi k_B T L) \qquad (8)$$

Here $(L - 2/K)$ is the effective thickness of normal metal at temperature T and $R_{NO} = \rho_N L/A$ is the resistance of the normal region without the presence of the proximity effect. We include a factor of 2 to account for proximity effects at both S/N interfaces in SNS structures. Evidence for the proximity effect is the $1/T$ dependence of residual resistance we have observed in similar films as shown in Fig. 7. The deviations from a $1/T$ dependence are caused by nonohmic critical current induced resistance changes near zero resistance.[8]

Extrapolation of (8) to zero resistance allows us to estimate the zero resistance temperature T_o and L, the length of the normal region. For Film 90220, $T_o = 57K$ and L = 4 nm if we take $V_N = 10^7$ cm/s, the Fermi velocity for normal $YBa_2Cu_3O_7$. This L = 4 nm normal metal barrier length corresponds to the thickness of the thickest normal metal region. Since L is much smaller than grain sizes in our films, this suggests that intragrain coupling dominates intergrain coupling as suggested by others.[14,15] This would support the evidence we have seen for SNS junctions in our films.

ACKNOWLEDGEMENTS

We are particularly grateful for Joe Rice, Dick Blaugher and their colleagues at Intermagnetics General Corporation for their assistance with the low temperature measurements. Honghua Li assisted with the growth and characterization of the films.

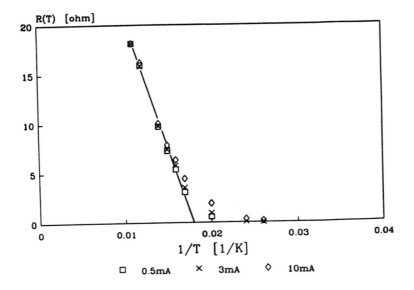

Fig. 7 Temperature dependence of dc resistance in a Y-Ba-Cu-O film (90220) at temperatures between 50 and 82K. This behavior is generally independent of current level.

REFERENCES

1. C.L. Bertin and K. Rose, J. Appl. Phys. 39, 2561 (1968).

2. C.L. Bertin and K. Rose, J. Appl. Phys. 42, 631 (1971).

3. W.J. Ayer and K. Rose, IEEE Trans. Magn. MAG-11, 61 (1975).

4. M. Leung et al., Appl. Phys. Lett. 50, 1691 (1987).

5. M. Itoh, H. Takigawa, And R. Ueda, IEEE Trans. Elec. Dev. ED-27, 150 (1980).

6. J.R. Tucker and M.J. Feldman, Rev. Mod. Phys. 57, 1055 (1985).

7. J.-Y. Jeong and K. Rose, Int. J. Infrared and MM Waves, 10, 1441 (1989).

8. J.-Y. Jeong and K. Rose, "Radiation Detection Mechanisms in High Temperature Superconductors", Proc. NYSIS Third Annual Conference on Superconductivity and Applications, September 1989.

9. H. Kanter and F.L. Vernon, J. Appl. Phys., 43, 3174 (1972).

10. D.E. McCumber, J. Appl. Phys., $\underline{39}$, 3113 (1968).

11. H. Kanter, Revue de Physique Applique, $\underline{9}$, 225 (1974).

12. R.B. van Dover, R.E. Howard, and M.R. Beasley, IEEE Trans. Magn. $\underline{\text{MAG-15}}$, 574 (1979).

13. G. Deutscher and P.G. deGennes, "Proximity Effects", Ch. 17 in "Superconductivity", edited by R.D. Parks (Marcel Dekker, New York, 1969).

14. B. Raveau and C. Michel, in "Novel Superconductors", edited by S.A. Wolf and V.Z. Kresin (Plenum Press, New York, 1987), pp. 599-610.

15. G. Deutscher and K.A. Muller, Phys. Rev. Lett., $\underline{59}$, 1745 (1987).

Subject Index

A

Absorption spectroscopy (*see* X-ray absorption spectroscopy)

ac losses, 639

  Nb-Ti alloy wire, 672

Acoustic vibration, 90–91, 94

Alignment, 121–23, 374–78, 414, 417–18, 447, 456, 478–85

  a-axis, on MgO substrate, 450

  annealing and, 577

  applied field and, 481–82

  biaxial, 375–78

  c-axis

    copper oxide planes, 679–80

    grain boundaries and, 121–29

    rf magnetron sputtered films, 433

    textured films, 446–52

  effect of curing time on, 481–82

  effect of particle size and shape on, 414–18

  effect of weak normal-aligning RE on, 482–84

  lattice mismatch and, 450–51

  morphology of BSCCO tape, 455

  nature of misorientation, 480–81

  shape orientation, 376–77

  symmetry matching and, 450–51

  twist boundaries, 125–28

Alloying, 673–74

Amorphous films, 555–56

Anderson-Kim theory, 85

Angle resolved photoemission spectra (ARPES), 161, 163

Anisotropy

  a–*b* plane, of normal state resistivity, 139–45

  effect on grain boundary structures, 121

  effect on crystal growth, 121, 124–25

  electronic, 27

  mechanical, 27

  of critical current density, 398–403, 453–460

  of susceptibility, 144

  of upper critical field, 401

resistivity, 368–69

  TlBaCaCuO, 368–69

Annealing, 466

  alignment and, 577

  contact properties, 520

  effect on oxygen stoichiometry in single crystals, 559

  effect on resistivity anomaly, 203

  optimization of sputtered film properties, 461–68

Anomalous behavior, LaSrCuO, 203–4

Antiferromagnetic phase, 32–33, 35–36

Antimony doping, 58

Applications

  energy applications, 635–45

  filters, microstrip, 652–55

  generators, 637

  maglev, 643

  microbridges, 661–63, 665

  microelectronic applications, 664

  microwave devices, 648

  motors, 636, 638–39, 643–44

  passive millimeter wave devices, 646

  power electronic applications, 643

  space applications, 646–57

  superconducting magnetic energy storage (SMES), 637

  superconducting quantum interference devices (SQUIDS), 661–64

  transportation applications, 636, 643

Auger electron spectroscopy (AES), 406–8, 473–74

B

$BaCO_3$, 611–12

$Ba_{1-x}K_xBiO_3$, 49, 52–53

Backreaction, 174

Band gap

  effect on carrier doping, 58, 166–68

  in fluorite perovskites, 166–68

$BaPb_{.75}Bi_{.25}O_3$, 55–56

$BaPb_{1-x}Bi_xO_3$, 47–49

697

Author Index

A

Alcock, C. B., 200
Allen, J. P., 429
Alp, E. E., 265
Anderson, W. A., 405, 518
Asaka, T., 355
Awaji, S., 398

B

Balachandran, U., 410, 439
Ballentine, P. H., 429
Barus, A. M. M., 413
Batlogg, B., 55
Bayya, S. S., 306
Berta, Y., 617
Bhargava, R. N., 93
Bichile, G. K., 183
Bielaczy, S., 461
Borisovskii, V. V., 153
Bourdillon, A. J., 165, 453
Brooks, K. C., 470
Budai, J. D., 336
Butcher, Jr., J. A., 569

C

Campuzano, J. C., 265
Cappelletti, R. L., 569
Cava, R. J., 55
Chen, B. J., 589
Chen, C. T., 206
Chen, F., 374, 478, 575
Chen, J., 478
Chen, J. W., 130
Chen, J. Y., 130
Chen, Y. Y., 130
Cheng, C. H., 130
Chien, F. Z., 100
Christen, D. K., 336
Chung, D. D. L., 410
Clapp, R. E., 272
Claus, H., 111
Cormack, A. N., 617
Crabtree, G. W., 111, 139

D

Dale, S. J., 635
Deshmukh, S., 183
Dhananjeyan, M. V. T., 234
Dong, S. Y., 446
Downey, J., 139
Duan, H.-M., 364
Durčok, S., 558
Durga Prasad, K. A., 234
Dwivedi, A., 617
Dynes, R. C., 46
Dyomin, A. V., 241, 249

E

Ebihara, K., 551
Eick, R. H., 317
Eschwei, M., 461

F

Faiz, M., 265
Fang, C. S., 130
Fang, Y., 111, 139
Feenstra, R., 336
Feng, A., 470
Fisher, L. M., 153
Fleshler, S., 111, 139
Fujishima, T., 551

G

Gammel, P., 55
Gao, Y., 439
Giessen, B. C., 374, 478, 486, 575
Gilmore, R., 173
Gonda, S., 326
Goodenough, J. B., 26
Gubser, D. U., 646
Gupta, A., 317

H

Habermeier, H.-U., 421
Haldar, P., 575
Hanada, T., 326

AIP Conference Proceedings

		L.C. Number	ISBN
No. 156	Advanced Accelerator Concepts (Madison, WI, 1986)	87-70635	0-88318-358-0
No. 157	Stability of Amorphous Silicon Alloy Materials and Devices (Palo Alto, CA, 1987)	87-70990	0-88318-359-9
No. 158	Production and Neutralization of Negative Ions and Beams (Brookhaven, NY, 1986)	87-71695	0-88318-358-7
No. 159	Applications of Radio-Frequency Power to Plasma: Seventh Topical Conference (Kissimmee, FL, 1987)	87-71812	0-88318-359-5
No. 160	Advances in Laser Science–II (Seattle, WA, 1986)	87-71962	0-88318-360-9
No. 161	Electron Scattering in Nuclear and Particle Science: In Commemoration of the 35th Anniversary of the Lyman-Hanson-Scott Experiment (Urbana, IL, 1986)	87-72403	0-88318-361-7
No. 162	Few-Body Systems and Multiparticle Dynamics (Crystal City, VA, 1987)	87-72594	0-88318-362-5
No. 163	Pion–Nucleus Physics: Future Directions and New Facilities at LAMPF (Los Alamos, NM, 1987)	87-72961	0-88318-363-3
No. 164	Nuclei Far from Stability: Fifth International Conference (Rosseau Lake, ON, 1987)	87-73214	0-88318-364-1
No. 165	Thin Film Processing and Characterization of High-Temperature Superconductors (Anaheim, CA, 1987)	87-73420	0-88318-365-X
No. 166	Photovoltaic Safety (Denver, CO, 1988)	88-42854	0-88318-366-8
No. 167	Deposition and Growth: Limits for Microelectronics (Anaheim, CA, 1987)	88-71432	0-88318-367-6
No. 168	Atomic Processes in Plasmas (Santa Fe, NM, 1987)	88-71273	0-88318-368-4
No. 169	Modern Physics in America: A Michelson-Morley Centennial Symposium (Cleveland, OH, 1987)	88-71348	0-88318-369-2
No. 170	Nuclear Spectroscopy of Astrophysical Sources (Washington, DC, 1987)	88-71625	0-88318-370-6
No. 171	Vacuum Design of Advanced and Compact Synchrotron Light Sources (Upton, NY, 1988)	88-71824	0-88318-371-4
No. 172	Advances in Laser Science–III: Proceedings of the International Laser Science Conference (Atlantic City, NJ, 1987)	88-71879	0-88318-372-2
No. 173	Cooperative Networks in Physics Education (Oaxtepec, Mexico, 1987)	88-72091	0-88318-373-0
No. 174	Radio Wave Scattering in the Interstellar Medium (San Diego, CA, 1988)	88-72092	0-88318-374-9